Structure & Function of the Body

Structure & Function OF THE BODY

GARY A. THIBODEAU, PhD
Chancellor and Professor of Biology,
University of Wisconsin—River Falls,
River Falls, Wisconsin

KEVIN T. PATTON, PhD
Professor, Department of Life Sciences,
St. Charles County Community College,
St. Peters, Missouri

TENTH EDITION

St. Louis Baltimore Boston Carlsbad Chicago Naples New York Philadelphia Portland
London Madrid Mexico City Singapore Sydney Tokyo Toronto Wiesbaden

Dedicated to Publishing Excellence

A Times Mirror
Company

Vice President and Publisher: James M. Smith
Senior Acquisitions Editor: Ronald E. Worthington, PhD
Developmental Editor: Kristin M. Shahane
Project Manager: Carol Sullivan Weis
Production Editor: Karen M. Rehwinkel
Design Manager: Sheilah Barrett
Design: Rose Design
Manufacturing Manager: Karen Lewis
Cover Illustration: William C. Andrea, CMI

TENTH EDITION
Copyright © 1997 by Mosby-Year Book, Inc.

Previous editions copyrighted 1960, 1964, 1968, 1972, 1976, 1980, 1984, 1988, 1992

Printed in the United States of America
Composition by The Clarinda Company
Lithography/color film by Color Associates, Inc.
Printing/binding by Von Hoffmann Press, Inc.

Mosby-Year Book, Inc.
11830 Westline Industrial Drive
St. Louis, Missouri 63146

Library of Congress Cataloging in Publication Data

Thibodeau, Gary A.,
    Structure & function of the body / Gary A. Thibodeau, Kevin
T. Patton.—10th ed.
      p.   cm.
    Includes index.
    ISBN 0-8151-8714-9 (HC). — ISBN 0-8151-8713-0 (SC)
    1. Human physiology.   2. Human anatomy.   I. Patton, Kevin T.
II. Title.
    [DNLM: 1. Anatomy.   2. Physiology.   QS 4 T427s 1997]
    QP34.5.T5   1997
    612—dc20
    DNLM/DLC
    for Library of Congress   96-17642
                                        CIP

98 99 00 / 9 8 7 6 5 4 3

Preface

We have been privileged to teach human anatomy and physiology for several decades. Each time we interact with students in the classroom or laboratory we become more aware of the challenges of teaching—and learning—about a body of knowledge that is not only fascinating and complex but ever-growing.

Advances in many areas of science have an impact on anatomy and physiology, perhaps more than other biological disciplines. Classroom instructors in the late 1990s are faced with the growing dilemma of selecting the most appropriate information for inclusion in introductory but nonetheless rigorous and demanding anatomy and physiology courses. New information is constantly being published describing advances in areas such as medicine, laboratory and clinical instrumentation, physiological chemistry, biotechnology, immunology, and molecular genetics. The result is an information explosion of incredible magnitude that makes it more and more difficult for teachers to select appropriate material and to give an up-to-date, contemporary presentation of anatomy and physiology that is accurate and user friendly for students. This edition of *Structure & Function of the Body* retains many features that have proved successful in over 35 years of classroom use; yet as a new text it presents a wealth of carefully selected new content, as well as pedagogical enhancements that will better serve the needs of today's instructors and students. The writing style and depth of coverage are intended to challenge, reward, and reinforce introductory students as they grasp and assimilate important concepts.

During the revision of this text, each change in the selection, sequencing, or method of presentation of material was evaluated by anatomy and physiology teachers working in the field—teachers currently assisting students to learn about human structure and function for the first time. The result is a text that students will read—one designed to help the teacher teach and the student learn. It is particularly suited to introductory anatomy and physiology courses in nursing and allied health-related programs. Emphasis is on material required for entry into more advanced courses, completion of professional licensing examinations, and successful application of information in a practical work-related environment.

Dr. Kevin Patton, who contributed significantly to the previous edition of this book, joins now as a coauthor to form a dynamic new writing team. We are confident that you will agree that our teamwork has resulted in a synergistic blend of talents evident throughout the book.

◉ *Special Features*

UNIFYING THEMES

Anatomy and physiology encompass a body of knowledge that, because of its sheer magnitude, can easily discourage and overwhelm the introductory student. There is no question, however, that competency in both anatomy and physiology is essential for student success in almost every clinical or advanced course in a health-related or science curriculum. If a textbook is to be successful as a teaching tool in such a complex and important learning environment, it must assist and complement the efforts of instructor and student. It must help unify information, stimulate critical thinking, and motivate students to master a new vocabulary as they learn about the beauty and connectedness of human structure and function.

Structure & Function of the Body is dominated by two major unifying themes: the *complementarity of normal structure and function* and *homeostasis.* In every chapter of the book the student is shown how organized anatomical structures of a particular size, shape, form, or placement serve unique and specialized functions. Repeated emphasis of this principle encourages students to integrate otherwise isolated factual information into a cohesive and understandable whole. As a result, anatomy and physiology emerge as living and dynamic topics of personal interest and importance to the student. The integrating principle of homeostasis is used to show how the "normal" interaction of structure and function is achieved and maintained by dynamic counterbalancing forces within the body.

PRESENTATION OF CLINICAL AND PATHOLOGICAL INFORMATION

Highly selective clinical and pathological examples are included in each chapter of the book to help students understand that the disease process is a disruption in homeostasis and a breakdown of the normal integration of form and function. In the use of clinical examples the intent is to reinforce the normal mechanisms of body defense and adaptation and to stimulate student interest. The unifying themes are based on the need for introductory students in anatomy and physiology to have a firm foundation in the normal state before primary learning emphasis is focused on the diseased state. This text deals primarily with normal anatomy and physiology. An alternate text, *The Human Body in Health & Disease,* emphasizes pathology in addition to normal structure and function and is intended as an introduction to structural anomaly and pathophysiology.

ORGANIZATION AND CONTENT

The 20 chapters of *Structure & Function of the Body* present the core material of anatomy and physiology most *important* for introductory students. The selection of appropriate information in both disciplines was designed to eliminate the confusing mix of nonessential and overly specialized material that unfortunately accompanies basic information in many introductory textbooks. Information is presented so that students know and understand what is *important.* Further, pedagogical aids in each chapter identify learning objectives and then reinforce successful mastery of this clearly identified core material. As a result, students and instructors can interact more effectively as active participants in the learning experience. The sequencing of chapters in the book follows a course organization most commonly used in teaching at the undergraduate level. Because each chapter is self-contained, instructors have the flexibility to alter the sequence of material to fit personal teaching preferences or the special content or time constraints of their courses or students.

At every level of organization, both within and between chapters, care has been taken to couple structural information with important functional concepts. Students are presented with a conceptual framework and the necessary information on which to build. In each chapter of the text, appropriate physiological content balances the anatomical information that is presented. As a result, the student has a more integrated understanding of human structure and function. Throughout the text, examples that stress the complementarity of structure and function have been consciously selected to emphasize the importance of homeostasis as a unifying concept.

Acquiring and *using* the terms so necessary for any study of anatomy and physiology can be difficult for many students. To assist students in this area, new terms are introduced, defined, and incorporated into a working vocabulary. In addition to vocabulary development, every chapter incorporates skillfully designed visuals to reinforce written information with sensory input. Information related to body structure and function is presented in ways designed to help students develop the ability to integrate conceptual material and reinforce the normal integration of form and function.

The style of presentation of material in this text and its readability, accuracy, and level of coverage have been carefully developed to meet the needs of undergraduate students taking an introductory course in anatomy and physiology. *Structure & Function of the Body* remains an introductory textbook—a teaching book and not a reference text. No textbook can replace the direction and stimulation provided by an enthusiastic teacher to a curious and involved student. A good textbook, however, can and should be enjoyable to read and helpful to both.

PEDAGOGICAL FEATURES

Structure & Function of the Body is a student-oriented text. Written in a very readable style, it has numerous pedagogical aids that maintain interest and motivation. Every chapter contains the following elements that facilitate learning and the retention of information in the most effective manner.

Chapter Outline: An overview outline introduces each chapter and enables the student to preview the content and direction of the chapter at the major concept level before the detailed reading.

Chapter Objectives: Each chapter opening page contains approximately five measurable objectives for the student. Each objective clearly identifies for the student, before he or she reads the chapter, what the key goals should be and what information should be mastered.

Key Terms and Pronunciation Guide: Key terms, when introduced and defined in the text body, are identified in **boldface** to highlight their importance. A pronunciation guide follows each new term that students may find difficult to pronounce correctly.

Boxed Inserts and Essays: Brief boxed inserts or longer essays appear in every chapter. These inserts include information ranging from clinical applications to sidelights on recent research or related topics to exercise and fitness. Pathological conditions are sometimes explained in essay format to help students better understand the relationship between normal structure and function. Examples include AIDS, ulcers, and shingles. New boxed essays on Home Health Care issues have been added to this edition and reflect the growing trend in home based health care. Due to this trend, many students using this text will at some point encounter some of the issues discussed in the boxes. Topics reflect the diversity and wide range of conditions and problems a home health care worker must be prepared to deal with, including procedures, assessments, and administration. All boxed material is highlighted with an easily recognized symbol so that students can see at a glance whether the box contains clinical, general, fitness, or home health care information.

Outline Summaries: Extensive and detailed end-of-chapter summaries in outline format provide excellent guides for students as they review the text materials when preparing for examinations. Many students also find these detailed guides useful as a chapter preview in conjunction with the chapter outline.

Chapter Tests: Objective-type Chapter Test questions are included at the end of each chapter. They serve as quick checks for the recall and mastery of important subject matter. They are also designed as aids to increase the retention of information. Answers to all Chapter Test questions are provided at the end of the text.

Review Questions: Subjective review questions at the end of each chapter allow students to use a narrative format to discuss concepts and synthesize important chapter information for review by the instructor. The answers to these review questions are available in the Instructor's Manual that accompanies the text.

Critical Thinking Questions: Review questions that encourage students to use critical thinking skills are highlighted at the end of the Review Questions section. Answers to these questions are found in the Instructor's Manual along with the answers to the other Review Questions.

Chemistry Appendix: Recognizing that some students may need to review basic chemistry, we have added a new fully-illustrated Chemistry of Life appendix. This section discusses in simple, straightforward terms the concepts needed to understand basic anatomy and physiology with *concept summaries* to emphasize key points. As an appendix, this information will be readily available to students who need it, but it won't burden the text itself with unnecessary detail for those who do not. References within the chapters point students exactly to the appropriate coverage in the Appendix.

Additional learning and study aids at the end of the text include **Common Medical Abbreviations, Prefixes,** and **Suffixes;** an extensive **Glossary** of terms to assist students in mastering the vocabulary of anatomy and physiology; and a detailed **Index** that serves as a ready reference for locating information.

ILLUSTRATIONS

A major strength of *Structure & Function of the Body* is the exceptional quality, accuracy, and beauty of the illustration program. The truest test of any illustration is how effectively it can complement and strengthen written information found in the text and how successfully it can be used by the student as a learning tool. Extensive use has been made of full-color illustrations, micrographs, and dissection photographs throughout the text. Illustrations proven pedagogically effective in previous editions of *Structure & Function of the Body* have been retained or updated in appearance to provide accurate information and visual appeal. Each illustration is carefully referred to in the text and is designed to support the text discussion. The illustrations are an integral part of the learning process and should be carefully studied by the student.

Supplements

The supplements package has been carefully planned and developed to assist instructors and to enhance their use of the text. Each supplement, including the test items and study guide, has been thoroughly reviewed by many of the same instructors who reviewed the text.

INSTRUCTOR'S RESOURCE MANUAL AND TEST BANK

The Instructor's Resource Manual and Test Bank, prepared by Judith Diehl of Reid State Technical College, provides text adopters with substantial support in teaching from the text. The following features are included in every chapter:

Sample lecture outlines based on student objectives

Transparency masters for duplication or overhead projection

Outlines and **worksheets** for class demonstrations

Suggestions for student activities and assignments

Sources of audiovisual support

Information on current topics for distribution to students

Answers to the Chapter Test and Review Questions in the textbook

Critical Thinking Exercises for each chapter provide a step-by-step problem-solving opportunity for students

A **computerized test bank** is also available for Windows and MacIntosh users.

TRANSPARENCY ACETATES

A set of full-color transparency acetates—all with large, easy-to-read labels—is available to adopters of the text for use as a teaching aid.

STUDY GUIDE

The Study Guide, written by Linda Swisher of Sarasota County Vocational Technical Center, provides students with additional self-study aids, including chapter overviews, topic reviews, review questions keyed to specific pages in the text, and application and labeling exercises, as well as answers to the questions in the Study Guide.

MULTIMEDIA

Exciting new software is now available to facilitate the learning of difficult concepts in anatomy and physiology using the power of interactive media. **The Dynamic Human CD-ROM** uses animations, interactive anatomy and histology images, clinical illustrations and movies, and three-dimensional rotating viewers to bring the 10 major body systems to life for the student to explore and learn.

Mosby is pleased to bring to the scientific community a remarkable series of multimedia and print products made possible the National Library of Medicine's Visible Human Project (VHP) and innovative software technology developed by EAI. The first of these products is the **Dissectable Human CD-ROM,** which uses segmented, volume rendered VHP data to present systems-based human anatomy and digital dissection of major organs, both for the first time. Body systems can be studied in isolation or in context with other organs and tissues, with rotational viewing in three dimensions. Labeling can be toggled on or off, and authoring capability allows users to create customized presentations and study programs.This software will change the way anatomy is taught, and the way it is understood. Images from the Mosby-EAI image library are also available in a four color atlas, **Mosby's Systems Atlas of Anatomy.**

In addition to these exciting products, Mosby offers **Body Spectrum** software. The software provides reinforcement of anatomy with computerized coloring and labeling exercises.

A WORD OF THANKS

Many people have contributed to the development and success of *Structure & Function of the Body.* We extend our thanks and deep appreciation to the various students and classroom instructors who have provided us with helpful suggestions following their use of earlier editions of this text.

A specific "thank you" goes to the following instructors who critiqued in detail the ninth and tenth editions of this text or various drafts of the revision. Their invaluable comments were instrumental in the development of this new edition.

Patricia Laing-Arie
Meridian Technology Center
Stillwater, Oklahoma

Bert Atsma
Union County College
Cranford, New Jersey

Ethel J. Avery
Trenholm State Technical College
Montgomery, Alabama

Barbara Barger
Clarion County Career Center
Shippenville, Pennsylvania

Lydia R. Chavana
South Texas Vo-Tech Institute
McAllen, Texas

Marie Conn
Mayo State Vo-Tech School
Pikeville, Kentucky

Joseph Devine
Allied Health Careers
Austin, Texas

Edna M. Dilmore
Bessemer State Technical College
Bessemer, Alabama

Sally Flesch
Black Hawk College
Moline, Illinois

Denise L. Kampfhenkel
Schreiner College
Kerrville, Texas

Anne Lilly
Santa Rosa Junior College
Santa Rosa, California

Richard E. McKeeby
Union County College
Cranford, New Jersey

Keith R. Orloff
California Paramedical and Technical College
Long Beach, California

Christine Payne
Sarasota County Vocational Technical Center
Sarasota, Florida

Ann Senisi Scott
Nassau Tech VOCES
Westbury, New York

Anna M. Strand
Gogebic Community College
Ironwood, Michigan

Eugene R. Volz
Sacramento City College
Sacramento, California

Iris Wilkelhake
Southeast Community College
Lincoln, Nebraska

At Mosby–Year Book, Inc., thanks are due all who have worked with us on this new edition. We wish especially to acknowledge the support and effort of Ronald E. Worthington, PhD, senior editor; Kristin Shahane, developmental editor; Carol Sullivan Weis, project manager; Karen Rehwinkel, production editor; and Sheilah Barrett, design manager, all of whom were instrumental in bringing this edition to successful completion.

GARY A. THIBODEAU
KEVIN T. PATTON

Contents in Brief

Contents

CHAPTER 6

 The Muscular System, 110

CHAPTER 7

 The Nervous System, 136

CHAPTER 8

 The Senses, 172

CHAPTER 9

 The Endocrine System, 190

APPENDIX A

 Chemistry of Life, 426

APPENDIX B

 Common Medical Abbreviations, Prefixes, and Suffixes, 441

An Introduction to the Structure and Function of the Body

1

OBJECTIVES

After you have completed this chapter, you should be able to:

1. Define the terms *anatomy* and *physiology.*

2. List and discuss in order of increasing complexity the levels of organization of the body.

3. Define the term *anatomical position.*

4. List and define the principal directional terms and sections (planes) used in describing the body and the relationship of body parts to one another.

5. List the nine abdominopelvic regions and the abdominopelvic quadrants.

6. List the major cavities of the body and the subdivisions of each.

7. Discuss and contrast the axial and the appendicular subdivisions of the body. Identify a number of specific anatomical regions in each area.

8. Explain the meaning of the term *homeostasis* and give an example of a typical homeostatic mechanism.

There are many wonders in our world, but none is more wondrous than the human body. This is a textbook about that incomparable structure. It deals with two very distinct and yet interrelated sciences: **anatomy** and **physiology.** As a science, anatomy is often defined as the study of the structure of an organism and the relationships of its parts. The word *anatomy* is derived from two Greek words that mean "a cutting up." Anatomists learn about the structure of the human body by cutting it apart. This process, called **dissection,** is still the principal technique used to isolate and study the structural components or parts of the human body. Physiology is the study of the functions of living organisms and their parts. It is a dynamic science that requires active experimentation. In the chapters that follow, you will see again and again that anatomical structures seem designed to perform specific functions. Each has a particular size, shape, form, or position in the body related directly to its ability to perform a unique and specialized activity.

Structural Levels of Organization

Before you begin the study of the structure and function of the human body and its many parts, it is important to think about how those parts are organized and how they might logically fit together into a functioning whole. Examine Figure 1-1. It illustrates the differing levels of organization that influence body structure and function. Note that the levels of organization progress from the least complex (chemical level) to the most complex (body as a whole).

Organization is one of the most important characteristics of body structure. Even the word *organism,* used to denote a living thing, implies organization.

Although the body is a single structure, it is made up of trillions of smaller structures. Atoms and molecules are often referred to as the **chemical level** of organization (see Appendix A). The existence of life depends on the proper levels and proportions of many chemical substances in the cells of the body. Many of the physical and chemical phenomena that play important roles in the

life process will be reviewed in the next chapter. Such information provides an understanding of the physical basis for life and for the study of the next levels of organization so important in the study of anatomy and physiology—cells, tissues, organs, and systems.

Cells are considered to be the smallest "living" units of structure and function in our body. Although long recognized as the simplest units of living matter, cells are far from simple. They are extremely complex, a fact you will discover in Chapter 2.

Tissues are somewhat more complex than cells. By definition a tissue is an organization of many similar cells that act together to perform a common function. Cells are held together and surrounded by varying amounts and varieties of gluelike, nonliving intercellular substances.

Organs are more complex than tissues. An organ is a group of several different kinds of tissues arranged so that they can together act as a unit to perform a special function. For instance, the lungs shown in Figure 1-1 are an example of organization at the organ level.

Systems are the most complex units that make up the body. A system is an organization of varying numbers and kinds of organs arranged so that they can together perform complex functions for the body. The organs of the respiratory system shown in Figure 1-1 permit air to enter the body and travel to the lungs, where the eventual exchange of oxygen and carbon dioxide occurs. Organs of the respiratory system include the nose, the windpipe or trachea, and the complex series of bronchial tubes that permit passage of air into the lungs.

The **body as a whole** is all the atoms, molecules, cells, tissues, organs, and systems that you will study in subsequent chapters of this text. Although capable of being dissected or broken down into many parts, the body is a unified and complex assembly of structurally and functionally interactive components, each working together to ensure healthy survival.

Anatomical Position

Discussions about the body, the way it moves, its posture, or the relationship of one area to another

FIGURE 1–1   *Structural Levels of Organization in the Body.*

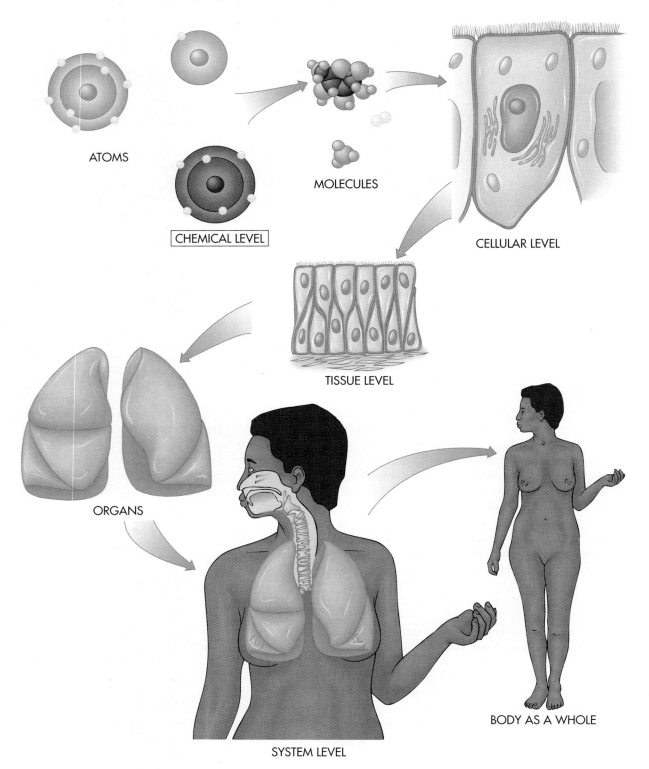

ATOMS

MOLECULES

CHEMICAL LEVEL

CELLULAR LEVEL

TISSUE LEVEL

ORGANS

SYSTEM LEVEL

BODY AS A WHOLE

FIGURE 1–2 *Anatomical Position.* The body is in an erect or standing posture with the arms at the sides and the palms forward. The head and feet also point forward.

assume that the body as a whole is in a specific position called the **anatomical position.** In this reference position (Figure 1-2) the body is in an erect or standing posture with the arms at the sides and palms turned forward. The head and feet also point forward. The anatomical position is a reference position that gives meaning to the directional terms used to describe the body parts and regions.

Supine and **prone** are terms used to describe the position of the body when it is not in the anatomical position. In the supine position the body is lying face upward, and in the prone position the body is lying face downward.

Anatomical Directions

When studying the body, it is often helpful to know where an organ is in relation to other structures. The following directional terms are used in describing relative positions of body parts:

1. Superior and **inferior** (Figure 1-3)—*superior* means "toward the head," and *inferior* means "toward the feet." *Superior* also means "upper" or "above," and *inferior* means "lower" or "below." For example, the lungs are located superior to the diaphragm, whereas the stomach is located inferior to it.

2. Anterior and **posterior** (Figure 1-3)—*anterior* means "front" or "in front of"; *posterior* means "back" or "in back of." In humans, who walk in an upright position, *ventral* (toward the belly) can be used in place of anterior, and *dorsal* (toward the back) can be used for posterior. For example, the nose is on the anterior surface of the body, and the shoulder blades are on its posterior surface.

3. Medial and **lateral** (Figure 1-3)—*medial* means "toward the midline of the body"; *lateral* means "toward the side of the body or away from its midline." For example, the great toe is at the medial side of the foot, and the little toe is at its lateral side. The heart lies medial to the lungs, and the lungs lie lateral to the heart.

4. Proximal and **distal** (Figure 1-3)—*proximal* means "toward or nearest the trunk of the body, or nearest the point of origin of one of its parts"; *distal* means "away from or farthest from the trunk or the point of origin of a body part." For example, the elbow lies at the proximal end of the lower arm, whereas the hand lies at its distal end.

5. Superficial and **deep**—*superficial* means nearer the surface; *deep* means farther away from the body surface. For example, the skin of the arm is superficial to the muscles below it, and the bone of the upper arm is deep to the muscles that surround and cover it.

Planes or Body Sections

To facilitate the study of individual organs or the body as a whole, it is often useful to subdivide or "cut" it into smaller segments. To do this, body planes or sections have been identified by special

FIGURE 1–3 *Directions and Planes of the Body.*

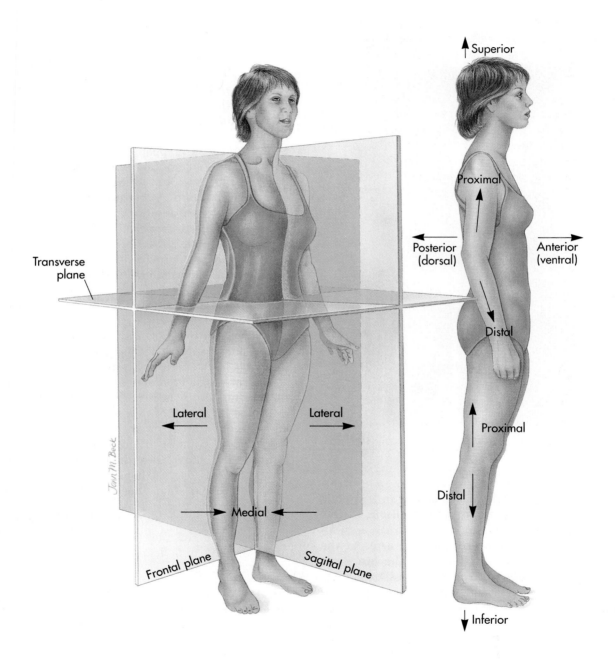

names. Read the following definitions and identify each term in Figure 1-3.

1. **Sagittal**—a sagittal cut or section is a lengthwise plane running from front to back. It divides the body or any of its parts into right and left sides. The sagittal plane shown in Figure 1-3 divides the body into two *equal halves.* This unique type of sagittal plane is called a **midsagittal plane.**

2. **Frontal**—a frontal (*coronal*) plane is a lengthwise plane running from side to side. As you can see in Figure 1-3, a frontal plane divides the body or any of its parts into anterior and posterior (front and back) portions.

3. **Transverse**—a transverse plane is a horizontal or crosswise plane. Such a plane (Figure 1-3) divides the body or any of its parts into upper and lower portions.

● *Body Cavities*

Contrary to its external appearance, the body is not a solid structure. It is made up of open spaces or cavities that in turn contain compact, well-ordered arrangements of internal organs. The two major body cavities are called the **ventral** and **dorsal body cavities.** The location and outlines of the body cavities are illustrated in Figure 1-4. The ventral cavity includes the **thoracic cavity,** a space that you may think of as your chest cavity. Its midportion is a subdivision of the thoracic cavity, called the **mediastinum;** its other subdivisions are called the right and left **pleural cavities.** The ventral cavity in Figure 1-4 is broken down into an **abdominal cavity** and a **pelvic cavity.** Actually, they form only one cavity, the **abdominopelvic cavity,** because no physical partition separates them. In Figure 1-4 a dotted line shows the approximate point of separation between the abdominal and pelvic subdivisions. Notice, however, that an actual physical partition, represented in the figure as a wide band, separates the thoracic cavity from the abdominal cavity. This muscular partition is the **diaphragm.** It is dome-shaped and is the most important muscle for breathing.

To make it easier to locate organs in the large abdominopelvic cavity, anatomists have divided the abdominopelvic cavity into the nine regions

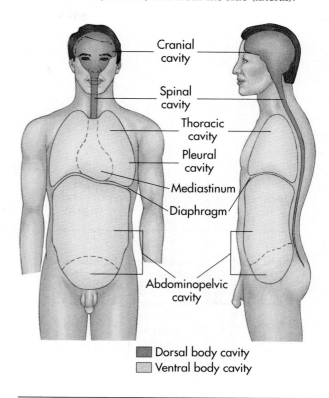

FIGURE 1–4 *Body Cavities.* Location and subdivisions of the dorsal and ventral body cavities as viewed from the front (anterior) and from the side (lateral).

Cranial cavity
Spinal cavity
Thoracic cavity
Pleural cavity
Mediastinum
Diaphragm
Abdominopelvic cavity

■ Dorsal body cavity
■ Ventral body cavity

shown in Figure 1-5 and defined them as follows:

1. **Upper abdominopelvic regions**—the right and left hypochondriac regions and the epigastric region lie above an imaginary line across the abdomen at the level of the ninth rib cartilages.

2. **Middle regions**—the right and left lumbar regions and the umbilical region lie below an imaginary line across the abdomen at the level of the ninth rib cartilages and above an imaginary line across the abdomen at the top of the hip bones.

3. **Lower regions**—the right and left iliac (or inguinal) regions and the hypogastric region lie below an imaginary line across the abdomen at the level of the top of the hip bones.

FIGURE 1–5   *The Nine Regions of the Abdominopelvic Cavity.* The most superficial organs are shown.

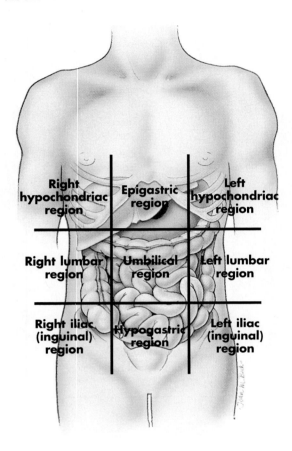

FIGURE 1–6   *Division of the Abdominopelvic Cavity into Four Quadrants.* Diagram showing relationship of internal organs to the four abdominal quadrants.

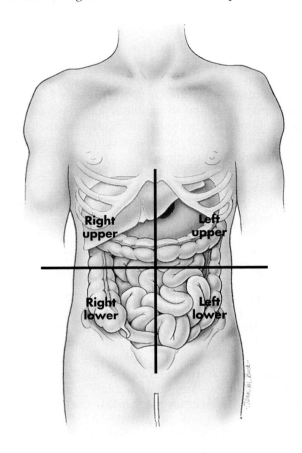

Another and perhaps easier way to divide the abdominopelvic cavity is shown in Figure 1-6. This method is frequently used by health professionals and is useful for locating pain or describing the location of a tumor. As you can see in Figure 1-6, the midsagittal and transverse planes, which were described in the previous section, pass through the navel (umbilicus) and divide the abdominopelvic region into the following **four quadrants:** right upper or superior, right lower or inferior, left upper or superior, and left lower or inferior.

The dorsal cavity shown in Figure 1-4 includes the space inside the skull that contains the brain; it is called the **cranial cavity.** The space inside the spinal column is called the **spinal cavity;** it contains the spinal cord. The cranial and spinal cavities are **dorsal cavities,** whereas the thoracic and abdominopelvic cavities are **ventral cavities.**

Some of the organs in the largest body cavities are visible in Figure 1-7 and are listed in Table 1-1. Find each body cavity in a model of the human body if you have access to one. Try to identify the organs in each cavity, and try to visualize their locations in your own body. Study Figures 1-4 and 1-7.

FIGURE 1–7   *Organs of the Major Body Cavities.*   A view from the front.

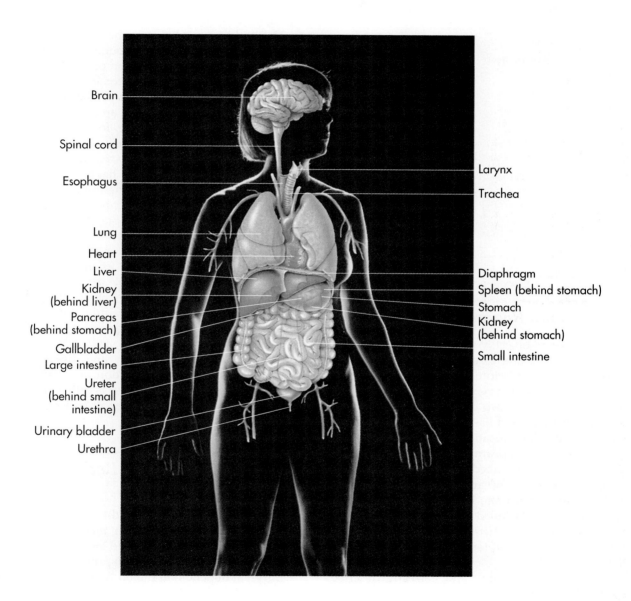

Brain

Spinal cord

Esophagus

Larynx

Trachea

Lung

Heart

Liver

Kidney
(behind liver)

Pancreas
(behind stomach)

Gallbladder

Large intestine

Ureter
(behind small
intestine)

Urinary bladder

Urethra

Diaphragm

Spleen (behind stomach)

Stomach

Kidney
(behind stomach)

Small intestine

TABLE 1–1   Body Cavities

BODY CAVITY	ORGAN(S)
Ventral Body Cavity	
Thoracic cavity	
Mediastinum	Trachea, heart, blood vessels
Pleural cavities	Lungs
Abdominopelvic cavity	
Abdominal cavity	Liver, gallbladder, stomach, spleen, pancreas, small intestine, parts of large intestine
Pelvic cavity	Lower (sigmoid) colon, rectum, urinary bladder, reproductive organs
Dorsal Body Cavity	
Cranial cavity	Brain
Spinal cavity	Spinal cord

⬤ Body Regions

To recognize an object, you usually first notice its overall structure and form. For example, a car is recognized as a car before the specific details of its tires, grill, or wheel covers are noted. Recognition of the human form also occurs as you first identify overall shape and basic outline. However, for more specific identification to occur, details of size, shape, and appearance of individual body areas must be described. Individuals differ in overall appearance because specific body areas such as the face or torso have unique identifying characteristics. Detailed descriptions of the human form require that specific regions be identified and appropriate terms be used to describe them.

The ability to identify and correctly describe specific body areas is particularly important in the health sciences. For a patient to complain of pain in the head is not as specific and therefore not as useful to a physician or nurse as a more specific and localized description. Saying that the pain is facial provides additional information and helps to more specifically identify the area of pain. By using correct anatomical terms such as forehead, cheek, or chin to describe the area of pain, attention can be focused even more quickly on the specific anatomical area that may need attention. Familiarize yourself with the more common terms used to describe specific body regions identified in Figure 1-8 and listed in Table 1-2.

The body as a whole can be subdivided into two major portions or components: **axial** and **appendicular.** The axial portion of the body consists of the head, neck, and torso or trunk; the appendicular portion consists of the upper and lower extremities. Each major area is subdivided as shown in Figure 1-8. Note, for example, that the torso is composed of thoracic, abdominal, and pelvic areas, and the upper extremity is divided into arm, forearm, wrist, and hand components. Although most terms used to describe gross body regions are well understood, misuse is common. The word *leg* is a good example: it refers to the area of the lower extremity between the knee and ankle and not to the entire lower extremity.

The structure of the body changes in many ways and at varying rates during a lifetime. Before young adulthood, it develops and grows; after young adulthood, it gradually undergoes degenerative changes. With advancing age, there is a generalized decrease in size or a wasting away of many body organs and tissues that affects the structure and function of many body areas. This degenerative process is called **atrophy.** Nearly every chapter of this book will refer to a few of these changes.

⬤ The Balance of Body Functions

Although they may have very different structures, all living organisms maintain mechanisms that ensure survival of the body and success in propagating its genes through its offspring.

Survival depends on the body maintaining relatively constant conditions within the body. **Homeostasis** is what physiologists call the relative constancy of the internal environment. The cells of the body live in an internal environment made up mostly of water combined with salts and other dissolved substances. Like fish in a fish-

FIGURE 1–8 *Axial and Appendicular Divisions of the Body.* Specific body regions are labeled.

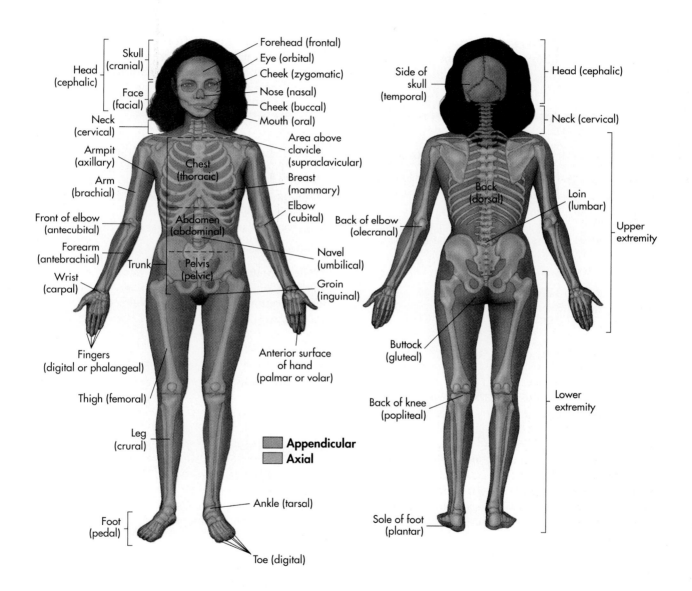

bowl, the cells are able to survive only if the conditions of their watery environment remain stable. The temperature, salt content, acid level (pH), fluid volume and pressure, oxygen concentration, and other vital conditions must remain within acceptable limits. To maintain constant water conditions in a fishbowl, one may add a heater, an air pump, and filters. Likewise, the body has mechanisms that act as heaters, air pumps, and the like, to maintain conditions of its internal fluid environment.

Because the activities of cells and external disturbances are always threatening internal stability, or homeostasis, the body must constantly work to maintain or restore that stability. To accomplish this self-regulation, a highly complex and integrated communication control system is required. The basic type of control system in the body is called a **feedback loop.** The idea of a feedback loop is borrowed from engineering. Figure 1-9, *A*, shows how an engineer would describe the feedback loop that maintains stability of tempera-

TABLE 1–2   Descriptive Terms for Body Regions

BODY REGION	AREA OR EXAMPLE	BODY REGION	AREA OR EXAMPLE
Abdominal (ab-DOM-in-al)	Anterior torso below diaphragm	**Facial, cont'd**	
Antebrachial (an-tee-BRAY-kee-al)	Forearm	**Zygomatic** (zye-go-MAT-ik)	Upper cheek
Antecubital (an-tee-KYOO-bi-tal)	Depressed area just in front of elbow	**Femoral** (FEM-or-al)	Thigh
		Gluteal (GLOO-tee-al)	Buttock
Axillary (AK-si-lair-ee)	Armpit	**Inguinal** (ING-gwi-nal)	Groin
Brachial (BRAY-kee-al)	Arm	**Lumbar** (LUM-bar)	Lower back between ribs and pelvis
Buccal (BUK-al)	Cheek		
Carpal (KAR-pal)	Wrist	**Mammary** (MAM-er-ee)	Breast
Cephalic (se-FAL-ik)	Head	**Occipital** (ok-SIP-i-tal)	Back of lower skull
Cervical (SER-vi-kal)	Neck	**Olecranal** (oh-LEK-kra-nal)	Back of elbow
Cranial (KRAY-nee-al)	Skull	**Palmar** (PAHL-mar)	Palm of hand
Crural (KROOR-al)	Leg	**Pedal** (PED-al)	Foot
Cubital (KYOO-bi-tal)	Elbow	**Pelvic** (PEL-vik)	Lower portion of torso
Cutaneous (kyoo-TANE-ee-us)	Skin (or body surface)	**Perineal** (pair-i-NEE-al)	Area (perineum) between anus and genitals
Digital (DIJ-i-tal)	Fingers or toes	**Plantar** (PLAN-tar)	Sole of foot
Dorsal (DOR-sal)	Back	**Popliteal** (pop-li-TEE-al)	Area behind knee
Facial (FAY-shal)	Face	**Supraclavicular** (soo-pra-kla-VIK-yoo-lar)	Area above clavicle
Frontal (FRON-tal)	Forehead		
Nasal (NAY-zal)	Nose	**Tarsal** (TAR-sal)	Ankle
Oral (OR-al)	Mouth	**Temporal** (TEM-por-al)	Side of skull
Orbital or **ophthalmic** (OR-bi-tal or op-THAL-mik)	Eyes	**Thoracic** (tho-RAS-ik)	Chest
		Umbilical (um-BILL-ih-kal)	Area around navel or umbilicus
		Volar (VO-lar)	Palm or sole

ture in a building. Cold winds outside a building may cause a decrease in building temperature below normal. A **sensor,** in this case a thermometer, detects the change in temperature. Information from the sensor *feeds back* to a **control center**—a thermostat in this example—that compares the actual temperature to the normal temperature and responds by activating the building's furnace. The furnace is called an **effector** because it has an effect on the controlled condi-

tion (temperature). Because the sensor continually feeds information back to the control center, the furnace will be automatically shut off when the temperature has returned to normal.

As you can see in Figure 1-9, *B*, the body uses a similar feedback loop in restoring body temperature when we become chilled. Nerve endings that act as temperature sensors feed information to a control center in the brain that compares actual body temperature to normal body temperature. In

FIGURE 1–9 *Negative Feedback Loops.* **A,** An engineer's diagram showing how a relatively constant room temperature (*controlled condition*) can be maintained. A thermostat (*control center*) receives feedback information from a thermometer (*sensor*) and responds by counteracting a change from normal by activating a furnace (*effector*). **B,** A physiologist's diagram showing how a relatively constant body temperature (*controlled condition*) can be maintained. The brain (*control center*) receives feedback information from nerve endings called cold receptors (*sensors*) and responds by counteracting a change from normal by activating shivering by muscles (*effectors*).

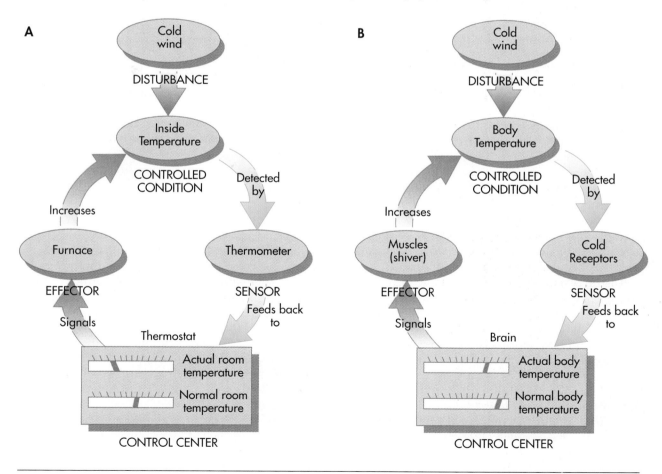

response to a chill, the brain sends nerve signals to muscles that shiver. Shivering produces heat that increases our body temperature. We stop shivering when feedback information tells the brain that body temperature has increased to normal.

Feedback loops such as those shown in Figure 1-9 are called **negative feedback** loops because they oppose, or negate, a change in a controlled condition. Most homeostatic control loops in the body involve negative feedback because reversing changes back toward a normal value tends to stabilize conditions—exactly what homeostasis is all about. An example of a negative feedback loop occurs when decreasing blood oxygen concentration caused by muscles using oxygen during exercise is counteracted by an increase in breathing to bring the blood oxygen level back up to normal. Another example is the excretion of larger than usual volumes of urine when the volume of fluid in the body is greater than the normal, ideal amount.

Although not common, **positive feedback** loops exist in the body and are involved in normal function. Positive feedback control loops are stimulatory. Instead of opposing a change in the internal environment and causing a "return to normal," positive feedback loops amplify or reinforce the change that is occurring. This type of feedback loop causes an ever-increasing rate of events to occur until something stops the process. An example of a positive feedback loop includes the events that cause rapid increases in uterine contractions before the birth of a baby. Another example is the increasingly rapid sticking together of blood cells called *platelets* to form a plug that begins formation of a blood clot.

It is important to realize that homeostatic control mechanisms can only maintain a *relative* constancy. All homeostatically controlled conditions in the body do not remain absolutely constant. Rather, conditions normally fluctuate near a normal, ideal value. Thus body temperature, for example, rarely remains exactly the same for very long; it usually fluctuates up and down near a person's normal body temperature.

Because all organs function to help maintain homeostatic balance, we will be discussing negative and positive feedback mechanisms often throughout the remaining chapters of this book.

Exercise Physiology

Exercise physiologists study the effects of exercise on the body organ systems. They are especially interested in the complex control mechanisms that preserve or restore homeostasis during or immediately after periods of strenuous physical activity. Exercise, defined as any significant use of skeletal muscles, is a normal activity with beneficial results. However, exercise disrupts homeostasis. For example, when muscles are worked, the core body temperature rises and blood CO_2 levels increase. These and many other body functions quickly deviate from "normal ranges" that exist at rest. Complex control mechanisms must then "kick in" to restore homeostasis.

As a scientific discipline, exercise physiology attempts to explain many body processes in terms of how they maintain homeostasis. Exercise physiology has many practical applications in therapy and rehabilitation, athletics, occupational health, and general wellness. This specialty concerns itself with the function of the whole body, not just one or two body systems.

Before leaving this brief introduction to physiology, we must pause to state the important principle that maintaining the balance of body functions is related to age. During childhood, homeostatic functions gradually become more and more efficient and effective. They operate with maximum efficiency and effectiveness during young adulthood. During late adulthood and old age, they gradually become less and less efficient and effective. Changes and functions occurring during the early years are called *developmental processes*; those occurring after young adulthood are called *aging processes*. In general, developmental processes improve efficiency of functions; aging processes usually diminish it.

OUTLINE SUMMARY

STRUCTURAL LEVELS OF ORGANIZATION
(Figure 1-1)

A Organization is an outstanding characteristic of body structure

B The body is a unit constructed of the following smaller units:

  1 Cells—the smallest structural units; organizations of various chemicals

  2 Tissues—organizations of similar cells

  3 Organs—organizations of different kinds of tissues

  4 Systems—organizations of many different kinds of organs

ANATOMICAL POSITION (Figure 1-2)
Standing erect with the arms at the sides and palms turned forward

ANATOMICAL DIRECTIONS

A Superior—toward the head, upper, above

  Inferior—toward the feet, lower, below

B Anterior—front, in front of (same as ventral in humans)

  Posterior—back, in back of (same as dorsal in humans)

C Medial—toward the midline of a structure

  Lateral—away from the midline or toward the side of a structure

D Proximal—toward or nearest the trunk, or nearest the point of origin of a structure

  Distal—away from or farthest from the trunk, or farthest from a structure's point of origin

E Superficial—nearer the body surface

  Deep—farther away from the body surface

PLANES OR BODY SECTIONS (Figure 1-3)

A Sagittal plane—lengthwise plane that divides a structure into right and left sections

B Midsagittal—sagittal plane that divides the body into two equal halves

C Frontal (coronal) plane—lengthwise plane that divides a structure into anterior and posterior sections

D Transverse plane—horizontal plane that divides a structure into upper and lower sections

BODY CAVITIES (Figure 1-4):

A Ventral cavity

  1 Thoracic cavity

    a Mediastinum—midportion of thoracic cavity; heart and trachea located in mediastinum

    b Pleural cavities—right lung located in right pleural cavity, left lung in left pleural cavity

  2 Abdominopelvic cavity

    a Abdominal cavity contains stomach, intestines, liver, gallbladder, pancreas, and spleen

    b Pelvic cavity contains reproductive organs, urinary bladder, and lowest part of intestine

    c Abdominopelvic regions (Figures 1-5 and 1-6)

      (1) Nine regions

      (2) Four quadrants

B Dorsal cavity

  1 Cranial cavity contains brain

  2 Spinal cavity contains spinal cord

BODY REGIONS (Figure 1-8)

A Axial region—head, neck, and torso or trunk

B Appendicular region—upper and lower extremities

THE BALANCE OF BODY FUNCTIONS

A Survival of the individual and of the genes is the body's most important business

B Survival depends on the maintenance or restoration of homeostasis (relative constancy of the internal environment; Figure 1-9); the body uses negative feedback loops and, less often, positive feedback loops to maintain or restore homeostas is

C All organs function to maintain homeostasis

D Body functions are related to age; peak efficiency is during young adulthood, diminishing efficiency occurs after young adulthood

---

NEW WORDS

abdominopelvic quadrants (4)	thoracic	superficial	organ
abdominopelvic regions (9)	control center	deep	system
anatomical position	directional terms	effector loop	physiology
anatomy	superior	feedback	planes of section
atrophy	inferior	homeostasis	sagittal
cavities	anterior	mediastinum	midsagittal
abdominal	posterior	negative feedback	frontal
cranial	ventral	organization	transverse
pelvic	dorsal	(structural levels)	positive feedback
pleural	medial	chemical	prone
spinal	lateral	cellular	sensor
	proximal	tissue	supine
	distal		

CHAPTER TEST

1. The study of the structure of an organism and the relationship of its parts is called _____; the study of the functions of that organism is called _____.

2. An organization of many similar cells that together perform a common function is called a _____.

3. The term _____ means "toward the side of the body."

4. A frontal plane divides the body or any of its parts into _____ and _____ portions.

5. The body as a whole can be subdivided into two major portions: _____ and _____.

6. The relative constancy of the body's internal environment is described by the term _____.

Match the body area in column B with the area or example in column A. (Only one answer is correct.)

COLUMN A

7. _____ Skull
8. _____ Groin
9. _____ Breast
10. _____ Sole of foot
11. _____ Chest
12. _____ Wrist
13. _____ Fingers or toes
14. _____ Armpit
15. _____ Mouth
16. _____ Buttock

COLUMN B

a. Mammary
b. Thoracic
c. Digital
d. Carpal
e. Inguinal
f. Axillary
g. Gluteal
h. Cranial
i. Oral
j. Plantar

17. The mediastinum is a subdivision of the:
    a. Thoracic cavity
    b. Pleural cavity
    c. Abdominal cavity
    d. Pelvic cavity

18. Which of the following is an example of a lower abdominal region?
    a. Epigastric region
    b. Umbilical region
    c. Right hypochondriac region
    d. Hypogastric region

19. Which of the following represents the least complex of the structural levels of organization?
    a. System
    b. Tissue
    c. Cell
    d. Organ

20. The diaphragm separates the:
    a. Right and left pleural cavities
    b. Abdominal and pelvic cavities
    c. Thoracic and abdominal cavities
    d. Mediastinum and pleural cavities

REVIEW QUESTIONS

1. Name the four kinds of structural units of the body. Define each briefly.
2. In one word, what is the one dominant function of the body or of any living thing?
3. Explain briefly what *homeostasis* means.
4. What do *proximal* and *distal* mean?
5. On what surface of the foot are the toenails located?
6. What structures lie lateral to the bridge of the nose?
7. Which joint—hip or knee—lies at the distal end of the thigh?
8. Define *anatomical position*.
9. List the major subdivisions of the axial and appendicular areas of the body.
10. List and define the four major planes of section.

CRITICAL THINKING

11. In what cavity could you find each of the following?
    brain
    esophagus
    gallbladder
    heart
    liver
12. Identify and discuss the mechanisms of negative and positive feedback loops in homeostasis.

Cells and Tissues

2

OBJECTIVES

After you have completed this chapter, you should be able to:

1. Identify and discuss the basic structure and function of the three major components of a cell.

2. List and briefly discuss the functions of the primary cellular organelles.

3. Compare the major passive and active transport processes that act to move substances through cell membranes.

4. Compare and discuss DNA and RNA and their function in protein synthesis.

5. Discuss the stages of mitosis and explain the importance of cellular reproduction.

6. Explain how epithelial tissue is grouped according to shape and arrangement of cells.

7. List and briefly discuss the major types of connective and muscle tissue.

8. List the three structural components of a neuron.

About 300 years ago Robert Hooke looked through his microscope—one of the very early, somewhat primitive ones—at some plant material. What he saw must have surprised him. Instead of a single magnified piece of plant material, he saw many small pieces. Because they reminded him of miniature prison cells, that is what he called them—cells. Since Hooke's time, thousands of individuals have examined thousands of plant and animal specimens and found them all, without exception, to be composed of cells. This fact, that cells are the smallest structural units of living things, has become the foundation of modern biology. Many living things are so simple that they consist of just one cell. The human body, however, is so complex that it consists not of a few thousand or millions or even billions of cells but of many trillions of them. This chapter discusses cells first and then tissues.

Cells

SIZE AND SHAPE

Human cells are microscopic in size; that is, they can be seen only when magnified by a microscope. However, they vary considerably in size. An ovum (female sex cell), for example, has a diameter of a little less than 1000 micrometers (about ¹⁄₂₅ of an inch), whereas red blood cells have a diameter of only 7.5 micrometers. Cells differ even more notably in shape than in size. Some are flat, some are brick shaped, some are thread-like, and some have irregular shapes.

COMPOSITION

Cells contain **cytoplasm** (SI-to-plazm), or "living matter," a substance that exists only in cells. The term *cyto-* is a Greek combining form and denotes a relationship to a cell. Each cell in the body is surrounded by a thin membrane, the **plasma membrane**. This membrane separates the cell contents from the dilute salt water solution called **interstitial** (in-ter-STISH-al) **fluid,** or simply **tissue fluid,** that bathes every cell in the body. Numerous specialized structures called **organelles** (or-gan-ELZ), which will be described in subsequent sections, are contained within the cytoplasm of each cell. A small, circular body called the **nucleus** (NOO-kle-us) is also inside the cell.

Important information related to body composition is included in Appendix A. You are encouraged to review this material, which includes a discussion of the chemical elements and compounds important to body structure and function.

PARTS OF THE CELL

The three main parts of a cell are:

1. Plasma membrane
2. Cytoplasm
3. Nucleus

The plasma membrane surrounds the entire cell, forming its outer boundary. The cytoplasm is all the living material inside the cell (except the nucleus). The nucleus is a large, membrane-bound structure in most cells that contains the genetic code.

Plasma Membrane

As the name suggests, the **plasma membrane** is the membrane that encloses the cytoplasm and forms the outer boundary of the cell. It is an incredibly delicate structure—only about 7 nm (nanometers) or 3/10,000,000 of an inch thick! Yet it has a precise, orderly structure (Figure 2-1). Two layers of phosphate-containing fat molecules called **phospholipids** form a fluid framework for the plasma membrane. Another kind of fat molecule called *cholesterol* is also a component of the plasma membrane. Cholesterol helps stabilize the phospholipid molecules to prevent breakage of the plasma membrane. Note in Figure 2-1 that protein molecules dot the surfaces of the membrane and extend all the way through the phospholipid framework.

Despite its seeming fragility, the plasma membrane is strong enough to keep the cell whole and intact. It also performs other life-preserving functions for the cell. It serves as a well-guarded gateway between the fluid inside the cell and the fluid around it. Certain substances move through it, but it bars the passage of others. The plasma membrane even functions as a communication device. How? Some protein molecules on the membrane's outer surface serve as receptors for certain other molecules when the other molecules contact them. In other words, certain molecules bind to certain receptor proteins. For example, some hormones (chemicals secreted into blood from ductless glands) bind to membrane receptors, and a change in cell functions follows. We might therefore think of such hormones as chemical messages, communicated to cells by binding to their cytoplasmic membrane receptors.

The plasma membrane also identifies a cell as coming from one particular individual. Its surface proteins serve as positive identification tags because they occur only in the cells of that individual. A practical application of this fact is made in *tissue typing,* a procedure performed before an organ from one individual is transplanted into another. Carbohydrate chains attached to the surface of cells often play a role in the identification of cell types.

Cytoplasm

Cytoplasm is the specialized living material of cells. It lies between the plasma membrane and the nucleus, which can be seen in Figure 2-2 as a round or spherical structure in the center of the cell. Numerous small structures are part of the cytoplasm, along with the fluid that serves as the interior environment of each cell. As a group, the small structures that make up much of the cytoplasm are called **organelles.** This name means "little organs," an appropriate name because they function like organs function for the body.

Look again at Figure 2-2. Notice how many different kinds of structures you can see in the cytoplasm of this cell. A little more than a generation ago, almost all of these organelles were

FIGURE 2–1 *Structure of the Plasma Membrane.* Note that protein molecules may penetrate completely through the two layers of phospholipid molecules.

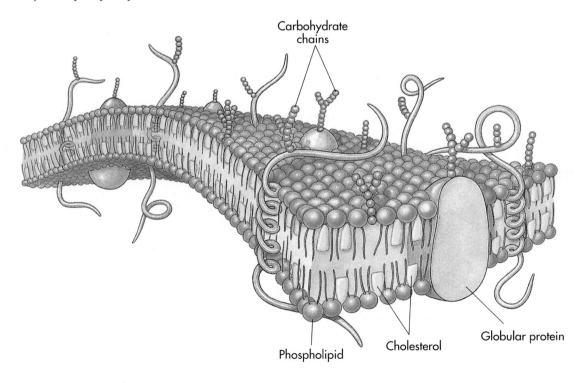

FIGURE 2–2 *General Characteristics of the Cell.* Artist's interpretation of cell structure.

unknown. They are so small that they are invisible even when magnified 1000 times by a light microscope. Electron microscopes brought them into view by magnifying them many thousands of times. We shall briefly discuss the following organelles, which are found in cytoplasm (see also Table 2-1):

1. Ribosomes
2. Endoplasmic reticulum
3. Golgi apparatus
4. Mitochondria
5. Lysosomes
6. Centrioles
7. Cilia
8. Flagella

Ribosomes. Organelles called **ribosomes** (RI-bo-sohms), shown as dots in Figure 2-2, are very tiny particles found throughout the cell. Some ribosomes are found temporarily attached to a network of membranous canals called *endoplasmic reticulum (ER)*—another type of organelle described in the next paragraph. Ribosomes may also be free in the cytoplasm. Ribosomes perform a very complex function; they make enzymes and other protein compounds. Their nickname, "protein factories," indicates this function.

Endoplasmic reticulum. An **endoplasmic reticulum** (en-doe-PLAZ-mik ree-TIK-yoo-lum) **(ER)** is a system of membranes forming a network of connecting sacs and canals that wind back and forth through a cell's cytoplasm, all the way from the nucleus and almost to the plasma membrane. The tubular passageways or canals in the ER carry proteins and other substances through the cytoplasm of the cell from one area to another. There are two types of ER: *rough* and *smooth*. Rough ER gets its name from the fact that many ribosomes are attached to its outer surface, giving it a rough texture similar to sandpaper. As ribosomes make their proteins, they may attach to the rough ER and drop the protein into the interior of the ER. The ER then transports the proteins to areas where chemical processing takes place. These areas of the ER are so full that ribosomes do not attach to pass their proteins, giving this type of ER a smooth texture. Fats, carbohydrates, and proteins that make up cellular membrane material are manufactured in smooth ER. Thus the smooth ER makes new membrane for the cell. To sum up:

TABLE 2–1   Some Major Cell Structures and Their Functions

CELL STRUCTURE	FUNCTION(S)
Plasma membrane	Serves as the boundary of the cell; protein and carbohydrate molecules on outer surface of plasma membrane perform various functions; for example, they serve as markers that identify cells of each individual or as receptor molecules for certain hormones
Ribosomes	Synthesize proteins; a cell's "protein factories"
Endoplasmic reticulum (ER)	Ribosomes attached to rough ER synthesize proteins; smooth ER synthesizes lipids and certain carbohydrates
Golgi apparatus	Chemically processes, then packages substances from the ER
Mitochondria	ATP synthesis; a cell's "powerhouses"
Lysosomes	A cell's "digestive system"
Centrioles	Function in cell reproduction
Cilia	Short, hairlike extensions on the free surfaces of some cells capable of movement
Flagella	Single and much larger projections of cell surfaces than cilia; the only example in humans is the "tail" of a sperm cell
Nucleus	Dictates protein synthesis, thereby playing an essential role in other cell activities, namely active transport, metabolism, growth, and heredity
Nucleoli	Play an essential role in the formation of ribosomes

rough ER receives and transports newly made proteins and smooth ER makes new membrane.

Golgi Apparatus. The **Golgi** (GOL-jee) **apparatus** consists of tiny, flattened sacs stacked on one another near the nucleus. Little bubbles, or sacs, break off the smooth ER and carry new proteins and other compounds to the sacs of the Golgi apparatus. These little sacs, also called **vesicles,** fuse with the Golgi sacs and allow the contents of both to mingle. The Golgi apparatus chemically processes the molecules from the ER, then packages them into little vesicles that break away from the Golgi apparatus and move slowly outward to the plasma membrane. Each vesicle fuses with the plasma membrane, opens to the outside of the cell, and releases its contents. An example of a Golgi apparatus product is the slippery substance called mucus. If we wanted to nickname the Golgi apparatus, we might call it the cell's "chemical processing and packaging center."

Mitochondria. Mitochondria (my-toe-KON-dree-ah) are another kind of organelle in all cells. Mitochondria are so tiny that a lineup of 15,000 or more of them would fill a space only about 2.5 cm or 1 inch long. Two membranous sacs, one inside the other, compose a mitochondrion. The inner membrane forms folds that look like miniature incomplete partitions. Within a mitochondrion's fragile walls, complex, energy-releasing chemical reactions occur continuously. Because these reactions supply most of the power for cellular work, mitochondria have been nicknamed the cell's "power plants." The survival of cells and therefore of the body depends on mitochondrial chemical reactions. Enzymes (molecules that promote specific chemical reactions), which are found in mitochondrial walls and inner substance, use oxygen to break down glucose and other nutrients to release energy required for cellular work. The process is called *aerobic* or *cellular respiration.*

Lysosomes. The **lysosomes** (LYE-so-sohms) are membranous-walled organelles that in their active stage look like small sacs, often with tiny particles in them (Figure 2-2). Because lysosomes contain chemicals (enzymes) that can digest food compounds, one of their nicknames is "digestive bags." Lysosomal enzymes can also digest substances other than foods. For example, they can digest and thereby destroy microbes that invade the cell. Thus lysosomes can protect cells against destruction by microbes. Yet, paradoxically, lysosomes sometimes kill cells instead of protecting them. If their powerful enzymes escape from the lysosome sacs into the cytoplasm, they kill the cell by digesting it. This fact has earned lysosomes their other nickname, which is "suicide bags."

Centrioles. The **centrioles** (SEN-tree-olz) are paired organelles. Two of these rod-shaped structures exist in every cell. They are arranged so that they lie at right angles to each other (Figure 2-2). Each centriole is composed of fine tubules that play an important role during cell division.

Cilia. Cilia (SIL-ee-ah) are extremely fine, almost hairlike extensions on the exposed or free surfaces of some cells. Cilia are organelles capable of movement. One cell may have a hundred or more cilia capable of moving together in a wave-like fashion over the surface of a cell. They often have highly specialized functions. For example, by moving as a group in one direction, they propel mucus upward over the cells that line the respiratory tract.

Flagella. A **flagellum** (flah-JEL-um) is a single projection extending from the cell surface. Flagella are much larger than cilia. In the human, the only example of a flagellum is the "tail" of the male sperm cell. Propulsive movements of the flagellum make it possible for sperm to "swim" or move toward the ovum after they are deposited in the female reproductive tract (Figure 2-3).

Nucleus

Viewed under a light microscope, the **nucleus** of a cell looks like a very simple structure—just a small sphere in the central portion of the cell. However, its simple appearance belies the complex and critical role it plays in cell function. The nucleus ultimately controls every organelle in the cytoplasm. It also controls the complex process of cell reproduction. In other words, the nucleus must function properly for a cell to accomplish its normal activities and be able to duplicate itself.

Note that the cell nucleus in Figure 2-2 is surrounded by a **nuclear envelope.** The envelope, made up of two separate membranes, encloses a special type of cell material in the nucleus called **nucleoplasm.** Nucleoplasm contains a number of specialized structures; two of the most important

FIGURE 2–3 *Human Sperm.* Note the tail-like flagellum on each sperm cell. The flagella are so long that they do not fit into the photograph at this magnification.

Flagellum

are shown in Figure 2-2. They are the **nucleolus** (noo-KLEE-oh-lus) and the **chromatin** (KRO-mah-tin) **granules.**

Nucleolus. The nucleolus is critical in protein formation because it "programs" the formation of ribosomes in the nucleus. The ribosomes then migrate through the nuclear envelope into the cytoplasm of the cell and produce proteins.

Chromatin and Chromosomes. Chromatin granules in the nucleus are threadlike structures made of proteins and hereditary material called **DNA** or **deoxyribonucleic** (dee-OK-see-rye-bo-noo-KLEE-ik) **acid.** DNA is the genetic material often described as the chemical "blueprint" of the body. It determines everything from gender to body build and hair color in every human being. During cell division, DNA molecules become tightly coiled. They then look like short, rodlike structures and are called **chromosomes.** The importance and function of DNA will be explained in greater detail in the section on cell reproduction later in this chapter. Information about DNA can also be found in *Appendix A: Chemistry of Life,* which begins on p. 426.

RELATIONSHIP OF CELL STRUCTURE AND FUNCTION

Every human cell performs certain functions; some maintain the cell's survival, and others help maintain the body's survival. In many instances, the number and type of organelles allow cells to differ dramatically in terms of their specialized functions. For example, cells that contain large numbers of mitochondria, such as heart muscle cells, are capable of sustained work. Why? Because the numerous mitochondria found in these cells supply the necessary energy required for rhythmic and ongoing contractions. Movement of the flagellum of a sperm cell is another example of the way a specialized organelle has a specialized function. The sperm's flagellum propels it through the reproductive tract of the female, thus increasing the chances of successful fertilization. This is how and why organizational structure at the cellular level is so important for function in living organisms. There are examples in every chapter of the text to illustrate how structure and function are intimately related at every level of body organization.

Movement of Substances Through Cell Membranes

The plasma membrane in every healthy cell separates the contents of the cell from the tissue fluid that surrounds it. At the same time the membrane must permit certain substances to enter the cell and allow others to leave. Heavy traffic moves continuously in both directions through cell membranes. Molecules of water, foods, gases, wastes, and many other substances stream in and out of all cells in endless procession. A number of processes allow this mass movement of substances into and out of cells. These transport processes are classified under two general headings:

1. Passive transport processes
2. Active transport processes

As implied by their name, active transport processes require the expenditure of energy by the cell, and passive transport processes do not. The energy required for active transport processes is obtained from a very important chemical substance called **adenosine triphosphate** (ah-DEN-o-sen tri-FOS-fate) or **ATP.** ATP is produced in the mitochondria using energy from nutrients and is capable of releasing that energy to do work in the cell. For active transport processes to occur, the breakdown of ATP and the use of the released energy are required.

The details of active and passive transport of substances across cell membranes are much easier to understand if you keep in mind the following two key facts: (1) in passive transport processes, no cellular energy is required to move substances from a high concentration to a low concentration; and (2) in active transport processes, cellular energy is required to move substances from a low concentration to a high concentration.

PASSIVE TRANSPORT PROCESSES

The primary passive transport processes that move substances through the cell membranes include the following:

1. Diffusion
   a. Osmosis
   b. Dialysis
2. Filtration

Scientists describe the movement of substances in passive systems as going "down a concentration gradient." This means that substances in passive systems move from a region of high concentration to a region of low concentration until they reach equal proportions on both sides of the membrane. As you read the next few paragraphs, refer to Table 2-2, which summarizes important information about passive transport processes.

Diffusion

Diffusion is a good example of a passive transport process. Diffusion is the process by which substances scatter themselves evenly throughout an available space. The system does not require additional energy for this movement. To demon-

TABLE 2–2	Passive Transport Processes		
PROCESS	**DESCRIPTION**		**EXAMPLES**
Diffusion	Movement of particles through a membrane from an area of high concentration to an area of low concentration—that is, down the concentration gradient		Movement of carbon dioxide out of all cells; movement of sodium ions into nerve cells as they conduct an impulse
Osmosis	Diffusion of water through a selectively permeable membrane in the presence of at least one impermeant solute		Diffusion of water molecules into and out of cells to correct imbalances in water concentration
Filtration	Movement of water and small solute particles, but not larger particles, through a filtration membrane; movement occurs from area of high pressure to area of low pressure		In the kidney, water and small solutes move from blood vessels but blood proteins and blood cells do not, thus beginning the formation of urine

FIGURE 2–4   *Diffusion.* Note that the membrane is permeable to glucose and water and that it separates a 10% glucose solution from a 20% glucose solution. The container on the left shows the two solutions separated by the membrane at the start of diffusion. The container on the right shows the result of diffusion after time.

 Tonicity

A salt (NaCl) solution is said to be **isotonic** (*iso* = equal) if it contains the same concentration of salt normally found in a living red blood cell. A 0.9% NaCl solution is isotonic; that is, it contains the same level of NaCl as found in red cells. A solution that contains a higher level of salt (above 0.9%) is said to be **hypertonic** (*hyper* = above) and one containing less (below 0.9%) is hypotonic (*hypo* = below). With what you now know about filtration, diffusion, and osmosis, can you predict what would occur if red blood cells were placed in isotonic, hypotonic, and hypertonic solutions?

Examine the figures. Note that red blood cells placed in isotonic solution remain unchanged because there is no effective difference in salt or water concentrations. The movement of water into and out of the cells is about equal. This is not the case

with red cells placed in hypertonic salt solution; they immediately lose water from their cytoplasm into the surrounding salty solution, and they shrink. This process is called **crenation.**

The opposite occurs if red cells are placed in a hypotonic solution; they swell as water enters the cell from the surrounding dilute solution. Eventually the cells break or **lyse,** and the hemoglobin they contain is released into the solution.

Hypotonic solution (cells lyse)   Isotonic solution   Hypertonic solution (cells crenate)

strate diffusion of particles throughout a fluid, perform this simple experiment the next time you pour yourself a cup of coffee or tea. Place a cube of sugar on a teaspoon and lower it gently to the bottom of the cup. Let it stand for 2 or 3 minutes, and then, holding the cup steady, take a sip off the top. It will taste sweet. Why? Because some of the sugar molecules will have diffused from the area of high concentration near the sugar cube at the bottom of the cup to the area of low concentration at the top of the cup.

The process of diffusion is shown in Figure 2-4. Note that both substances diffuse rapidly through the membrane in both directions. However, as indicated by the green arrows, more glucose moves out of the 20% solution, where the concentration is higher, into the 10% solution, where the concentration is lower, than in the opposite direction. This is an example of movement down a concentration gradient. Simultaneously, more water moves from the 10% solution, where there are more water molecules, into the 20% solution, where there are fewer water molecules. This is also an example of movement down a concentration gradient. Water moves from high to low concentration. The result? Equilibration (balancing) of the concentrations of the two solutions after an interval of time. From then on, equal amounts of glucose will diffuse in both directions, as will equal amounts of water.

Osmosis and dialysis. Osmosis (os-MO-sis) and **dialysis** (dye-AL-i-sis) are specialized examples of diffusion. In both cases, diffusion occurs across a selectively permeable membrane. The plasma membrane of a cell is said to be selectively permeable because it permits the passage of certain substances but not others. This is a necessary property if the cell is to permit some substances, such as nutrients, to gain entrance to the cell while excluding others. Osmosis is the diffusion of *water* across a selectively permeable membrane. However, in the case of dialysis, substances called **solutes,** which are dissolved particles in water, move across a selectively permeable membrane by diffusion.

Filtration

Filtration is the movement of water and solutes through a membrane because of a greater pushing force on one side of the membrane than on the other side. The force is called *hydrostatic pressure,* which is simply the force or weight of a fluid pushing against some surface. A principle about filtration that is of great physiological importance is that it always occurs *down* a hydrostatic pressure gradient. This means that when two fluids have unequal hydrostatic pressures and are separated by a membrane, water and diffusible solutes or particles (those to which the membrane is permeable) filter out of the solution that has the higher hydrostatic pressure into the solution that has the lower hydrostatic pressure. Filtration is the process responsible for urine formation in the kidney; wastes are filtered out of the blood into the kidney tubules because of a difference in hydrostatic pressure.

ACTIVE TRANSPORT PROCESSES

Active transport is the uphill movement of a substance through a living cell membrane. *Uphill* means "up a concentration gradient" (that is, from a lower to a higher concentration). The energy required for this movement is obtained from ATP. Because the formation and breakdown of ATP requires complex cellular activity, active transport mechanisms can take place only through living membranes. Table 2-3 summarizes active transport processes.

Ion Pumps

A specialized cellular component called the ion pump makes possible a number of active transport mechanisms. An ion pump is a protein structure in the cell membrane called a *carrier.* The ion pump uses energy from ATP to actively move ions across cell membranes *against* their concentration gradients. "Pump" is an appropriate term because it suggests that active transport moves a substance in an uphill direction just as a water pump, for example, moves water uphill.

Ion pumps are very specific, and different ion pumps are required to move different types of ions. For example, sodium pumps move sodium ions only. Likewise, calcium pumps move calcium ions and potassium pumps move potassium ions.

Some ion pumps are "coupled" to one another so that two or more different substances may be moved through the cell membrane at one time. For example, the **sodium-potassium pump** is an ion pump that pumps sodium ions out of a cell

TABLE 2–3   Active Transport Processes

PROCESS	DESCRIPTION		EXAMPLES
Ion Pump	Movement of solute particles from an area of low concentration to an area of high concentration (up the concentration gradient) by means of a carrier protein structure		In muscle cells, pumping of nearly all calcium ions to special compartments—or out of the cell
Phagocytosis	Movement of cells or other large particles into cell by trapping it in a section of plasma membrane that pinches off inside the cell		Trapping of bacterial cells by phagocytic white blood cells
Pinocytosis	Movement of fluid and dissolved molecules into a cell by trapping them in a section of plasma membrane that pinches off inside the cell		Trapping of large protein molecules by some body cells

while it pumps potassium ions into the cell. Because both ions are moved against their concentration gradients, this pump creates a high sodium concentration outside the cell and a high potassium concentration inside the cell. Such a pump is required to remove sodium from the inside of a nerve cell after it has rushed in as a result of the passage of a nerve impulse. Some ion pumps are coupled with other specific carriers that transport glucose, amino acids, and other substances.

Phagocytosis and Pinocytosis

Phagocytosis (fag-o-sye-TOE-sis) is another example of how a cell can use its active transport mechanism to move an object or substance through the plasma membrane and into the cytoplasm. The term *phagocytosis* comes from a Greek word meaning "to eat." The word is appropriate because the process permits a cell to engulf and literally "eat" foreign material (Figure 2-5). Cer-

tain white blood cells destroy bacteria in the body by phagocytosis. During this process the cell membrane forms a pocket around the material to be moved into the cell and, by expenditure of energy from ATP, the object is moved to the interior of the cell. Once inside the cytoplasm, the bacterium fuses with a lysosome and is destroyed.

Pinocytosis (pin-o-sye-TOE-sis) is an active transport mechanism used to incorporate fluids or dissolved substances into cells. Again, the term is appropriate because it comes from the Greek word meaning "drink."

◗ Cell Reproduction and Heredity

All human cells that reproduce do so by a process called **mitosis** (my-TOE-sis). During this process a cell divides to multiply; one cell divides to form two cells. Cell reproduction and ultimately the transfer of heritable traits is closely tied to the production of proteins. Two *nucleic acids*, **ribonucleic acid** or **RNA** in the cytoplasm and **deoxyribonucleic acid** or **DNA** in the nucleus play crucial roles in protein synthesis.

DNA MOLECULE AND GENETIC INFORMATION

Chromosomes, which are composed largely of DNA, make heredity possible. The "genetic information" contained in DNA molecules ultimately determines the transmission and expression of heritable traits such as skin color and blood group from each generation of parents to their children.

FIGURE 2–5 *Phagocytosis of Bacteria by a White Blood Cell.* Phagocytosis is an active transport mechanism that requires expenditure of energy. Note how an extension of cytoplasm envelops the bacteria, which are drawn through the cell membrane and into the cytoplasm.

Structurally, the DNA molecule resembles a long, narrow ladder made of a pliable material. It is twisted round and round its axis, taking on the shape of a double helix. Each DNA molecule is made of many smaller units, namely, a sugar, bases, and phosphate units. The bases are adenine, thymine, guanine, and cytosine. As you can see in Figure 2-6, *A*, each step in the DNA ladder consists of a pair of bases. Only two combinations of bases occur, and the same two bases invariably pair off with each other in a DNA molecule. Adenine always binds to thymine, and cytosine always binds to guanine. This characteristic of DNA structure is called **complementary base pairing.**

A **gene** is a specific segment of base pairs in a chromosome. Although the types of base pairs in all chromosomes are the same, the order or *sequence* of base pairs is not the same. This fact has tremendous functional importance because it is the sequence of base pairs in each gene of each chromosome that determines heredity. Each gene directs the synthesis of one kind of protein molecule that may function, for example, as an enzyme, a structural component of a cell, or a specific hormone. In humans having 46 chromosomes in each body cell, the nuclear DNA represents a total genetic information package or **genome** (JEE-nohm) of over 3 *billion* base pairs in about 100,000 genes that make up the 23 pairs of chromosomes received from both parents. Is it any wonder, then, with such vast amounts of genetic information in each of our cells, that no two of us inherit exactly the same traits?

Genetic Code

How do genes bring about heredity? There is, of course, no short and easy answer to that question. We know that the genetic information contained in each gene is capable of "directing" the synthesis of a specific protein. The unique sequence of a thousand or so base pairs in a gene determines the sequence of specific building blocks required to form a particular protein. This store of information in each gene is called the *genetic code.* In summary, the coded information in genes controls protein and enzyme production, enzymes facilitate cellular chemical reactions, and cellular chemical reactions determine cell structure and function and therefore heredity.

FIGURE 2–6 *Protein Synthesis.* **A,** The DNA molecule contains a sequence of base pairs, or gene, that represents a sequence of amino acids. **B,** During transcription, the DNA code is "transcribed" as an mRNA molecule forms. **C,** During translation, the mRNA code is "translated" at the ribosome and the proper sequence of amino acids is assembled. The amino acid strand coils or folds as it is formed. **D,** The coiled amino acid strand folds again to form a protein molecule with a specific, complex shape.

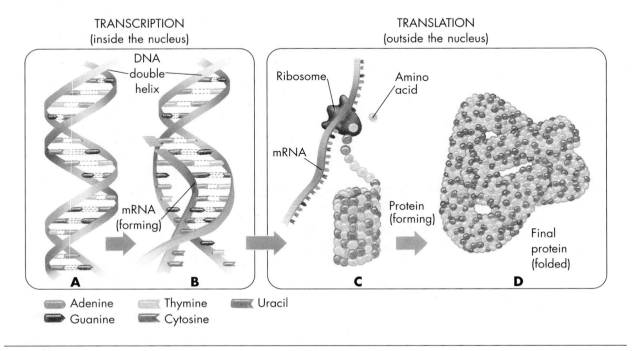

RNA Molecules and Protein Synthesis

DNA, with its genetic code that dictates directions for protein synthesis, is contained in the nucleus of the cell. The actual process of protein synthesis, however, occurs in ribosomes and on ER. Another specialized nucleic acid, ribonucleic acid or RNA, transfers this genetic information from the nucleus to the cytoplasm. (Note: if you are not familiar with the chemical structure of proteins or nucleic acids, you may want to review *Appendix A: Chemistry of Life,* p. 426, before reading further.)

Both RNA and DNA are composed of four bases, a sugar, and phosphate. RNA, however, is a single- rather than a double-stranded molecule, and it contains a different sugar and base component. The base uracil replaces thymine.

The process of transferring genetic information from the nucleus into the cytoplasm where pro-

teins are actually produced requires completion of two specialized steps called **transcription** and **translation.**

Transcription. During *transcription* the double-stranded DNA molecule separates or unwinds, and a special type of RNA called **messenger RNA** or **mRNA** is formed (Figure 2-6, *B*). Each strand of mRNA is a duplicate or copy of a particular gene sequence along one of the newly separated DNA spirals. The messenger RNA is said to have been "transcribed" or copied from its DNA mold or template. The mRNA molecules pass from the nucleus to the cytoplasm to direct protein synthesis in the ribosomes and ER.

Translation. *Translation* is the synthesis of a protein by ribosomes, which use the information contained in an mRNA molecule to direct the choice and sequencing of the appropriate chemical building blocks called *amino acids* (Figure 2-6, *C*). As amino acids are assembled into their proper sequence by a ribosome, a protein strand forms. This strand then folds on itself and perhaps even combines with another strand to form a complete protein molecule. The specific, complex shape of each type of protein molecule allows the molecule to perform specific functions in the cell. It is clear that because DNA directs the shape of each protein, DNA also directs the function of each protein in a cell.

CELL DIVISION

The process of cell reproduction involves the division of the nucleus (mitosis) and the cytoplasm. After the process is complete, two daughter cells result; both have the same genetic material as the cell that preceded them. As you can see in Figure 2-7, the specific and visible stages of cell division are preceded by a period called **interphase** (IN-ter-faze). During interphase the cell is said to be "resting." However, it is resting only from the standpoint of active cell division. In all other

FIGURE 2–7 *Mitosis.* For simplicity, only four chromosomes are shown in the diagram.

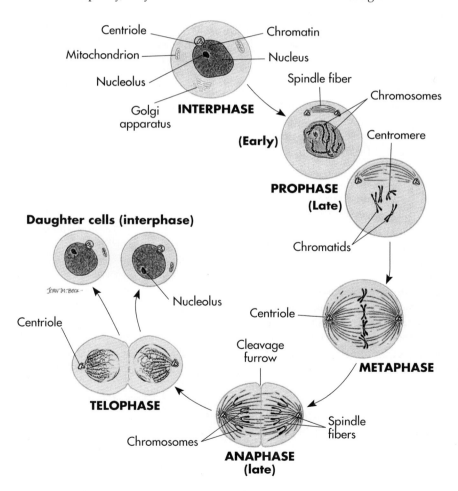

aspects it is exceedingly active. During interphase and just before mitosis begins, the DNA of each chromosome replicates itself.

DNA Replication

DNA molecules possess a unique ability that no other molecules in the world have. They can make copies of themselves, a process called **DNA replication.** Before a cell divides to form two new cells, each DNA molecule in its nucleus forms another DNA molecule just like itself. When a DNA molecule is not replicating, it has the shape of a tightly coiled double helix. As it begins replication, short segments of the DNA molecule uncoil and the two strands of the molecule pull apart between their base pairs. The separated strands therefore contain unpaired bases. Each unpaired base in each of the two separated strands attracts its complementary base (in the nucleoplasm) and binds to it. Specifically, each adenine attracts and binds to a thymine, and each cytosine attracts and binds to a guanine. These steps are repeated over and over throughout the length of the DNA molecule. Thus each half of a DNA molecule becomes a whole DNA molecule identical to the original DNA molecule. After DNA replication is complete, the cell continues to grow until it is ready for the first phase of mitosis.

Prophase

Look at Figure 2-7 and note the changes that identify the first stage of mitosis, prophase (PRO-faze). The chromatin becomes "organized." Chromosomes in the nucleus have formed two strands called **chromatids** (KRO-mah-tids). Note that the two chromatids are held together by a beadlike structure called the **centromere** (SEN-tro-meer). In the cytoplasm the centrioles are moving away from each other as a network of tubules called **spindle fibers** forms between them. These spindle fibers serve as "guidewires" and assist the chromosomes to move toward opposite ends of the cell later in mitosis.

Metaphase

By the time metaphase (MET-ah-faze) begins, the nuclear envelope and nucleolus have disappeared. Note in Figure 2-7 the chromosomes have aligned themselves across the center of the cell. Also, the centrioles have migrated to opposite ends of the cell, and spindle fibers are attached to each chromatid.

Anaphase

As anaphase (AN-ah-faze) begins, the beadlike centromeres, which were holding the paired chromatids together, break apart. As a result, the individual chromatids, identified once again as chromosomes, move away from the center of the cell. Movement of chromosomes occurs along spindle fibers toward the centrioles. Note in Figure 2-7 that chromosomes are being pulled to opposite ends of the cell. A **cleavage furrow** that begins to divide the cell into two daughter cells can be seen for the first time at the end of anaphase.

Telophase

During telophase (TEL-o-faze) cell division is completed. Two nuclei appear, and chromosomes become less distinct and appear to break up. As the nuclear envelope forms around the chromatin,

TABLE 2–4	Stages of Cell Division
STAGE	**CHARACTERISTICS**
Prophase	The chromatin condenses into visible chromosomes
	Chromatids become attached at the centromere
	Spindle fibers appear
	The nucleolus and nuclear envelope disappear
Metaphase	Spindle fibers attach to each chromatid
	Chromosomes align across the center of the cell
Anaphase	Centromeres break apart
	Chromosomes move away from the center of the cell
	The cleavage furrow appears
Telophase	The nuclear envelope and both nuclei appear
	The cytoplasm and organelles divide equally
	The process of cell division is completed

the cleavage furrow completely divides the cell into two parts. Before division is complete, each nucleus is surrounded by cytoplasm in which organelles have been equally distributed. By the end of telophase, two separate daughter cells, each having identical genetic characteristics, are formed. Each cell is fully functional and will perhaps itself undergo mitosis in the future.

Results of Cell Division

Mitosis results in the production of identical new cells. In the adult, mitosis replaces cells that have become less functional with age or have been damaged or destroyed by illness or injury. During periods of body growth, mitosis allows groups of similar cells to develop into **tissues.**

If the body loses its ability to control mitosis, an abnormal mass of proliferating cells develops. This mass is a **neoplasm** (NEE-o-plazm). Neoplasms may be relatively harmless growths called *benign* (be-NINE) *tumors* or dangerous and *malignant* (mah-LIG-nant) cancerous growths. The stages of mitosis are listed in Table 2-4 with descriptions of changes that occur during each stage.

 Tissues

The four main kinds of tissues that compose the body's many organs follow:

1. Epithelial tissue
2. Connective tissue
3. Muscle tissue
4. Nervous tissue

Tissues differ from each other in the size and shape of their cells, in the amount and kind of material between the cells, and in the special functions they perform to help maintain the body's survival. In Table 2-5, you will find a listing of the four major tissues and the various subtypes of each. The table also includes examples of the location of the tissues and a primary function of each tissue type.

EPITHELIAL TISSUE

Epithelial (ep-i-THEE-lee-al) **tissue** covers the body and many of its parts. It also lines various parts of the body. Because epithelial cells are packed close together with little or no intercellular material between them, they form continuous sheets that contain no blood vessels. Examine Figure 2-8. It illustrates how this large group of tissues can be subdivided according to the **shape** and **arrangement** of the cells found in each type.

Shape of Cells

If classified according to *shape,* epithelial cells are:

1. Squamous (flat and scalelike)
2. Cuboidal (cube shaped)
3. Columnar (higher than they are wide)

Arrangement of Cells

If categorized according to *arrangement* of cells, epithelial tissue can be classified as the following:

1. Simple (a single layer of cells of the same shape)
2. Stratified (many layers of cells of the same shape)
3. Transitional (several layers of cells of differing shapes)

Several types of epithelium are described in the paragraphs that follow and are illustrated in Figures 2-9 to 2-12.

Simple Squamous Epithelium

Simple squamous (SKWAY-mus) **epithelium** consists of a single layer of very thin and irregularly shaped cells. Because of its structure, substances can readily pass through simple squamous epithelial tissue, making absorption its special function. Absorption of oxygen into the blood, for example, takes place through the simple squamous epithelium that forms the tiny air sacs in the lungs (Figure 2-9).

Stratified Squamous Epithelium

Stratified squamous epithelium (Figure 2-10) consists of several layers of closely packed cells, an arrangement that makes this tissue a specialist at protection. For instance, stratified squamous epithelial tissue protects the body against invasion by microorganisms. Most microbes cannot work their way through a barrier of stratified squamous tissue such as that which composes the surface of skin and of mucous membranes.

TABLE 2–5 Tissues

TISSUE	LOCATION(S)	FUNCTION(S)
Epithelial		
Simple squamous	Alveoli of lungs	Absorption by diffusion of respiratory gases between alveolar air and blood
	Lining of blood and lymphatic vessels	Absorption by diffusion, filtration, and osmosis
Stratified squamous	Surface of lining of mouth and esophagus	Protection
	Surface of skin (epidermis)	Protection
Simple columnar	Surface layer of lining of stomach, intestines, parts of respiratory tract	Protection; secretion; absorption
Stratified transitional	Urinary bladder	Protection
Pseudostratified	Surface of lining of trachea	Protection
Simple cuboidal	Glands; kidney tubules	Secretion, absorption
Connective (most widely distributed of all tissues)		
Areolar	Area between other tissues and organs	Connection
Adipose (fat)	Area under skin	Protection
	Padding at various points	Insulation; support; nutrient reserve
Dense fibrous	Tendons; ligaments; fascia; scar tissue	Flexible but strong connection
Bone	Skeleton	Support; protection
Cartilage	Part of nasal septum; area covering articular surfaces of bones; larynx; rings in trachea and bronchi	Firm but flexible support
	Disks between vertebrae	Withstand pressure
	External ear	Flexible support
Blood	Blood vessels	Transportation
Hemopoietic	Red bone marrow	Blood cell formation
Muscle		
Skeletal (striated voluntary)	Muscles that attach to bones	Maintenance of posture; movement of bones
	Eyeball muscles	Eye movements
	Upper third of esophagus	First part of swallowing
Cardiac (striated involuntary)	Wall of heart	Contraction of heart
Smooth (nonstriated involuntary or visceral)	Walls of tubular viscera of digestive, respiratory, and genitourinary tracts	Movement of substances along respective tracts
	Walls of blood vessels and large lymphatic vessels	Changing of diameter of vessels
	Ducts of glands	Movement of substances along ducts
	Intrinsic eye muscles (iris and ciliary body)	Changing of diameter of pupils and shape of lens
	Arrector muscles of hairs	Erection of hairs (goose pimples)
Nervous		
	Brain; spinal cord; nerves	Irritability; conduction

FIGURE 2–8 *Classification of Epithelial Tissues.* The tissues are classified according to the shape and arrangement of cells.

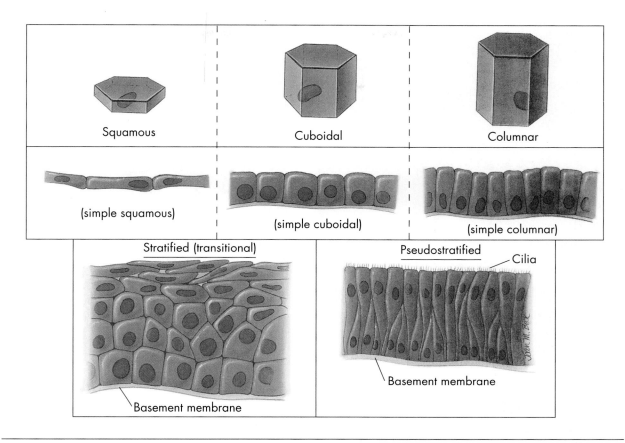

FIGURE 2–9 *Simple Squamous Epithelium.* **A,** Photomicrograph of lung tissue shows thin simple squamous epithelium lining the alveolar air sacs. **B,** Sketch of photomicrograph.

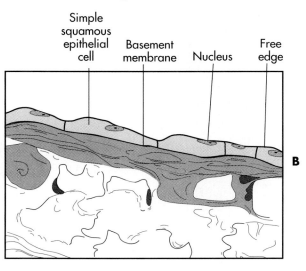

FIGURE 2–10 *Stratified Squamous Epithelium.* **A,** Photomicrograph. **B,** Sketch of the photomicrograph. Note the many layers of epithelial cells that have been stained yellow.

A

Superficial squamous cell

B

Basal squamous cell   Basement membrane

FIGURE 2–11 *Simple Columnar Epithelium.* **A,** Photomicrograph. **B,** Sketch of the photomicrograph. Note the goblet or mucus-producing cells that are present.

A

Goblet cells

B

Columnar epithelial cell

One way of preventing infections, therefore, is to take good care of your skin. Don't let it become cracked from chapping, and guard against cuts and scratches.

Simple Columnar Epithelium

Simple columnar epithelium can be found lining the inner surface of the stomach, intestines, and some areas of the respiratory and reproductive tracts. In Figure 2-11 the simple columnar cells are arranged in a single layer lining the inner surface of the colon or large intestine. These epithelial cells are higher than they are wide, and the nuclei are located toward the bottom of each cell. The "open spaces" among the cells are specialized **goblet cells** that produce mucus. The regular columnar-shaped cells specialize in absorption.

Stratified Transitional Epithelium

Stratified transitional epithelium is typically found in body areas subjected to stress and must be able to stretch; an example would be the wall of the urinary bladder. In many instances, up to 10 layers of cuboidal-shaped cells of varying sizes are present in the absence of stretching. When stretching occurs, the epithelial sheet expands, the number of cell layers decreases, and cell shape changes from cuboidal to squamous (flat) in appearance. The fact that transitional epithelium has this ability keeps the bladder wall from tearing under the pressures of stretching. Stratified transitional epithelium is shown in Figures 2-8 and 2-12.

Pseudostratified Epithelium

Pseudostratified epithelium, illustrated in Figure 2-8, is typical of that which lines the trachea or windpipe. Look carefully at the illustration. Note that each cell actually touches the gluelike **basement membrane** that lies under all epithelial tissues. Although the epithelium in Figure 2-8 (pseudostratified) appears to be two cell layers thick, it is not. This is the reason it is called *pseudo* (or false) stratified epithelium. The cilia that extend from the cells are capable of moving in unison. In doing so, they move mucus along the lining surface of the trachea, thus affording protection against entry of dust or foreign particles into the lungs.

Cuboidal Epithelium

Simple cuboidal epithelium does not form protective coverings but instead forms tubules or other groupings specialized for secretory activity

---

FIGURE 2–12 *Stratified Transitional Epithelium.* **A,** Photomicrograph of tissue lining the urinary bladder wall. **B,** Sketch of the photomicrograph. Note the many layers of epithelial cells of various shapes.

FIGURE 2–13 *Simple Cuboidal Epithelium.* This scanning electron micrograph shows how a single layer of cuboidal cells can form glands. The secreting cells arrange themselves into single or branched tubules that open onto a surface—the lining of the stomach in this case.

FIGURE 2–14 *Adipose Tissue.* Photomicrograph showing the large storage spaces for fat inside the adipose tissue cells.

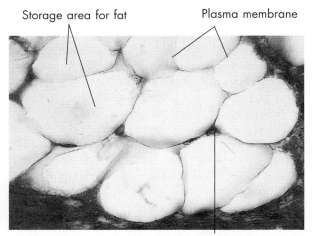

Storage area for fat　　Plasma membrane

Nucleus of adipose cell

Tubular gland　　Cuboidal cells forming wall of gland

(Figure 2-13). Secretory cuboidal cells usually function in clusters or tubes of secretory cells commonly called glands. Glands of the body may be classified as exocrine if they release their secretion through a duct or as endocrine if they release their secretion directly into the bloodstream. Examples of glandular secretions include saliva produced by the salivary glands, digestive juices, sweat or perspiration, and hormones such as those secreted by the pituitary or thyroid glands. Simple cuboidal epithelium also forms the tubules that form urine in the kidneys.

CONNECTIVE TISSUE

Connective tissue is the most abundant and widely distributed tissue in the body. It also exists in more varied forms than any of the other tissue types. It is found in skin, membranes, muscles, bones, nerves, and all internal organs. Connective tissue exists as delicate, paper-thin webs that hold internal organs together and give them shape. It also exists as strong and tough cords, rigid bones, and even in the form of a fluid—blood.

The functions of connective tissue are as varied as its structure and appearance. It connects tissues to each other and forms a supporting framework for the body as a whole and for its individual organs. As blood, it transports substances throughout the body. Several other kinds of connective tissue function to defend us against microbes and other invaders.

Connective tissue differs from epithelial tissue in the arrangement and variety of its cells and in the amount and kinds of intercellular material, called **matrix,** found between its cells. In addition to the relatively few cells embedded in the matrix of most types of connective tissue, varying numbers and kinds of fibers are also present. The structural quality and appearance of the matrix and fibers determine the qualities of each type of connective tissue. The matrix of blood, for example, is a liquid, but other types of connective tissue, such as cartilage, have the consistency of firm rubber. The matrix of bone is hard and rigid, although the matrix of connective tissues such as tendons and ligaments is strong and flexible.

The following list identifies a number of the major types of connective tissue in the body. Photomicrographs of several are also shown.

1. Areolar connective tissue
2. Adipose or fat tissue
3. Fibrous connective tissue
4. Bone
5. Cartilage
6. Blood
7. Hemopoietic tissue

Areolar and Adipose Connective Tissue

Areolar (ah-REE-o-lar) **connective tissue** is the most widely distributed of all connective tissue types. It is the "glue" that gives form to the internal organs. It consists of delicate webs of fibers and of a variety of cells embedded in a loose matrix of soft, sticky gel.

Adipose (AD-i-pose) or **fat tissue** is specialized to store lipids. In Figure 2-14, numerous spaces have formed in the tissue so that large quantities of fat can accumulate inside cells.

Fibrous Connective Tissue

Fibrous connective tissue (Figure 2-15) consists mainly of bundles of strong, white **collagen** fibers arranged in parallel rows. This type of connective tissue composes tendons. It provides great strength and flexibility but it does not stretch. Such characteristics are ideal for these structures that anchor our muscles to our bones.

Bone and Cartilage

Bone is one of the most highly specialized forms of connective tissue. The matrix of bone is hard and calcified. It forms numerous structural building blocks called **osteons** (AHS-tee-onz), or *Haversian* (ha-VER-shan) *systems*. When bone is viewed under a microscope, we can see these circular arrangements of calcified matrix and cells that give bone its characteristic appearance (Figure 2-16). Bones are a storage area for calcium and provide support and protection for the body.

Cartilage differs from bone in that its matrix is the consistency of a firm plastic or a gristlelike gel. Cartilage cells, which are called **chondrocytes** (KON-dro-sites), are located in many tiny spaces distributed throughout the matrix (Figure 2-17).

FIGURE 2–15 *Dense Fibrous Connective Tissue.* Photomicrograph of tissue in the tendon. Note the multiple bundles of collagenous fibers in a parallel arrangement.

Bundle of collagen fibers

Fiber-producing cells

FIGURE 2–16 *Bone Tissue.* Photomicrograph of dried, ground bone. Many wheel-like structural units of bone, known as osteons (Haversian systems), are apparent in this section.

Osteon (Haversian system)

FIGURE 2–17 *Cartilage.* Photomicrograph showing the chondrocytes distributed throughout the gel-like matrix.

Matrix   Chondrocyte

FIGURE 2–18 *Blood.* Photomicrograph of a human blood smear. This smear shows two white blood cells surrounded by a number of smaller red blood cells. The liquid matrix of this tissue is also called *plasma.*

Red blood cells

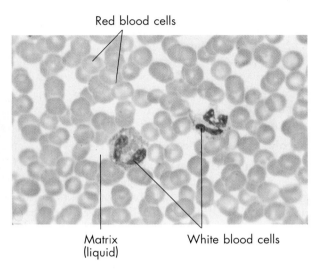

Matrix   White blood cells
(liquid)

Blood and Hemopoietic Tissue

Because its matrix is liquid, blood is perhaps the most unusual form of connective tissue. It has transportational and protective functions in the body. Red and white blood cells are the cell types common to blood (Figure 2-18).

Hemopoietic (hee-mo-poy-ET-ik) **tissue** is the connective tissue found in the red marrow cavities of bones and in organs such as the spleen, tonsils, and lymph nodes. This type of tissue is responsible for the formation of blood cells and lymphatic system cells important in our defense against disease.

MUSCLE TISSUE

Muscle cells are the movement specialists of the body. They have a higher degree of contractility (ability to shorten or contract) than any other tissue cells. Unfortunately, injured muscle cells are often slow to heal and are frequently replaced by scar tissue if injured. There are three kinds of muscle tissue: **skeletal, cardiac,** and **smooth.**

Skeletal Muscle Tissue

Skeletal or striated muscle is called *voluntary* because willed or voluntary control of skeletal muscle contractions is possible. Note in Figure 2-19 that, when viewed under a microscope, skeletal muscle is characterized by many cross striations and many nuclei per cell. Individual cells are long and threadlike and are often called *fibers.* Skeletal muscles are attached to bones and when contracted produce voluntary and controlled body movements.

Cardiac Muscle Tissue

Cardiac muscle forms the walls of the heart, and the regular but involuntary contractions of cardiac muscle produce the heartbeat. Under the light microscope (Figure 2-20), cardiac muscle fibers have cross striations (like skeletal muscle) and thicker dark bands called *intercalated disks.* Cardiac muscle fibers branch and reform to produce an interlocking mass of contractile tissue.

FIGURE 2–19 *Skeletal Muscle.* Photomicrograph showing the striations of the muscle cell fibers in longitudinal section.

Cross striations of muscle cell

Nuclei of muscle cell   Muscle fiber

FIGURE 2–20 *Cardiac Muscle.* Photomicrograph showing the branched, lightly striated fibers. The darker bands, called *intercalated disks,* which are characteristic of cardiac muscle, are easily identified in this tissue section.

Nucleus of muscle cell

Intercalated disks

 Tissues and Fitness

Achieving and maintaining an ideal body weight is a health-conscious goal. However, a better indicator of health and fitness is **body composition.** Exercise physiologists assess body composition to identify the percentage of the body made of lean tissue and the percentage made of fat. Body-fat percentage is often determined by using calipers to measure the thickness of skin folds at certain places on the body. A person with low body weight may still have a high ratio of fat to muscle, an unhealthy condition. In this case the individual is "underweight" but "overfat." In other words, fitness depends more on the percentage and ratio of specific tissue types than the overall amount of tissue present.

Therefore one goal of a good fitness program is a desirable body-fat percentage. For men, the ideal is 15% to 18%, and for women, the ideal is 20% to 22%. Because fat contains stored energy (measured in calories), a low-fat percentage means a low-energy reserve. High body-fat percentages are associated with several life-threatening conditions, including cardiovascular disease. A balanced diet and an exercise program ensure that the ratio of fat to muscle tissue stays at a level appropriate for maintaining homeostasis.

FIGURE 2–21 *Smooth Muscle.* Photomicrograph, longitudinal section. Note the central placement of nuclei in the spindle-shaped smooth muscle fibers.

Smooth muscle cells

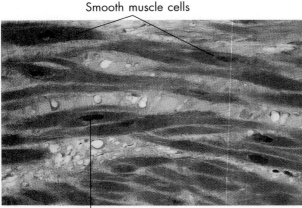

Nucleus of muscle cell

FIGURE 2–22 *Nervous Tissue.* Photomicrograph of neurons in a smear of the spinal cord. Both neurons in this slide show characteristic cell bodies and multiple cell processes.

Nerve cell body   Axon

Dendrites   Glial cells

Smooth Muscle Tissue

Smooth (visceral) muscle is said to be *involuntary* because it is not under conscious or willful control. Under a microscope (Figure 2-21), smooth muscle cells are seen as long, narrow fibers but not nearly as long as skeletal or striated fibers. Individual smooth muscle cells appear smooth (that is, without cross striations) and have only one nucleus per fiber. Smooth muscle helps form the walls of blood vessels and hollow organs such as the intestines and other tube-shaped structures in the body. Contractions of smooth (visceral) muscle propel food material through the digestive tract and help regulate the diameter of blood vessels. Contraction of smooth muscle in the tubes of the respiratory system, such as the bronchioles in the lungs, can impair breathing and result in asthma attacks and labored respiration.

NERVOUS TISSUE

The function of **nervous tissue** is rapid communication between body structures and control of body functions. Nervous tissue consists of two kinds of cells: nerve cells, or **neurons** (NOO-rons), which are the functional or conducting units of the system, and special connecting and supporting cells called **glia** (GLEE-ah) or *neuroglia.*

All neurons are characterized by a **cell body** and two types of processes: one **axon,** which transmits a nerve impulse away from the cell body, and one or more **dendrites** (DEN-drites), which carry impulses toward the cell body. Both neurons in Figure 2-22 have many dendrites extending from the cell body.

CELLS

A Size and shape
  1 Human cells vary considerably in size
  2 All are microscopic
  3 Cells differ notably in shape
B Composition
  1 Cytoplasm containing specialized organelles surrounded by a plasma membrane
  2 Organization of cytoplasmic substances important for life
C Structural parts
  1 Plasma membrane (Figure 2-1)
    a Forms outer boundary of cell
    b Thin, two-layered membrane of phospholipids containing proteins
    c Is selectively permeable
  2 Cytoplasm (Figure 2-2)
    a Organelles
      (1) Ribosomes
        (a) May attach to rough ER or lie free in cytoplasm
        (b) Manufacture proteins
        (c) Often called *protein factories*
      (2) Endoplasmic reticulum (ER)
        (a) Network of connecting sacs and canals
        (b) Carry substances through cytoplasm
        (c) Types are rough and smooth
        (d) Rough ER collects and transports proteins made by ribosomes
        (e) Smooth ER synthesizes chemicals; makes new membrane
      (3) Golgi apparatus
        (a) Group of flattened sacs near nucleus
        (b) Collect chemicals that move from the smooth ER in vesicles
        (c) Called the *chemical processing and packaging center*
      (4) Mitochondria
        (a) Composed of inner and outer membranes
        (b) Involved with energy-releasing chemical reactions
        (c) Often called *power plants* of the cell
      (5) Lysosomes
        (a) Membranous-walled organelles
        (b) Contain digestive enzymes
        (c) Have protective function (eat microbes)
        (d) Often called *suicide bags*
      (6) Centrioles
        (a) Paired organelles
        (b) Lie at right angles to each other near nucleus
        (c) Function in cell reproduction
      (7) Cilia
        (a) Fine, hairlike extensions found on free or exposed surfaces of some cells
        (b) Capable of moving in unison in a wave-like fashion
      (8) Flagella
        (a) Single projections extending from cell surfaces
        (b) Much larger than cilia
        (c) "Tails" of sperm cells only example of flagella in humans
  3 Nucleus
    a Controls cell because it contains the genetic code—instructions for making proteins, which in turn determine cell structure and function
    b Component structures include nuclear envelope, nucleoplasm, nucleolus, and chromatin granules
    c Chromosomes contain DNA
D Relationship of cell structure and function
  1 Regulation of life processes
  2 Survival of species through reproduction of the individual
  3 Relationship of structure to function apparent in number and type of organelles seen in different cells
    a Heart muscle cells contain many mitochondria required to produce adequate energy needed for continued contractions
    b Flagellum of sperm cell gives motility, allowing movement of sperm through female reproductive tract, thus increasing chances for fertilization (Figure 2-3)

MOVEMENTS OF SUBSTANCES THROUGH CELL MEMBRANES

A Passive transport processes do not require added energy and result in movement "down a concentration gradient"
  1 Diffusion (Figure 2-4)
    a Substances scatter themselves evenly throughout an available space
    b It is unnecessary to add energy to the system
    c Movement is from high to low concentration
    d Osmosis and dialysis are specialized examples of diffusion across a selectively permeable membrane
    e Osmosis is diffusion of water
    f Dialysis is diffusion of solutes
  2 Filtration
    a Movement of water and solutes caused by hydrostatic pressure on one side of membrane
    b Responsible for urine formation

B Active transport processes occur only in living cells; movement of substances is "up the concentration gradient"; requires energy from ATP

  1 Ion pumps

    a An ion pump is protein complex in cell membrane

    b Ion pumps use energy from ATP to move substances across cell membranes against their concentration gradients

    c Examples: sodium-potassium pump, calcium pump

    d Some ion pumps work with other carriers so that glucose or amino acids are transported along with ions

  2 Phagocytosis and pinocytosis

    a Both are active transport mechanisms because they require cell energy

    b Phagocytosis is a protective mechanism often used to destroy bacteria (Figure 2-5)

    c Pinocytosis is used to incorporate fluids or dissolved substances into cells

CELL REPRODUCTION

A DNA structure—large molecule shaped like a spiral staircase; sugar (deoxyribose), and phosphate units compose sides of the molecule; base pairs (adenine-thymine or guanine-cytosine) compose "steps"; base pairs always the same but sequence of base pairs differs in different DNA molecules; a gene is a specific sequence of base pairs within a DNA molecule; genes dictate formation of enzymes and other proteins by ribosomes, thereby indirectly determining a cell's structure and functions; in short, genes are heredity determinants (Figure 2-6)

B Genetic code

  1 Genetic information—stored in base-pair sequences on genes—expressed through protein synthesis

  2 RNA molecules and protein synthesis

    a DNA—contained in cell nucleus

    b Protein synthesis—occurs in cytoplasm, thus genetic information must pass from the nucleus to the cytoplasm

    c Process of transferring genetic information from nucleus to cytoplasm where proteins are produced requires completion of *transcription* and *translation* (Figure 2-6).

  3 Transcription

    a Double-stranded DNA separates to form messenger RNA or mRNA

    b Each strand of mRNA duplicates a particular gene (base-pair sequence) from a segment of DNA

    c mRNA molecules pass from the nucleus to the cytoplasm where they direct protein synthesis in ribosomes and ER

  4 Translation

    a Involves synthesis of proteins in cytoplasm by ribosomes

    b Requires use of information contained in mRNA

C Cell Division—reproduction of cell involving division of the nucleus (mitosis) and the cytoplasm; period when the cell is not actively dividing is called interphase

D DNA replication—process by which each half of a DNA molecule becomes a whole molecule identical to the original DNA molecule; precedes mitosis

E Mitosis—process in cell division that distributes identical chromosomes (DNA molecules) to each new cell formed when the original cell divides; enables cells to reproduce their own kind; makes heredity possible

F Stages of mitosis (Figure 2-7)

  1 Prophase—first stage

    a Chromatin granules become organized

    b Chromosomes (pairs of linked chromatids) appear

    c Centrioles move away from nucleus

    d Nuclear envelope disappears, freeing genetic material

    e Spindle fibers appear

  2 Metaphase—second stage

    a Chromosomes align across center of cell

    b Spindle fibers attach themselves to each chromatid

  3 Anaphase—third stage

    a Centromeres break apart

    b Separated chromatids now called *chromosomes*

    c Chromosomes are pulled to opposite ends of cell

    d Cleavage furrow develops at end of anaphase

  4 Telophase—fourth stage

    a Cell division is completed

    b Nuclei appear in daughter cells

    c Nuclear envelope and nucleoli appear

    d Cytoplasm is divided (cytokinesis)

    e Daughter cells become fully functional

TISSUES

A Epithelial tissue

  1 Covers body and lines body cavities

  2 Cells packed closely together with little matrix

  3 Classified by shape of cells (Figure 2-8)

    a Squamous

    b Cuboidal

    c Columnar

4 Classified by arrangement of cells
  a Simple
  b Stratified
  c Transitional
5 Simple squamous epithelium (Figure 2-9)
  a Single layer of scalelike cells
  b Absorption is function
6 Stratified squamous epithelium (Figure 2-10)
  a Several layers of closely packed cells
  b Protection is primary function
7 Simple columnar epithelium (Figure 2–11)
  a Columnar cells arranged in a single layer
  b Line stomach and intestines
  c Contain mucus-producing goblet cells
  d Specialized for absorption
8 Stratified transitional epithelium (Figure 2-12)
  a Found in body areas, such as urinary bladder, that stretch
  b Up to 10 layers of cuboidal-shaped cells that distort to squamous shape when stretched
9 Pseudostratified epithelium
  a Each cell touches basement membrane
  b Lines the trachea
10 Simple cuboidal epithelium (Figure 2-13)
  a Often specialized for secretory activity
  b Cuboidal cells may be grouped into glands
  c May secrete into ducts, directly into blood, and on body surface
  d Examples of secretions include saliva, digestive juice, and hormones
  e Cuboidal epithelium also forms the urine-producing tubules of the kidney
B Connective tissue
  1 Most abundant tissue in body
  2 Most widely distributed tissue in body
  3 Multiple types, appearances, and functions
  4 Relatively few cells in intercellular matrix
  5 Types
    a Areolar—glue that holds organs together
    b Adipose (fat)—lipid storage is primary function (Figure 2-14)
    c Fibrous—strong fibers; example is tendon (Figure 2-15)
    d Bone—matrix is calcified; function in support and protection (Figure 2-16)
    e Cartilage—chondrocyte is cell type (Figure 2-17)
    f Blood—matrix is fluid; function is transportation (Figure 2-18)
C Muscle tissue (Figures 2-19 to 2-21)
  1 Types
    a Skeletal—attaches to bones; also called *striated* or *voluntary*; control is voluntary; striations apparent when viewed under a microscope (Figure 2-19)
    b Cardiac—also called *striated involuntary*; composes heart wall; ordinarily cannot control contractions (Figure 2-20)
    c Smooth—also called *nonstriated (visceral)* or *involuntary*; no cross striations; found in blood vessels and other tube-shaped organs (Figure 2-21)
D Nervous tissue (Figure 2-22)
  1 Cell types
    a Neurons—conducting cells
    b Glia (neuroglia)—supportive and connecting cells
  2 Neurons
    a Cell components
      (1) Cell body
      (2) Axon (one) carries nerve impulse away from cell body
      (3) Dendrites (one or more) carry nerve impulse toward the cell body
  3 Function—rapid communication between body structures and control of body functions

---

N E W   W O R D S

adenosine triphosphate (ATP)	dendrite	telophase	osteon
adipose	deoxyribonucleic acid (DNA)	neuron	ribonucleic acid (RNA)
areolar	glia	nucleoplasm	spindle fiber
axon	goblet cell	organelle	squamous
centriole	hemopoietic	cilia	transcription
centromere	hypertonic	endoplasmic reticulum (ER)	translation
chondrocyte	hypotonic	flagellum	transport
chromatid	interphase	Golgi apparatus	dialysis
chromatin	lyse	lysosome	diffusion
collagen	matrix	mitochondria	filtration
columnar	mitosis	nucleolus	osmosis
crenation	prophase	nucleus	phagocytosis
cuboidal	metaphase	plasma membrane	pinocytosis
cytoplasm	anaphase	ribosome	

C H A P T E R   T E S T

1. Which of the following groups represents the three principal parts of a cell?
   a. Plasma membrane, nucleus, cytoplasm
   b. Ribosomes, plasma membrane, mitochondria
   c. Lysosomes, centrioles, plasma membrane
   d. Nucleus, Golgi apparatus, mitochondria
2. Which of the following is considered an active transport process?
   a. Osmosis
   b. Ion pump
   c. Filtration
   d. Dialysis
3. Deoxyribonucleic acid or DNA:
   a. Is exactly the same as RNA
   b. Contains the compound uracil
   c. Is sometimes called the "heredity molecule"
   d. Is not involved with genes or chromosomes
4. The stage of mitosis in which the chromosomes align themselves across the center of the cell is called:
   a. Prophase
   b. Anaphase
   c. Telophase
   d. Metaphase
5. Which of the following terms describes flat, plate-like cells that permit substances to readily pass through them?
   a. Squamous
   b. Stratified
   c. Transitional
   d. Columnar
6. Osteons are associated with which of the following?
   a. Adipose or fat tissue
   b. Bone
   c. Fibrous connective tissue
   d. Cartilage
7. Which of the following types of muscle is considered voluntary?
   a. Cardiac
   b. Smooth
   c. Visceral
   d. Skeletal

Select the most appropriate answer in column B for each item in Column A. (Only one answer is correct.)

COLUMN A	COLUMN B
8. _____Cellular organelles	a. Osmosis
9. _____Example of diffusion	b. Blood
10. _____Active transport mechanism	c. Mitochondria
11. _____Characteristic of DNA	d. Anaphase
12. _____Stage of mitosis	e. Complementary base pairing
13. _____Type of connective tissue	f. Pinocytosis
14. _____Composed of epithelial tissue	g. Dendrite
15. _____Component of a nerve cell	h. Glands

16. The nickname, "protein factories," is often used to describe _____.
17. Substances move "up a concentration gradient" in _____ transport processes.
18. The first stage of mitosis is called _____.
19. Absorption, protection, and secretion are important functions of _____ tissue.
20. No blood vessels are found in _____ tissue.

REVIEW QUESTIONS

1. Identify the three main parts of a cell.
2. Describe the structure and functions of the plasma membrane.
3. What and where are the following? What functions do they perform?
   nucleolus
   endoplasmic reticulum
   ribosomes
   chromatin
4. Discuss the genetic code and its relationship to protein synthesis.
5. What is a gene? How does a gene differ from a chromosome?
6. Compare and contrast active and passive transport systems. List by name the major active and passive transport processes.
7. List the stages of mitosis and briefly describe what occurs during each period.
8. When does DNA replication occur with respect to mitosis?
9. Identify the stage of mitosis that could be described as "prophase in reverse."
10. Explain how epithelial tissue can be classified according to the shape and arrangement of the cells.
11. Compare the matrix found in bone, areolar connective tissue, and blood.
12. What is the most widely distributed tissue in the body? What tissue exists in more varied forms than any other tissue type? In what type of tissue is appearance most determined by the nature of the matrix?
13. Compare and contrast the three major types of muscle tissue.
14. Identify the two basic types of cells found in nervous tissue. What is the difference between an axon and a dendrite?

CRITICAL THINKING

15. Discuss the importance of the following three chemical substances: DNA, RNA, and ATP. Where in the cell would you expect to find the largest quantity of each?
16. What actually moves during osmosis? During dialysis? What is required for both osmosis and dialysis to occur that is not required for diffusion?
17. How is ATP involved in active transport processes?
18. Explain how the structure of stratified transitional epithelium is related to its function.

Organ Systems of the Body

OBJECTIVES

*After you have completed this chapter,
you should be able to:*

1. Define and contrast the terms *organ* and *organ system.*
2. List the 11 major organ systems of the body.
3. Identify and locate the major organs of each major organ system.
4. Briefly describe the major functions of each major organ system.
5. Identify and discuss the major subdivisions of the reproductive system.

The words *organ* and *system* were discussed in Chapter 1 as having special meanings when applied to the body. An **organ** is a structure made up of two or more kinds of tissues organized in such a way that the tissues can together perform a more complex function than can any tissue alone. A **system** is a group of organs arranged in such a way that they can together perform a more complex function than can any organ alone. This chapter gives an overview of the 11 major organ systems of the body.

In the chapters that follow, the presentation of information on individual organs and an explanation of how they work together to accomplish complex body functions will form the basis for the discussion of each organ system. For example, a detailed description of the skin as the primary organ of the integumentary system will be covered in Chapter 4, and information on the bones of the body as organs of the skeletal system will be presented in Chapter 5. A knowledge of individual organs and how they are organized into groups makes much more meaningful the understanding of how a particular organ system functions as a unit in the body.

When you have completed your study of the major organ systems in the chapters that follow, it will be possible to view the body not as an assembly of individual parts but as an integrated and functioning whole. This chapter names the systems of the body and the major organs that compose them, and it briefly describes the functions of each system. It is intended to provide a basic "road map" to help you anticipate and prepare for the more detailed information that follows in the remainder of the text.

◯ *Organ Systems of the Body*

In contrast to cells, which are the smallest structural units of the body, organ systems are its largest and most complex structural units. The 11 major organ systems that compose the human body are listed below.

1. Integumentary
2. Skeletal
3. Muscular
4. Nervous
5. Endocrine
6. Circulatory
7. Lymphatic
8. Respiratory
9. Digestive
10. Urinary
11. Reproductive
    a. Male subdivision
    b. Female subdivision

Examine Figure 3-1 to find a diagrammatical listing of the body systems and the major organs in each. In addition to the information contained in Figure 3-1, each system is presented in visual form in Figures 3-2 through 3-13. Visual presentation of material is often useful in understanding the interrelationships that are so important in anatomy and physiology.

INTEGUMENTARY SYSTEM

Note in Figure 3-2 that the skin is the largest and most important organ in the **integumentary** (in-teg-yoo-MEN-tar-ee) **system.** Its weight in most adults is 20 pounds or more, accounting for about 16% of total body weight and making it the body's heaviest organ. The integumentary system includes the skin and its **appendages,** which include the hair, nails, and specialized sweat- and oil-producing glands. In addition, a number of microscopic and highly specialized sense organs are embedded in the skin. They permit the body to respond to pain, pressure, touch, and changes in temperature.

The integumentary system is crucial to survival. Its primary function is *protection.* The skin protects underlying tissue against invasion by harmful bacteria, bars entry of most chemicals, and minimizes the chances of mechanical injury to underlying structures. In addition, the skin regulates body temperature by sweating, synthesizes important chemicals, and functions as a sophisticated sense organ.

SKELETAL SYSTEM

The sternum or breastbone, the humerus, and the femur shown in Figure 3-3 are examples of the 206 individual organs (bones) found in the **skeletal system.** The system includes not only bones but also related tissues such as cartilage and liga-

FIGURE 3–1 *Body Systems and Their Organs.*

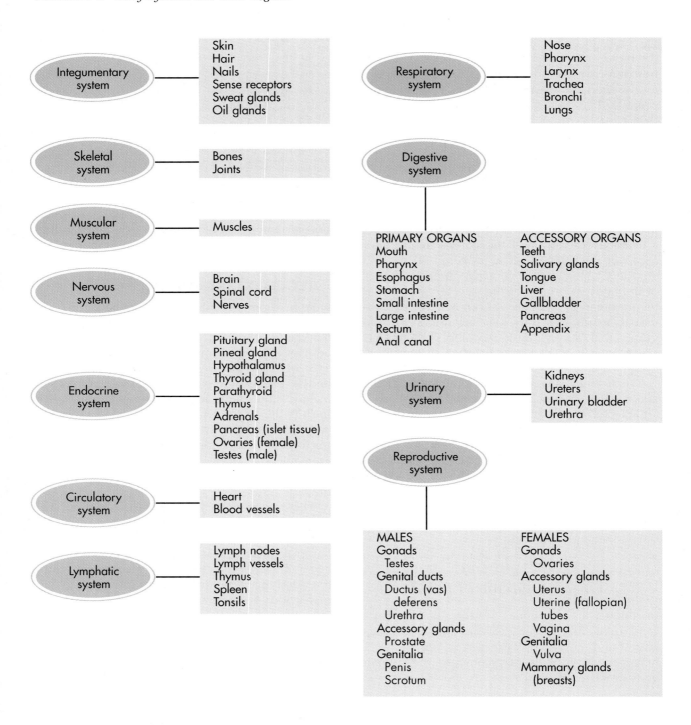

Integumentary system
- Skin
- Hair
- Nails
- Sense receptors
- Sweat glands
- Oil glands

Skeletal system
- Bones
- Joints

Muscular system
- Muscles

Nervous system
- Brain
- Spinal cord
- Nerves

Endocrine system
- Pituitary gland
- Pineal gland
- Hypothalamus
- Thyroid gland
- Parathyroid
- Thymus
- Adrenals
- Pancreas (islet tissue)
- Ovaries (female)
- Testes (male)

Circulatory system
- Heart
- Blood vessels

Lymphatic system
- Lymph nodes
- Lymph vessels
- Thymus
- Spleen
- Tonsils

Respiratory system
- Nose
- Pharynx
- Larynx
- Trachea
- Bronchi
- Lungs

Digestive system

PRIMARY ORGANS	ACCESSORY ORGANS
Mouth	Teeth
Pharynx	Salivary glands
Esophagus	Tongue
Stomach	Liver
Small intestine	Gallbladder
Large intestine	Pancreas
Rectum	Appendix
Anal canal	

Urinary system
- Kidneys
- Ureters
- Urinary bladder
- Urethra

Reproductive system

MALES	FEMALES
Gonads	Gonads
Testes	Ovaries
Genital ducts	Accessory glands
Ductus (vas)	Uterus
deferens	Uterine (fallopian)
Urethra	tubes
Accessory glands	Vagina
Prostate	Genitalia
Genitalia	Vulva
Penis	Mammary glands
Scrotum	(breasts)

FIGURE 3–2   *Integumentary System.*

FIGURE 3–3   *Skeletal System.*

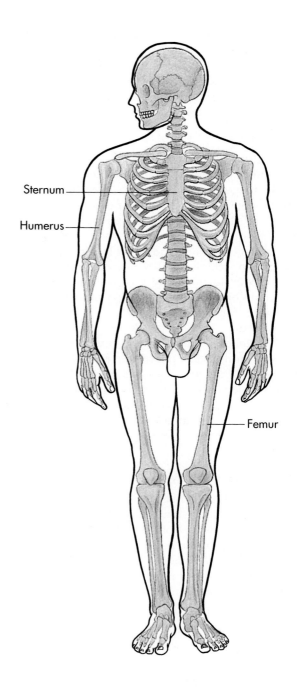

ments that together provide the body with a rigid framework for support and protection. In addition, the skeletal system, through the existence of **joints** between bones, makes possible the movements of body parts. Without joints, we could make no movements; our bodies would be rigid, immobile hulks. Bones also serve as storage areas for important minerals such as calcium and phosphorus. The formation of blood cells in the red marrow of certain bones is another crucial function of the skeletal system.

MUSCULAR SYSTEM

Individual skeletal muscles are the organs of the **muscular system.** In addition to **voluntary** or **skeletal** muscles, which have the ability to contract when stimulated and are under conscious control, the muscular system also contains **smooth** or **involuntary** muscles found in organs such as the stomach and small intestine. The third type of muscle tissue is the **cardiac muscle** of the heart. Muscles not only produce movement and maintain body posture but also generate the heat required for maintaining a constant core body temperature.

The tendon labeled in Figure 3-4 represents how tendons attach muscles to bones. When stimulated by a nervous impulse, muscle tissue shortens or contracts. Voluntary movement occurs when skeletal muscles contract because of the way muscles are attached to bones and the way bones articulate or join together with one another in joints.

NERVOUS SYSTEM

The brain, spinal cord, and nerves are the organs of the **nervous system.** As you can see in Figure 3-5, nerves extend from the brain and spinal cord to every area of the body. The extensive networking of the components of the nervous system makes it possible for this complex system to perform its primary functions. These include the following:

1. Communication between body functions
2. Integration of body functions
3. Control of body functions
4. Recognition of sensory stimuli

These functions are accomplished by specialized signals called **nerve impulses.** In general, the

FIGURE 3–4   *Muscular System.*

Muscle

Tendon

FIGURE 3–5 *Nervous System.*

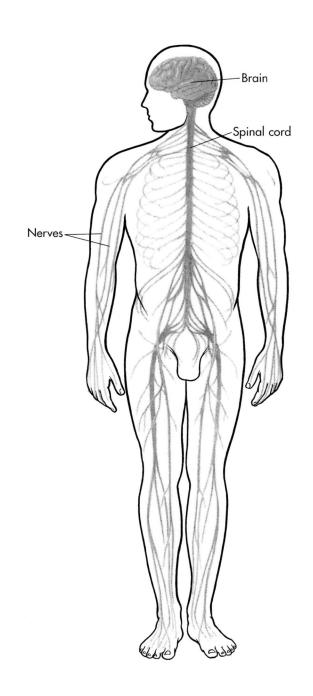

functions of the nervous system result in rapid activity that lasts usually for a short duration. For example, we can chew our food normally, walk, and perform coordinated muscular movements only if our nervous system functions properly. The nerve impulse permits the rapid and precise control of diverse body functions. Other types of nerve impulses cause glands to secrete fluids. In addition, elements of the nervous system can recognize certain **stimuli** (STIM-yoo-lye), such as heat, light, pressure, or temperature, that affect the body. When stimulated, these specialized components of the nervous system, called **sense organs** (discussed in Chapter 8), generate nervous impulses that travel to the brain or spinal cord where analysis or relay occurs and, if needed, appropriate action is initiated.

ENDOCRINE SYSTEM

The **endocrine system** is composed of specialized glands that secrete chemicals known as **hormones** directly into the blood. Sometimes called *ductless glands,* the organs of the endocrine system perform the same general functions as the nervous system: communication, integration, and control. The nervous system provides rapid, brief control by fast-traveling nerve impulses. The endocrine system provides slower but longer-lasting control by hormone secretion; for example, secretion of growth hormone controls the rate of development over long periods of gradual growth.

In addition to controlling growth, hormones are the main regulators of metabolism, reproduction, and other body activities. They play important roles in fluid and electrolyte balance, acid-base balance, and energy metabolism.

As you can see in Figure 3-6, the endocrine glands are widely distributed throughout the body. The **pituitary** (pi-TOO-i-TAIR-ee) **gland, pineal** (PIN-e-al) **gland,** and **hypothalamus** (hi-po-THAL-ah-mus) are located in the skull. The **thyroid** (THY-roid) and **parathyroid** (PAIR-ah-THY-roid) **glands** are in the neck, and the **thymus** (THY-mus) **gland** is in the thoracic cavity, specifically in the mediastinum (see Figure 1-4, p. 5). The **adrenal** (ah-DRE-nal) **glands** and **pancreas** (PAN-kree-as) are found in the abdominal cavity. Note in Figure 3-6 that the ovaries in the female and the testes in the male also function as endocrine glands.

FIGURE 3–6   *Endocrine System.*

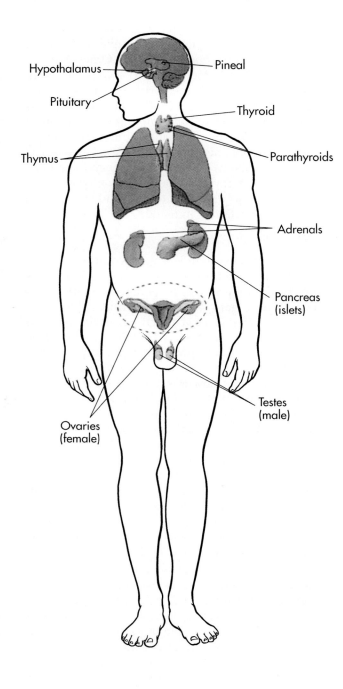

Hypothalamus

Pituitary

Pineal

Thyroid

Thymus

Parathyroids

Adrenals

Pancreas
(islets)

Testes
(male)

Ovaries
(female)

CIRCULATORY SYSTEM

The **circulatory system** consists of the heart, which is a muscular pumping device as shown in Figure 3-7, and a closed system of vessels made up of **arteries, veins,** and **capillaries.** As the name implies, blood contained in the circulatory system is pumped by the heart around a closed circle or circuit of vessels as it passes through the body. The term **cardiovascular** refers to the system of the heart and blood vessels.

The primary function of the circulatory system is *transportation.* The need for an efficient transportation system in the body is critical. Transportation needs include continuous movement of oxygen and carbon dioxide, nutrients, hormones, and other important substances. Wastes produced by the cells are released into the bloodstream on an ongoing basis and are transported by the blood to the excretory organs. The circulatory system also helps regulate body temperature by distributing heat throughout the body and by assisting in retaining or releasing heat from the body by regulating blood flow near the body surface. Certain cells of the circulatory system can also become involved in the defense of the body or immunity.

LYMPHATIC SYSTEM

The **lymphatic system** is composed of **lymph nodes, lymphatic vessels,** and specialized lymphatic organs such as the **tonsils, thymus,** and **spleen.** Note that the thymus in Figure 3-8 functions as an endocrine and as a lymphatic gland. Instead of containing blood, the lymphatic vessels are filled with lymph, a whitish, watery fluid that contains lymphocytes, proteins, and some fatty molecules. No red blood cells are present. The lymph is formed from the fluid around the body cells and diffuses into the lymph vessels. However, unlike blood, lymph does not circulate repeatedly through a closed circuit or loop of vessels. Instead, lymph flowing through lymphatic vessels eventually enters the circulatory system by passing through large ducts, including the **thoracic duct** shown in Figure 3-8, which in turn connect with veins in the upper area of the thoracic cavity. Collections of lymph nodes can be seen in the axillary (armpit) and in the inguinal (groin) areas of the body in Figure 3-8. The formation and movement of lymph are discussed in Chapter 11.

FIGURE 3–7 *Circulatory System.*

FIGURE 3–8 *Lymphatic System.*

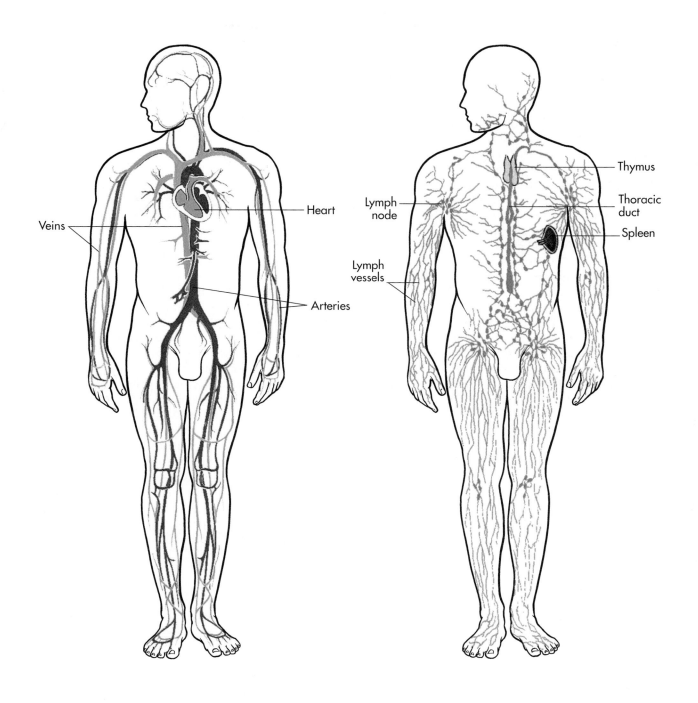

The functions of the lymphatic system include movement of fluids and certain large molecules from the tissue spaces around the cells and movement of fat-related nutrients from the digestive tract back to the blood. The lymphatic system is also involved in the functioning of the immune system, which plays a critical role in the defense mechanism of the body against disease.

RESPIRATORY SYSTEM

The organs of the **respiratory system** include the nose, **pharynx** (FAIR-inks), **larynx** (LAR-inks), **trachea** (TRAY-kee-ah), **bronchi** (BRON-ki), and lungs (Figure 3-9). Together these organs permit the movement of air into the tiny, thin-walled sacs of the lungs called **alveoli** (al-VE-o-li). In the alveoli, oxygen from the air is exchanged for the waste product carbon dioxide, which is carried to the lungs by the blood so that it can be eliminated from the body. The organs of the respiratory system perform a number of functions in addition to permitting movement of air into the alveoli. For example, if you live in a cold or dry environment, incoming air can be warmed and humidified as it passes over the lining of the respiratory air passages. In addition, inhaled irritants such as pollen or dust passing through the respiratory tubes can be trapped in the sticky mucus that covers the lining of many respiratory passages and then eliminated from the body. The respiratory system is also involved in regulating the acid-base balance of the body—a function that will be discussed in Chapter 18.

DIGESTIVE SYSTEM

The organs of the **digestive system** (Figure 3-10) are often separated into two groups: the *primary organs* and the *secondary* or *accessory organs* (Figure 3-1). They work together to ensure proper digestion and absorption of nutrients. The primary organs include the mouth, pharynx, esophagus, stomach, small intestine, large intestine, rectum, and anal canal. The accessory organs of digestion include the teeth, salivary glands, tongue, liver, gallbladder, pancreas, and appendix.

The primary organs of the digestive system form a tube, open at both ends, called the **gastrointestinal** (GAS-tro-in-TES-ti-nal) or **GI tract.** Food that enters the tract is digested, its nutrients are absorbed, and the undigested residue is eliminated from the body as waste material called

feces (FEE-seez). The accessory organs assist in the mechanical or chemical breakdown of ingested food. The appendix, although classified as an accessory organ of digestion and physically attached to the digestive tube, is not functionally important in the digestive process. However, inflammation of the appendix, called **appendicitis** (ah-PEN-di-SYE-tis) is a very serious clinical condition and frequently requires surgery.

URINARY SYSTEM

The organs of the **urinary** system include the **kidneys, ureters** (u-REE-ters), **bladder,** and **urethra** (yoo-RE-thrah).

The kidneys (Figure 13-11) "clear" or clean the blood of the waste products continually produced by the metabolism of foodstuff in the body cells. The kidneys also play an important role in maintaining the electrolyte, water, and acid-base balances in the body.

The waste product produced by the kidneys is called **urine** (YOOR-in). After it is produced by the kidneys, it flows out of the kidneys through the ureters into the urinary bladder, where it is stored. Urine passes from the bladder to the outside of the body through the urethra. In the male the urethra passes through the penis, which has a double function; it transports urine and semen or seminal fluid. Therefore it has urinary and reproductive purposes. In the female the urinary and reproductive passages are completely separate, so the urethra performs only a urinary function.

In addition to the organs of the urinary system, other organs are also involved in the elimination of body wastes. Undigested food residues and metabolic wastes are eliminated from the intestinal tract as feces, and the lungs rid the body of carbon dioxide. The skin also serves an excretory function by eliminating water and some salts in sweat.

REPRODUCTIVE SYSTEM

The normal function of the **reproductive system** is different from the normal function of other organ systems of the body. The proper functioning of the reproductive systems ensures survival, not of the individual but of the species—the human race. In addition, production of the hormones that permit the development of sexual characteristics occurs as a result of normal reproductive system activity.

FIGURE 3–9 *Respiratory System.*

FIGURE 3–10 *Digestive System.*

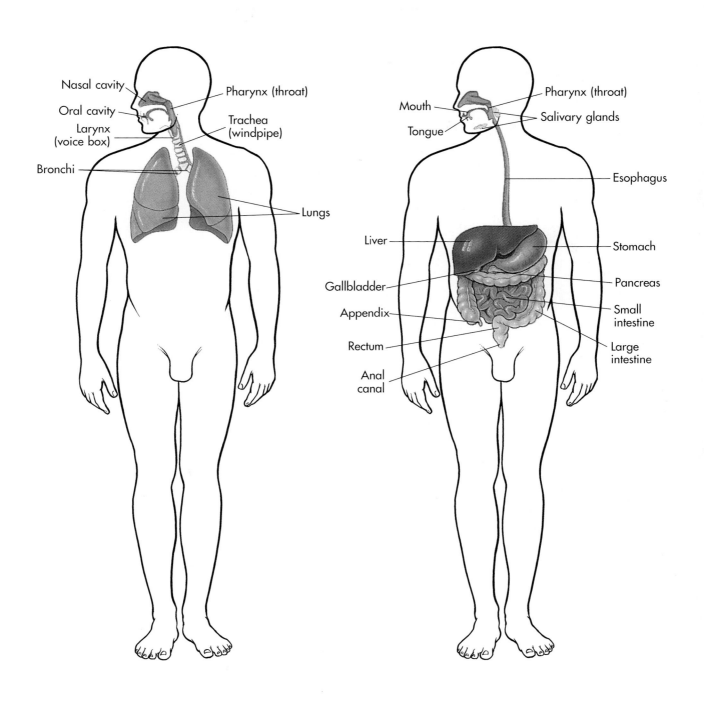

FIGURE 3–11   *Urinary System.*

FIGURE 3–12   *Male Reproductive System.*

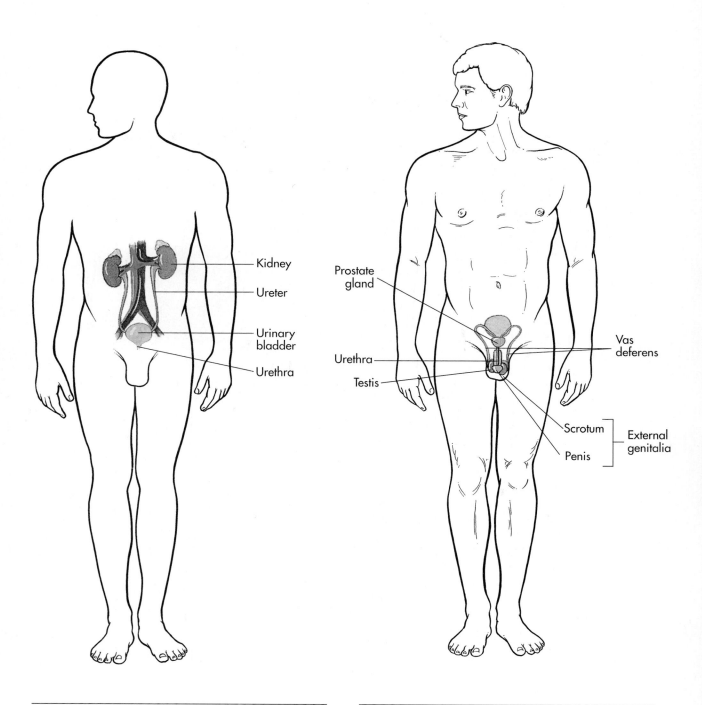

Kidney

Ureter

Urinary
bladder

Urethra

Prostate
gland

Urethra

Testis

Vas
deferens

Scrotum

Penis

External
genitalia

Male Reproductive System

The male reproductive structures shown in Figure 3-12 include the **gonads** (GO-nads), called **testes** (TES-teez), which produce the sex cells or **sperm**; one of the important **genital ducts,** called the **vas deferens** (vas DEF-er-enz); and the **prostate** (PROSS-tate), which is classified as an **accessory gland** in the male. The **penis** (PEE-nis) and **scrotum** (SKRO-tum) are supporting structures and together are known as the **genitalia** (jen-i-TAIL-yah). The urethra, which is identified in Figure 3-11 as part of the urinary system, passes through the penis. It serves as a genital duct that carries sperm to the exterior and as a passageway for the elimination of urine. Functioning together, these structures produce, transfer, and ultimately introduce sperm into the female reproductive tract, where fertilization can occur. Sperm produced by the testes travel through a number of genital ducts, including the vas deferens, to exit the body. The prostate and other accessory glands, which add fluid and nutrients to the sex cells as they pass through the ducts and the supporting structures (especially the penis), permit transfer of sex cells into the female reproductive tract.

Female Reproductive System

The female **gonads** are the **ovaries.** The **accessory organs** shown in Figure 3-13 include the **uterus** (YOO-ter-us), **uterine** (YOO-ter-in) or **fallopian tubes,** and the **vagina** (vah-JYE-nah). In the female the term **vulva** (VUL-vah) is used to describe the external genitalia. The breasts or **mammary glands** are also classified as external accessory sex organs in the female.

　　The reproductive organs in the female produce the sex cells or **ova;** receive the male sex cells (sperm); permit fertilization and transfer of the sex cells to the uterus; and allow for the development, birth, and nourishment of offspring.

　　As you study the more detailed structure and function of the organ systems in the chapters that follow, always relate the system and its component organs to the body as a whole. No one body system functions entirely independently of other systems. Instead, you will find that they are structurally and functionally interrelated and interdependent.

FIGURE 3–13 *Female Reproductive System.*

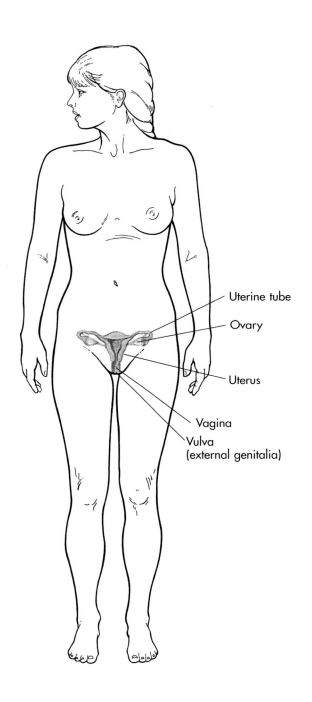

Uterine tube

Ovary

Uterus

Vagina

Vulva
(external genitalia)

Cancer Screening Tests

A knowledge of the structure and function of the body organ systems is a critically important "first step" in understanding and using information that empowers us to become more sophisticated guardians of our own health and well-being. For example, a better understanding of the reproductive system helps individuals participate in a more direct and personal way in cancer prevention screening techniques.

Breast and testicular self-examinations to detect cancer are two important ways that women and men can participate directly in protecting their own health. Instruction in these techniques is an important part of many home health care educational outreach services. Although specific instructions for these self-tests lie beyond the scope of a textbook of normal anatomy and physiology, it is important to note that this information is readily available from the American Cancer Society and from most hospitals, clinics, and health care providers.

 ## Paired Organs

Have you ever wondered what advantage there might be in having two kidneys, two lungs, two eyes, and two of many other organs? Although the body could function well with only one of each, most of us are born with a pair of these organs. For paired organs that are vital to survival, such as the kidneys, this arrangement allows for the accidental loss of one organ without immediate threat to the survival of the individual. Athletes who have lost one vital organ through injury or disease are often counseled against participating in contact sports that carry the risk of damaging the remaining organ. If the second organ is damaged, total loss of a vital function, such as sight, or even death may result.

DEFINITIONS AND CONCEPTS

A Organ—a structure made up of two or more kinds of tissues organized in such a way that they can together perform a more complex function than can any tissue alone

B Organ system—a group of organs arranged in such a way that they can together perform a more complex function than can any organ alone

C A knowledge of individual organs and how they are organized into groups makes more meaningful the understanding of how a particular organ system functions as a whole

ORGAN SYSTEMS

A Integumentary system
  1 Structure—organs
    a Skin
    b Hair
    c Nails
    d Sense receptors
    e Sweat glands
    f Oil glands
  2 Functions
    a Protection
    b Regulation of body temperature
    c Synthesis of chemicals
    d Sense organ

B Skeletal system
  1 Structure
    a Bones
    b Joints
  2 Functions
    a Support
    b Movement (with joints and muscles)
    c Storage of minerals
    d Blood cell formation

C Muscular system
  1 Structure
    a Muscles
      (1) Voluntary or striated
      (2) Involuntary or smooth
      (3) Cardiac
  2 Functions
    a Movement
    b Maintenance of body posture
    c Production of heat

D Nervous system
  1 Structure
    a Brain
    b Spinal cord
    c Nerves
    d Sense organs
  2 Functions
    a Communication
    b Integration
    c Control
    d Recognition of sensory stimuli
  3 System functions by production of nerve impulses caused by stimuli of various types
  4 Control is fast-acting and of short duration

E Endocrine system
  1 Structure
    a Pituitary gland
    b Pineal gland
    c Hypothalamus
    d Thyroid gland
    e Parathyroid glands
    f Thymus gland
    g Adrenal glands
    h Pancreas
    i Ovaries (female)
    j Testes (male)
  2 Functions
    a Secretion of special substances called *hormones* directly into the blood
    b Same as nervous system—communication, integration, control
    c Control is slow and of long duration
    d Examples of hormone regulation:
      (1) Growth
      (2) Metabolism
      (3) Reproduction
      (4) Fluid and electrolyte balance

F Circulatory system
  1 Structure
    a Heart
    b Blood vessels
  2 Functions
    a Transportation
    b Regulation of body temperature
    c Immunity (body defense)

G Lymphatic system
  1 Structure
    a Lymph nodes
    b Lymphatic vessels
    c Thymus
    d Spleen
  2 Functions
    a Transportation
    b Immunity (body defense)

H Respiratory system
  1 Structure
    a Nose
    b Pharynx

    c Larynx
    d Trachea
    e Bronchi
    f Lungs
  2 Functions
    a Exchange of waste gas (carbon dioxide) for oxygen in the lungs
    b Area of gas exchange in the lungs called *alveoli*
    c Filtration of irritants from inspired air
    d Regulation of acid-base balance

I Digestive system
  1 Structure
    a Primary organs
      (1) Mouth
      (2) Pharynx
      (3) Esophagus
      (4) Stomach
      (5) Small intestine
      (6) Large intestine
      (7) Rectum
      (8) Anal canal
    b Accessory organs
      (1) Teeth
      (2) Salivary glands
      (3) Tongue
      (4) Liver
      (5) Gallbladder
      (6) Pancreas
      (7) Appendix
  2 Functions
    a Mechanical and chemical breakdown (digestion) of food
    b Absorption of nutrients
    c Undigested waste product that is eliminated is called *feces*
    d Appendix is a structural but not a functional part of digestive system
    e Inflammation of appendix is called *appendicitis*

J Urinary system
  1 Structure
    a Kidneys
    b Ureters
    c Urinary bladder
    d Urethra
  2 Functions
    a "Clearing" or cleaning blood of waste products—waste product excreted from body is called *urine*
    b Electrolyte balance
    c Water balance
    d Acid-base balance
    e In male, urethra has urinary and reproductive functions

K Reproductive system
  1 Structure
    a Male
      (1) Gonads—testes
      (2) Genital ducts—vas deferens, urethra
      (3) Accessory gland—prostate
      (4) Supporting structures—genitalia (penis and scrotum)
    b Female
      (1) Gonads—ovaries
      (2) Accessory organs—uterus, uterine (fallopian) tubes, vagina
      (3) Supporting structures—genitalia (vulva), mammary glands (breasts)
  2 Functions
    a Survival of species
    b Production of sex cells (male: sperm; female: ova)
    c Transfer and fertilization of sex cells
    d Development and birth of offspring
    e Nourishment of offspring
    f Production of sex hormones

N E W   W O R D S

Review the names of organ systems and individual organs in Figures 3-1 through 3-13.

appendix	gastrointestinal (GI)	hormone	nerve impulse
cardiovascular	tract	integumentary	stimuli
endocrine	genitalia	lymphatic	urine
feces			

CHAPTER TEST

1. The largest and most important organ of the integumentary system is the _____.
2. The 206 individual organs of the skeletal system are called _____.
3. The four primary functions of the nervous system are _____, _____, _____, and _____.
4. The organs of the endocrine system secrete substances called _____ into the blood.
5. The thymus and spleen are classified as _____ organs.
6. The waste product produced by the kidneys is called _____.
7. The tongue, liver, and gallbladder are classified as _____ organs of digestion.
8. Oxygen from the air is exchanged for the waste product carbon dioxide in thin-walled sacs in the lungs called _____.

Select the most appropriate answer in column B for each item in column A. (Only one answer is correct.)

COLUMN A	COLUMN B
9. _____Appendage of skin	a. Lymphatic system
10. _____Storage site for calcium	b. Muscles
11. _____Function to generate heat	c. Feces
12. _____Permit rapid communication	d. Hair
13. _____Important in body defense mechanism	e. Vulva
14. _____Undigested residue in GI tract	f. Testes
15. _____Female external genitalia	g. Bones
16. _____Male gonads	h. Nerve impulses

REVIEW QUESTIONS

1. Give brief definitions of the terms *organ* and *organ system.*
2. List the 11 major organ systems of the body.
3. Discuss the structure and generalized functions of the integumentary system.
4. What is the relationship between a stimulus and a nerve impulse?
5. Define the term *hormone.*
6. List and discuss the structural and functional components of the circulatory system.
7. Identify the waste products associated with the urinary and digestive systems.
8. List the primary and accessory organs of the digestive system.
9. Compare and contrast the structure and function of the male and female reproductive systems.

CRITICAL THINKING

10. Explain the functional interaction that occurs between the skeletal, muscular, and nervous systems.
11. Compare the generalized functions of the nervous and endocrine systems. How are they similar? How do they differ?

The Integumentary System and Body Membranes

OBJECTIVES

*After you have completed this chapter,
you should be able to:*

1. Classify, compare the structure of, and give examples of each type of body membrane.
2. Describe the structure and function of the epidermis and dermis.
3. List and briefly describe each accessory organ of the skin.
4. List and discuss the three primary functions of the integumentary system.
5. Classify burns and describe how to estimate the extent of a burn injury.

n Chapter 1 the concept of progressive organization of body structures from simple to complex was established. Complexity in body structure and function progresses from cells to tissues and then to organs and organ systems. This chapter discusses the skin and its appendages—the hair, the nails, and the skin glands—as an organ system. This system is called the **integumentary system. Integument** (in-TEG-yoo-ment) is another name for the skin, and the skin itself is the principal organ of the integumentary system. The skin is one of a group of anatomically simple but functionally important sheetlike structures called **membranes.** This chapter will begin with classification and discussion of the important body membranes. Study of the structure and function of the integument will follow. Ideally, you should study the skin and its appendages before proceeding to the more traditional organ systems in the chapters that follow to improve your understanding of how structure is related to function.

⃝ *Classification of Body Membranes*

The term **membrane** refers to a thin, sheetlike structure that may have many important functions in the body. Membranes cover and protect the body surface, line body cavities, and cover the inner surfaces of the hollow organs such as the digestive, reproductive, and respiratory passageways. Some membranes anchor organs to each other or to bones, and others cover the internal organs. In certain areas of the body, membranes secrete lubricating fluids that reduce friction during organ movements such as the beating of the heart or lung expansion and contraction. Membrane lubricants also decrease friction between bones in joints. There are two major categories or types of body membranes:

1. **Epithelial membranes,** composed of epithelial tissue and an underlying layer of specialized connective tissue
2. **Connective tissue membranes,** composed exclusively of various types of connective tissue; no epithelial cells are present in this type of membrane

EPITHELIAL MEMBRANES

There are three types of epithelial tissue membranes in the body:

1. Cutaneous membrane
2. Serous membranes
3. Mucous membranes

Cutaneous Membrane

The **cutaneous** (kyoo-TAY-nee-us) **membrane** or **skin** is the primary organ of the integumentary system. It is one of the most important and certainly one of the largest and most visible organs. In most individuals the skin composes some 16% of the body weight. It fulfills the requirements necessary for an epithelial tissue membrane in that it has a superficial layer of epithelial cells and an underlying layer of supportive connective tissue. Its structure is uniquely suited to its many functions. The skin will be discussed in depth later in the chapter.

Serous Membranes

Like all epithelial membranes, a **serous** (SE-rus) **membrane** is composed of two distinct layers of tissue. The epithelial sheet is a thin layer of simple squamous epithelium. The connective tissue layer forms a very thin, gluelike **basement membrane** that holds and supports the epithelial cells.

The serous membrane that lines body cavities and covers the surfaces of organs in those cavities is in reality a single, continuous sheet of tissue covering two different surfaces. The name of the serous membrane is determined by its location. Using this criterion results in two types of serous membranes; the first type lines body cavities, and the second type covers the organs in those cavities. The serous membrane, which lines the walls of a body cavity much like wallpaper covers the walls of a room, is called the **parietal** (pah-RYE-i-tal) **portion.** The other type of serous membrane, which covers the surface of organs found in body cavities, is called the **visceral** (VIS-er-al) **portion.**

The serous membranes of the thoracic and abdominal cavities are identified in Figure 4-1. In the thoracic cavity the serous membranes are called **pleura** (PLOOR-ah), and in the abdominal cavity, they are called **peritoneum** (pair-i-toe-NEE-um). Look again at Figure 4-1 to note the placement of the **parietal** and **visceral pleura** and

FIGURE 4–1   *Types of Body Membranes.* A, Epithelial membranes, including cutaneous membrane (skin), serous membranes (parietal and visceral pleura and peritoneum), and mucous membranes. **B,** Connective tissue membranes, including synovial membranes. See text for explanation.

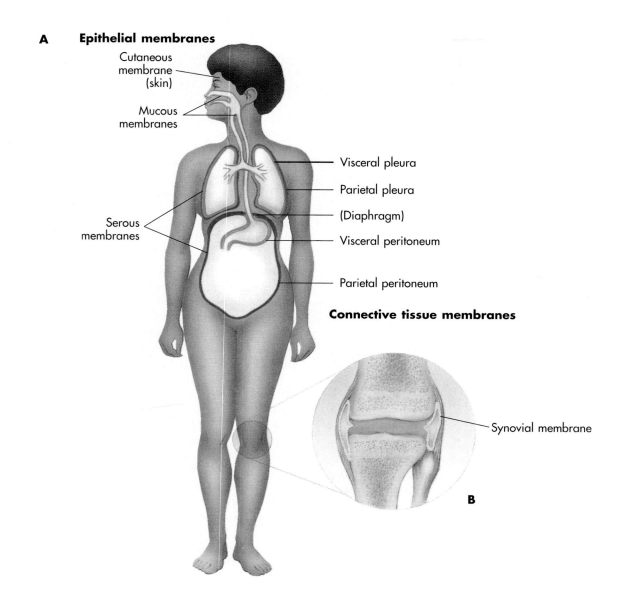

A   **Epithelial membranes**

Cutaneous membrane (skin)

Mucous membranes

Serous membranes

Visceral pleura

Parietal pleura

(Diaphragm)

Visceral peritoneum

Parietal peritoneum

Connective tissue membranes

Synovial membrane

B

the **parietal** and **visceral peritoneum.** In both cases the parietal layer forms the lining of the bodycavity, and the visceral layer covers the organs found in that cavity.

Serous membranes secrete a thin, watery fluid that helps reduce friction and serves as a lubricant when organs rub against one another and against the walls of the cavities that contain them. **Pleurisy** (PLOOR-i-see) is a very painful pathological condition characterized by inflammation of the serous membranes (pleura) that line the chest cavity and cover the lungs. Pain is caused by irritation and friction as the lungs rub against the walls of the chest cavity. In severe cases the inflamed surfaces of the pleura fuse, and permanent damage may develop. The term **peritonitis** (pair-i-toe-NYE-tis) is used to describe inflammation of the serous membranes in the abdominal cavity. Peritonitis is sometimes a serious complication of an infected appendix.

Mucous Membranes

Mucous (MYOO-kus) **membranes** are epithelial membranes that line body surfaces opening directly to the exterior. Examples of mucous membranes include those lining the respiratory, digestive, urinary, and reproductive tracts. The epithelial component of a mucous membrane varies, depending on its location and function. In the esophagus, for example, a tough, abrasion-resistant stratified squamous epithelium is found. A thin layer of simple columnar epithelium covers the walls of the lower segments of the digestive tract.

The epithelial cells of most mucous membranes secrete a thick, slimy material called **mucus** that keeps the membranes moist and soft.

The term **mucocutaneous** (myoo-ko-kyoo-TAY-nee-us) **junction** is used to describe the transitional area that serves as a point of "fusion" where skin and mucous membranes meet. Such junctions lack accessory organs such as hair or sweat glands that characterize skin. These transitional areas are generally moistened by mucous glands within the body orifices or openings where these junctions are located. The eyelids, nasal openings, vulva, and anus have mucocutaneous junctions that may become sites of infection or irritation.

CONNECTIVE TISSUE MEMBRANES

Unlike cutaneous, serous, and mucous membranes, connective tissue membranes do not contain epithelial components. The **synovial** (si-NO-vee-al) **membranes** lining the spaces between bones and joints that move are classified as connective tissue membranes. These membranes are smooth and slick and secrete a thick and colorless lubricating fluid called **synovial fluid.** The membrane itself, with its specialized fluid, helps reduce friction between the opposing surfaces of bones in movable joints. Synovial membranes also line the small, cushionlike sacs called **bursae** (BER-see) found between moving body parts.

● *The Skin*

The brief description of the skin in Chapter 3 (see p. 47) identified it not only as the primary organ of the integumentary system but also as the largest and one of the most important organs of the body. Architecturally the skin is a marvel. Consider the incredible number of structures fitting into 1 square inch of skin: 500 sweat glands; over 1000 nerve endings; yards of tiny blood vessels; nearly 100 oil or **sebaceous** (se-BAY-shus) glands; 150 sensors for pressure, 75 for heat, 10 for cold; and millions of cells.

STRUCTURE OF THE SKIN

The skin or cutaneous membrane is a sheetlike organ composed of the following layers of distinct tissue (Figure 4-2):

1. The **epidermis** is the outermost layer of the skin. It is a relatively thin sheet of stratified squamous epithelium.
2. The **dermis** is the deeper of the two layers. It is thicker than the epidermis and is made up largely of connective tissue.

As you can see in Figure 4-2, the layers of the skin are supported by a thick layer of loose connective tissue and fat called **subcutaneous** (sub-kyoo-TAY-nee-us) **tissue.** Fat in the subcutaneous layer insulates the body from extremes of heat and cold. It also serves as a stored source of energy for the body and can be used as a food source if required. In addition, the subcutaneous

FIGURE 4–2 *Microscopic View of the Skin.* The epidermis, shown in longitudinal section, is raised at one corner to reveal the ridges in the dermis.

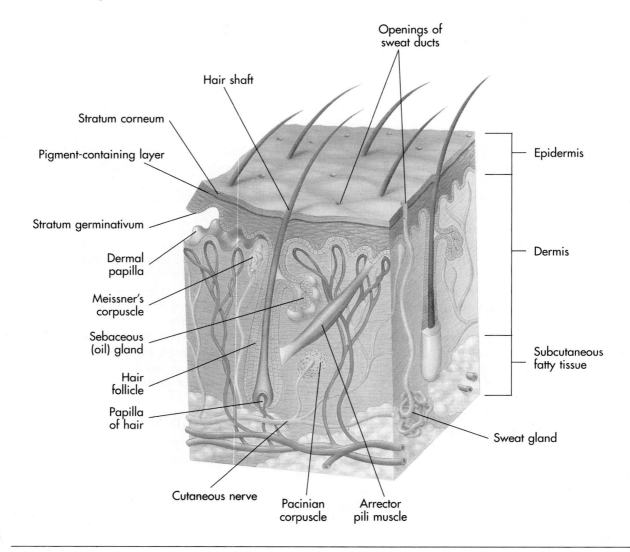

tissue acts as a shock-absorbing pad and helps protect underlying tissues from injury caused by bumps and blows to the body surface.

Epidermis

The tightly packed epithelial cells of the epidermis are arranged in many distinct layers. The cells of the innermost layer, called the **stratum germinativum,** undergo mitosis and reproduce themselves (Figure 4-2). As they move toward the surface of the skin, these new cells "specialize" in ways that increase their ability to provide protection for the body tissues that lie below them. This ability is of critical clinical significance. It enables the skin to repair itself if it is injured. The self-repairing characteristic of normal skin makes it possible for the body to maintain an effective barrier against infection, even when it is subjected to injury and normal wear and tear. As new cells are produced in the deep layer of the epidermis, they

FIGURE 4–3 *Photomicrograph of the Skin.* Many dead cells of the stratum corneum have flaked off from the surface of the epidermis. Note that the epidermis is very cellular. The dermis has fewer cells and more connective tissue.

"Flaked" cells from stratum corneum

Epidermis

Dermis

move upward through additional layers, or "strata" of cells. As they approach the surface, the cytoplasm is replaced by one of nature's most unique proteins, a substance called **keratin** (KARE-ah-tin). Keratin is a tough, waterproof material that provides cells in the outer layer of the skin with a horny, abrasion-resistant, and protective quality. The tough outer layer of the epidermis is called the **stratum corneum** (KOR-nee-um). Cells filled with keratin are continually pushed to the surface of the epidermis. In the photomicrograph of the skin shown in Figure 4-3, many of the outermost cells of the stratum corneum have been dislodged. These dry, dead cells filled with keratin "flake off" by the thousands onto our clothes, our bathwater, and things we handle. Millions of epithelial cells reproduce daily to replace the millions shed—just one example of the work our bodies do without our knowledge, even when we seem to be resting.

The deepest cell layer of the epidermis identified in Figure 4-2 is responsible for the production of a specialized **pigment** substance that gives color to the skin. The term *pigment* comes from a Latin word meaning "paint." It is this epidermal layer that gives color to the skin. The brown pigment **melanin** (MEL-ah-nin) is produced by specialized cells in this layer. These cells are called **melanocytes** (MEL-ah-no-sites). The higher the concentration of melanin, the deeper the color of skin. The amount of melanin in your skin depends first on the skin color genes you have inherited. That is, heredity determines how dark or light your basic skin color is. However, other factors such as sunlight can modify this hereditary effect. Prolonged exposure to sunlight in light-skinned people darkens the exposed area because it leads to increased melanin deposits in the epidermis. If the skin contains little melanin, a change in color can occur if the volume of blood in the skin changes significantly or if the amount of oxygen in the blood is increased or decreased. In these individuals increased blood flow to the skin or increased blood oxygen levels can cause a pink flush to appear. However, if blood oxygen levels decrease or if actual blood flow is reduced dramatically, the skin turns a bluish gray color—a condition called **cyanosis** (SYE-ah-NO-sis). In general, the less abundant the melanin deposits in the skin, the more visible the changes in color caused by the change in skin blood volume or oxygen level. Conversely, the richer the skin's pigmentation, the less noticeable such changes will be.

The cells of the epidermis are packed tightly together. They are held firmly to one another and to the dermis below by specialized junctions between the membranes of adjacent cells. If these specialized links, sometimes described as "spot welds," are weakened or destroyed, the skin falls apart. When this occurs because of burns, friction injuries, or exposure to irritants, **blisters** may result.

The junction that exists between the thin epidermal layer of the skin above and the dermal layer below is called the **dermal-epidermal junction.** The area of contact between dermis and epidermis "glues" them together and provides support for the epidermis, which is attached to its upper surface. Blister formation also occurs if this junction is damaged or destroyed. The junction is visible in Figure 4-2, which shows the epidermis raised on one corner to reveal the underlying dermis more clearly.

Dermis

The dermis is the deeper of the two primary skin layers and is much thicker than the epidermis. It is composed largely of connective tissue. Instead of cells being crowded close together like the epithelial cells of the epidermis, they are scattered far apart, with many fibers in between. Some of the fibers are tough and strong (collagen or white fibers), and others are stretchable and elastic (elastic or yellow fibers).

The upper region of the dermis is characterized by parallel rows of peglike projections called **dermal papillae** (pah-PIL-ee), which are visible in

Figure 4-2. These upward projections are interesting and useful features. They form an important part of the dermal-epidermal junction that helps bind the skin layers together. In addition, they form the ridges and grooves that make possible fingerprinting as a means of identification.

You can observe these ridges on the tips of the fingers and on the skin covering the palms of your hands. Observe in Figure 4-2 how the epidermis follows the contours of the dermal papillae. These ridges develop sometime before birth. Not only is their pattern unique in each individual, but also it never changes except to grow larger—two facts that explain why our fingerprints or footprints positively identify us. Many hospitals identify newborn babies by footprinting them soon after birth.

The deeper area of the dermis is filled with a dense network of interlacing fibers. Most of the fibers in this area are collagen that gives toughness to the skin. However, elastic fibers are also present. These make the skin stretchable and elastic (able to rebound). As we age, the number of elastic fibers in the dermis decreases, and the amount of fat stored in the subcutaneous tissue is reduced. Wrinkles develop as the skin loses elasticity, sags, and becomes less soft and pliant.

In addition to connective tissue elements, the dermis contains a specialized network of nerves and nerve endings to process sensory information such as pain, pressure, touch, and temperature. At various levels of the dermis, there are muscle fibers, hair follicles, sweat and sebaceous glands, and many blood vessels.

APPENDAGES OF THE SKIN

Hair

The human body is covered with millions of hairs. Indeed, at the time of birth most of the specialized structures called **follicles** (FOL-li-kuls) that are required for hair growth are already present. They develop early in fetal life and by birth are present in most parts of the skin. The hair of a newborn infant is extremely fine and soft; it is called **lanugo** (lah-NOO-go) from the Latin word meaning "down." In premature infants, lanugo may be noticeable over most of the body, but soon after birth the lanugo is lost and replaced by new hair that is stronger and more pigmented. Although only a few areas of the skin are hair-

Subcutaneous Injection

Although the subcutaneous layer is not part of the skin, it carries the major blood vessels and nerves to the skin above it. The rich blood supply and loose, spongy texture of the subcutaneous layer make it an ideal site for the rapid and relatively pain-free absorption of injected material. Liquid medicines such as insulin and pelleted implant materials are often administered by **subcutaneous injection** into this spongy and porous layer beneath the skin.

FIGURE 4–4 *Hair Follicle.* Relationship of a hair follicle and related structures to the epidermal and dermal layers of the skin.

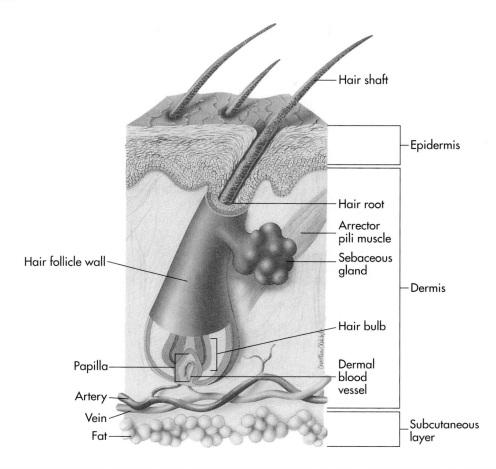

less—notably the lips, the palms of the hands, and the soles of the feet—most body hair remains almost invisible. Hair is most visible on the scalp, eyelids, and eyebrows. The coarse hair that first appears in the pubic and axillary regions at the time of puberty develops in response to the secretion of hormones.

Hair growth begins when cells of the epidermal layer of the skin grow down into the dermis, forming a small tube called the **hair follicle.** The relationship of a hair follicle and its related structures to the epidermal and dermal layers of the skin is shown in Figure 4-4. Hair growth begins from a small, cap-shaped cluster of cells called the

hair papilla (pah-PIL-ah), which is located at the base of the follicle. The papilla is nourished by a dermal blood vessel. Note in Figure 4-4 that part of the hair, namely the **root,** lies hidden in the follicle. The visible part of a hair is called the **shaft.** Figure 4-5 shows shafts of hair extending from their follicles.

As long as cells in the papilla of the hair follicle remain alive, new hair will replace any that is cut or plucked. Contrary to popular belief, frequent cutting or shaving does not make hair grow faster or become coarser. Why? Because neither process affects the epithelial cells that form the hairs, since they are embedded in the dermis.

FIGURE 4–5 *Hair Shaft and Follicle.* Scanning electron micrograph showing shafts of hair extending from their follicles.

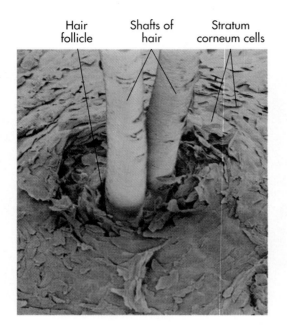

Hair follicle   Shafts of hair   Stratum corneum cells

A tiny, smooth (involuntary) muscle can be seen in Figure 4-4. It is called an **arrector pili** (ah-REK-tor PYE-lie) muscle. It is attached to the base of a dermal papilla above and to the side of a hair follicle below. Generally, these muscles contract only when we are frightened or cold. When contraction occurs, each muscle simultaneously pulls on its two points of attachment (that is, up on a hair follicle but down on a part of the skin). This produces little raised places, called *goose pimples,* between the depressed points of the skin and at the same time pulls the hairs up until they are more or less straight. The name *arrector pili* describes the function of these muscles; it is Latin for "erectors of the hair." We unconsciously recognize these facts in expressions such as "I was so frightened my hair stood on end."

Receptors

Receptors in the skin make it possible for the body surface to act as a sense organ, relaying messages to the brain concerning sensations such as touch, pain, temperature, and pressure. Receptors differ in structure from the highly complex to the very simple. Figure 4-6 shows enlarged views of a **Meissner's** (MIZE-ners) **corpuscle** and a **Pacin-**

FIGURE 4–6 *Skin Receptors.* Receptors are specialized nerve endings that make it possible for the skin to act as a sense organ. **A,** Meissner's corpuscle. **B,** Pacinian corpuscle. (See also Figure 4-2.)

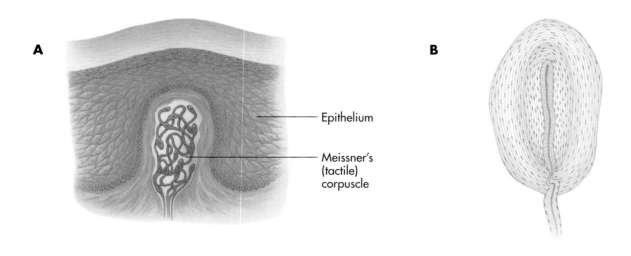

A

Epithelium

Meissner's (tactile) corpuscle

B

ian (pah-SIN-ee-an) **corpuscle.** Look again at Figure 4-2 and find these receptors. The Pacinian corpuscle is deep in the dermis. It is capable of detecting *pressure* on the skin surface. The Meissner's corpuscle is generally located close to the skin surface. It is capable of detecting sensations of *light touch.* Both specialized receptors are widely distributed in skin. Additional receptors in the skin respond to other types of stimuli. For example, **free nerve endings** respond to pain, and receptors called **Krause's end bulbs** detect sensations of touch, low frequency vibration, and possibly cold. Other receptors mediate sensations of heat, crude touch, and vibration.

Nails

Nails are classified as accessory organs of the skin and are produced by cells in the epidermis. They form when epidermal cells over the terminal ends of the fingers and toes fill with keratin and become hard and platelike. The components of a typical fingernail and its associated structures are shown in Figure 4-7. In this illustration the fingernail of the index finger is viewed from above and in sagittal section. (Recall that a sagittal section divides a body part into right and left portions.) Look first at the nail as seen from above. The visible part of the nail is called the **nail body.** The rest of the nail, namely, the **root,** lies in a groove and is hidden by a fold of skin called the **cuticle** (KYOO-ti-kul). In the sagittal section you can see the nail root from the side and note its relationship to the cuticle, which is folded back over its upper surface. The nail body nearest the root has a crescent-shaped white area known as the **lunula** (LOO-nyoo-lah), or "little moon." You should be able to identify this area easily on your own nails; it is most noticeable on the thumbnail. Under the nail lies a layer of epithelium called the **nail bed,** which is labeled on the sagittal section in Figure 4- 7. Because it contains abundant blood vessels, it appears pink in color through the translucent nail bodies. If blood oxygen levels drop and cyanosis develops, the nail bed will turn blue.

FIGURE 4–7 *Structure of Nails.* **A,** Fingernail viewed from above. **B,** Sagittal section of fingernail and associated structures.

Depilatories

All **depilatories** (de-PIL-ah-toe-rees) are used to remove unwanted hair. They dissolve the protein in hair shafts that extend above the skin surface. Because the follicle is not affected, regrowth of hair continues at a normal rate. Unfortunately, use of depilatories sometimes results in allergic reactions producing irritation and rash. If used to remove body hair before surgery, reactions to depilatories may delay healing of the surgical incision.

Skin Glands

The skin glands include the two varieties of **sweat** or **sudoriferous** (soo-doe-RIF-er-us) **glands** and the microscopic **sebaceous glands.**

Sweat (sudoriferous) glands. Sweat glands are the most numerous of the skin glands. They can be classified into two groups—**eccrine** (EK-rin) and **apocrine** (AP-o-krin)—based on type of secretion and location. **Eccrine sweat glands** are by far the more numerous, important, and widespread sweat glands in the body. They are quite small and, with few exceptions, are distributed over the total body surface. Throughout life they produce a transparent, watery liquid called **perspiration,** or **sweat.** Sweat assists in the elimination of waste products such as ammonia and uric acid. In addition to elimination of waste, sweat plays a critical role in helping the body maintain a constant temperature. Anatomists estimate that a single square inch of skin on the palms of the hands contains about 3000 eccrine sweat glands. With a magnifying glass you can locate the pinpoint-size openings on the skin that you probably call **pores.** The pores are outlets of small ducts from the eccrine sweat glands.

Apocrine sweat glands are found primarily in the skin in the armpit (axilla) and in the pigmented skin areas around the genitals. They are larger than the eccrine glands, and instead of watery sweat, they secrete a thicker, milky secretion. The odor associated with apocrine gland secretion is not caused by the secretion itself. Instead, it is caused by the contamination and decomposition of the secretion by skin bacteria. Apocrine glands enlarge and begin to function at puberty.

Sebaceous glands. Sebaceous glands secrete oil for the hair and skin. Oil or sebaceous glands grow where hairs grow. Their tiny ducts open into hair follicles (Figure 4-4) so that their secretion, called **sebum** (SEE-bum), lubricates the hair and skin. Someone aptly described sebum as "nature's skin cream" because it prevents drying and cracking of the skin. Sebum secretion increases during adolescence, stimulated by the increased blood levels of the sex hormones. Frequently sebum

Decubitus Ulcers

Family members, nurses, or other professionals who provide home health care services for bedridden or otherwise immobilized individuals need to be aware of the causes and nature of **decubitus** (de-KU-bi-tus) **ulcers** or pressure sores. Decubitus means "lying down," a name that hints at a common cause of pressure sores: lying in one position for long periods. Also called *bedsores*, these lesions appear after blood flow to a local area of skin slows because of pressure on skin covering a bony prominence such as the heel (see illustration). Ulcers form and infections develop as lack of blood flow causes tissue damage. Frequent changes in body position and soft support cushions help prevent decubitus ulcers.

accumulates in and enlarges some of the ducts of the sebaceous glands, forming white pimples. This sebum often darkens, forming a **blackhead.** Sebum secretion decreases in late adulthood, contributing to increased wrinkling and cracking of the skin.

FUNCTIONS OF THE SKIN

The skin or cutaneous membrane serves three important functions that contribute to survival. The most important functions are:

1. Protection
2. Temperature regulation
3. Sense organ activity

Protection

The skin as a whole is often described as our "first line of defense" against a multitude of hazards. It protects us against the daily invasion of deadly microbes. The tough, keratin-filled cells of the stratum corneum also resist the entry of harmful chemicals and protect against physical tears and cuts. Because it is waterproof, **keratin** also protects the body from excessive fluid loss. Melanin in the pigment layer of the skin prevents the sun's harmful ultraviolet rays from penetrating the interior of the body.

Temperature Regulation

The skin plays a key role in regulating the body's temperature. Incredible as it seems, on a hot and humid day the skin can serve as a means for releasing almost 3000 calories of body heat— enough heat energy to boil over 20 liters of water! It accomplishes this feat by regulating sweat secretion and by regulating the flow of blood close to the body surface. When sweat evaporates from the body surface, heat is also lost. The principle of heat loss through evaporation is basic to many cooling systems. When increased quantities of blood are allowed to fill the vessels close to the skin, heat is also lost by radiation. Blood supply to the skin far exceeds the amount needed by the skin. Such an abundant blood supply primarily enables the regulation of body temperature.

Sense Organ Activity

The skin functions as an enormous sense organ. Its millions of nerve endings serve as antennas or receivers for the body, keeping it informed of changes in its environment. The specialized receptors shown in Figures 4-2 and 4-6 make it possible for the body to detect sensations of light touch (Meissner's corpuscles) and pressures (Pacinian corpuscles). Other receptors make it possible for us to respond to the sensations of pain, heat, and cold.

BURNS

Burns constitute one of the most serious and frequent problems that affect the skin. Typically, we think of a burn as an injury caused by fire or by contact of the skin with a hot surface. However, overexposure to ultraviolet light (sunburn) or contact of the skin with an electric current or a harmful chemical such as an acid can also cause burns.

 Exercise and the Skin

Excess heat produced by the skeletal muscles during exercise increases the core body temperature far beyond the normal range. Because blood in vessels near the skin's surface dissipates heat well, the body's control centers adjust blood flow so that more warm blood from the body's core is sent to the skin for cooling. During exercise, blood flow in the skin can be so high that the skin takes on a redder coloration.

To help dissipate even more heat, sweat production increases to as high as 3 L per hour during exercise. Although each sweat gland produces very little of this total, over 3 million individual sweat glands are found throughout the skin. Sweat evaporation is essential to keeping body temperature in balance, but excessive sweating can lead to a dangerous loss of fluid. Because normal drinking may not replace the water lost through sweating, it is important to increase fluid consumption during and after any type of exercise to avoid **dehydration.**

FIGURE 4–8   *The "Rule of Nines."* Dividing the body into 11 areas of 9% each helps one to estimate the amount of skin surface burned in an adult.

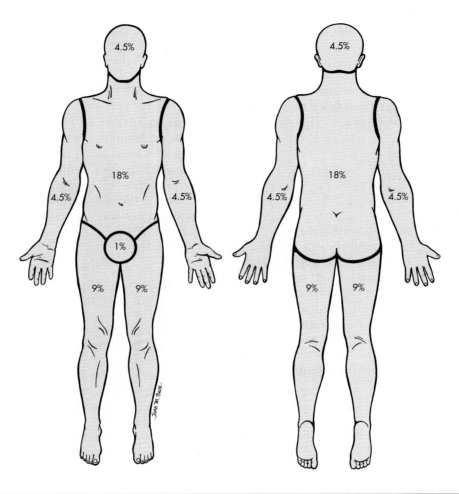

Estimating Body Surface Area

When burns involve large areas of the skin, treatment and the possibility for recovery depend in large part on the **total area involved** and the **severity of the burn.** The severity of a burn is determined by the depth of the injury, as well as by the amount of body surface area affected.

The "**rule of nines**" is one of the most frequently used methods of determining the extent of a burn injury. With this technique (Figure 4-8) the body is divided into 11 areas of 9% each, with the area around the genitals representing the additional 1% of body surface area. As you can see in Figure 4-8, in the adult 9% of the skin covers the head and each upper extremity, including front and back surfaces. Twice as much, or 18%, of the total skin area covers the front and back of the trunk and each lower extremity, including front and back surfaces.

Classification of Burns

The classification system used to describe the severity of burns is based on the number of tissue layers involved. The most severe burns destroy not only layers of the skin and subcutaneous tissue but underlying tissues, as well.

First-degree burns. A **first-degree burn** (for example, a typical sunburn) causes minor discomfort and some reddening of the skin. Although the surface layers of the epidermis may peel in 1 to 3 days, no blistering occurs, and actual tissue destruction is minimal.

Second-degree burns. A **second-degree burn** involves the deep epidermal layers and always causes injury to the upper layers of the dermis. Although deep second-degree burns damage sweat glands, hair follicles, and sebaceous glands, complete destruction of the dermis does not occur. Blisters, severe pain, generalized swelling, and fluid loss characterize this type of burn. Scarring is common. First- and second-degree burns are called **partial-thickness burns.**

Third-degree burns. A **third-degree,** or **full-thickness burn** is characterized by complete destruction of the epidermis and dermis. In addition, tissue death extends below the primary skin layers into the subcutaneous tissue. Third-degree burns often involve underlying muscles and even bone. One distinction between second- and third-degree burns is that third-degree lesions are insensitive to pain immediately after injury because of the destruction of nerve endings. The fluid loss that results from third-degree burns is a very serious problem.

O U T L I N E S U M M A R Y

CLASSIFICATION OF BODY MEMBRANES

A Classification of body membranes (Figure 4-1)
  1 Epithelial membranes—composed of epithelial tissue and an underlying layer of connective tissue
  2 Connective tissue membranes—composed largely of various types of connective tissue

B Epithelial membranes
  1 Cutaneous membrane—the skin
  2 Serous membranes—simple squamous epithelium on a connective tissue basement membrane
    a Types
      (1) Parietal—line walls of body cavities
      (2) Visceral—cover organs found in body cavities
    b Examples
      (1) Pleura—parietal and visceral layers line walls of thoracic cavity and cover the lungs
      (2) Peritoneum—parietal and visceral layers line walls of abdominal cavity and cover the organs in that cavity
    c Diseases
      (1) Pleurisy—inflammation of the serous membranes that line the chest cavity and cover the lungs
      (2) Peritonitis—inflammation of the serous membranes in the abdominal cavity that line the walls and cover the abdominal organs
  3 Mucous membranes
    a Line body surfaces that open directly to the exterior
    b Produce mucus, a thick secretion that keeps the membranes soft and moist

C Connective tissue membranes
  1 Do not contain epithelial components
  2 Produce a lubricant called *synovial fluid*
  3 Examples are the synovial membranes in the spaces between joints and in the lining of bursal sacs

THE SKIN

A Structure (Figure 4-2)—two primary layers called *epidermis* and *dermis*
  1 Epidermis
    a Outermost and thinnest primary layer of skin
    b Composed of several layers of stratified squamous epithelium
    c Stratum germinativum—innermost layer of cells that continually reproduce, and new cells move toward the surface
    d As cells approach surface, they are filled with a tough, waterproof protein called *keratin* and eventually flake off
    e Stratum corneum—outermost layer of keratin-filled cells
    f Pigment-containing layer—epidermal layer that contains pigment cells called *melanocytes,* which produce the brown pigment melanin
    g Blisters—caused by breakdown of union between cells or primary layers of skin
    h Dermal-epidermal junction—specialized area between two primary skin layers
  2 Dermis
    a Deeper and thicker of the two primary skin layers and composed largely of connective tissue
    b Upper area of dermis characterized by parallel rows of peglike dermal papillae
    c Ridges and grooves in dermis form pattern unique to each individual (basis of fingerprinting)
    d Deeper areas of dermis filled with network of tough collagenous and stretchable elastic fibers
    e Number of elastic fibers decreases with age and contributes to wrinkle formation
    f Dermis also contains nerve endings, muscle fibers, hair follicles, sweat and sebaceous glands, and many blood vessels

B Appendages of the skin
  1 Hair (Figures 4-4 and 4-5)
    a Soft hair of fetus and newborn called *lanugo*
    b Hair growth requires epidermal tubelike structure called *hair follicle*
    c Hair growth begins from hair papilla
    d Hair root lies hidden in follicle and visible part of hair called *shaft*
    e Arrector pili—specialized smooth muscle that produces "goose pimples" and causes hair to stand up straight
  2 Receptors (Figure 4-6)
    a Specialized nerve endings—make it possible for skin to act as a sense organ
    b Meissner's corpuscle—capable of detecting light touch
    c Pacinian corpuscle—capable of detecting pressure
  3 Nails (Figure 4-7)
    a Produced by epidermal cells over terminal ends of fingers and toes
    b Visible part called *nail body*
    c Root lies in a groove and is hidden by cuticle
    d Crescent-shaped area nearest root called *lunula*
    e Nail bed may change color with change in blood flow

4 Skin glands
 a Types
  (1) Sweat or sudoriferous
  (2) Sebaceous
 b Sweat or sudoriferous glands
  (1) Types
   (a) Eccrine sweat glands
    • Most numerous, important, and wide-spread of the sweat glands
    • Produce perspiration or sweat, which flows out through pores on skin surface
    • Function throughout life and assist in body heat regulation
   (b) Apocrine sweat glands
    • Found primarily in axilla and around genitalia
    • Secrete a thicker, milky secretion quite different from eccrine perspiration
    • Breakdown of secretion by skin bacteria produces odor
 c Sebaceous glands
  (1) Secrete oil or sebum for hair and skin
  (2) Level of secretion increases during adolescence
  (3) Amount of secretion regulated by sex hormones
  (4) Sebum in sebaceous gland ducts may darken to form a blackhead
C Functions of the skin
 1 Protection—first line of defense
  a Against infection by microbes
  b Against ultraviolet rays from sun
  c Against harmful chemicals
  d Against cuts and tears

2 Temperature regulation
 a Skin can release almost 3000 calories of body heat per day
  (1) Mechanisms of temperature regulation
   (a) Regulation of sweat secretion
   (b) Regulation of flow of blood close to the body surface
3 Sense organ activity
 a Skin functions as an enormous sense organ
 b Receptors serve as receivers for the body, keeping it informed of changes in its environment
D Burns
 1 Treatment and recovery or survival depend on total area involved and severity or depth of the burn
 2 Estimating body surface area using the "rule of nines" (Figure 4-8) in adults
  a Body divided into 11 areas of 9% each
  b Additional 1% of body surface area around genitals
 3 Classification of burns
  a First-degree (partial-thickness) burns—only surface layers of epidermis involved
  b Second-degree (partial-thickness) burns—involve the deep epidermal layers and always cause injury to the upper layers of the dermis
  c Third-degree (full-thickness) burns—characterized by complete destruction of the epidermis and dermis
   (1) May involve underlying muscle and bone
   (2) Lesion is insensitive to pain because of destruction of nerve endings immediately after injury—intense pain is soon experienced

---

N E W   W O R D S

apocrine sweat gland	eccrine sweat gland	mucocutaneous junction	pleura
arrector pili	epidermis	mucous membrane	pleurisy
blister	follicle	mucus	sebaceous gland
bursa	keratin	Pacinian corpuscle	serous membrane
cutaneous	Krause's end bulb	papilla	stratum corneum
cuticle	lanugo	parietal	subcutaneous
cyanosis	lunula	peritoneum	sudoriferous gland
dehydration	Meissner's corpuscle	peritonitis	synovial membrane
depilatories	melanin		visceral portion
dermis	melanocyte		

CHAPTER TEST

1. There are two major categories of body membranes called _____ membranes and _____ membranes.
2. The membrane that lines the walls of a body cavity is called the _____ portion of the membrane, and the portion that covers the surface of organs found in body cavities is called the _____ portion of that membrane.
3. The connective tissue membranes that line joint spaces are called _____ membranes.
4. The principal organ of the integumentary system is the _____.
5. The skin is classified as a _____ membrane.
6. The two principal layers of the skin are called the _____ and the _____.
7. The tough waterproof material that provides cells in the outer layer of the skin with a protective quality is called _____.
8. The brown pigment that gives color to the skin is known as _____.
9. The part of a hair that lies hidden in the follicle is called the _____.
10. The Pacinian corpuscle is capable of detecting _____ on the skin surface.
11. Sweat or sudoriferous glands are classified into two groups, _____ and _____, based on type of secretion and location.
12. Specialized glands that secrete oil for the hair and skin are known as _____ glands.
13. The "rule of nines" is used in the treatment and prognosis of _____.

Select the most correct answer from Column B for each statement in Column A. (Only one answer is correct.)

COLUMN A	COLUMN B
14. _____ Another name for skin	a. Lanugo
15. _____ Outermost layer of skin	b. Arrector pili
16. _____ Contains melanocytes	c. Shaft
17. _____ Fine, soft hair of newborn	d. Integument
18. _____ Visible part of a hair	e. Pigment-containing layer
19. _____ Produce "goose pimples"	f. Bursae
20. _____ Lined with synovial membrane	g. Epidermis

R E V I E W   Q U E S T I O N S

1. Define the term *membrane* and discuss a variety of functions that membranes serve in the body.
2. Classify body membranes.
3. Discuss the two types of serous membranes and give examples of each.
4. Compare mucous and synovial membranes.
5. How do the terms *integument* and *integumentary system* differ in meaning?
6. List the appendages of the skin.
7. Discuss the three primary functions of the skin.
8. What are the two major factors that determine the treatment of and possibility for recovery from burn injuries?

C R I T I C A L   T H I N K I N G

9. Identify and compare the two main layers of the skin. How are these layers related to the subcutaneous layer?
10. Classify the skin glands. Locate each type of gland and compare the secretions of each.
11. Classify and compare the three major types of burn injuries.

The Skeletal System

OBJECTIVES

After you have completed this chapter, you should be able to:

1. List and discuss the generalized functions of the skeletal system.

2. Identify the major anatomical structures found in a typical long bone.

3. Discuss the microscopic structure of bone and cartilage, including the identification of specific cell types and structural features.

4. Explain how bones are formed, how they grow, and how they are remodeled.

5. Identify the two major subdivisions of the skeleton and list the bones found in each area.

6. List and compare the major types of joints in the body and give an example of each.

The primary organs of the skeletal system, bones, lie buried within the muscles and other soft tissues, providing a rigid framework and support structure for the whole body. In this respect the skeletal system functions like steel girders in a building; however, unlike steel girders, bones can be moved. Bones are also living organs. They can change and help the body respond to a changing environment. This ability of bones to change allows our bodies to grow and change.

Our study of the skeletal system will begin with an overview of its function. We will then classify bones by their structure and describe the characteristics of a typical bone. After discussing the microscopic structure of skeletal tissues, we will briefly outline bone growth and formation. With this information, the study of specific bones and the way they are assembled in the skeleton will be more meaningful. The chapter will end with a discussion of skeletal functions and an overview of joints or **articulations** (ar-tick-yoo-LAY-shuns).

An understanding of how bones articulate with one another in joints and how they relate to other body structures provides a basis for understanding the functions of many other organ systems. Coordinated movement, for example, is possible only because of the way bones are joined to one another and because of the way muscles are attached to those bones. In addition, knowing where specific bones are in the body will assist you in locating other body structures that will be discussed later.

Functions of the Skeletal System

SUPPORT

Bones form the body's supporting framework. All the softer tissues of the body literally hang from the skeletal framework.

PROTECTION

Hard, bony "boxes" protect delicate structures enclosed within them. For example, the skull protects the brain. The breastbone and ribs protect vital organs (heart and lungs) and also a vital tissue (red bone marrow, the blood cell–forming tissue).

MOVEMENT

Muscles are anchored firmly to bones. As muscles contract and shorten, they pull on bones and thereby move them.

STORAGE

Bones play an important part in maintaining homeostasis of blood calcium, a vital substance required for normal nerve and muscle function. They serve as a safety-deposit box for calcium. When the amount of calcium in blood increases above normal, calcium moves out of the blood and into the bones for storage. Conversely, when blood calcium decreases below normal, calcium moves in the opposite direction. It comes out of storage in bones and enters the blood.

HEMOPOIESIS

The term **hemopoiesis** (hee-mo-poy-EE-sis) is used to describe the process of blood cell formation. It is a combination of two Greek words: *hemo* (HEE-mo) meaning "blood" and *poiesis* (poy-EE-sis) meaning "to make." Blood cell formation is a vital process carried on in **red bone marrow.** Red bone marrow is soft connective tissue inside the hard walls of some bones.

Types of Bones

There are four types of bones. Their names suggest their shapes: *long* (for example, humerus or upper arm bone), *short* (for example, carpals or wrist bones), *flat* (for example, frontal or skull bone), and *irregular* (for example, vertebrae or spinal bones). Many important bones in the skeleton are classified as long bones, and all have several common characteristics. By studying a typical long bone, you can become familiar with the structural features of the entire group.

● *Structure of Long Bones*

Figure 5-1 will help you learn the names of the main parts of a long bone. Identify each of the following:

1. **Diaphysis** (dye-AF-i-sis) or shaft—a hollow tube made of hard compact bone, hence a rigid and strong structure light enough in weight to permit easy movement
2. **Medullary cavity**—the hollow area inside the diaphysis of a bone; contains soft **yellow bone marrow,** an inactive, fatty form of marrow found in the adult skeleton
3. **Epiphyses** (e-PIF-i-sees) or the ends of the bone—red bone marrow fills in small spaces in the spongy bone composing the epiphyses
4. **Articular cartilage**—a thin layer of cartilage covering each epiphysis; functions like a small rubber cushion would if it were placed over the ends of bones where they form a joint
5. **Periosteum**—a strong fibrous membrane covering a long bone except at joint surfaces, where it is covered by articular cartilage
6. **Endosteum**—a fibrous membrane that lines the medullary cavity

● *Microscopic Structure of Bone and Cartilage*

The skeletal system contains two major types of connective tissue: **bone** and **cartilage.** Bone has different appearances and textures, depending on its location. In Figure 5-2, *A,* the outer layer of bone is hard and dense. Bone of this type is called **dense** or **compact bone.** The porous bone in the end of the long bone is called *spongy bone.* As the name implies, spongy bone contains many spaces that may be filled with marrow. Compact or dense bone appears solid to the naked eye. Figure 5-2, *B,* shows the microscopic appearance of spongy and compact bone. The needlelike threads of spongy bone that surround a network of spaces are called **trabeculae** (trah-BEK-yoo-lee).

As you can see in Figures 5-2 and 5-3, compact or dense bone does not contain a network of open spaces. Instead, the matrix is organized into

FIGURE 5–1   *Longitudinal Section of a Long Bone.*

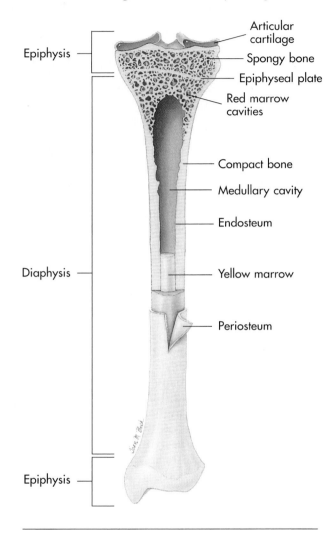

numerous structural units called **osteons** or *Haversian systems.* Each circular and tubelike osteon is composed of calcified matrix arranged in multiple layers resembling the rings of an onion. Each ring is called a **concentric lamella** (lah-MEL-ah). The circular rings or lamellae surround the **central canal,** which contains a blood vessel.

Bones are not lifeless structures. Within their hard, seemingly lifeless matrix are many living bone cells called **osteocytes** (OS-tee-o-sites).

FIGURE 5–2 *Microscopic Structure of Bone.* Longitudinal section of a long bone **(A)** shows the location of the microscopic section illustrated in **B.** Note that the compact bone forming the hard shell of the bone is constructed of cylindrical units called *osteons.* Spongy bone is constructed of bony projections called *trabeculae.*

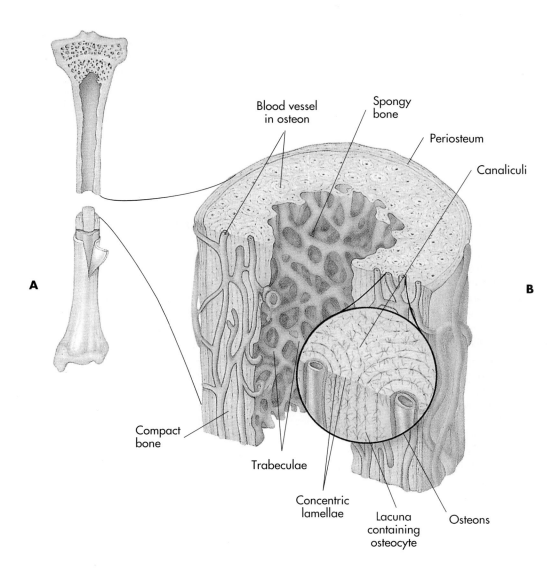

FIGURE 5–3   *Compact Bone.* Photomicrograph shows osteon system of organization.

Osteon (Haversian system)

Osteocyte in lacuna

Canaliculi

Lamella

Central canal

FIGURE 5–4   *Cartilage Tissue.* Photomicrograph shows chondrocytes scattered around the tissue in spaces called lacunae.

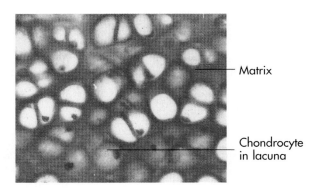

Matrix

Chondrocyte in lacuna

Osteocytes lie between the hard layers of the lamellae in little spaces called **lacunae** (lah-KOO-nee). In Figures 5-2, *B*, and 5-3, note that tiny passageways or canals called **canaliculi** (kan-ah-LIK-yoo-lye) connect the lacunae with one other and with the central canal in each Haversian system. Nutrients pass from the blood vessel in the Haversian canal through the canaliculi to the osteocytes. Note also in Figure 5-2, *B*, that numerous blood vessels from the outer **periosteum** (pair-ee-OS-tee-um) enter the bone and eventually pass through the Haversian canals.

   Cartilage both resembles and differs from bone. Like bone, it consists more of intercellular substance than of cells. Innumerable collagenous fibers reinforce the matrix of both tissues. However, in cartilage the fibers are embedded in a firm gel instead of in a calcified cement substance like they are in bone; hence cartilage has the flexibility of a firm plastic rather than the rigidity of bone. Note in Figure 5-4 that cartilage cells, called **chondrocytes** (KON-dro-sites), like the osteocytes of bone, are located in lacunae. In cartilage, lacunae are suspended in the cartilage matrix much like air bubbles in a block of firm gelatin. Because there are no blood vessels in cartilage, nutrients must diffuse through the matrix to reach the cells. Because of this lack of blood vessels, cartilage rebuilds itself very slowly after an injury.

Bone Formation and Growth

When the skeleton begins to form in a baby before its birth, it consists not of bones but of cartilage and fibrous structures shaped like bones. Gradually these cartilage "models" become transformed into real bones when the cartilage is replaced with calcified bone matrix. This process of constantly "remodeling" a growing bone as it changes from a small cartilage model to the characteristic shape and proportion of the adult bone requires continuous activity by bone-forming cells called **osteoblasts** (OS-tee-o-blasts) and bone-resorbing cells called **osteoclasts** (OS-tee-o-clasts). The laying down of calcium salts in the gel-like matrix of the forming bones is an ongoing process. This calcification process is what makes bones as "hard as bone." The combined action of the osteoblasts and osteoclasts sculpts bones into their adult shapes (Figure 5-5). The process of "sculpting" by the bone-forming and -resorbing cells allows bones to respond to stress or injury by changing size, shape, and density. The stresses placed on certain bones during exercise increase the rate of bone deposition. For this reason, athletes or dancers may have denser, stronger bones than less active people.

   Most bones of the body are formed from cartilage models as illustrated in Figures 5-5 and 5-6.

FIGURE 5–5   *Endochondral Ossification.* **A,** Bone formation begins with a cartilage model. **B** and **C,** Invasion of the diaphysis (shaft) by blood vessels and the combined action of osteoblast and osteoclast cells result in cavity formation, calcification, and the appearance of bone tissue. **D** and **E,** centers of ossification also appear in the epiphyses (ends) of the bone. **F,** Note the epiphyseal plate, indications that this bone is not yet mature and that additional growth is possible. **G,** In a mature bone, only a faint epiphyseal line marks where the cartilage has disappeared and the centers of ossification have fused together.

FIGURE 5–6 *Bone Development in a Newborn.* An infant's skeleton has many bones that are not yet completely ossified.

Osteoporosis

Osteoporosis (os-tee-o-po-RO-sis) is one of the most common and serious of all bone diseases. It is characterized by excessive loss of calcified matrix and collagenous fibers from bone. Osteoporosis occurs most frequently in elderly white females. Although white and black males are also susceptible, black women are seldom affected by it.

Because sex hormones play important roles in stimulating osteoblast activity after puberty, decreasing levels of these hormones in the blood of elderly persons reduces new bone growth and the maintenance of existing bone mass. Therefore some resorption of bone and subsequent loss of bone mass is an accepted consequence of advancing years. However, bone loss in osteoporosis goes far beyond the modest decrease normally seen in old age. The result is a dangerous pathological condition resulting in bone degeneration, increased susceptibility to "spontaneous fractures," and pathological curvature of the spine. Treatment may include sex hormone therapy and dietary supplements of calcium and vitamin D to replace deficiencies or to offset intestinal malabsorption.

This process is called **endochondral** (en-doe-KON-dral) **ossification** (os-i-fi-KAY-shun), meaning "formed in cartilage." A few flat bones, such as the skull bones illustrated in Figure 5-6, are formed by another process in connective tissue membranes.

As you can see in Figure 5-5, a long bone grows and ultimately becomes "ossified" from small centers located in both ends of the bone, called **epiphyses,** and from a larger center located in the shaft or the **diaphysis** of the bone. As long as any cartilage, called an **epiphyseal plate,** remains between the epiphyses and the diaphysis, growth continues. Growth ceases when all epiphyseal cartilage is transformed into bone. All that remains is an *epiphyseal line* that marks the location where the two centers of ossification have fused together. Physicians sometimes use this knowledge to determine whether a child is going to grow any more. They have an x-ray study performed on the child's wrist, and if it shows a layer of epiphyseal cartilage, they know that additional growth will occur. However, if it shows no epiphyseal cartilage, they know that growth has stopped and that the individual has attained adult height.

Epiphyseal Fracture

The point of articulation between the epiphysis and diaphysis of a growing long bone is susceptible to injury if overstressed, especially in the young child or preadolescent athlete. In these individuals the epiphyseal plate can be separated from the diaphysis or epiphysis, causing an **epiphyseal fracture.** This x-ray study shows such a fracture in a young boy

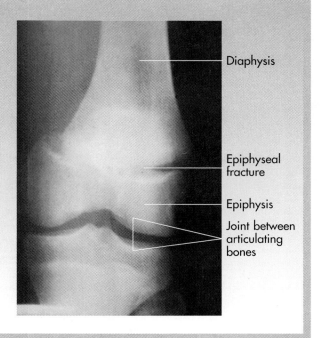

Diaphysis

Epiphyseal fracture

Epiphysis

Joint between articulating bones

Divisions of Skeleton

The human skeleton has two divisions: the **axial skeleton** and the **appendicular skeleton.** Bones of the center or axis of the body make up the axial skeleton. The bones of the skull, spine, and chest and the hyoid bone in the neck are all in the axial skeleton. The bones of the upper and lower extremities or appendages make up the appendicular skeleton. The appendicular skeleton consists of the bones of the upper extremities (shoulder, pectoral girdles, arms, wrists, and hands) and the lower extremities (hip, pelvic girdles, legs, ankles, and feet) (Table 5-1). Locate the various parts of the axial skeleton and the appendicular skeleton in Figure 5-7.

AXIAL SKELETON
Skull

The skull consists of 8 bones that form the **cranium,** 14 bones that form the **face,** and 6 tiny bones in the **middle ear.** You will probably want to learn the names and locations of these bones.

TABLE 5–1 Main Parts of the Skeleton*

AXIAL SKELETON[†]	APPENDICULAR SKELETON[‡]
Skull	Upper Extremities
Cranium	Shoulder (pectoral) girdle
Ear bones	Arm
Face	Wrists
	Hands
Spine	Lower Extremities
Vertebrae	Hip (pelvic) girdle
Thorax	Legs
Ribs	Ankles
Sternum	Feet
Hyoid Bone	

*Total bones = 206.
[†]Total = 80 bones.
[‡]Total = 126 bones.

FIGURE 5–7   *Human Skeleton.* The axial skeleton is distinguished by its bluer tint. **A,** Anterior view.

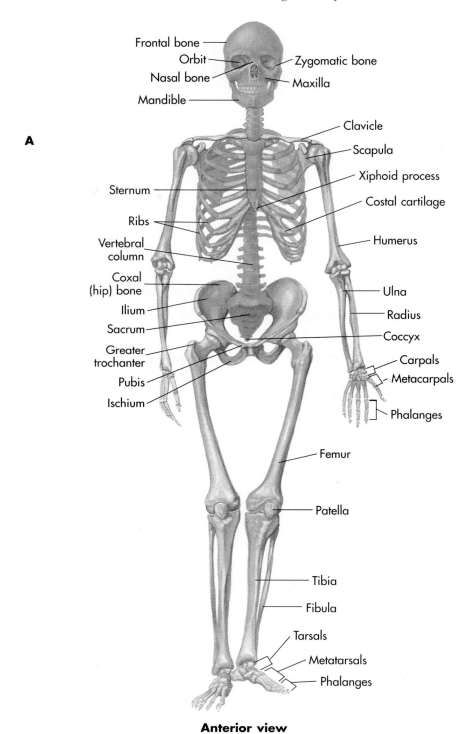

A

Frontal bone
Orbit
Nasal bone
Mandible

Zygomatic bone
Maxilla

Clavicle
Scapula
Xiphoid process
Costal cartilage

Sternum
Ribs
Vertebral column
Coxal (hip) bone
Ilium
Sacrum
Greater trochanter
Pubis
Ischium

Humerus

Ulna
Radius
Coccyx

Carpals
Metacarpals

Phalanges

Femur

Patella

Tibia
Fibula
Tarsals
Metatarsals
Phalanges

Anterior view

FIGURE 5–7 cont'd **B,** Posterior view

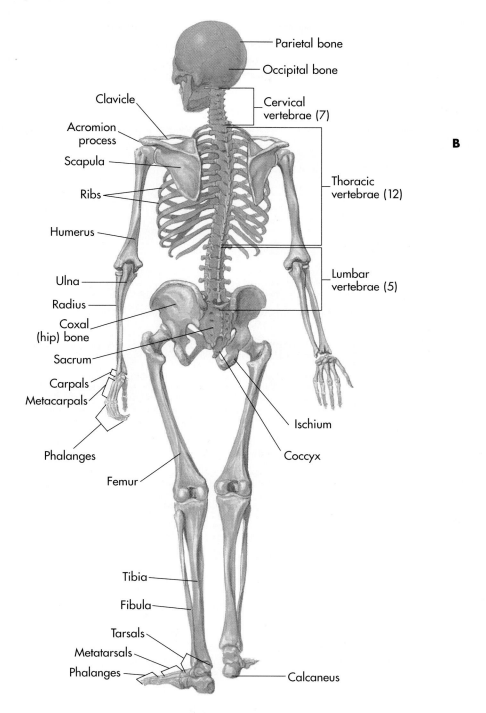

B

Posterior view

These are given in Table 5-2. Find as many of them as you can on Figure 5-8. Feel their outlines in your own body where possible. Examine them on a skeleton if you have access to one.

"My sinuses give me so much trouble." Have you ever heard this complaint or perhaps uttered it yourself? **Sinuses** are spaces or cavities inside some of the cranial bones. Four pairs of them (those in the frontal, maxillary, sphenoid, and ethmoid bones) have openings into the nose and thus are referred to as **paranasal sinuses.** Sinuses give trouble when the mucous membrane that lines them becomes inflamed, swollen, and painful. For example, inflammation in the frontal sinus (*frontal*

TABLE 5–2   Bones of the Skull

NAME	NUMBER	DESCRIPTION
Cranial Bones		
Frontal	1	Forehead bone; also forms front part of floor of cranium and most of upper part of eye sockets; cavity inside bone above upper margins of eye sockets (orbits) called *frontal sinus;* lined with mucous membrane
Parietal	2	Form bulging topsides of cranium
Temporal	2	Form lower sides of cranium; contain *middle* and *inner ear* structures; *mastoid sinuses* are mucosa-lined spaces in *mastoid process,* the protuberance behind ear; *external auditory canal* is tube leading into temporal bone; muscles attach to *styloid process*
Occipital	1	Forms back of skull; spinal cord enters cranium through large hole *(foramen magnum)* in occipital bone
Sphenoid	1	Forms central part of floor of cranium; pituitary gland located in small depression in sphenoid called *sella turcica (Turkish saddle);* muscles attach to *pterygoid process*
Ethmoid	1	Complicated bone that helps form floor of cranium, side walls and roof of nose and part of its middle partition (nasal septum—made up of the *vomer* and the *perpendicular plate*), and part of orbit; contains honeycomb-like spaces, the *ethmoid sinuses; superior* and *middle conchae* are projections of ethmoid bone; form "ledges" along side wall of each nasal cavity
Face Bones		
Nasal	2	Small bones that form upper part of bridge of nose
Maxilla	2	Upper jawbones; also help form roof of mouth, floor, and side walls of nose and floor of orbit; large cavity in maxillary bone is *maxillary sinus*
Zygomatic	2	Cheek bones; also help form orbit
Mandible	1	Lower jawbone articulates with temporal bone at *condyloid process;* small anterior hole for passage of nerves and vessels is the *mental foramen*
Lacrimal	2	Small bones; help form medial wall of eye socket and side wall of nasal cavity
Palatine	2	Form back part of roof of mouth and floor and side walls of nose and part of floor of orbit
Inferior concha	2	Form curved "ledge" along inside of side wall of nose, below middle concha
Vomer	1	Forms lower, back part of nasal septum
Ear Bones		
Malleus	2	Malleus, incus, and stapes are tiny bones in middle ear cavity in temporal bone; *malleus* means "hammer"—shape of bone
Incus	2	*Incus* means "anvil"—shape of bone
Stapes	2	*Stapes* means "stirrup"—shape of bone

FIGURE 5–8 *The Skull.* **A,** Right side. **B,** Front.

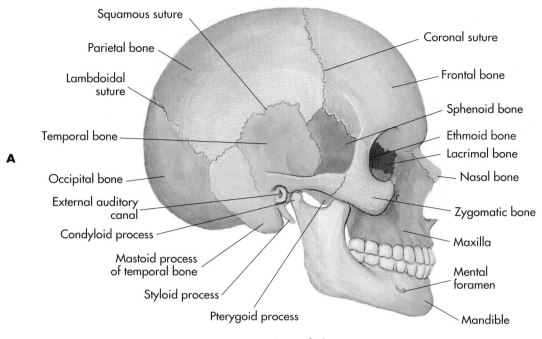

Squamous suture

Parietal bone

Lambdoidal suture

Temporal bone

Occipital bone

External auditory canal

Condyloid process

Mastoid process of temporal bone

Styloid process

Pterygoid process

Coronal suture

Frontal bone

Sphenoid bone

Ethmoid bone

Lacrimal bone

Nasal bone

Zygomatic bone

Maxilla

Mental foramen

Mandible

A

Lateral view

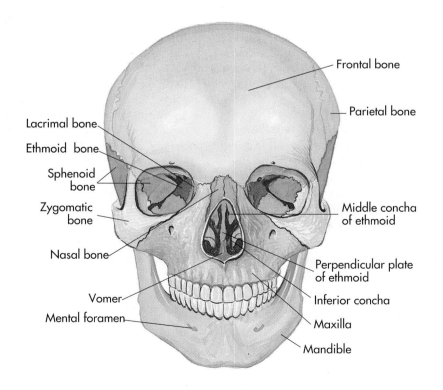

Frontal bone

Parietal bone

Lacrimal bone

Ethmoid bone

Sphenoid bone

Zygomatic bone

Nasal bone

Vomer

Mental foramen

Middle concha of ethmoid

Perpendicular plate of ethmoid

Inferior concha

Maxilla

Mandible

B

Anterior view

sinusitis) often starts from a common cold. The letters *-itis* added to a word mean "inflammation of."

Note in Figure 5-8 that the two parietal bones, which give shape to the bulging topside of the skull, form immovable joints called **sutures** with several bones: the *lambdoidal suture* with the occipital bone, the *squamous suture* with the temporal bone and part of the sphenoid, and the *coronal suture* with the frontal bone.

You may be familiar with the "soft spots" on a baby's skull. These are six **fontanels,** or areas where ossification is incomplete at birth. You can see them in Figure 5-6. Fontanels allow some compression of the skull during birth without much risk of breaking the skull bones. They may also be important in determining the position of the baby's head before delivery. The fontanels fuse to form sutures before a baby is 2 years old.

Spine (Vertebral Column)

The term *vertebral column* may conjure up a mental picture of the spine as a single long bone shaped like a column in a building, but this is far from true. The vertebral column consists of a series of separate bones or **vertebrae** connected in such a way that they form a flexible curved rod (see Figure 5-9). Different sections of the spine have different names: cervical region, thoracic region, lumbar region, sacrum, and coccyx. They are illustrated in Figure 5-9 and described in Table 5-3.

Although individual vertebrae are small bones, irregular in shape, they have several well-defined parts. Note, for example, in Figure 5-10, the body of the lumbar vertebra shown there, its spinous process (or spine), its two transverse processes, and the hole in its center, called the *vertebral foramen*. The superior and inferior articular processes permit limited and controlled movement between adjacent vertebrae. To feel the tip of the spinous process of one of your vertebrae, simply bend your head forward and run your fingers down the back of your neck until you feel a projection of bone at shoulder level. This is the tip of the seventh cervical vertebra's long spinous process. The seven cervical vertebrae form the supporting framework of the neck.

Have you ever noticed the four curves in your spine? Your neck and the small of your back curve slightly inward or forward, whereas the chest region of the spine and the lowermost portion curve in the opposite direction (Figure 5-9). The cervical and lumbar curves of the spine are called *concave curves,* and the thoracic and sacral curves are called *convex curves.* This is not true, however, of a newborn baby's spine. It forms a continuous convex curve from top to bottom (Figure 5-11). Gradually, as the baby learns to hold up his or her head, a reverse or concave curve develops in his or her neck, (cervical region). Later, as the baby learns to stand, the lumbar region of his or her spine also becomes concave.

The normal curves of the spine have important functions. They give it enough strength to support the weight of the rest of the body. They also provide the balance necessary for us to stand and walk on two feet instead of having to crawl on all fours. A curved structure has more strength than a straight one of the same size and materials. (The next time you pass a bridge, look to see whether or not its supports form a curve.) Clearly the spine needs to be a strong structure. It supports the head that is balanced on top of it, the ribs and internal organs that are suspended from it in front, and the hips and legs that are attached to it below.

TABLE 5–3 Bones of the Vertebral Column		
NAME	**NUMBER**	**DESCRIPTION**
Cervical	7	Upper seven vertebrae, in neck region; first cervical vertebra called *atlas;* second, *axis*
Thoracic vertebrae	12	Next twelve vertebrae; ribs attach to these
Lumbar vertebrae	5	Next five vertebrae; are in small of back
Sacrum	1	In child, five separate vertebrae; in adult, fused into one
Coccyx	1	In child, three to five separate vertebrae; in adult, fused into one

FIGURE 5–9 *The Spinal Column.* View shows the 7 cervical vertebrae, the 12 thoracic vertebrae, the 5 lumbar vertebrae, the sacrum, and the coccyx. **A,** Lateral view. **B,** Anterior view. **C,** Posterior view.

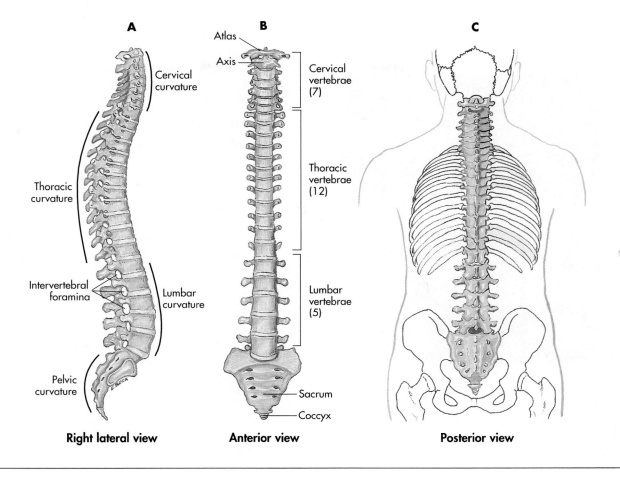

FIGURE 5–10 *The Third Lumbar Vertebra.* **A,** From above. **B,** From the side.

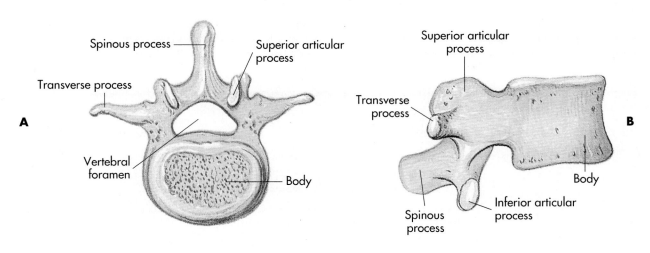

FIGURE 5–11 *Spinal Curvature of an Infant.* The spine of the newborn baby forms continuous convex curve.

Thorax

Twelve pairs of ribs, the sternum (breastbone), and the thoracic vertebrae form the bony cage known as the **thorax** or **chest.** Each of the 12 pairs of ribs is attached posteriorly to a vertebra. Also, all the ribs except the lower two pairs are attached to the sternum and so have anterior and posterior anchors. Look closely at Figure 5-12, and you can see that the first seven pairs of ribs (sometimes referred to as the *true ribs*) are attached to the ster-num by costal cartilage. The eighth, ninth, and tenth pairs of ribs are attached to the cartilage of the seventh ribs and are sometimes called *false ribs.* The last two pairs of ribs, in contrast, are not attached to any costal cartilage but seem to float free in front, hence their descriptive name, *floating ribs* (Table 5-4).

APPENDICULAR SKELETON

Of the 206 bones that form the skeleton as a whole, 126 are contained in the appendicular sub-division. Look again at Figure 5-7 to identify the appendicular components of the skeleton. Note that the bones in the shoulder or pectoral girdle connect the bones of the arm, forearm, wrist, and hands to the axial skeleton of the thorax, and the hip or pelvic girdle connects the bones of the thigh, leg, ankle, and foot to the axial skeleton of the pelvis.

Upper Extremity

The **scapula** (SKAP-yoo-lah) or shoulder blade and the **clavicle** (KLAV-ik-kul) or collar bone compose the *shoulder* or *pectoral girdle.* This connects the upper extremity to the axial skeleton. The only direct point of attachment between bones occurs at the **sternoclavicular** (ster-no-klah-VIK-yoo-lar) **joint** between the clavicle and the sternum or breastbone. As you can see in Figures 5-7 and 5-12, this joint is very small. Because the upper extremity is capable of a wide range of motion, great pressures can occur at or near the joint. As a result, fractures of the clavicle are very common.

The **humerus** (HYOO-mer-us) is the long bone of the arm and the second longest bone in the body. It is attached to the scapula at its proximal end and articulates with the two bones of the fore-

TABLE 5–4	Bones of the Thorax	
NAME	**NUMBER**	**DESCRIPTION**
True ribs	14	Upper seven pairs; attached to sternum by *costal cartilages*
False ribs	10	Lower five pairs; lowest two pairs do not attach to sternum, therefore, called *floating ribs;* next three pairs attached to sternum by costal cartilage of seventh ribs
Sternum	1	Breastbone; shaped like a dagger; piece of cartilage at lower end of bone called *xiphoid process*

FIGURE 5–12 *Bones of the Thorax.* Rib pairs 1 through 7, the true ribs, are attached by cartilage to the sternum. Rib pairs 8 through 10, the false ribs, are attached to the cartilage of the seventh pair. Rib pairs 11 and 12 are called floating ribs because they have no anterior cartilage attachments.

arm at the elbow joint. The bones of the forearm are the **radius** and the **ulna.** The anatomy of the elbow is a good example of how structure determines function. Note in Figure 5-13 that the large bony process of the ulna, called the **olecranon** (o-LEK-rah-non) **process,** fits nicely into a large depression on the posterior surface of the humerus, called the **olecranon fossa.** This structural relationship makes possible movement at the joint.

The radius and the ulna of the forearm articulate with each other and with the distal end of the humerus at the elbow joint. In addition, they also touch each another distally where they articulate with the bones of the wrist. In the anatomical position, with the arm at the side and the palm facing forward, the radius runs along the lateral

side of the forearm, and the ulna is located along the medial border.

The wrist and the hand have more bones in them for their size than any other part of the body—8 **carpal** (KAR-pal) or wrist bones, 5 **metacarpal** (met-ah-KAR-pal) bones that form the support structure for the palm of the hand, and 14 **phalanges** (fah-LAN-jeez) or finger bones—27 bones in all (Table 5-5). This composition is very important structurally. The presence of many small bones in the hand and wrist and the many movable joints between them makes the human hand highly maneuverable. Some anatomists refer to the hand and wrist as the functional "reason" for the upper extremity. Figure 5-14 shows the relationships between the bones of the wrist and hand.

FIGURE 5–13 *Bones of the Arm, Elbow Joint, and Forearm.* Posterior aspect of right humerus **(A)**, right radius and ulna **(B)**, and right elbow **(C)**.

FIGURE 5–14 *Bones of the Right Hand and Wrist.* There are 14 phalanges in each hand. Each of these bones is called a phalanx.

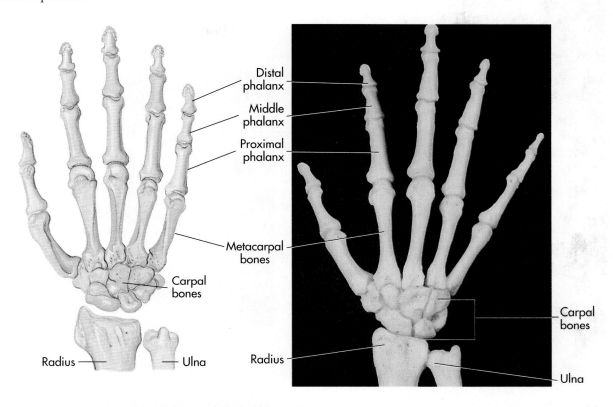

TABLE 5–5 Bones of the Upper Extremities

NAME	NUMBER	DESCRIPTION
Clavicle	2	Collarbones; only joints between shoulder girdle and axial skeleton are those between each clavicle and sternum *(sternoclavicular joints)*
Scapula	2	Shoulder blades; scapula plus clavicle forms *shoulder girdle; acromion process*—tip of shoulder that forms joint with clavicle; *glenoid cavity*—arm socket
Humerus	2	Upper arm bone (Muscles are attached to the *greater tubercle* and to the *medial* and *lateral epicondyles;* the *trochlea* articulates with the ulna; the *surgical neck* is a common fracture site.)
Radius	2	Bone on thumb side of lower arm (Muscles are attached to the *radial tuberosity* and to the *styloid process.*)
Ulna	2	Bone on little finger side of lower arm; *olecranon process*—projection of ulna known as elbow or "funny bone" (Muscles are attached to the *coronoid process* and to the *styloid process.*)
Carpal bones	16	Irregular bones at upper end of hand; anatomical wrist
Metacarpals	10	Form framework of palm of hand
Phalanges	28	Finger bones; three in each finger, two in each thumb

FIGURE 5–15 *Bones of the Thigh, Knee Joint, and Leg.* Anterior aspect of right femur **(A),** anterior aspect of the knee **(B),** right tibia and fibula **(C),** and posterior aspect of the right knee **(D).**

Lower Extremity

The *hip* or *pelvic girdle* connects the legs to the trunk. The hip girdle as a whole consists of two large **coxal** or pelvic bones, one located on each side of the pelvis. These two bones, with the sacrum and coccyx behind, provide a strong base of support for the torso and connect the lower extremities to the axial skeleton. In an infant's body each coxal bone consists of three separate bones—the **ilium** (ILL-ee-um), the **ischium** (IS-kee-um), and the **pubis** (PYOO-bis) (Figure 5-6). These bones grow together to become one bone in an adult (Figures 5-7 and 5-18).

Just as the humerus is the only bone in the arm, the **femur** (FEE-mur) is the only bone in the thigh (Figure 5-15). It is the longest bone in the body and articulates proximally (toward the hip) with the coxal bone in a deep, cup-shaped socket called the **acetabulum** (as-e-TAB-yoo-lum). The articulation of the head of the femur in the acetabulum is more stable than the articulation of the head of the humerus with the scapula in the upper extremity. As a result, dislocation of the hip occurs less often than does disarticulation of the shoulder. Distally, the femur articulates with the knee cap or **patella** (pah-TEL-ah) and the **tibia** or "shinbone." The tibia forms a rather sharp edge or crest along the front of your lower leg. A slender, non–weight-bearing, and rather fragile bone named the **fibula** lies along the outer or lateral border of the lower leg.

Toe bones have the same name as finger bones—**phalanges.** There are the same number of toe bones as finger bones, a fact that might surprise you because toes are shorter than fingers. Foot bones comparable to the metacarpals and carpals of the hand have slightly different names. They are called **metatarsals** and **tarsals** in the foot (Figure 5-16). Just as each hand contains five metacarpal bones, each foot contains five metatarsal bones. However, the foot has only seven tarsal bones, in contrast to the hand's eight carpals. The largest tarsal bone is the **calcaneus** or heel bone. The bones of the lower extremities are summarized in Table 5-6.

You stand on your feet, so certain features of their structure make them able to support the body's weight. The great toe, for example, is considerably more solid and less mobile than the thumb. The foot bones are held together in such a

way as to form springy lengthwise and crosswise arches. These provide great supporting strength and a highly stable base. Strong ligaments and leg muscle tendons normally hold the foot bones firmly in their arched positions. Frequently, however, the foot ligaments and tendons weaken. The arches then flatten, a condition appropriately called *fallen arches* or *flatfeet.*

Two arches extend in a lengthwise direction in the foot (Figure 5-17, *A*). One lies on the inside part of the foot and is called the **medial longitudinal arch.** The other lies along the outer edge of the foot and is named the **lateral longitudinal arch.** Another arch extends across the ball of the foot; this arch is called the **transverse** or **metatarsal arch** (Figure 5-17, *B*).

FIGURE 5–16 *Bones of the Right Foot.* Compare the names and numbers of foot bones (viewed here from above) with those of the hand bones shown in Figure 5–14.

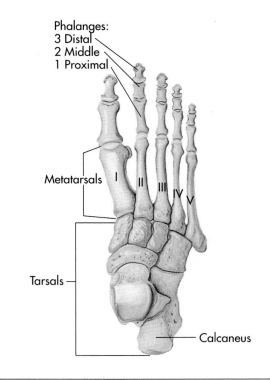

Phalanges:
3 Distal
2 Middle
1 Proximal

Metatarsals I II III IV V

Tarsals

Calcaneus

TABLE 5–6   Bones of the Lower Extremities

NAME	NUMBER	DESCRIPTION
Coxal bone	2	Hipbones; *ilium*—upper flaring part of pelvic bone; *ischium*—lower back part; *pubic bone*—lower front part; *acetabulum*—hip socket; *symphysis pubis*—joint in midline between two pubic bones; *pelvic inlet*—opening into *true pelvis* or pelvic cavity; if pelvic inlet is misshapen or too small, infant skull cannot enter true pelvis for natural birth
Femur	2	Thigh or upper leg bones; *head of femur*—ball-shaped upper end of bone; fits into acetabulum (Muscles are attached to the *greater* and *lesser trochanters* and to the *lateral* and *medial epicondyles*; the *lateral* and *medial condyles* form articulations at the knee.)
Patella	2	Kneecap
Tibia	2	Shinbone; *medial malleolus*—rounded projection at lower end of tibia commonly called *inner anklebone*; muscles are attached to the *tibial tuberosity*
Fibula	2	Long slender bone of lateral side of lower leg; *lateral malleolus*—rounded projection at lower end of fibula commonly called *outer anklebone*
Tarsal bones	14	Form heel and back part of foot; anatomical ankle; largest is the *calcaneus*
Metatarsals	10	Form part of foot to which toes are attached; tarsal and metatarsal bones arranged so that they form three arches in foot; *inner longitudinal arch* and *outer longitudinal arch,* which extend from front to back of foot, and transverse or *metatarsal arch,* which extends across foot
Phalanges	28	Toe bones; three in each toe, two in each great toe

FIGURE 5–17   *Arches of the Foot.* **A,** Medial and lateral longitudinal arches. **B,** Transverse arch.

A

Medial longitudinal arch

Lateral longitudinal arch

Longitudinal arches

B

Transverse arch

Differences Between a Man's and a Woman's Skeleton

A man's skeleton and a woman's skeleton differ in several ways. If you were to examine a male skeleton and a female skeleton placed side by side, you would probably first notice the difference in their sizes. Most male skeletons are larger than most female skeletons, a structural difference that seems to have no great functional importance. Structural differences between the male and female hipbones, however, do have functional importance. The female pelvis is made so that the body of a baby can be cradled in it before birth and can pass through it during birth. Although the individual male hipbones (coxal bones) are generally larger than the individual female hipbones, together the male hipbones form a narrower structure than do the female hipbones. A man's pelvis is shaped something like a funnel, but a woman's pelvis has a broader, shallower shape, more like a basin. (Incidentally, the word *pelvis* means "basin.") Another difference is that the pelvic inlet and pelvic outlet are both normally much wider in the female than in the male. Figure 5-18 shows this difference clearly. The angle at the front of the female pelvis where the two pubic bones join is wider than it is in the male.

FIGURE 5–18 *Comparison of the Male and Female Pelvis.* Notice the narrower width of the male pelvis, giving it a more funnel-like shape than the female pelvis.

Male

Female

 Joints (Articulations)

Every bone in the body, except one, connects to at least one other bone. In other words, every bone but one forms a joint with some other bone. (The exception is the hyoid bone in the neck, to which the tongue anchors.) Most of us probably never think much about our joints unless something goes wrong with them and they do not function properly. Then their tremendous importance becomes painfully clear. Joints hold our bones together securely and at the same time make it possible for movement to occur between the bones—between most of them, that is. Without joints we could not move our arms, legs, or any other of our body parts. Our bodies would, in short, be rigid, immobile hulks. Try, for example, to move your arm at your shoulder joint in as many directions as you can. Try to do the same thing at your elbow joint. Now examine the shape of the bones at each of these joints on a skeleton or in Figure 5-7. Do you see why you cannot move your arm at your elbow in nearly as many directions as you can at your shoulder?

KINDS OF JOINTS

One method classifies joints into three types according to the degree of movement they allow:

1. Synarthroses (no movement)
2. Amphiarthroses (slight movement)
3. Diarthroses (free movement)

Differences in joint structure account for differences in the degree of movement that is possible.

Synarthroses

A synarthrosis is a joint in which fibrous connective tissue grows between the articulating (join-

ing) bones holding them close together. The joints between cranial bones are synarthroses, commonly called *sutures* (Figure 5-19, *A*).

Amphiarthroses

An amphiarthrosis is a joint in which cartilage connects the articulating bones. The symphysis pubis, the joint between the two pubic bones, is an amphiarthrosis (Figure 5-19, *B*).

Joints between the bodies of the vertebrae are also amphiarthroses. These joints make it possible to flex the trunk forward or sideways and even to circumduct and rotate it. Strong ligaments connect the bodies of the vertebrae, and fibrous disks lie between them. The central core of these intervertebral disks consists of a pulpy, elastic substance that loses some of its resiliency with age.

Diarthroses

Fortunately most of our joints by far are diarthroses. Such joints allow considerable move-

FIGURE 5–19 *Joints of the Skeleton.* **A,** Synarthrotic joint. **B,** Amphiarthrotic joint.

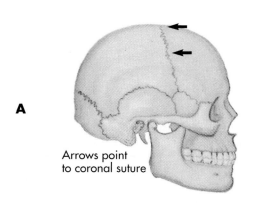

A

Arrows point to coronal suture

B

Symphysis pubis

Palpable Bony Landmarks

Health professionals often identify externally palpable bony landmarks when dealing with the sick and injured. **Palpable** bony landmarks are bones that can be touched and identified through the skin. They serve as reference points in identifying other body structures.

There are externally palpable bony landmarks throughout the body. Many skull bones such as the zygomatic bone, can be palpated. The medial and lateral epicondyles of the humerus, the olecranon process of the ulna, and the styloid process of the ulna and the radius at the wrist can be palpated on the upper extremity. The highest corner of the shoulder is the acromion process of the scapula.

When you put your hands on your hips, you can feel the superior edge of the ilium called the *iliac crest.* The anterior end of the crest, called the *anterior superior iliac spine,* is a prominent landmark used often as a clinical reference. The medial malleolus of the tibia and the lateral malleolus of the fibula are prominent at the ankle. The calcaneus or heel bone is easily palpated on the posterior aspect of the foot. On the anterior aspect of the lower extremity, examples of palpable bony landmarks include the patella or knee cap, the anterior border of the tibia or shin bone, and the metatarsals and phalanges of the toes. Try to identify as many of the externally palpable bones of the skeleton as possible on

Zygomatic bone

Acromion process of scapula

Medial epicondyle of humerus

Lateral epicondyle of humerus

Iliac crest

Styloid process of radius

Styloid process of ulna

Patella

Anterior border of tibia

Lateral malleolus of fibula

Medial malleolus of tibia

Calcaneus

your own body. Using these as points of reference will make it easier for you to visualize the placement of other bones that cannot be touched or palpated through the skin.

---

ment, sometimes in many directions and sometimes in only one or two directions.

Structure. Diarthroses (freely movable joints) are made alike in certain ways. All have a joint capsule, a joint cavity, and a layer of cartilage over the ends of two joining bones (Figure 5-20). The **joint capsule** is made of the body's strongest and toughest material, fibrous connective tissue, and is lined with a smooth, slippery synovial mem-

brane. The capsule fits over the ends of the two bones somewhat like a sleeve. Because it attaches firmly to the shaft of each bone to form its covering (called the *periosteum; peri* means "around," and *osteon* means "bone"), the joint capsule holds the bones securely together but at the same time permits movement at the joint. The structure of the joint capsule, in other words, helps make possible the joint's function.

FIGURE 5–20   *Structure of a Diarthrotic Joint.* Each diarthrosis has a joint capsule, a joint cavity, and a layer of cartilage over the ends of the joined bones.

Joint cavity

Bone

Bursa

Synovial membrane

Blood vessel

Nerve

Joint capsule

Articular cartilage

Bone

Periosteum

Total Hip Replacement

Because total hip replacement (THR) is the most common orthopedic operation performed on older persons (over 200,000 procedures per year in the U.S.), home health care professionals often work with patients recovering from THR surgery.

The THR procedure involves replacement of the femoral head by a metal prosthesis and the acetabular socket by a polyethylene cup. The prostheses are usually coated with a porous material that allows natural growth of bone to mesh with the artificial material. Such meshing of tissue and prostheses ensures stability of the parts without the loosening that the use of glues in the past often allowed. First introduced in 1953, THR technique has advanced to the state that the procedure has a success rate of about 85%.

Patients at home after THR surgery should progress through proper surgical healing and recovery, including stabilization of the prostheses as new tissue grows into their porous surfaces. THR patients should also expect some improvement in regained use of the affected hip, including weight-bearing and walking movements.

Ligaments (cords or bands made of the same strong fibrous connective tissue as the joint capsule) also grow out of the periosteum and lash the two bones together even more firmly.

The layer of **articular cartilage** over the joint ends of bones acts like a rubber heel on a shoe—it absorbs jolts. The **synovial membrane** secretes a lubricating fluid (synovial fluid) that allows easier movement with less friction.

There are several types of diarthroses, namely, ball-and-socket, hinge, pivot, saddle, gliding, and condyloid (Figure 5-21). Because they differ in structure, they differ also in their possible range of movement. In a ball-and-socket joint, a ball-shaped head of one bone fits into a concave socket of another bone. Shoulder and hip joints, for example, are ball-and-socket joints. Of all the joints in our bodies, these permit the widest range of movements. Think for a moment about how many ways you can move your upper arms. You can move them forward, you can move them backward, you can move them away from the sides of your body, and you can move them back down to your sides. You can also move them around so as to describe a circle with your hands.

FIGURE 5–21 *Types of Diarthrotic Joints.* Notice that the structure of each type dictates its function (movement).

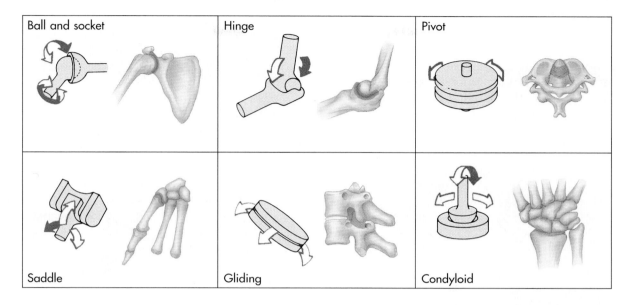

Hinge joints, like the hinges on a door, allow movements in only two directions, namely, flexion and extension. Flexion is bending a part; extension is straightening it out. Elbow and knee joints and the joints in the fingers are hinge joints.

Pivot joints are those in which a small projection of one bone pivots in an arch of another bone. For example, a projection of the axis, the second vertebra in the neck, pivots in an arch of the atlas, the first vertebra in the neck. This rotates the head, which rests on the atlas.

Only one pair of saddle joints exists in the body—between the metacarpal bone of each thumb and a carpal bone of the wrist (the name of this carpal bone is the *trapezium*). Because the articulating surfaces of these bones are saddle-shaped, they make possible the human thumb's great mobility, a mobility no animal's thumb pos-

sesses. We can flex, extend, abduct, adduct, and circumduct our thumbs, and most important of all, we can move our thumbs to touch the tip of any one of our fingers. (This movement is called *opposing the thumb to the fingers.*) Without the saddle joints at the base of each of our thumbs, we could not do such a simple act as picking up a pin or grasping a pencil between thumb and forefinger.

Gliding joints are the least movable diarthrotic joints. Their flat articulating surfaces allow limited gliding movements, such as that at the superior and inferior articulating processes between successive vertebrae.

Condyloid joints are those in which a condyle (an oval projection) fits into an elliptical socket. An example is the fit of the distal end of the radius into depressions in the carpal bones.

The Knee Joint

The knee is the largest and most vulnerable joint. Because the knee is often subjected to sudden, strong forces during athletic activity, knee injuries are among the most common type of athletic injury. Sometimes, the articular cartilages on the tibia become torn when the knee twists while bearing weight. The ligaments holding the tibia and femur together can also be injured in this way. Knee injuries may also occur when a weight-bearing knee is hit by another person.

FUNCTIONS OF BONE

A Supports and gives shape to the body
B Protects internal organs
C Helps make movements possible
D Stores calcium
E Hemopoiesis or blood cell formation

TYPES OF BONES

A Long—Example: humerus (upper arm)
B Short—Example: carpals (wrist)
C Flat—Example: frontal (skull)
D Irregular—Example: vertebrae (spinal cord)

STRUCTURE OF LONG BONES

A Structural components (Figure 5-1)
 1 Diaphysis or shaft
 2 Medullary cavity containing yellow marrow
 3 Epiphyses or ends of the bone; spongy bone contains red bone marrow
 4 Articular cartilage—covers epiphyses as a cushion
 5 Periosteum—strong membrane covering bone except at joint surfaces
 6 Endosteum—lines medullary cavity

MICROSCOPIC STRUCTURE OF BONE AND CARTILAGE

A Bone types (Figure 5-2)
 1 Spongy
  a Texture results from needlelike threads of bone called *trabeculae* surrounded by a network of open spaces
  b Found in epiphyses of bones
  c Spaces contain red bone marrow
 2 Compact
  a Structural unit is Haversian system—composed of concentric lamella, lacunae containing osteocytes, and canaliculi, all covered by periosteum
B Cartilage (Figure 5-4)
 1 Cell type called *chondrocyte*
 2 Matrix is gel-like and lacks blood vessels

BONE FORMATION AND GROWTH

(Figures 5-5 and 5-6)
A Sequence of development early—cartilage models replaced by calcified bone matrix
B Osteoblasts form new bone, and osteoclasts resorb bone

DIVISIONS OF SKELETON

Skeleton composed of the following divisions and their subdivisions:
A Axial skeleton
 1 Skull
 2 Spine
 3 Thorax
 4 Hyoid bone
B Appendicular skeleton
 1 Upper extremities, including shoulder girdle
 2 Lower extremities, including hip girdle
C Location and description of bones—see Figures 5-7 to 5-17 and Tables 5-2 to 5-6

DIFFERENCES BETWEEN A MAN'S AND A WOMAN'S SKELETON

A Size—male skeleton generally larger
B Shape of pelvis—male pelvis deep and narrow, female pelvis broad and shallow
C Size of pelvic inlet—female pelvic inlet generally wider, normally large enough for baby's head to pass through it (Figure 5-18)
D Pubic angle—angle between pubic bones of female generally wider

JOINT (ARTICULATIONS)

A Kinds of joints (Figures 5-19 to 5-21)
 1 Synarthroses (no movement)—fibrous connective tissue grows between articulating bones; for example, sutures of skull
 2 Amphiarthroses (slight movement)—cartilage connects articulating bones; for example, symphysis pubis
 3 Diarthroses (free movement)—most joints belong to this class
  a Structures of freely movable joints—joint capsule and ligaments hold adjoining bones together but permit movement at joint
  b Articular cartilage—covers joint ends of bones and absorbs joints
  c Synovial membrane—lines joint capsule and secretes lubricating fluid
  d Joint cavity—space between joint ends of bones
B Types of freely movable joints—ball-and-socket, hinge, pivot, saddle, gliding, and condyloid

---

N E W · W O R D S

amphiarthroses
appendicular skeleton
articular cartilage
articulation
axial skeleton
canaliculi
chondrocytes
compact bone

diaphysis
diarthroses
epiphyses
fontanels
hemopoiesis
lacunae
lamella
medullary cavity

osteoblasts
osteoclasts
osteocytes
osteon
pectoral girdle
pelvic girdle
periosteum
red bone marrow

sinus
skull
spine
synarthroses
synovial membrane
thorax
trabeculae
yellow bone marrow

C H A P T E R   T E S T

1. Spongy bone contains needlelike threads of bone known as _____.
2. The structural units of compact bone are called _____.
3. Bone-forming cells are called _____, whereas bone-resorbing cells are called _____.
4. The hollow shaft of a long bone is also known as the_____.
5. The human skeleton has two main divisions, the _____ skeleton and the _____ skeleton.
6. Spaces or cavities located inside some of the cranial bones are called _____.
7. The supporting framework of the neck is formed by the seven _____ vertebrae.
8. The shoulder blade is also known as the _____, and the collar bone is called the _____.
9. The long bones of the forearm are the _____ and the _____.
10. The large bony process of the ulna that forms the elbow is called the _____ process.
11. There are 8 _____ bones in the wrist and 14 _____ or finger bones in each hand.
12. Each hip bone consists of three separate bones called the _____, _____, and _____, which fuse in the adult to become one bone.

13. The femur of the leg articulates distally with the knee cap or _____ and the shinbone or _____.
14. The metatarsal and tarsal bones are located in the _____.
15. Joints that permit free movement are called _____ joints.

Circle the "T" before each true statement and the "F" before each false statement.

T   F   16. Adult bones do not contain living cells.
T   F   17. Red bone marrow functions in hemopoiesis or blood cell formation.
T   F   18. The hollow cylindrical portion of a long bone is called the *epiphysis*.
T   F   19. Haversian systems are components of spongy bone.
T   F   20. Lacunae are found in compact bone and cartilage.
T   F   21. There are more bones in the axial than in the appendicular skeleton.
T   F   22. The term *phalanges* is used to describe the bones of the fingers and the toes.
T   F   23. "Soft spots" in the skull at birth are called *fontanels*.
T   F   24. Synarthroses are freely movable joints.
T   F   25. Synovial membranes are found in diarthrotic joints.

REVIEW QUESTIONS

1. Discuss the mechanism of bone formation and growth.
2. Compare the structures of compact and spongy bone.
3. Discuss the following cell types: osteocyte, osteoblast, osteoclast, chondrocyte.
4. List the major structural components of a typical long bone, and briefly describe the function of each.
5. List and discuss the generalized functions of the skeletal system.
6. What are the two major subdivisions of the skeleton? What are the major body areas in each subdivision?
7. Give the correct anatomical name for each of the following: collarbone, breastbone, wrist bones, finger bones, forearm bones, thigh bone, hip bone, knee cap, ankle bones, and neck vertebrae.
8. Compare the structures and functions of the arms and legs, pectoral and pelvic girdles, shoulder and hip joints, and hands and feet.
9. Classify the major types of joints.
10. List and compare the types of freely movable joints.

CRITICAL THINKING

11. Is it possible to tell whether a child is going to grow taller? If so, how?
12. Describe one functionally important difference between a male and a female adult skeleton. Why is this difference important?

The Muscular System

6

BOXED ESSAYS

Carpal Tunnel Syndrome
Intramuscular Injections

OBJECTIVES

*After you have completed this chapter,
you should be able to:*

1. List, locate in the body, and compare the structure and function of the three major types of muscle tissue.

2. Discuss the microscopic structure of a skeletal muscle sarcomere and motor unit.

3. Discuss how a muscle is stimulated and compare the major types of skeletal muscle contractions.

4. Name, identify on a model or diagram, and give the function of the major muscles of the body discussed in this chapter.

5. List and explain the most common types of movement produced by skeletal muscles.

Although we will initially review the three types of muscle tissue introduced earlier (see Chapter 3), the plan for this chapter is to focus on skeletal or voluntary muscle—those muscle masses that attach to bones and actually move them about when contraction or shortening of muscle cells, or muscle fibers, occurs. If you weigh 120 pounds, about 50 pounds of your weight comes from your skeletal muscles, the "red meat" of the body that is attached to your bones.

Movements caused by skeletal muscle contraction vary in complexity from blinking an eye to the coordinated and fluid movements of a gifted athlete. Not many of our body structures can claim as great an importance for happy, useful living as can our voluntary muscles, and only a few can boast of greater importance for life itself. Our ability to survive often depends on our ability to adjust to the changing conditions of our environment. Movements frequently constitute a major part of this adjustment.

Muscle Tissue

Under the microscope, threadlike and cylindrical skeletal muscle cells appear in bundles. They are characterized by many crosswise stripes and multiple nuclei (Figure 6-1, *A*). Each fine thread is a muscle cell or, as it is usually called, a *muscle fiber.* This type of muscle tissue has three names: *skeletal muscle,* because it attaches to bone; *striated muscle,* because of its cross stripes or striations; and *voluntary muscle,* because its contractions can be controlled voluntarily.

In addition to **skeletal muscle,** the body also contains two other kinds of muscle tissue: cardiac muscle and nonstriated, smooth, or involuntary muscle. **Cardiac muscle** composes the bulk of the heart. Cells in this type of muscle are also cylindrical, branch frequently, (Figure 6-1, *B*) and then recombine into a continuous mass of interconnected tissue. Like skeletal muscle cells, they have cross striations. They also have unique dark bands called *intercalated disks* where the plasma membranes of adjacent cardiac fibers come in contact with each other. Cardiac muscle tissue demonstrates the principle that "form follows function."

The interconnected nature of cardiac muscle fibers helps the tissue to contract as a unit and increases the efficiency of the heart muscle in pumping blood.

Nonstriated or **smooth muscle** cells are tapered at each end, have a single nucleus, and lack the cross stripes or striations of skeletal muscle cells (Figure 6-1, *C*). They have a smooth, even appearance when viewed through a microscope. They are called *involuntary* because we normally do not have control over their contractions. Smooth or involuntary muscle forms an important part of blood vessel walls and of many hollow internal organs (viscera) such as the gut, urethra, and ureters. Because of its location in many visceral structures, it is sometimes called *visceral muscle.* Although we cannot willfully control the action of smooth muscle, its contractions are highly regulated so that, for example, food is passed through the digestive tract or urine is pushed through the ureters into the bladder.

Muscle cells specialize in contraction or shortening. Every movement we make is produced by contractions of skeletal muscle cells. Contractions of cardiac muscle cells keep the blood circulating, and smooth muscle contractions do many things; for instance, they move food into and through the stomach and intestines and make a major contribution to the maintenance of normal blood pressure.

Structure of Skeletal Muscle

A skeletal muscle is an organ composed mainly of striated muscle cells and connective tissue. Most skeletal muscles attach to two bones that have a movable joint between them. In other words, most muscles extend from one bone across a joint to another bone. Also, one of the two bones is usually more stationary in a given movement than the other. The muscle's attachment to this more stationary bone is called its **origin.** Its attachment to the more movable bone is called the muscle's **insertion.** The rest of the muscle (all of it except its two ends) is called the *body* of the muscle (Figure 6-2).

Tendons anchor muscles firmly to bones. Made of dense fibrous connective tissue in the shape of

FIGURE 6–1   *Muscle Tissue.* **A,** Skeletal muscle. **B,** Cardiac muscle. **C,** Smooth muscle.

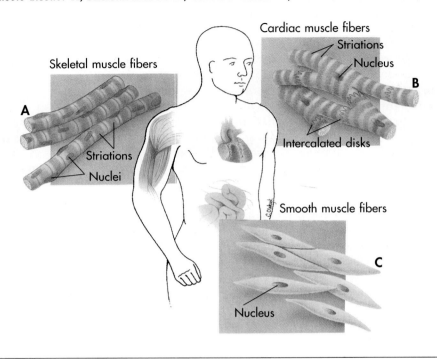

FIGURE 6–2   *Attachments of a Skeletal Muscle.* A muscle originates at a relatively stable part of the skeleton (origin) and inserts at the skeletal part that is moved when the muscle contracts (insertion).

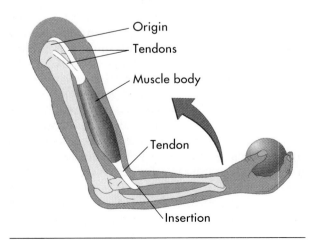

heavy cords, tendons have great strength. They do not tear or pull away from bone easily. Yet any emergency room nurse or physician sees many tendon injuries—severed tendons and tendons torn loose from bones.

Small fluid-filled sacs called **bursae** lie between some tendons and the bones beneath them (see Figure 5-20). These small sacs are made of connective tissue and are lined with **synovial membrane.** The synovial membrane secretes a slippery lubricating fluid (synovial fluid) that fills the bursa. Like a small, flexible cushion, a bursa makes it easier for a tendon to slide over a bone when the tendon's muscle shortens. **Tendon sheaths** enclose some tendons. Because these tube-shaped structures are also lined with synovial membrane and are moistened with synovial fluid, they, like the bursae, facilitate body movement.

MICROSCOPIC STRUCTURE

Muscle tissue consists of specialized contractile cells or **muscle fibers** that are grouped together and arranged in a highly organized way. Each skeletal muscle fiber is itself filled with two kinds of very fine and threadlike structures called **thick** and **thin myofilaments** (my-o-FIL-a-ments). The thick myofilaments are formed from a protein called **myosin,** and the thin myofilaments are composed mostly of the protein **actin.** Find the label **sarcomere** (SAR-ko-meer) in Figure 6-3. Think of the sarcomere as the basic functional or *contractile unit* of skeletal muscle. Recall that the osteon (Haversian system) serves as the basic building block of compact bone; the sarcomere serves that function in skeletal muscle. The submicroscopic structure of a sarcomere consists of numerous actin and myosin myofilaments arranged so that, when viewed under a microscope, dark and light stripes or cross striations are seen. The repeating units or sarcomeres are separated from each other by dark bands called *Z lines.*

Although the sarcomeres in the upper portion (Figure 6-3, *A*) and in the electron photomicrograph (EM) of Figure 6-3, *B,* are in a relaxed state, the thick and thin myofilaments, which are lying parallel to each other, still overlap. Now look at the diagrams in the lower portion of Figure 6-3, *A.* Note that contraction of the muscle causes the two types of myofilaments to slide toward each other and shorten the sarcomere and thus the entire muscle. When the muscle relaxes, the sarcomeres can return to resting length, and the filaments resume their resting positions.

An explanation of how a skeletal muscle contracts is provided by the **sliding filament model.** According to this model, during contraction, the thick and thin myofilaments in a muscle fiber first attach to one another by forming "bridges" that then act as levers to ratchet or pull the myofilaments past each other. The connecting bridges between the myofilaments form properly only if calcium is present. During the relaxed state, calcium is within the endoplasmic reticulum (see Chapter 2) in the muscle cell. It is released into the cytoplasm when the muscle is stimulated to contract. The shortening of a muscle cell also requires energy. This is supplied by the breakdown of adenosine triphosphate (ATP) molecules, the energy storage molecules of the cell.

Functions of Skeletal Muscle

The three primary functions of the muscular system are:

1. Movement
2. Posture or muscle tone
3. Heat production

MOVEMENT

Muscles move bones by pulling on them. Because the length of a skeletal muscle becomes shorter as its fibers contract, the bones to which the muscle attaches move closer together. As a rule, only the insertion bone moves. Look again at Figure 6-2. As the ball is lifted, the shortening of the muscle body pulls the insertion bone toward the origin bone. The origin bone stays put, holding firm, while the insertion bone moves toward it. One tremendously important function of skeletal muscle contractions therefore is to produce body movements. Remember this simple rule: a muscle's insertion bone moves toward its origin bone. It can help you understand muscle actions.

Voluntary muscular movement is normally smooth and free of jerks and tremors because skeletal muscles generally work in coordinated teams, not singly. Several muscles contract while others relax to produce almost any movement that you can imagine. Of all the muscles contracting simultaneously, the one that is mainly responsible for producing a particular movement is called the **prime mover** for that movement. The other muscles that help in producing the movement are called **synergists** (SIN-er-jists). As prime movers and synergist muscles at a joint contract, other muscles, called **antagonists** (an-TAG-o-nists), relax. When antagonist muscles contract, they produce a movement opposite to that of the prime movers and their synergist muscles.

Locate the biceps brachii, brachialis, and triceps brachii muscles in Figure 6-6. All the muscles in these figures are involved in bending and straightening the forearm at the elbow joint. The biceps brachii is the prime mover during bending, and the brachialis is its helper or synergist muscle. When the biceps brachii and brachialis muscles bend the forearm, the triceps brachii relaxes. Therefore while the forearm bends, the triceps brachii is the antagonistic muscle. While the fore-

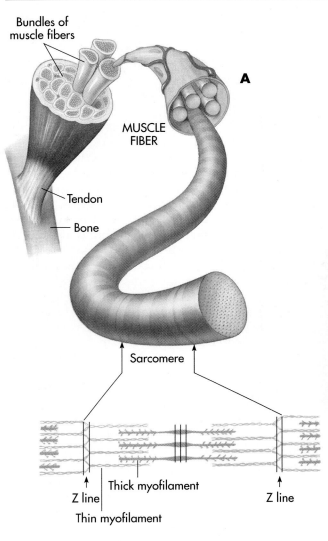

Bundles of
muscle fibers

MUSCLE
FIBER

Tendon

Bone

A

Sarcomere

Z line

Thick myofilament

Thin myofilament

Z line

FIGURE 6–3   *Structure of Skeletal Muscle.* A, Each muscle organ has many muscle fibers, each containing many bundles of thick and thin filaments. The diagrams show the overlapping thick and thin filaments arranged to form adjacent segments called *sarcomeres.* During contraction, the thin filaments are pulled toward the center of each sarcomere, shortening the whole muscle. **B,** This electron micrograph shows that the overlapping thick and thin filaments within each sarcomere create a pattern of dark striations in the muscle. The extreme magnification allowed by electron microscopy has revolutionized our concept of the structure and function of skeletal muscle and other tissues.

B

Z line   Sarcomere   Z line

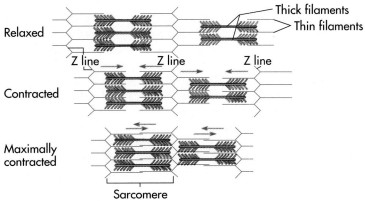

Relaxed

Z line   Z line   Z line

Thick filaments

Thin filaments

Contracted

Maximally
contracted

Sarcomere

arm straightens, these three muscles continue to work as a team. However, during straightening, the triceps brachii becomes the prime mover and the biceps brachii and brachialis become the antagonistic muscles. This combined and coordinated activity is what makes our muscular movements smooth and graceful.

POSTURE

We are able to maintain our body position because of a specialized type of skeletal muscle contraction called **tonic contraction.** Because relatively few of a muscle's fibers shorten at one time in a tonic contraction, the muscle as a whole does not shorten, and no movement occurs. Consequently, tonic contractions do not move any body parts. They do hold muscles in position, however. In other words, muscle tone maintains **posture.** Good posture means that body parts are held in the positions that favor best function. These positions balance the distribution of weight and therefore put the least strain on muscles, tendons, ligaments, and bones. To have good posture in a standing position, for example, you must stand with your head and chest held high, your chin, abdomen, and buttocks pulled in, and your knees bent slightly.

To judge for yourself how important good posture is, consider some of the effects of poor posture. Besides detracting from appearance, poor posture makes a person tire more quickly. It puts an abnormal pull on ligaments, joints, and bones and sometimes leads to deformities. Poor posture crowds the heart, making it harder for it to contract. Poor posture also crowds the lungs, decreasing their breathing capacity.

Skeletal muscle tone maintains posture by counteracting the pull of gravity. Gravity tends to pull the head and trunk down and forward, but the tone in certain back and neck muscles pulls just hard enough in the opposite direction to overcome the force of gravity and hold the head and trunk erect. The tone in thigh and leg muscles puts just enough pull on thigh and leg bones to counteract the pull of gravity on them that would otherwise collapse the hip and knee joints and cause us to fall in a heap.

HEAT PRODUCTION

Healthy survival depends on our ability to maintain a constant body temperature. A fever or elevation in body temperature of only a degree or two above 37° C (98.6° F) is almost always a sign of illness. Just as serious is a fall in body temperature. Any decrease below normal, a condition called **hypothermia** (hy-po-THER-mee-ah), drastically affects cellular activity and normal body function. The contraction of muscle fibers produces most of the heat required to maintain body temperature. Energy required to produce a muscle contraction is obtained from ATP. Most of the energy released during the breakdown of ATP during a muscular contraction is used to shorten the muscle fibers; however, some of the energy is lost as heat during the reaction. This heat helps us to maintain our body temperature at a constant level.

Fatigue

If muscle cells are stimulated repeatedly without adequate periods of rest, the strength of the muscle contraction decreases, resulting in **fatigue.** If repeated stimulation occurs, the strength of the contraction continues to decrease, and eventually the muscle loses its ability to contract.

During exercise, the stored ATP required for muscle contraction becomes depleted. Formation of more ATP results in a rapid consumption of oxygen and nutrients, often outstripping the ability of the muscle's blood supply to replenish them. When oxygen supplies run low, the muscle cells switch to a type of energy conversion that does not require oxygen. This process produces lactic acid that may result in muscle soreness after exercise. The term **oxygen debt** describes the continued increased metabolism that must occur in a cell to remove excess lactic acid that accumulates during prolonged exercise. Thus the depleted energy reserves are replaced. Labored breathing after the cessation of exercise is required to "pay the debt" of oxygen required for the metabolic effort. This mechanism is a good example of homeostasis at work. The body returns the cells' energy and oxygen reserves to normal, resting levels.

FIGURE 6–4 *Motor Neuron.* A motor unit consists of one motor neuron and the muscle fibers supplied by its branches.

- Myelin sheath
- Schwann cell
- Motor neuron
- Neuromuscular junction
- Nucleus
- Muscle fibers
- Myofibrils

Role of Other Body Systems in Movement

Remember that muscles do not function alone. Other structures such as bones and joints must function with them. Most skeletal muscles cause movements by pulling on bones across movable joints.

The respiratory, circulatory, nervous, muscular, and skeletal systems play essential roles in producing normal movements. This fact has great practical importance. For example, a person might have perfectly normal muscles and still not be able to move normally. He might have a nervous system disorder that shuts off impulses to certain skeletal muscles and thereby results in **paralysis.** Multiple sclerosis acts in this way, but so do some other conditions such as a brain hemorrhage, a brain tumor, or a spinal cord injury. Skeletal system disorders, especially arthritis, have disabling effects on body movement. Muscle functioning, then, depends on the functioning of many other parts of the body. This fact illustrates a principle that is repeated often in this book. It can be simply stated: Each part of the body is one of many components in a large, interactive system. The normal function of one part depends on the normal function of the other parts.

Motor Unit

Before a skeletal muscle can contract and pull on a bone to move it, the muscle must first be stimulated by nerve impulses. Muscle cells are stimulated by a nerve fiber called a **motor neuron** (Figure 6-4). The point of contact between the nerve ending and the muscle fiber is called a **neuromuscular junction.** Specialized chemicals are released by the motor neuron in response to a nervous impulse. These chemicals then generate events within the muscle cell that result in contraction or shortening of the muscle cell. A single motor neuron, with the muscle cells it innervates, is called a **motor unit** (Figure 6-4).

Muscle Stimulus

In a laboratory setting a single muscle fiber can be isolated and subjected to stimuli of varying intensities so that it can be studied. Such experiments show that a muscle fiber does not contract until an applied stimulus reaches a certain level of intensity. The minimal level of stimulation required to cause a fiber to contract is called the **threshold stimulus.**

When a muscle fiber is subjected to a threshold stimulus, it contracts completely. Because of this, muscle cells are said to respond **"all or none."** However, a muscle is composed of many muscle cells that are controlled by different motor units and that have different threshold-stimulus levels. Although each fiber in a muscle such as the biceps brachii responds all or none when subjected to a threshold stimulus, the muscle as a whole does not. This fact has tremendous importance in everyday life. It allows you to pick up a 2-liter bottle of soda or a 20 kg weight by stimulating the same muscle but executing contractions of varying strength. This is possible because different numbers of motor units can be activated for different loads. Once activated, however, each fiber always responds all or none.

Types of Skeletal Muscle Contraction

In addition to the specialized tonic contraction of muscle that maintains muscle tone and posture, other types of contraction also occur. Additional types of muscle contraction include the following:

1. **Twitch contraction**
2. **Tetanic contraction**
3. **Isotonic contraction**
4. **Isometric contraction**

TWITCH AND TETANIC CONTRACTIONS

A **twitch** is a quick, jerky response to a stimulus. Twitch contractions can be seen in isolated muscles during research, but they play a minimal role in normal muscle activity. To accomplish the coordinated and fluid muscular movements needed for most daily tasks, muscles must contract not in a jerky but in a smooth and sustained way.

A **tetanic contraction** is a more sustained and steady response than a twitch. It is produced by a series of stimuli bombarding the muscle in rapid succession. Contractions "melt" together to produce a sustained contraction or *tetanus.* About 30 stimuli per second, for example, evoke a tetanic contraction in certain types of skeletal muscle. Tetanic contraction is not necessarily a maximal contraction in which each muscle fiber responds at the same time. In most cases, only a few areas of the muscle undergo contractions at any time.

ISOTONIC CONTRACTION

In most cases, isotonic contraction of muscle produces movement at a joint. With this type of contraction the muscle shortens, and the insertion end moves toward the point of origin (Figure 6-5, *A*). Walking, running, breathing, lifting, and twisting are examples of isotonic contraction.

ISOMETRIC CONTRACTION

Contraction of a skeletal muscle does not always produce movement. Sometimes, it increases the tension within a muscle but does not shorten the muscle. When the muscle does not shorten and no movement results, it is called an *isometric contraction.* The word *isometric* comes from Greek words that mean "equal measure." In other words, a muscle's length during an isometric contraction and during relaxation is about equal. Although muscles do not shorten (and thus produce no movement) during isometric contractions, tension within them increases (Figure 6-5, *B*). Because of this, repeated isometric contractions make muscles grow larger and stronger—hence the popularity in recent years of isometric exercises. Pushing against a wall or other immovable object is a good example of isometric exercise. Although no movement occurs and the muscle does not shorten, its internal tension increases dramatically.

Effects of Exercise on Skeletal Muscles

We know that exercise is good for us. Some of the benefits of regular, properly practiced exercise are greatly improved muscle tone, better posture, more efficient heart and lung function, less fatigue, and looking and feeling better.

FIGURE 6–5 *Types of Muscle Contraction.* **A,** In isotonic contraction the muscle shortens, producing movement. **B,** In isometric contraction the muscle pulls forcefully against a load but does not shorten.

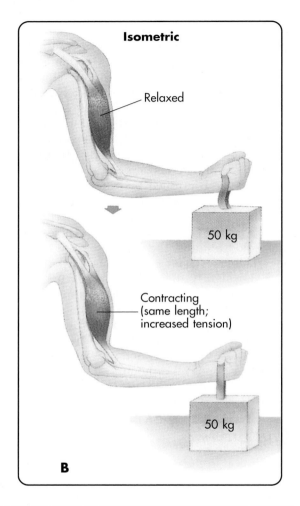

Skeletal muscles undergo changes that correspond to the amount of work that they normally do. During prolonged inactivity, muscles usually shrink in mass, a condition called **disuse atrophy.** Exercise, on the other hand, may cause an increase in muscle size called **hypertrophy**.

Muscle hypertrophy can be enhanced by **strength training,** which involves contracting muscles against heavy resistance. Isometric exercises and weight lifting are common strength-training activities. This type of training results in increased numbers of myofilaments in each muscle fiber. Although the number of muscle fibers

stays the same, the increased number of myofilaments greatly increases the mass of the muscle.

Endurance training, often called **aerobic training,** does not usually result in muscle hypertrophy. Instead, this type of exercise program increases a muscle's ability to sustain moderate exercise over a long period. Aerobic activities such as running, bicycling, or other primarily isotonic movements increase the number of blood vessels in a muscle without significantly increasing its size. The increased blood flow allows a more efficient delivery of oxygen and glucose to muscle fibers during exercise. Aerobic training

Carpal Tunnel Syndrome

Some physicians specialize in the field of occupational health, the study of health matters related to work or the workplace. Many problems seen by occupational health experts are caused by repetitive motions of the wrists or other joints. Word processors (typists) and meat cutters, for example, are at risk of developing conditions caused by repetitive motion injuries.

One common problem often caused by such repetitive motion is **tenosynovitis** (ten-o-sin-o-VYE-tis)—inflammation of the tendon sheath. Tenosynovitis can be painful, and the swelling characteristic of this condition can limit movement in affected parts of the body. For example, swelling of the tendon sheath around tendons in an area of the wrist known as the *carpal tunnel* can limit movement of the wrist, hand, and fingers. The figure shows the relative positions of the tendon sheath and medial nerve within the carpal tunnel. If this swelling, or any other lesion in the carpal tunnel, presses on the *median nerve,* a condition called **carpal tunnel syndrome** may result. Because the median nerve connects to the palm and radial side (thumb side) of the hand, carpal tunnel syndrome is characterized by weakness, pain, and tingling in this part of the hand. The pain and tingling may also radiate to the forearm and shoulder. Prolonged and severe

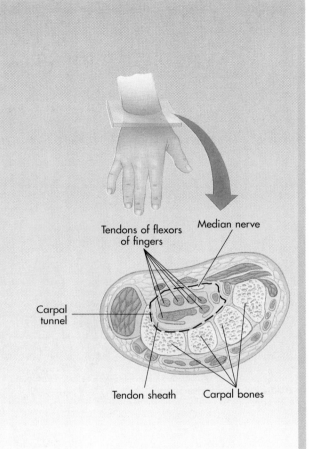

cases of carpal tunnel syndrome may be relieved by injection of inflammatory agents. A permanent cure is sometimes accomplished by surgical cutting or removal of the swollen tissue pressing on the median nerve.

also causes an increase in the number of mitochondria in muscle fibers. This allows production of more ATP as a rapid energy source.

Skeletal Muscle Groups

In the paragraphs that follow, representative muscles from the most important skeletal muscle groups will be discussed. Refer to Figure 6-6 often so that you will be able to see a muscle as you read about its placement on the body and its function. Table 6-1 identifies and groups muscles according to function and provides information about muscle action and points of origin and insertion. Keep in mind that muscles move bones, and the bones that they move are their insertion bones.

FIGURE 6–6 *General Overview of the Body Musculature.* **A,** Anterior view.

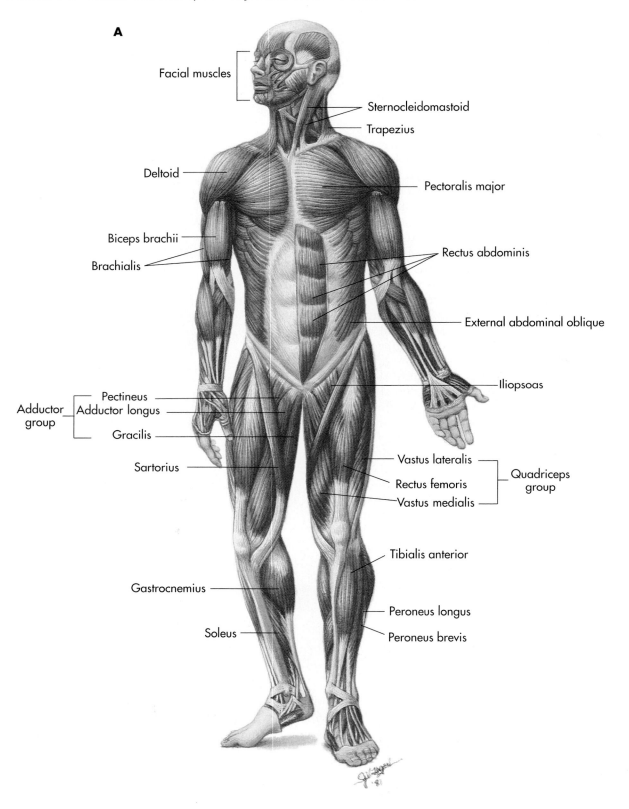

A

Facial muscles

Sternocleidomastoid

Trapezius

Deltoid

Pectoralis major

Biceps brachii

Brachialis

Rectus abdominis

External abdominal oblique

Iliopsoas

Pectineus

Adductor longus

Adductor group

Gracilis

Sartorius

Vastus lateralis

Rectus femoris

Quadriceps group

Vastus medialis

Tibialis anterior

Gastrocnemius

Peroneus longus

Soleus

Peroneus brevis

FIGURE 6–6, cont'd. B, Posterior view.

B

Sternocleidomastoid

Trapezius

Deltoid

Triceps brachii

Latissimus dorsi

External abdominal oblique

Gluteus maximus

Hamstring group
- Semitendinosus
- Biceps femoris
- Semimembranosus

Adductor magnus

Gastrocnemius

Peroneus longus

Peroneus brevis

Soleus

TABLE 6–1   Principal Muscles of the Body

MUSCLE	FUNCTION	INSERTION	ORIGIN
Muscles of the Head and Neck			
Frontal	Raises eyebrow	Skin of eyebrow	Occipital bone
Orbicularis oculi	Closes eye	Maxilla and frontal bone	Maxilla and frontal bone (encircles eye)
Orbicularis oris	Draws lips together	Encircles lips	Encircles lips
Zygomaticus	Elevates corners of mouth and lips	Angle of mouth and upper lip	Zygomatic
Masseter	Closes jaws	Mandible	Zygomatic arch
Temporal	Closes jaw	Mandible	Temporal region of the skull
Sternocleidomastoid	Rotates and extends head	Mastoid process	Sternum and clavicle
Trapezius	Extends head and neck	Scapula	Skull and upper vertebrae
Muscles That Move the Upper Extremities			
Pectoralis major	Flexes and helps adduct upper arms	Humerus	Sternum, clavicle, and upper rib cartilages
Latissimus dorsi	Extends and helps adduct upper arm	Humerus	Vertebrae and ilium
Deltoid	Abducts upper arm	Humerus	Clavicle and scapula
Biceps brachii	Flexes lower arm	Radius	Scapula
Triceps brachii	Extends lower arm	Ulna	Scapula and humerus
Muscles of the Trunk			
External oblique	Compresses abdomen	Midline of abdomen	Lower thoracic cage
Internal oblique	Compresses abdomen	Midline of abdomen	Pelvis
Transversus abdominis	Compresses abdomen	Midline of abdomen	Ribs, vertebrae, and pelvis
Rectus abdominis	Flexes trunk	Lower rib cage	Pubis
Muscles That Move the Lower Extremities			
Iliopsoas	Flexes thigh or trunk	Femur	Ilium and vertebrae
Sartorius	Flexes thigh and rotates lower leg	Tibia	Ilium
Gluteus maximus	Extends thigh	Femur	Ilium, sacrum, and coccyx
Adductor group			
Adductor longus	Adducts thigh	Femur	Pubis
Gracilis	Adducts thigh	Tibia	Pubis
Pectineus	Adducts thigh	Femur	Pubis

TABLE 6–1 Principal Muscles of the Body—Cont'd

MUSCLE	FUNCTION	INSERTION	ORIGIN
Muscles That Move the Lower Extremities (cont'd)			
Hamstring group			
Semimembranosus	Flexes lower leg	Tibia	Ischium
Semitendinosus	Flexes lower leg	Tibia	Ischium
Biceps femoris	Flexes lower leg	Fibula	Ischium and femur
Quadriceps group			
Rectus femoris	Extends lower leg	Tibia	Ilium
Vastus lateralis, intermedius, and medialus	Extend lower leg	Tibia	Femur
Tibialis anterior	Dorsiflexes foot	Metatarsals (foot)	Tibia
Gastrocnemius	Plantar flexes foot	Calcaneus (heel)	Femur
Soleus	Plantar flexes foot	Calcaneus (heel)	Tibia and fibula
Peroneus group			
Peroneus longus and brevis	Plantar flex foot	Tarsal and metatarsals (ankle and foot)	Tibia and fibula

MUSCLES OF THE HEAD AND NECK

The **muscles of facial expression** (Figure 6-7) allow us to communicate many different emotions nonverbally. Contraction of the **frontal** muscle, for example, allows you to raise your eyebrows in surprise and furrow the skin of your forehead into a frown. The **orbicularis** (or-bik-yoo-LAIR-is) **oris** (OR-iss), called the *kissing muscle,* puckers the lips. The **zygomaticus** (zye-go-MAT-ik-us) elevates the corners of the mouth and lips and has been called the *smiling muscle.*

The **muscles of mastication** are responsible for closing the mouth and producing chewing movements. As a group, they are among the strongest muscles in the body. The two largest muscles of the group, identified in Figure 6-7, are the **masseter** (mas-SEE-ter), which elevates the mandible, and the **temporal** (TEM-po-ral), which assists the masseter in closing the jaw.

The **sternocleidomastoid** (stern-o-kli-doe-MAS-toyd) and **trapezius** (tra-PEE-zee-us) muscles are easily identified in Figures 6-6 and 6-7.

The two sternocleidomastoid muscles are located on the anterior surface of the neck. They originate on the sternum and then pass up and cross the neck to insert on the mastoid process of the skull. Working together, they flex the head on the chest. If only one contracts, the head is both flexed and tilted to the opposite side. The triangular-shaped trapezius muscles form the line from each shoulder to the neck on its posterior surface. They have a wide line of origin extending from the base of the skull down the spinal column to the last thoracic vertebra. When contracted, the trapezius muscles help elevate the shoulders and extend the head backwards.

MUSCLES THAT MOVE THE UPPER EXTREMITIES

The upper extremity is attached to the thorax by the fan-shaped **pectoralis** (pek-tor-RAL-is) **major** muscle, which covers the upper chest, and by the **latissimus** (la-TIS-i-mus) **dorsi** muscle, which takes its origin from structures over the lower

FIGURE 6–7 *Muscles of the Head and Neck.* Muscles that produce most facial expressions surround the eyes, nose, and mouth. Large muscles of mastication stretch from the upper skull to the lower jaw. These powerful muscles produce chewing movements. The neck muscles connect the skull to the trunk of the body, rotating the head or bending the neck.

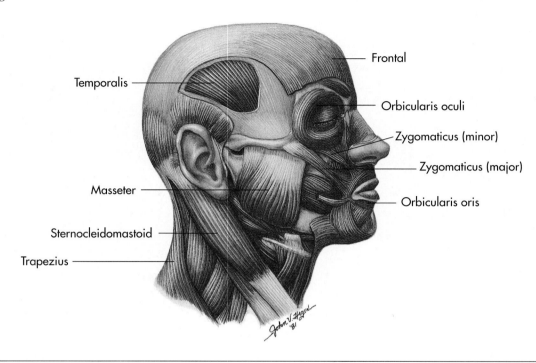

back (Figures 6-6 and 6-8). Both muscles insert on the humerus. The pectoralis major is a flexor, and the latissimus dorsi is an extensor of the upper arm.

The deltoid muscle forms the thick, rounded prominence over the shoulder and upper arm (Figure 6-6). The muscle takes its origin from the scapula and clavicle and inserts on the humerus. It is a powerful abductor of the upper arm.

As the name implies, the biceps brachii (BRAY-kee-eye) is a two-headed muscle that serves as a primary flexor of the forearm (Figure 6-6). It originates from the bones of the shoulder girdle and inserts on the radius in the forearm.

The **triceps brachii** is on the posterior or back surface of the upper arm. It has three heads of origin from the shoulder girdle and inserts into the olecranon process of the ulna. The triceps is an extensor of the elbow and thus performs a

straightening function. Because this muscle is responsible for delivering blows during fights, it is often called the *boxer's muscle.*

MUSCLES OF THE TRUNK

The muscles of the anterior or front side of the abdomen are arranged in three layers, with the fibers in each layer running in different directions much like the layers of wood in a sheet of plywood (Figure 6-8). The result is a very strong "girdle" of muscle that covers and supports the abdominal cavity and its internal organs.

The three layers of muscle in the anterolateral (side) abdominal walls are arranged as follows: the outermost layer or **external oblique;** a middle layer or **internal oblique;** and the innermost layer or **transversus abdominis.** In addition to these sheetlike muscles, the band- or strap-shaped **rectus abdominis** muscle runs down the midline of

FIGURE 6–8 *Muscles of the Trunk.* **A,** Anterior view showing superficial muscles. **B,** Anterior view showing deeper muscles.

the abdomen from the thorax to the pubis. The rectus abdominis and external oblique muscles can be seen in Figure 6-8. In addition to protecting the abdominal viscera, the rectus abdominis flexes the spinal column.

The *respiratory muscles* will be discussed in Chapter 13. **Intercostal muscles,** located between the ribs, and the sheetlike **diaphragm** separating the thoracic and abdominal cavities change the size and shape of the chest during breathing. As a result, air is moved into or out of the lungs.

MUSCLES THAT MOVE THE LOWER EXTREMITIES

The **iliopsoas** (il-ee-o-SO-us) originates from deep within the pelvis and the lower vertebrae to insert on the lesser trochanter of the femur and capsule of the hip joint. It is generally classified as a flexor of the thigh and an important postural muscle that stabilizes and keeps the trunk from falling over backward when you stand. However, if the thigh is fixed so that it cannot move, the iliopsoas flexes the *trunk.* An example would be doing sit-ups.

The **gluteus** (GLOO-tee-us) **maximus** (MAX-i-mus) forms the outer contour and much of the substance of the buttock. It is an important extensor of the thigh (Figure 6-6) and supports the torso in the erect position.

The **adductor muscles** originate on the bony pelvis and insert on the femur. They are located on the inner or medial side of the thighs. These muscles adduct or press the thighs together.

The three **hamstring muscles** are called the *semimembranosus, semitendinosus,* and *biceps femoris.* Acting together, they serve as powerful flexors of the lower leg (Figure 6-6). They originate on the ischium and insert on the tibia or fibula.

The **quadriceps** (KWOD-re-seps) **femoris** muscle group covers the upper thigh. The four thigh muscles—the *rectus femoris* and three *vastus* muscles—extend the lower leg (Figure 6-6 and Table 6-1). One component of the quadriceps group has its origin on the pelvis, and the remaining three originate on the femur; all four insert on the tibia. Only two of the vastus muscles are visible in Figure 6-6. The vastus intermedius is covered by the rectus femoris and is not visible.

Intramuscular Injections

Many drugs are administered by intramuscular injection. If the amount to be injected is 2 ml or less, the deltoid muscle is often selected as the site of injection. Note in Figure *A* that the needle is inserted into the muscle about two-fingers' breadth below the acromion process of the scapula and lateral to the tip of the acromion. If the amount of medication to be injected is 2 to 3 ml, the gluteal area shown in Figure *B* is often used. Injections are made into the gluteus medius muscle near the center of the upper outer quadrant, as shown in the illustration. Another technique of locating the proper injection site is to draw an imaginary diagonal line from a point of reference on the back of the bony pelvis (posterior superior iliac spine) to the greater trochanter of the femur. The injection is given about three-fingers' breadth above and one third of the way down the line. It is important that the sciatic nerve and the

superior gluteal blood vessels be avoided during the injection. Proper technique requires a knowledge of the underlying anatomy.

In addition to intramuscular injections, which are generally administered by a health care provider in an institutional setting, many individuals must self-administer injections of needed medications, such as insulin, on a regular basis in their homes. Educating these patients or their care givers on how to correctly administer medication by injection is an important issue in the delivery of home health care services. Topics that must be covered include instruction on proper injection techniques, selection of needle length and gauge, identification of important anatomic landmarks in making injection site selections, and the preparation and rotation of selected injection sites.

Trapezius

Acromion process of scapula

Deltoid

A

Superior gluteal artery and vein

Posterior superior iliac spine

Sciatic nerve

Gluteus maximus

Iliac crest

Gluteus medius

Greater trochanter

B

The **tibialis** (tib-ee-AL-is) **anterior** muscle (Figure 6-6) is located on the anterior or front surface of the leg. It dorsiflexes the foot. The **gastrocnemius** (gas-trok-NEE-mee-us) is the primary calf muscle. Note in Figure 6-6 that it has two fleshy components arising from both sides of the femur. It inserts through the Achilles tendon into the heel bone or calcaneus. The gastrocnemius is responsible for plantar flexion of the foot; because it is used to stand on tiptoe, it is sometimes called the *toe dancer's muscle.* A group of three muscles called the **peroneus** (pair-o-NEE-us) **group** (Figure 6-6) is found along the sides of the lower leg. As a group, these muscles plantar flex the foot. A long tendon from one component of the group—the *peroneus longus* muscle tendon—forms a support arch for the foot (see Figure 5-17).

◉ *Movements Produced by Skeletal Muscle Contractions*

The types of movement that may produce a muscle contraction at any joint depend largely on the shapes of the bones involved and the joint type (see Chapter 5). Muscles acting on some joints produce movement in several directions, whereas only limited movement is possible at other joints. The terms most often used to describe body movements are:

1. **Flexion**
2. **Extension**
3. **Abduction**
4. **Adduction**
5. **Rotation**
6. **Supination** and **pronation**
7. **Dorsiflexion** and **plantar flexion**

Flexion is a movement that makes the angle between two bones at their joint smaller than it was at the beginning of the movement. Most flexions are movements commonly described as bending. If you bend your lower arm at the elbow or your lower leg at the knee, you flex the lower arm and leg. **Extension** movements are the opposite of flexions. They make the angle between two bones at their joint larger than it was at the beginning of the movement. Therefore, extensions are straightening or stretching movements rather than bending movements. Figures 6-9 and 6-10 illustrate flexion and extension of the lower arm and leg.

Abduction means moving a part away from the midline of the body, such as moving your arm out to the side. **Adduction** means moving a part toward the midline, such as bringing your arms down to your sides from an elevated position. Figure 6-11, *A,* shows abduction and adduction.

Rotation is movement around a longitudinal axis. You rotate your head by moving your skull from side to side as in shaking your head "no" (Figure 6-11, *B*).

Supination and **pronation** refer to hand positions that result from rotation of the forearm. (The term *prone* refers to the body as a whole lying face down. *Supine* means lying face up.) Supination results in a hand position with the palm turned to the anterior position (as in the anatomical position), and pronation occurs when you turn the palm of your hand so that it faces posteriorly (Figure 6-11, *C*).

Dorsiflexion and **plantar flexion** refer to foot movements. In dorsiflexion the dorsum or top of the foot is elevated with the toes pointing upward. In plantar flexion the bottom of the foot is directed downward so that you are in effect standing on your toes (Figure 6-11, *D*).

As you study the illustrations and learn to recognize the muscles discussed in this chapter, you should attempt to group them according to function, as in Table 6-2. You will note, for example, that flexors produce many of the movements used for walking, sitting, swimming, typing, and many other activities. Extensors also function in these activities but perhaps play their most important role in maintaining an upright posture.

FIGURE 6–9 *Flexion and Extension of the Lower Arm.* **A** and **B,** When the lower arm is flexed at the elbow, the biceps brachii contracts while its antagonist, the triceps brachii, relaxes. **B** and **C,** When the lower arm is extended, the biceps brachii relaxes while the triceps brachii contracts.

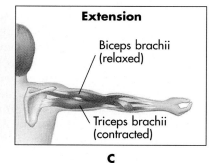

A   **B**   **C**

FIGURE 6–10 *Flexion and Extension of the Lower Leg.* **A** and **B,** When the lower leg flexes at the knee, muscles of the hamstring group contract while their antagonists in the quadriceps femoris group relax. **B** and **C,** When the lower leg extends, the hamstring muscles relax while the quadriceps femoris muscle contracts.

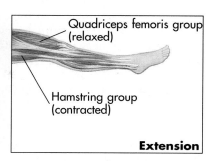

A   **B**   **C**

TABLE 6–2   Muscles Grouped According to Function

PART MOVED	FLEXORS	EXTENSORS	ABDUCTORS	ADDUCTORS
Upper arm	Pectoralis major	Latissimus dorsi	Deltoid	Pectoralis major and latissimus dorsi contracting together
Lower arm	Biceps brachii	Triceps brachii	None	None
Thigh	Iliopsoas and sartorius	Gluteus maximus	Gluteus medius	Adductor group
Lower leg	Hamstrings	Quadriceps group	None	None
Foot	Tibialis anterior	Gastrocnemius and soleus	Peroneus longus	Tibialis anterior

FIGURE 6–11 *Examples of Body Movements.* **A,** Adduction and abduction. **B,** Rotation. **C,** Pronation and supination. **D,** Dorsiflexion and plantar flexion.

INTRODUCTION

A Muscular tissue enables the body and its parts to move
  1 Movement caused by ability of muscle cells (called *fibers*) to shorten or contract
  2 Muscle cells shorten by converting chemical energy (obtained from food) into mechanical energy, which causes movement
  3 Three types of muscle tissue exist in body (see Chapter 2)

MUSCLE TISSUE

A Types of muscle tissue (Figure 6-1)
  1 Skeletal muscle—also called *striated* or *voluntary muscle*
    a Is 40% to 50% of body weight ("red meat" attached to bones)
    b Microscope reveals crosswise stripes or striations
    c Contractions can be voluntarily controlled
  2 Cardiac muscle—composes bulk of heart
    a Cardiac muscle cells branch frequently
    b Characterized by unique dark bands called *intercalated disks*
    c Interconnected nature of cardiac muscle cells allows heart to contract efficiently as a unit
  3 Nonstriated muscle or involuntary muscle—also called *smooth* or *visceral* muscle
    a Lacks cross stripes or striations when seen under a microscope; appears smooth
    b Found in walls of hollow visceral structures such as digestive tract, blood vessels, and ureters
    c Contractions not under voluntary control; movement caused by contraction is involuntary
B Function—all muscle cells specialize in contraction (shortening)

STRUCTURE OF SKELETAL MUSCLE

A Structure
  1 Each skeletal muscle is an organ composed mainly of skeletal muscle cells and connective tissue
  2 Most skeletal muscles extend from one bone across a joint to another bone
  3 Parts of a skeletal muscle
    a Origin—attachment to the bone that remains relatively stationary or fixed when movement at the joint occurs
    b Insertion—point of attachment to the bone that moves when a muscle contracts
    c Body—main part of the muscle

4 Muscles attach to bone by tendons—strong cords of fibrous connective tissue; some tendons enclosed in synovial-lined tubes and are lubricated by synovial fluid; tubes called *tendon sheaths*
5 Bursae—small synovial-lined sacs containing a small amount of synovial fluid; located between some tendons and underlying bones
B Microscopic structure (Figure 6-3)
  1 Contractile cells called *fibers*—grouped into bundles
  2 Fibers contain thick myofilaments (containing protein myosin) and thin myofilaments (composed of actin)
  3 Basic functional (contractile) unit called *sarcomere*; sarcomeres separated from each other by dark bands called *Z lines*
    a Sliding filament model explains mechanism of contraction
      (1) Thick and thin myofilaments slide past each other as a muscle contracts
      (2) Contraction requires calcium and energy-rich ATP molecules

FUNCTIONS OF SKELETAL MUSCLE

A Movement
  1 Muscles produce movement; as a muscle contracts, it pulls the insertion bone closer to the origin bone; movement occurs at the joint between the origin and the insertion
    a Groups of muscles usually contract to produce a single movement
      (1) Prime mover—muscle whose contraction is mainly responsible for producing a given movement
      (2) Synergist—muscle whose contractions help the prime mover produce a given movement
      (3) Antagonist—muscle whose actions oppose the action of a prime mover in any given movement
B Posture
  1 A specialized type of muscle contraction, called *tonic contraction*, enables us to maintain body position
    a In tonic contraction, only a few of a muscle's fibers shorten at one time
    b Tonic contractions produce no movement of body parts
    c Tonic contractions maintain muscle tone called *posture*

(1) Good posture reduces strain on muscles, tendons, ligaments, and bones

(2) Poor posture causes fatigue and may lead to deformity

C Heat production

  1 Survival depends on the body's ability to maintain a constant body temperature

    a Fever—an elevated body temperature—often a sign of illness

    b Hypothermia—a reduced body temperature

  2 Contraction of muscle fibers produces most of the heat required to maintain normal body temperature

FATIGUE

A Reduced strength of muscle contraction

B Caused by repeated muscle stimulation without adequate periods of rest

C Repeated muscular contraction depletes cellular ATP stores and outstrips the ability of the blood supply to replenish oxygen and nutrients

D Contraction in the absence of adequate oxygen produces lactic acid, which contributes to muscle soreness

E *Oxygen debt*—term used to describe the metabolic effort required to burn excess lactic acid that may accumulate during prolonged periods of exercise; the body is attempting to return the cells' energy and oxygen reserves to pre-exercise levels

ROLE OF OTHER BODY SYSTEMS IN MOVEMENT

A Muscle functioning depends on the functioning of many other parts of the body

  1 Most muscles cause movements by pulling on bones across movable joints

  2 Respiratory, circulatory, nervous, muscular, and skeletal systems play essential roles in producing normal movements

  3 Multiple sclerosis, brain hemorrhage, and spinal cord injury are examples of how pathological conditions in other body organ systems can dramatically affect movement

MOTOR UNIT (Figure 6-4)

A Stimulation of a muscle by a nerve impulse is required before a muscle can shorten and produce movement

B A motor neuron is the specialized nerve that transmits an impulse to a muscle, causing contraction

C A neuromuscular junction is the specialized point of contact between a nerve ending and the muscle fiber it innervates

D A motor unit is the combination of a motor neuron with the muscle cell or cells it innervates

MUSCLE STIMULUS

A A muscle will contract only if an applied stimulus reaches a certain level of intensity

  1 A threshold stimulus is the minimal level of stimulation required to cause a muscle fiber to contract

B Once stimulated by a threshold stimulus, a muscle fiber will contract completely, a response called *all or none*

C Different muscle fibers in a muscle are controlled by different motor units having different threshold-stimulus levels

  1 Although individual muscle fibers always respond all or none to a threshold stimulus, the muscle as a whole does not

  2 Different motor units responding to different threshold stimuli permit a muscle as a whole to execute contractions of graded force

TYPES OF SKELETAL MUSCLE CONTRACTION

A Twitch and tetanic contractions

  1 Twitch contractions are laboratory phenomena and do not play a significant role in normal muscular activity; they are a single contraction of muscle fibers caused by a single threshold stimulus

  2 Tetanic contractions are sustained and steady muscular contractions caused by a series of stimuli bombarding a muscle in rapid succession

B Isotonic contractions

  1 Contraction of a muscle that produces movement at a joint

  2 During isotonic contractions, the muscle shortens, causing the insertion end of the muscle to move toward the point of origin

  3 Most types of body movements such as walking and running are caused by isotonic contractions

C Isometric contractions

  1 Isometric contractions are muscle contractions that do not produce movement; the muscle as a whole does not shorten

  2 Although no movement occurs during isometric contractions, tension within the muscle increases

EFFECTS OF EXERCISE ON SKELETAL MUSCLES

A Exercise, if regular and properly practiced, improves muscle tone and posture, results in more efficient heart and lung functioning, and reduces fatigue

B Effects of exercise on skeletal muscles

  1 Muscles undergo changes related to the amount of work they normally do

    a Prolonged inactivity causes disuse atrophy

    b Regular exercise increases muscle size, called *hypertrophy*

2 Strength training is exercise involving contraction of muscles against heavy resistance
  a Strength training increases the numbers of myofilaments in each muscle fiber, and as a result, the total mass of the muscle increases
  b Strength training does not increase the number of muscle fibers
3 Endurance training is exercise that increases a muscle's ability to sustain moderate exercise over a long period; it is sometimes called *aerobic training*
  a Endurance training allows more efficient delivery of oxygen and nutrients to a muscle via increased blood flow
  b Endurance training does not usually result in muscle hypertrophy

SKELETAL MUSCLE GROUPS (Table 6-1)
A Muscles of the head and neck (Figure 6-7)
  1 Facial muscles
    a Orbicularis oculi
    b Orbicularis oris
    c Zygomaticus
  2 Muscles of mastication
    a Masseter
    b Temporal
  3 Sternocleidomastoid—flexes head
  4 Trapezius—elevates shoulders and extends head
B Muscles that move the upper extremities
  1 Pectoralis major—flexes upper arm
  2 Latissimus dorsi—extends upper arm
  3 Deltoid—abducts upper arm
  4 Biceps brachii—flexes forearm
  5 Triceps brachii—extends forearm
C Muscles of the trunk (Figure 6-8)
  1 Abdominal muscles
    a Rectus abdominis
    b External oblique
    c Internal oblique
    d Transversus abdominis

2 Respiratory muscles
  a Intercostal muscles
  b Diaphragm
D Muscles that move the lower extremities
  1 Iliopsoas—flexes thigh
  2 Gluteus maximus—extends thigh
  3 Adductor muscles—adduct thighs
  4 Hamstring muscles—flex lower leg
    a Semimembranosus
    b Semitendinosus
    c Biceps femoris
  5 Quadriceps femoris group—extend lower leg
    a Rectus femoris
    b Vastus muscles
  6 Tibialis anterior—dorsiflexes foot
  7 Gastrocnemius—plantar flexes foot
  8 Peroneus group—flex foot

TYPES OF MOVEMENTS PRODUCED BY SKELETAL MUSCLE CONTRACTIONS
(Figures 6-9 through 6-11)
A Flexion—movement that decreases the angle between two bones at their joint: bending
B Extension—movement that increases the angle between two bones at their joint: straightening
C Abduction—movement of a part away from the midline of the body
D Adduction—movement of a part toward the midline of the body
E Rotation—movement around a longitudinal axis
F Supination and pronation—hand positions that result from rotation of the forearm; supination results in a hand position with the palm turned to the anterior position; pronation occurs when the palm faces posteriorly
G Dorsiflexion and plantar flexion—foot movements; dorsiflexion results in elevation of the dorsum or top of the foot; during plantar flexion, the bottom of the foot is directed downward

NEW WORDS

abduction	flexion	neuromuscular junction	sarcomere
actin	hypertrophy	origin	sliding filament theory
adduction	hypothermia	oxygen debt	stimulus
all or none	insertion	paralysis	supination
antagonist	isometric	plantar flexion	synergist
atrophy	isotonic	posture	tendon
bursa	motor neuron	prime mover	tenosynovitis
dorsiflexion	motor unit	pronation	tetanic contraction
extension	myofilaments	rotation	tonic contraction
fatigue	myosin		

CHAPTER TEST

1. Cardiac muscle:
   a. Cells have no nuclei
   b. Is composed of cells that branch frequently
   c. Lines many hollow internal organs
   d. All of the above are correct
2. The term *origin* refers to:
   a. The attachment of a muscle to a bone that does not move when contraction occurs
   b. Attachment of a muscle to a bone that moves when contraction occurs
   c. The body of a muscle
   d. None of the above are correct
3. A sarcomere:
   a. Is the basic functional or contractile unit of skeletal muscle
   b. Contains only actin myofilaments
   c. Contains only myosin myofilaments
   d. Is found only in smooth or involuntary muscle
4. The muscle mainly responsible for producing a particular movement is called a:
   a. Synergist
   b. Prime mover
   c. Antagonist
   d. Fixator
5. A motor unit consists of:
   a. Only a motor neuron
   b. A motor neuron and the muscle cells it innervates
   c. Only contracting muscle cells
   d. None of the above are correct
6. Which of the following types of muscle contraction does not produce movement?
   a. Isotonic contraction
   b. Tetanic contraction
   c. Twitch contraction
   d. Isometric contraction
7. Which of the following terms refers to moving a part away from the midline of the body?
   a. Abduction
   b. Adduction
   c. Supination
   d. Flexion

8. Many hollow internal organs and blood vessel walls contain _____ muscle.

9. Muscles are anchored firmly to bones by _____.

10. Thick myofilaments are formed from a protein called _____.

11. The explanation of muscle contraction resulting from movement of thick and thin myofibrils toward one another is called the _____ theory.

12. The muscles that assist prime movers in producing a particular movement are called _____.

13. Skeletal muscle tone maintains _____ by counteracting the pull of gravity.

14. Depletion of oxygen in a muscle during vigorous and prolonged exercise is called _____.

15. Walking and running are examples of _____ muscle contraction.

16. Standing on your toes is an example of _____ flexion.

17. The term _____ is used to describe shrinkage or decrease in the size of a muscle.

18. The major flexor muscles of the lower leg are the _____.

Select the most correct answer from Column B for each statement in Column A. (Only one answer is correct.)

COLUMN A	COLUMN B
19. _____Contractile unit of muscle	a. Biceps brachii
20. _____Flexes lower arm	b. Masseter
21. _____Extends lower leg	c. Sarcomere
22. _____Closes jaw	d. Quadriceps group
23. _____Movable point of attachment	e. Pectoralis major
24. _____Flexes upper arm	f. Gluteus maximus
25. _____Extends thigh	g. Insertion

1. Compare the three kinds of muscle tissue regarding location, microscopic appearance, and nerve control.
2. Explain flexion, extension, abduction, and adduction; give an example of each.
3. Explain how skeletal muscles, bones, and joints work together to produce movements.
4. What two kinds of muscle contractions do not produce movement?
5. What is the name of the main muscle or muscles that
   a. Flex the upper arm?
   b. Flex the lower leg?
   c. Extend the lower arm?
   d. Extend the thigh?
   e. Abduct the upper arm?
   f. Elevate the mandible?
   g. Flex the spinal column?
   h. Dorsiflex the foot?
6. Give the approximate location of the biceps brachii, hamstrings, deltoid, pectoralis major, quadriceps femoris group, latissimus dorsi, trapezius, rectus abdominis, and gastrocnemius; tell what movement each produces.
7. What is meant by the term *tetanus?*
8. Discuss the microscopic structure of skeletal muscle tissue.
9. Explain good posture, hypertrophy, sarcomere, muscle tone, isometric contractions, isotonic contractions, tonic contractions, tetanus, threshold, and all or none.

C R I T I C A L   T H I N K I N G

10. Why can a spinal cord injury be followed by muscle paralysis?
11. Can a muscle contract very long if its blood supply is shut off? Give a reason for your answer.
12. Briefly describe changes that gradually take place in bones, joints, and muscles if a person habitually gets too little exercise.

The Nervous System

BOXED ESSAYS

Multiple Sclerosis (MS)
Suppressing Pain During Exercise
Lumbar Puncture
Herpes Zoster or Shingles

OBJECTIVES

*After you have completed this chapter,
you should be able to:*

1. List the organs and divisions of the nervous
   system and describe the generalized functions
   of the system as a whole.

2. Identify the major types of cells in the nervous
   system and discuss the function of each.

3. Identify the anatomical and functional components of a three-neuron reflex arc. Compare and
   contrast the propagation of a nerve impulse
   along a nerve fiber and across a synaptic cleft.

4. Identify the major anatomical components of
   the brain and spinal cord and briefly comment
   on the function of each.

5. Compare and contrast spinal and cranial nerves.

6. Discuss the anatomical and functional characteristics of the two divisions of the autonomic
   nervous system.

The normal body must accomplish a gigantic and enormously complex job—keeping itself alive and healthy. Each one of its billions of cells performs some activity that is a part of this function. Control of the body's billions of cells is accomplished mainly by two communication systems: the nervous system and the endocrine system. Both systems transmit information from one part of the body to another, but they do it in different ways. The nervous system transmits information very rapidly by nerve impulses conducted from one body area to another. The endocrine system transmits information more slowly by chemicals secreted by duct-less glands into the bloodstream and circulated from the glands to other parts of the body. Nerve impulses and hormones communicate information to body structures, increasing or decreasing their activities as needed for healthy survival. In other words, the communication systems of the body are also its control and integrating systems. They weld the body's hundreds of functions into its one overall function of keeping itself alive and healthy.

Recall that homeostasis is the balanced and controlled internal environment of the body that is basic to life itself. Homeostasis is possible only if our physiological control and integration systems function properly. Our plan for this chapter is to name the cells, organs, and divisions of the nervous system; discuss the generation of nervous impulses; and then discover how these impulses move between one area of the body and another. We will study not only the major components of the nervous system, such as the brain, spinal cord, and nerves, but also learn about how they function to maintain and regulate homeostasis. In Chapter 8, we will consider the special senses.

Organs and Divisions of the Nervous System

The organs of the nervous system as a whole include the brain and spinal cord, the numerous nerves of the body, the specialized sense organs such as the eyes and ears, and the microscopic sense organs such as those found in the skin. The system as a whole consists of two principal divisions called the central nervous system and the peripheral nervous system (Figure 7-1). Because the brain and spinal cord occupy a midline or central location in the body, they are together called the **central nervous system** or **CNS.** Similarly, the usual designation for the nerves of the body is the **peripheral nervous system** or **PNS.** Use of the term *peripheral* is appropriate because nerves extend to outlying or peripheral parts of the body. A subdivision of the peripheral nervous system, called the **autonomic nervous system** or **ANS,** consists of structures that regulate the body's automatic or involuntary functions (for example, the heart rate, the contractions of the stomach and intestines, and the secretion of chemical compounds by glands).

Cells of the Nervous System

The two types of cells found in the nervous system are called **neurons** (NOO-rons) or nerve cells and **glia** (GLEE-ah), which are specialized connective tissue cells. Neurons conduct impulses, whereas glia support neurons.

NEURONS

Each neuron consists of three parts: a main part called the neuron **cell body,** one or more branching projections called **dendrites** (DEN-drites), and one elongated projection known as an **axon.** Identify each part on the neuron shown in Figure 7-2. Dendrites are the processes or projections that transmit impulses to the neuron cell bodies or axons, and axons are the processes that transmit impulses away from the neuron cell bodies or dendrites.

The three types of neurons are classified according to the direction in which they transmit impulses: **sensory neurons, motor neurons,** and **interneurons.** Sensory neurons transmit impulses to the spinal cord and brain from all parts of the body. Motor neurons transmit impulses in the opposite direction—away from the brain and spinal cord. They do not conduct impulses to all parts of the body but only to two kinds of tissue—muscle and glandular epithelial tissue. Interneurons conduct impulses from sensory neurons to motor neurons. Sensory neurons are also called *afferent* neurons; motor neurons are called *efferent* neurons, and interneurons are called *central* or *connecting* neurons.

FIGURE 7–1 *Divisions of the Nervous System.*

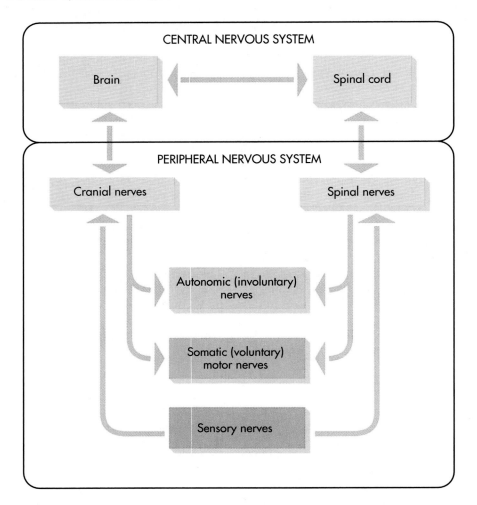

The axon shown in Figure 7-2, *B*, is surrounded by a segmented wrapping of a material called **myelin** (MY-e-lin). Myelin is a white, fatty substance formed by **Schwann cells** that wrap around some axons outside the central nervous system. Such fibers are called **myelinated fibers.** In Figure 7-2, *B*, one such axon has been enlarged to show additional detail. **Nodes of Ranvier** (rahn-vee-AY) are indentations between adjacent Schwann cells.

The outer cell membrane of a Schwann cell is called the **neurilemma** (noo-ri-LEM-mah). The fact that axons in the brain and cord have no neurilemma is clinically significant because it plays an essential part in the regeneration of cut and injured axons. Therefore the potential for regeneration in the brain and spinal cord is far less than it is in the peripheral nervous system.

GLIA

Glia—or *neuroglia* (noo-ROG-lee-ah)—do not specialize in transmitting impulses. Instead, they are special types of connective tissue cells. Their name is appropriate because it is derived from the Greek word *glia* meaning "glue." One function of glia cells is to hold the functioning neurons

FIGURE 7–2 *Neuron.* **A,** Diagram of a typical neuron showing dendrites, a cell body, and an axon. **B,** Segment of a myelinated axon cut to show detail of the concentric layers of the Schwann cell filled with myelin. **C,** Photomicrograph of neuron.

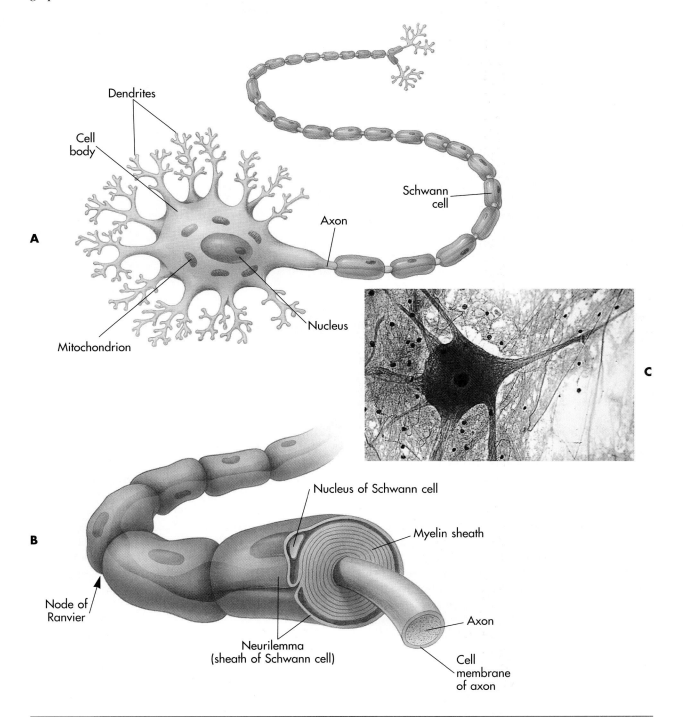

FIGURE 7–3 *Glia.* **A,** Astrocytes have extensions attached to blood vessels in the brain. **B,** Microglia within the central nervous system can enlarge and consume microbes by phagocytosis. **C,** Oligodendroglia have extensions that form myelin sheaths around axons in the central nervous system.

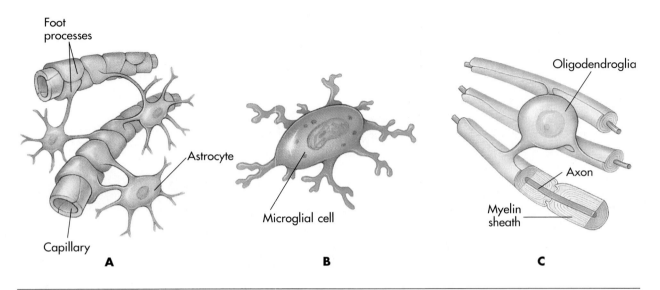

together and protect them. An important reason for discussing glia is that one of the most common types of brain tumor—called **glioma** (glee-O-mah)—develops from them.

Glia vary in size and shape (Figure 7-3). Some are relatively large cells that look somewhat like stars because of the threadlike extensions that jut out from their surfaces. These glia cells are called **astrocytes** (AS-tro-sites), a word that means "star cells" (Figure 7-3). Their threadlike branches attach to neurons and to small blood vessels, holding these structures close to each other. Along with the walls of the blood vessels, astrocyte branches form a two-layer structure called the *blood-brain barrier (BBB).* As its name implies, the blood-brain barrier separates the blood tissue and nervous tissue to protect vital brain tissue from harmful chemicals that might be in the blood.

Microglia (my-KROG-lee-ah) are smaller than astrocytes. They usually remain stationary, but in inflamed or degenerating brain tissue, they enlarge, move about, and act as microbe-eating scavengers. They surround the microbes, draw them into their cytoplasm, and digest them. Recall from Chapter 2 that phagocytosis is the scientific name for this important cellular process.

The **oligodendroglia** (ol-i-go-den-DROG-lee-ah) help to hold nerve fibers together and also serve another and probably more important function; they produce the fatty myelin sheath that envelops nerve fibers located in the brain and spinal cord.

● *Nerves*

A **nerve** is a group of peripheral nerve fibers (axons) bundled together like the strands of a cable. Because nerve fibers usually have a myelin sheath and myelin is white, nerves are called the **white matter** of the PNS. Bundles of axons in the CNS, called **tracts,** may also be myelinated and thus form this system's white matter. Tissue comprised of cell bodies and unmyelinated axons and dendrites is called **gray matter** because of its characteristic gray appearance.

Figure 7-4 shows that each axon in a nerve is surrounded by a thin wrapping of fibrous connective tissue called the **endoneurium** (en-doe-NOO-ree-um). Groups of these wrapped axons are called **fascicles.** Each fascicle is surrounded by a thin, fibrous **perineurium** (pair-i-NOO-ree-um). A

FIGURE 7–4 *The Nerve.* Each nerve contains axons bundled into fascicles. A connective tissue epineurium wraps the entire nerve. Perineurium surrounds each fascicle. Inset shows a scanning electron micrograph of a cross section of a nerve. From Kessel RG, Kardong RH: Tissues and organs: A text-atlas of scanning electron microscopy, WH Freeman, 1979.

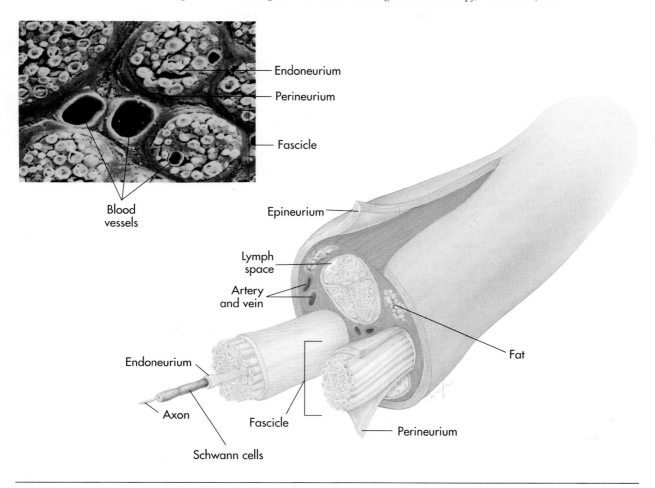

tough, fibrous sheath called the **epineurium** (ep-i-NOO-ree-um) covers the whole nerve.

Reflex Arcs

During every moment of our lives, nerve impulses speed over neurons to and from our spinal cords and brains. If all impulse conduction ceases, life itself ceases. Only neurons can provide the rapid communication between cells that is necessary for maintaining life. Hormonal messages are the only other kind of communications

the body can send, and they travel much more slowly than impulses. They can move from one part of the body to another only via circulating blood. Compared with impulse conduction, circulation is a very slow process.

Nerve impulses, sometimes called *action potentials*, can travel over trillions of routes—routes made up of neurons because they are the cells that conduct impulses. Hence the routes traveled by nerve impulses are sometimes spoken of as *neuron pathways*. A specialized type of neuron pathway, called a **reflex arc,** is important to nervous system functioning. The simplest kind of reflex arc is a two-neuron arc, so called because it consists of only two types of neurons: sensory neurons and motor neurons. Three-neuron arcs are the next simplest kind. They, of course, consist of all three kinds of neurons: sensory neurons, interneurons, and motor neurons. Reflex arcs are like one-way streets; they allow impulse conduction in only one direction. The next paragraph describes this direction in detail. Look frequently at Figure 7-5 as you read it.

Impulse conduction normally starts in receptors. **Receptors** are the beginnings of dendrites of sensory neurons. They are often located at some distance from the spinal cord (in tendons, skin, or mucous membranes, for example). In Figure 7-5 the sensory receptors are located in the quadriceps muscle group and in the patellar tendon. In the reflex that is illustrated there, stretch receptors are stimulated as a result of a tap on the patellar tendon from a rubber hammer used by a physician to elicit a reflex during a physical examination. The nerve impulse that is generated, its neu-

rologic pathway, and its ultimate "knee-jerk" effect is an example of the simplest form of a two-neuron reflex arc. In this reflex, only sensory and motor neurons are involved. The nerve impulse that is generated by stimulation of the stretch receptors travels along the length of the sensory neuron's dendrite to its cell body located in the **posterior (dorsal) root ganglion** (GANG-lee-on). A **ganglion** is a group of nerve-cell bodies located in the PNS. This ganglion is located near the spinal cord. Each spinal ganglion contains not one sensory neuron cell body as shown in Figure 7-5 but hundreds of them. The axon of the sensory neuron travels from the cell body in the dorsal root ganglion and ends near the dendrites of another neuron located in the gray matter of the spinal cord. A microscopic space separates the axon ending of one neuron from the dendrites of another neuron. This space is called a **synapse**. The nerve impulse stops at the synapse, chemical signals are sent across the gap, and the impulse then continues along the dendrites, cell body, and

FIGURE 7–5 *Patellar Reflex.* The neural pathway involved in the patellar ("knee-jerk") reflex.

Multiple Sclerosis (MS)

Many diseases are associated with disorders of the oligodendroglia. Because these glia cells are involved in myelin formation, these diseases are called **myelin disorders.** The most common primary disease of the CNS is a myelin disorder called **multiple sclerosis** or **MS.** It is characterized by myelin loss and destruction accompanied by varying degrees of oligodendroglial cell injury and death. The result is demyelination of the white matter of the CNS. Hard, plaquelike lesions replace the destroyed myelin, and affected areas are invaded by inflammatory cells. As the myelin around axons is lost, nerve conduction is impaired, and weakness, incoordination, visual impairment, and speech disturbances occur. Although the disease occurs in both sexes and all age groups, it is most common in women between 20 and 40 years.

The cause may be related to autoimmunity and viral infections in some individuals. It is relapsing and chronic in nature, but some cases of acute and unremitting disease have been reported. In most cases, MS is prolonged, with remissions and relapses occurring over many years. There is no known cure.

axon of the motor neuron. The motor neuron axon forms a synapse with a structure called an *effector,* an organ that puts nerve signals "into effect."

Effectors are muscles or glands, and muscle contractions and gland secretion are the only kinds of reflexes. The response to impulse conduction over a reflex arc is called a **reflex.** In short, impulse conduction by a reflex arc causes a reflex to occur. In a patellar reflex, the nerve impulses that reach the quadriceps muscle (the effector) result in the "knee-jerk" response.

Now turn your attention to the *interneuron* shown in Figure 7-5. Some reflexes involve three rather than two neurons. In these more complex types of responses, an interneuron, in addition to a sensory and motor neuron, is involved. In three-neuron reflexes, the end of the sensory neuron's axon synapses first with an interneuron before chemical signals are sent across a second synapse, resulting in conduction through the motor neuron. For example, application of an irritating stimulus to the skin of the thigh initiates a three-neuron reflex response that causes contraction of muscles to pull the leg away from the irritant—a three-neuron arc reaction called the *withdrawal reflex.* All interneurons lie entirely within the gray matter of the brain or spinal cord. Gray matter forms the H-shaped inner core of the spinal cord. Because of the presence of an interneuron, three-neuron reflex arcs have two synapses. A two-neuron reflex arc, however, has only a sensory neuron and a motor neuron with one synapse between them.

Identify the motor neuron in Figure 7-5. Observe that its dendrites and cell body, like those of an interneuron, are located in the spinal cord's gray matter. The axon of this motor neuron, however, runs through the anterior (ventral) root of the spinal nerve and terminates in a muscle.

 ## Nerve Impulses

What are nerve impulses? Here is one widely accepted definition: a nerve impulse is a self-propagating wave of electrical disturbance that travels along the surface of a neuron's plasma membrane. You might visualize this as a tiny spark sizzling its way along a fuse. Nerve impulses do not continually race along every nerve cell's surface. First they have to be initiated by a stimulus, a change in the neuron's environment. Pressure, temperature, and chemical changes are the usual stimuli. The membrane of

FIGURE 7–6 *Conduction of Nerve Impulses.* **A,** In an unmyelinated fiber, a nerve impulse (action potential) is a self-propagating wave of electrical disturbance. **B,** In a myelinated fiber, the action potential "jumps" around the insulating myelin in a rapid type of conduction called *saltatory conduction.*

each resting neuron has a slight positive charge on the outside and a negative charge on the inside, as shown in Figure 7-6. This occurs because there is normally an excess of sodium ions (Na$^+$) on the outside of the membrane. When a section of the membrane is stimulated, its Na$^+$ channels suddenly open, and Na$^+$ rushes inward. The inside of the membrane temporarily becomes positive, and the outside becomes negative. Although this section of the membrane immediately recovers, the electrical disturbance stimulates Na$^+$ channels in the next section of the membrane to open. Thus a self-propagating wave of disturbance—a nerve impulse—travels in one direction across the neuron's surface (Figure 7-6, *A*). If the traveling impulse encounters a section of membrane covered with insulating myelin, it simply "jumps" around the myelin. Called **saltatory conduction,** this type of impulse travel is

much faster than is possible in nonmyelinated sections. Saltatory conduction is illustrated in Figure 7-6, *B*.

The Synapse

Transmission of signals from one neuron to the next—across the synapse—is an important part of the nerve conduction process. By definition, a synapse is the place where impulses are transmitted from one neuron, called the **presynaptic neuron,** to another neuron, called the **postsynaptic neuron.** Three structures make up a synapse: a synaptic knob, a synaptic cleft, and the plasma membrane of a postsynaptic neuron. A **synaptic knob** is a tiny bulge at the end of a terminal branch of a presynaptic neuron's axon (Figure 7-7). Each synaptic knob contains many small sacs or vesi-

FIGURE 7–7   *Components of a Synapse.* Diagram shows synaptic knob or axon terminal of presynaptic neuron, the plasma membrane of a postsynaptic neuron, and a synaptic cleft. On the arrival of an action potential at a synaptic knob, neurotransmitter molecules are released from vesicles in the knob into the synaptic cleft. The combining of neurotransmitter and receptor molecules in the plasma membrane of the postsynaptic neuron opens ion channels and thereby initiates impulse conduction in the postsynaptic neuron.

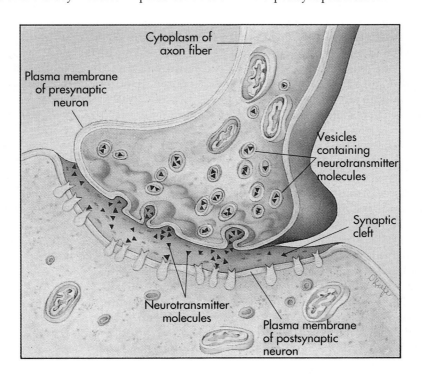

cles. Each vesicle contains a very small quantity of a chemical compound called a **neurotransmitter.** When a nerve impulse arrives at the synaptic knob, neurotransmitter molecules are released from the vesicles into the **synaptic cleft.** The synaptic cleft is the space between a synaptic knob and the plasma membrane of a *postsynaptic neuron.* It is an incredibly narrow space—only about two millionths of a centimeter in width. Identify the synaptic cleft in Figure 7-7. The plasma membrane of a postsynaptic neuron has protein molecules embedded in it opposite each synaptic knob. These serve as receptors to which neurotransmitter molecules bind. This binding can initiate an impulse in the postsynaptic neuron by opening ion channels in the postsynaptic membrane.

After impulse conduction by postsynaptic neurons is initiated, neurotransmitter activity is rapidly terminated. Either one or both of two mechanisms cause this. Some neurotransmitter molecules diffuse out of the synaptic cleft back into synaptic knobs. Other neurotransmitter molecules are metabolized into inactive compounds by specific enzymes.

Neurotransmitters are chemicals by which neurons communicate. As previously noted, at trillions of synapses in the CNS, presynaptic neurons release neurotransmitters that assist, stimulate, or inhibit postsynaptic neurons. At least 30 different compounds have been identified as neurotransmitters. They are not distributed randomly through the spinal cord and brain. Instead, specific neurotransmitters are localized in discrete groups of neurons and released in specific pathways.

For example, the substance named **acetylcholine** (as-e-til-KO-leen) is released at some of the synapses in the spinal cord and at neuromuscular (nerve-muscle) junctions. Other well-known neurotransmitters include **norepinephrine** (nor-ep-i-NEF-rin), **dopamine** (DOE-pa-meen), and **serotonin** (sair-o-TOE-nin). They belong to a group of compounds called **catecholamines** (kat-e-kol-AM-eens), which may play a role in sleep, motor function, mood, and pleasure recognition.

Two morphinelike neurotransmitters called **endorphins** (en-DOR-fins) and **enkephalins** (en-KEF-a-lins) are released at various spinal cord and brain synapses in the pain conduction pathway. These neurotransmitters inhibit conduction of pain impulses. They are natural pain killers.

Suppressing Pain During Exercise

Research shows that the release of endorphins increases during heavy exercise. Endorphins inhibit pain, so it is no wonder that pain associated with muscle fatigue decreases when endorphins are present. Normally, pain is a warning signal that calls attention to injuries or dangerous circumstances. However, it is better to inhibit severe pain if it stops us from continuing an activity that may be necessary for survival. Athletes and others who exercise heavily have even reported a unique feeling of well-being or euphoria associated with elevated endorphin levels.

Central Nervous System

The CNS, as its name implies, is centrally located. Its two major structures, the brain and spinal cord, are found along the midsagittal plane of the body (Figure 7-8). The brain is protected in the cranial cavity of the skull, and the spinal cord is surrounded in the spinal cavity by the vertebral column. In addition, the brain and spinal cord are covered by protective membranes called **meninges** (me-NIN-jeez), which are discussed in a later section of the chapter.

DIVISIONS OF THE BRAIN

The brain, one of our largest organs, consists of the following major divisions, named in ascending order beginning with most inferior part:

I **Brain stem**
  A **Medulla oblongata**
  B **Pons**
  C **Midbrain**
II **Cerebellum**
III **Diencephalon**
  A **Hypothalamus**
  B **Thalamus**
IV **Cerebrum**

Observe in Figure 7-9 the location and relative sizes of the medulla, pons, cerebellum, and cerebrum. Also identify the midbrain.

Brain Stem

The lowest part of the brain stem is the medulla oblongata. Immediately above the medulla lies the pons and above that the midbrain. Together these three structures are called the *brain stem* (Figure 7-9).

The **medulla oblongata** (ob-long-GAH-tah) is an enlarged, upward extension of the spinal cord. It lies just inside the cranial cavity above the large hole in the occipital bone called the *foramen magnum.* Like the spinal cord, the medulla consists of gray and white matter, but their arrangement differs in the two organs. In the medulla, bits of gray matter mix closely and intricately with white matter to form the *reticular formation* (*reticular* means "netlike"). In the spinal cord, gray and white matter do not intermingle; gray matter forms the interior core of the cord, and white matter surrounds it. The **pons** and **midbrain,** like the medulla, consist of white matter and scattered bits of gray matter.

All three parts of the brain stem function as two-way conduction paths. Sensory fibers conduct impulses up from the cord to other parts of the brain, and motor fibers conduct impulses down from the brain to the cord. In addition, many important reflex centers lie in the brain stem. The cardiac, respiratory, and vasomotor centers (collectively called the *vital centers*), for example, are located in the medulla. Impulses from these centers control heartbeat, respirations, and blood vessel diameter (which is important in regulating blood pressure).

Diencephalon

The **diencephalon** (dye-en-SEF-ah-lon) is a small but important part of the brain located between the midbrain below and the cerebrum above. It consists of two major structures: the hypothalamus and the thalamus.

Hypothalamus. The **hypothalamus** (hye-po-THAL-ah-mus), as its name suggests, is located below the thalamus. The posterior pituitary gland, the stalk that attaches it to the undersurface of the brain, and areas of gray matter located in the side walls of a fluid-filled space called the

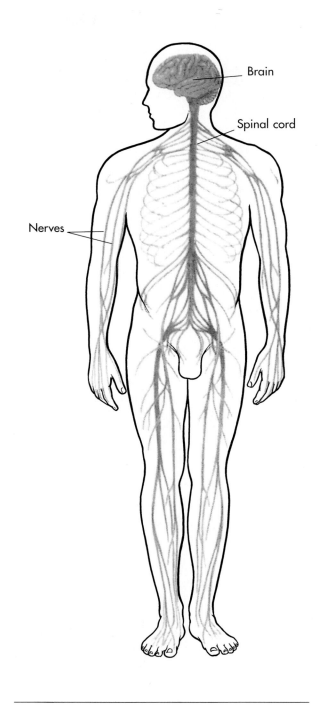

FIGURE 7–8 *The Nervous System.* The brain and spinal cord constitute the *central nervous system* (CNS), and the nerves make up the *peripheral nervous system* (PNS).

Brain

Spinal cord

Nerves

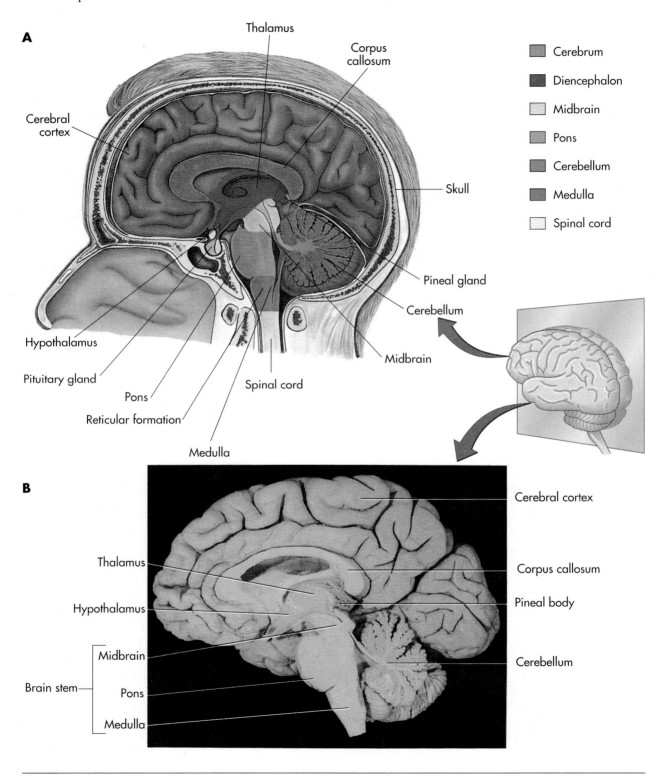

FIGURE 7–9 *Major Regions of the Central Nervous System.* **A,** Sagittal sections of the brain and spinal cord. **B,** Section of preserved brain.

A

Thalamus
Corpus callosum
Cerebral cortex
Skull
Pineal gland
Cerebellum
Midbrain
Hypothalamus
Pituitary gland
Pons
Spinal cord
Reticular formation
Medulla

Cerebrum
Diencephalon
Midbrain
Pons
Cerebellum
Medulla
Spinal cord

B

Cerebral cortex
Thalamus
Corpus callosum
Hypothalamus
Pineal body
Midbrain
Brain stem
Pons
Cerebellum
Medulla

third ventricle are extensions of the hypothalamus. Identify the pituitary gland and the hypothalamus in Figure 7-9.

The old adage, "Don't judge by appearances," applies well to appraising the importance of the hypothalamus. Measured by size, it is one of the least significant parts of the brain, but measured by its contribution to healthy survival, it is one of the most important brain structures. Impulses from neurons whose dendrites and cell bodies lie in the hypothalamus are conducted by their axons to neurons located in the spinal cord, and many of these impulses are then relayed to muscles and glands all over the body. Thus the hypothalamus exerts a major control over virtually all internal organs. Among the vital functions that it helps control are the heart beat, constriction and dilation of blood vessels, and contractions of the stomach and intestines.

Some neurons in the hypothalamus function in a surprising way; they make the hormones that the posterior pituitary gland secretes into the blood. Because one of these hormones (called *antidiuretic hormone* or *ADH*) affects the volume of urine excreted, the hypothalamus plays an essential role in maintaining the body's water balance.

Some of the neurons in the hypothalamus function as endocrine (ductless) glands. Their axons secrete chemicals called *releasing hormones* into the blood, which then carries them to the anterior pituitary gland. Releasing hormones, as their name suggests, control the release of certain anterior pituitary hormones. These in turn influence the hormone secretion of other endocrine glands. Thus the hypothalamus indirectly helps control the functioning of every cell in the body.

The hypothalamus is a crucial part of the mechanism for maintaining body temperature. Therefore marked elevation in body temperature in the absence of disease frequently characterizes injuries or other abnormalities of the hypothalamus. In addition, this important center is involved in functions such as the regulation of water balance, sleep cycles, and the control of appetite and many emotions involved in pleasure, fear, anger, sexual arousal, and pain.

Thalamus. Just above the hypothalamus is a dumbbell-shaped section of gray matter called the **thalamus** (THAL-ah-mus). Each enlarged end of the dumbbell lies in a lateral wall of the third ventricle. The thin center section of the thalamus passes from left to right through the third ventricle. The thalamus is composed chiefly of dendrites and cell bodies of neurons that have axons extending up toward the sensory areas of the cerebrum. The thalamus performs the following functions:

1. It helps produce sensations. Its neurons relay impulses to the cerebral cortex from the sense organs of the body.
2. It associates sensations with emotions. Almost all sensations are accompanied by a feeling of some degree of pleasantness or unpleasantness. The way that these pleasant and unpleasant feelings are produced is unknown except that they seem to be associated with the arrival of sensory impulses in the thalamus.
3. It plays a part in the so-called arousal or alerting mechanism.

Cerebellum

Structure. Look at Figure 7-9 to find the location, appearance, and size of the cerebellum. The cerebellum is the second largest part of the human brain. It lies under the occipital lobe of the cerebrum. In the cerebellum, gray matter composes the outer layer, and white matter composes the bulk of the interior.

Function. Most of our knowledge about cerebellar functions has come from observing patients who have some sort of disease of the cerebellum and from animals who have had the cerebellum removed. From such observations, we know that the cerebellum plays an essential part in the production of normal movements. Perhaps a few examples will make this clear. A patient who has a tumor of the cerebellum frequently loses his balance and topples over; he may reel like a drunken man when he walks. He cannot coordinate his muscles normally. He may complain, for instance, that he is clumsy about everything he does—that he cannot even drive a nail or draw a straight line. With the loss of normal cerebellar functioning, he has lost the ability to make precise movements. The general functions of the cerebellum, then, are to produce smooth coordinated movements, maintain equilibrium, and sustain normal postures.

Cerebrum

The **cerebrum** (SAIR-e-brum) is the largest and uppermost part of the brain. If you were to look at the outer surface of the cerebrum, the first features you would notice might be its many ridges and grooves. The ridges are called *convolutions* or *gyri* (JYE-rye), and the grooves are called *sulci* (SUL-kye). The deepest sulci are called *fissures;* the longitudinal fissure divides the cerebrum into right and left halves or hemispheres. These halves are almost separate structures except for their lower midportions, which are connected by a structure called the **corpus callosum** (COR-pus kal-LO-sum) (Figure 7-9). Two deep sulci subdivide each cerebral hemisphere into four major lobes and each lobe into numerous convolutions. The lobes are named for the bones that lie over them: the frontal lobe, the parietal lobe, the temporal lobe, and the occipital lobe. Identify these in Figure 7-10, *A.*

A thin layer of gray matter, made up of neuron dendrites and cell bodies, composes the surface of the cerebrum. Its name is the *cerebral cortex.* White matter, made up of bundles of nerve fibers (tracts), composes most of the interior of the cerebrum. Within this white matter, however, are a few islands of gray matter known as the **cerebral nuclei** or *basal ganglia,* whose functioning is essential for producing automatic movements and postures. Parkinson's disease is a disease of the cerebral nuclei. Because shaking or tremors are common symptoms of Parkinson's disease, it is also called "shaking palsy."

What functions does the cerebrum perform? This is a hard question to answer briefly because the neurons of the cerebrum do not function alone. They function with many other neurons in many other parts of the brain and in the spinal cord. Neurons of these structures continually bring impulses to cerebral neurons and continually transmit impulses away from them. If all other neurons were functioning normally and only cerebral neurons were not functioning, here are some of the things that you could not do. You could not think or use your will. You could not remember anything that has ever happened to you. You could not decide to make the smallest movement, nor could you make it. You would not see or hear. You could not experience any of the sensations that make life so rich and varied. Noth-

ing would anger or frighten you, and nothing would bring you joy or sorrow. You would, in short, be unconscious. These terms sum up cerebral functions: consciousness, thinking, memory, sensations, emotions, and willed movements. Figure 7-10, *B,* shows the areas of the cerebral cortex essential for willed movements, general sensations, vision, hearing, and normal speech.

Injury or disease can destroy neurons. A common example is the destruction of neurons of the motor area of the cerebrum that results from a **cerebrovascular accident (CVA),** which is a hemorrhage from or cessation of blood flow through cerebral blood vessels. When this happens, the victim can no longer voluntarily move the parts of his body on the side opposite to the side on which the CVA occurred. In nontechnical language, he has suffered a stroke. Note in Figure 7-10, *B,* the location of the motor area in the frontal lobe of the cerebrum.

It is important to understand that very specific areas of the cortex have very specific functions. For example, the temporal lobe's auditory areas interpret incoming nervous signals from the ear as very specific sounds. The visual area of the cortex in the occipital lobe helps you identify and understand specific images. Localized areas of the cortex are directly related to specific functions, as shown in Figure 7-10, *B.* This explains the very specific symptoms associated with an injury to localized areas of the cerebral cortex after a stroke or traumatic injury to the head. Table 7-1 summarizes the major components of the brain and their main functions.

SPINAL CORD

Structure

If you are of average height, your spinal cord is about 17 or 18 inches long (Figure 7-11). It lies inside the spinal column in the spinal cavity and extends from the occipital bone down to the bottom of the first lumbar vertebra. Place your hands on your hips, and they will line up with your fourth lumbar vertebra. Your spinal cord ends just above this level.

Look now at Figure 7-12. Notice the H-shaped core of the spinal cord. It consists of gray matter and so is composed mainly of dendrites and cell bodies of neurons. Columns of white matter form the outer portion of the spinal cord, and bundles

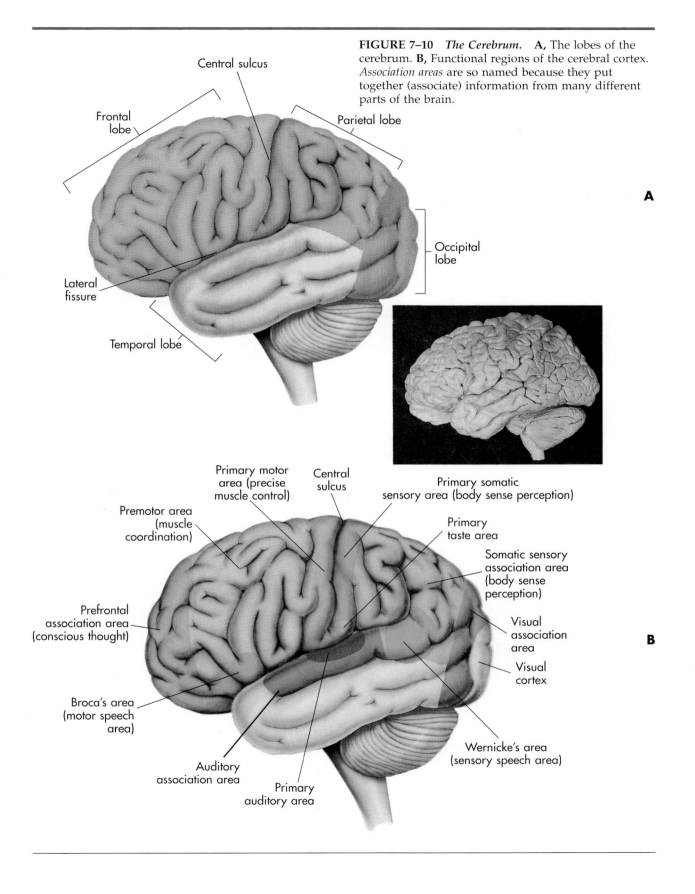

FIGURE 7–10 *The Cerebrum.* **A,** The lobes of the cerebrum. **B,** Functional regions of the cerebral cortex. *Association areas* are so named because they put together (associate) information from many different parts of the brain.

Central sulcus

Frontal lobe

Parietal lobe

Occipital lobe

Lateral fissure

Temporal lobe

A

Primary motor area (precise muscle control)

Central sulcus

Primary somatic sensory area (body sense perception)

Premotor area (muscle coordination)

Primary taste area

Somatic sensory association area (body sense perception)

Prefrontal association area (conscious thought)

Visual association area

Visual cortex

Broca's area (motor speech area)

Auditory association area

Primary auditory area

Wernicke's area (sensory speech area)

B

FIGURE 7–11 *Spinal Cord and Spinal Nerves.* Inset is a dissection of the cervical segment of the spinal cord showing emerging cervical nerves. The spinal cord is viewed from behind (posterior aspect).

of myelinated nerve fibers—the **spinal tracts**—make up the white columns.

Spinal cord tracts provide two-way conduction paths to and from the brain. **Ascending tracts** conduct impulses up the cord to the brain. **Descending tracts** conduct impulses down the cord from the brain. Tracts are functional organizations in that all axons composing one tract serve one general function. For instance, fibers of the spinothalamic tracts serve a sensory function. They transmit impulses that produce sensations of crude touch, pain, and temperature. Other

TABLE 7–1 Functions of Major Divisions of the Brain

BRAIN AREA	FUNCTION
Brain stem	
Medulla oblongata	Two-way conduction pathway between the spinal cord and higher brain centers; cardiac, respiratory, and vasomotor control center
Pons	Two-way conduction pathway between areas of the brain and other regions of the body; influences respiration
Midbrain	Two-way conduction pathway; relay for visual and auditory impulses
Diencephalon	
Hypothalamus	Regulation of body temperature, water balance, sleep-cycle control, appetite, and sexual arousal
Thalamus	Sensory relay station from various body areas to cerebral cortex; emotions and alerting or arousal mechanisms
Cerebellum	Muscle coordination; maintenance of equilibrium and posture
Cerebrum	Sensory perception, emotions, willed movements, consciousness, and memory

FIGURE 7–12 *Spinal Cord.* Cross section of the spinal cord showing the horns, pathways (nerve tracts), and roots.

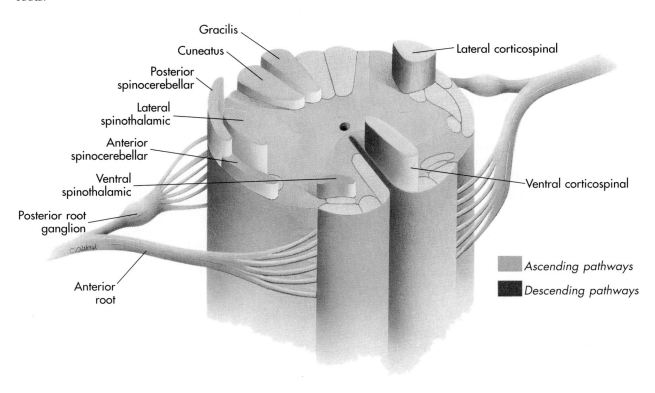

ascending tracts shown in Figure 7-12 include the gracilis and cuneatus tracts, which transmit sensations of touch and pressure up to the brain, and the anterior and posterior spinocerebellar tracts, which transmit information about muscle length to the cerebellum. Descending tracts include the lateral and ventral corticospinal tracts, which transmit impulses controlling many voluntary movements.

Functions

To try to understand spinal cord functions, think about a hotel telephone switching system. Suppose a guest in Room 108 calls the switching system and keys in the extension number for Room 520, and in a second or so, someone in that room answers. Very briefly, three events took place: a message traveled into the switching system, the system routed the message along the proper path, and the message traveled out from the switching system toward Room 520. The telephone switching system provided the network of connections that made possible the completion of the call. We might say that the switching system transferred the incoming call to an outgoing line. The spinal cord functions similarly. It contains the centers for thousands and thousands of reflex arcs. Look back at Figure 7-5. The interneuron shown there is an example of a spinal cord reflex center. It switches or transfers incoming sensory impulses to outgoing motor impulses, thereby making it possible for a reflex to occur. Reflexes that result from conduction over arcs whose centers lie in the spinal cord are called *spinal cord reflexes*. Two common kinds of spinal cord reflexes are withdrawal and jerk reflexes. An example of a withdrawal reflex is pulling one's hand away from a hot surface. The familiar knee jerk is an example of a jerk reflex.

In addition to functioning as the primary reflex center of the body, the spinal cord tracts, as previously noted, carry impulses to and from the brain. Sensory impulses travel up to the brain in ascending tracts, and motor impulses travel down from the brain in descending tracts. Therefore if an injury cuts the cord all the way across, impulses can no longer travel to the brain from any part of the body located below the injury, nor can they travel from the brain down to these parts. In short, this kind of spinal cord injury produces a loss of sensation, which is called **anesthesia** (an-es-THEE-zee-ah), and a loss of the ability to make voluntary movements, which is called **paralysis** (pah-RAL-i-sis).

COVERINGS AND FLUID SPACES OF THE BRAIN AND SPINAL CORD

Nervous tissue is not a sturdy tissue. Even moderate pressure can kill nerve cells, so nature safeguards the chief organs made of this tissue—the spinal cord and the brain—by surrounding them with a tough, fluid-containing membrane called the **meninges** (me-NIN-jeez). The meninges are then surrounded by bone. The spinal meninges

FIGURE 7–13   *Spinal Cord.* The meninges, spinal nerves, and sympathetic trunk are visible.

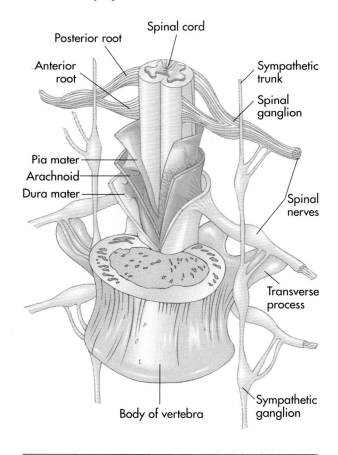

form a tubelike covering around the spinal cord and line the bony vertebral foramen of the vertebrae that surround the cord. Look at Figure 7-13, and you can identify the three layers of the spinal meninges. They are the **dura mater** (DOO-rah MA-ter), which is the tough outer layer that lines the vertebral canal, the **pia** (PEE-ah) **mater,** which is the innermost membrane covering the spinal cord itself, and the **arachnoid** (ah-RAK-noyd), which is the membrane between the dura and the pia mater. The arachnoid resembles a cobweb with fluid in its spaces. The word *arachnoid* means "cobweblike." It comes from *arachne,* the Greek word for spider. Arachne is the name of the girl who was changed into a spider by Athena because she boasted of the fineness of her weaving—at least, so an ancient Greek myth tells us.

The meninges that form the protective covering around the spinal cord also extend up and around the brain to enclose it completely (Figure 7-13). Fluid fills the subarachnoid spaces between the pia mater and arachnoid in the brain and spinal cord. This fluid is called **cerebrospinal fluid (CSF).**

Cerebrospinal (ser-e-bro-SPI-nal) **fluid** also fills spaces in the brain called cerebral **ventricles.** In Figure 7-14, you can see the irregular shapes of the ventricles of the brain. These illustrations can also help you visualize the location of the ventricles if you remember that these large spaces lie deep inside the brain and that there are two lateral ventricles. One lies inside the right half of the cerebrum (the largest part of the human brain), and the other lies inside the left half.

FIGURE 7–14 *Fluid Spaces of the Brain.* The large figure shows the ventricles highlighted within the brain in a left lateral view. The small figure shows the ventricles from above.

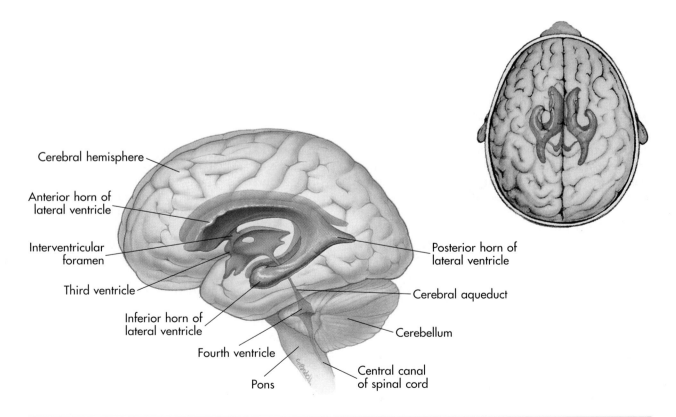

Cerebral hemisphere

Anterior horn of lateral ventricle

Interventricular foramen

Third ventricle

Inferior horn of lateral ventricle

Fourth ventricle

Pons

Posterior horn of lateral ventricle

Cerebral aqueduct

Cerebellum

Central canal of spinal cord

Lumbar Puncture

The extension of the meninges beyond the spinal cord is convenient for performing lumbar punctures without danger of injuring the spinal cord. A **lumbar puncture** is the withdrawal of some CSF from the subarachnoid space in the lumbar region of the spinal cord. The physician inserts a needle just above or below the fourth lumbar vertebra, knowing that the spinal cord ends an inch or more above the level. The fourth lumbar vertebra can be easily located because it lies on a line with the iliac crest. Placing a patient on his side and arching his back by drawing the knees and chest together separates the vertebrae sufficiently to introduce the needle. Lumbar punctures are often performed to withdraw CSF for analysis or to reduce pressure caused by swelling of the brain or spinal cord after injury or disease.

CSF is one of the body's circulating fluids. It forms continually from fluid filtering out of the blood in a network of brain capillaries known as the **choroid plexus** (KO-royd PLEK-sus) and into the ventricles. CSF seeps from the lateral ventricles into the third ventricle and flows down through the cerebral aqueduct (find this in Figure 7-14 and 7-15) into the fourth ventricle. Most of the CSF moves from the fourth ventricle into the subarachnoid space near the cerebellum. Some of it moves into the small, tubelike central canal of the cord and then out into the subarachnoid spaces. Then it moves leisurely down and around the cord and up and around the brain (in the subarachnoid spaces of their meninges) and returns to the blood (in the veins of the brain).

Remembering that this fluid forms continually from blood, circulates, and is resorbed into blood can be useful. It can help you understand certain abnormalities. Suppose a person has a brain tumor that presses on the cerebral aqueduct. This blocks the way for the return of CSF to the blood. Because the fluid continues to form but cannot drain away, it accumulates in the ventricles or in the meninges. Other conditions can cause an accumulation of CSF in the ventricles. An example is **hydrocephalus** (hye-dro-SEF-ah-lus) or "water on the brain." One form of treatment involves surgical placement of a hollow tube or catheter through the blocked channel so that CSF can drain into another location in the body.

Peripheral Nervous System

The nerves connecting the brain and spinal cord to other parts of the body constitute the **peripheral nervous system (PNS)**. This system includes **cranial** and **spinal nerves** that connect the brain and spinal cord, respectively, to peripheral structures such as the skin surface and the skeletal muscles. In addition, other structures in the **autonomic nervous system** or **ANS** are considered part of the PNS. These connect the brain and spinal cord to various glands in the body and to the cardiac and smooth muscle in the thorax and abdomen.

CRANIAL NERVES

Twelve pairs of cranial nerves are attached to the undersurface of the brain, mostly from the brain stem. Figure 7-16 shows the attachments of these nerves. Their fibers conduct impulses between the brain and structures in the head and neck and in the thoracic and abdominal cavities. For instance, the second cranial nerve (optic nerve) conducts impulses from the eye to the brain, where these impulses produce vision. The third cranial nerve (oculomotor nerve) conducts impulses from the brain to muscles in the eye, where they cause contractions that move the eye. The tenth cranial nerve (vagus nerve) conducts impulses between the medulla oblongata and structures in the neck and thoracic and abdominal cavities. The names of each cranial nerve and a brief description of their functions are listed in Table 7-2.

SPINAL NERVES
Structure

Thirty-one pairs of nerves are attached to the spinal cord in the following order: eight pairs are attached

FIGURE 7–15 *Flow of Cerebrospinal Fluid.* The fluid produced by filtration of blood by the choroid plexus of each ventricle flows inferiorly through the lateral ventricles, interventricular foramen, third ventricle, cerebral aqueduct, fourth ventricle, and subarachnoid space and to the blood.

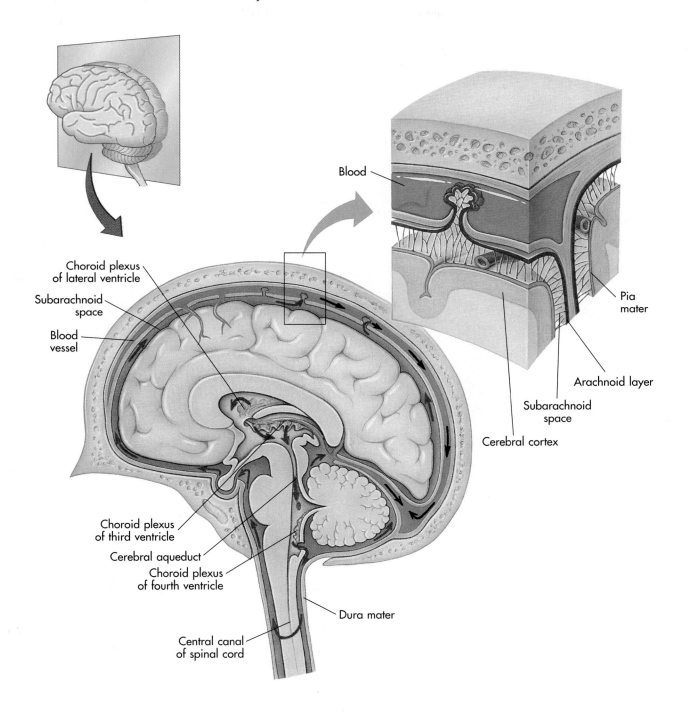

FIGURE 7–16 *Cranial Nerves.* View of the undersurface of the brain shows attachments of the cranial nerves.

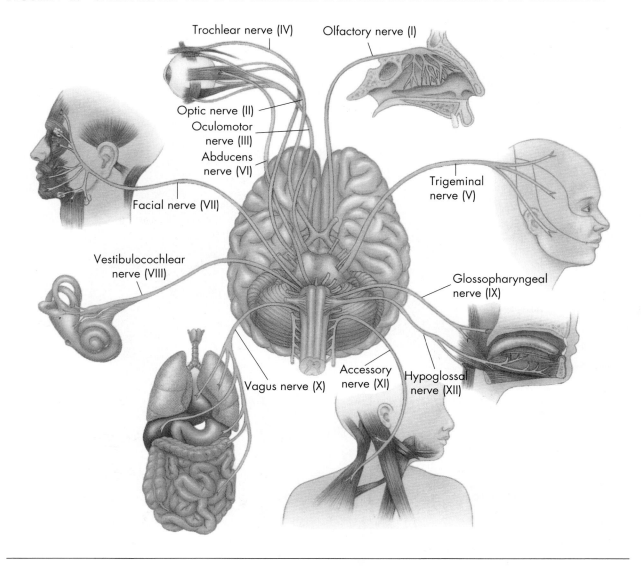

Trochlear nerve (IV)

Olfactory nerve (I)

Optic nerve (II)

Oculomotor nerve (III)

Abducens nerve (VI)

Facial nerve (VII)

Trigeminal nerve (V)

Vestibulocochlear nerve (VIII)

Glossopharyngeal nerve (IX)

Vagus nerve (X)

Accessory nerve (XI)

Hypoglossal nerve (XII)

to the cervical segments, twelve pairs are attached to the thoracic segments, five pairs are attached to the lumbar segments, five pairs are attached to the sacrospinal segments, and one pair is attached to the coccygeal segment (Figure 7-11). Unlike cranial nerves, spinal nerves have no special names; instead, a letter and number identify each one. C1, for example, indicates the pair of spinal nerves attached to the first segment of the cervical part of the cord, and T8 indicates nerves attached to the eighth segment of the thoracic part of the spinal cord. In Figure 7-11 the cervical area of the spine has been dissected to show the emerging spinal nerves in that area. After spinal nerves exit from the spinal cord, they branch to form the many peripheral nerves of the trunk and limbs. Sometimes, nerve fibers from several spinal nerves are reorganized to form a single peripheral nerve.

TABLE 7–2   Cranial Nerves

NERVE*	CONDUCT IMPULSES	FUNCTIONS
I Olfactory	From nose to brain	Sense of smell
II Optic	From eye to brain	Vision
III Oculomotor	From brain to eye muscles	Eye movements
IV Trochlear	From brain to external eye muscles	Eye movements
V Trigeminal	From skin and mucous membrane of head and from teeth to brain; also from brain to chewing muscles	Sensations of face, scalp, and teeth, chewing movements
VI Abducens	From brain to external eye muscles	Turning eyes outward
VII Facial	From taste buds of tongue to brain; from brain to face muscles	Sense of taste; contraction of muscles of facial expression
VIII Vestibulocochlear	From ear to brain	Hearing; sense of balance
IX Glossopharyngeal	From throat and taste buds of tongue to brain; also from brain to throat muscles and salivary glands	Sensations of throat, taste, swallowing movements, secretion of saliva
X Vagus	From throat, larynx, and organs in thoracic and abdominal cavities to brain; also from brain to muscles of throat and to organs in thoracic and abdominal cavities	Sensations of throat and larynx and of thoracic and abdominal organs; swallowing, voice production, slowing of heartbeat, acceleration of peristalsis (gut movements)
XI Accessory	From brain to certain shoulder and neck muscles	Shoulder movements; turning movements of head
XII Hypoglossal	From brain to muscles of tongue	Tongue movements

*The first letter of the words in the following sentence are the first letters of the names of the cranial nerves, in the correct order. Many anatomy students find that using this sentence, or one like it, helps in memorizing the names and numbers of the cranial nerves. It is "**O**n **O**ld **O**lympus' **T**iny **T**ops, **A** **F**riendly **V**iking **G**rew **V**ines **A**nd **H**ops."

This reorganization can be seen as a network of intersecting or "braided" branches called a **plexus.** Figure 7-11 shows several plexuses.

Functions

Spinal nerves conduct impulses between the spinal cord and the parts of the body not supplied by cranial nerves. The spinal nerves shown in Figure 7-11 contain, as do all spinal nerves, sensory and motor fibers. Spinal nerves therefore function to make possible sensations and movements. A disease or injury that prevents conduction by a spinal nerve thus results in a loss of feeling and a loss of movement in the part supplied by that nerve.

Detailed mapping of the skin's surface reveals a close relationship between the source on the spinal cord of each spinal nerve and the part of the body that it innervates (Figure 7-17). Knowledge of the segmental arrangement of spinal nerves is useful to physicians. For instance, a neurologist can identify the site of a spinal cord or nerve abnormality from the area of the body that is insensitive to a pinprick. Skin surface areas that are supplied by a single spinal nerve are called **dermatomes** (DER-mah-tomes). A dermatome "map" of the body is shown in Figure 7-17.

 Autonomic Nervous System

The **autonomic nervous system (ANS)** consists of certain motor neurons that conduct impulses from

FIGURE 7–17   *Dermatomes.* Segmental dermatome distribution of spinal nerves to the front, back, and side of the body. *C*, Cervical segments; *T*, thoracic segments; *L*, lumbar segments; *S*, sacral segments; *CX*, coccygeal segment.

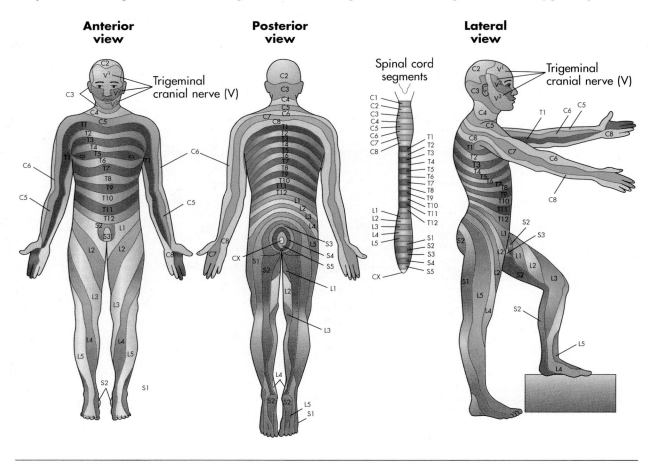

the spinal cord or brain stem to the following kinds of tissues:

1. Cardiac muscle tissue
2. Smooth muscle tissue
3. Glandular epithelial tissue

The ANS consists of the parts of the nervous system that regulate involuntary functions (for example, the heartbeat, contractions of the stomach and intestines, and secretions by glands). On the other hand, motor nerves that control the voluntary actions of skeletal muscles are sometimes called the *somatic nervous system.*

The autonomic nervous system consists of two divisions called the **sympathetic nervous system** and the **parasympathetic nervous system** (Figure 7-18).

FUNCTIONAL ANATOMY

Autonomic neurons are the motor neurons that make up the ANS. The dendrites and cell bodies of some autonomic neurons are located in the gray matter of the spinal cord or brain stem. Their axons extend from these structures and terminate in ganglia. These autonomic neurons are called **preganglionic neurons** because they conduct

FIGURE 7–18   *Innervation of the Major Target Organs by the Autonomic Nervous System.* The sympathetic pathways are highlighted with red, and the parasympathetic pathways are highlighted with blue.

impulses between the spinal cord and a ganglion. In the ganglia the axon endings of preganglionic neurons synapse with the dendrites or cell bodies of postganglionic neurons. **Postganglionic neurons,** as their name suggests, conduct impulses from a ganglion to cardiac muscle, smooth muscle, or glandular epithelial tissue.

Autonomic or **visceral effectors** are the tissues to which autonomic neurons conduct impulses. Specifically, visceral effectors are cardiac muscle that makes up the wall of the heart, smooth muscle that partially makes up the walls of blood vessels and other hollow internal organs, and glandular epithelial tissue that makes up the secreting part of glands.

AUTONOMIC CONDUCTION PATHS

Conduction paths to visceral and somatic effectors from the CNS (spinal cord or brain stem) differ somewhat. Autonomic paths to visceral effectors, as the right side of Figure 7-19 shows, consist of two-neuron relays. Impulses travel over preganglionic neurons from the spinal cord or brain stem to autonomic ganglia. There, they are relayed across synapses to postganglionic neurons, which then conduct the impulses from the ganglia to visceral effectors. Compare the autonomic conduction path with the somatic conduction path illustrated on the left side of Figure 7-19. Somatic motor neurons, like the ones shown here, conduct all the way from the spinal cord or brain

FIGURE 7–19 *Autonomic Conduction Paths.* The left side of the diagram shows that one somatic motor neuron conducts impulses all the way from the spinal cord to a somatic effector. Conduction from the spinal cord to any visceral effector, however, requires a relay of at least two autonomic motor neurons—a preganglionic and a postganglionic neuron, shown on the right side of the diagram.

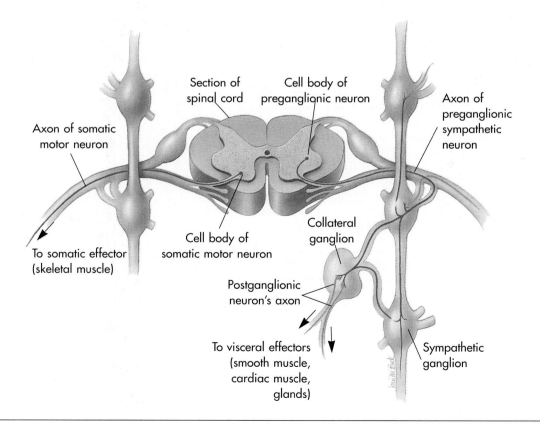

stem to somatic effectors with no intervening synapses.

SYMPATHETIC NERVOUS SYSTEM

Structure

Sympathetic preganglionic neurons have dendrites and cell bodies in the gray matter of the thoracic and upper lumbar segments of the spinal cord. The sympathetic system has also been referred to as the *thoracolumbar system.* Look now at the right side of Figure 7-19. Follow the course of the axon of the sympathetic preganglionic neuron shown there. It leaves the spinal cord in the anterior (ventral) root of a spinal nerve. It next enters the spinal nerve but soon leaves it to extend to and through a sympathetic ganglion and terminate in a collateral ganglion. There, it synapses with several postganglionic neurons whose axons extend to terminate in visceral effectors. Also shown in Figure 7-19, branches of the preganglionic axon may ascend or descend to terminate in ganglia above and below their point of origin. All sympathetic preganglionic axons therefore synapse with many postganglionic neurons, and these frequently terminate in widely separated organs. Hence sympathetic responses are usually wide-spread, involving many organs and not just one.

Sympathetic postganglionic neurons have dendrites and cell bodies in sympathetic ganglia. Sympathetic ganglia are located in front of and at each side of the spinal column. Because short fibers extend between the sympathetic ganglia, they look a little like two chains of beads and are often referred to as the *sympathetic chain ganglia.* Axons of sympathetic postganglionic neurons travel in spinal nerves to blood vessels, sweat glands, and arrector hair muscles all over the body. Separate autonomic nerves distribute many sympathetic postganglionic axons to various internal organs.

Functions of the Sympathetic Nervous System

The sympathetic nervous system functions as an emergency system. Impulses over sympathetic fibers take control of many internal organs when we exercise strenuously and when strong emotions—anger, fear, hate, anxiety—are elicited. In short, when we must cope with stress of any kind, sympathetic impulses increase to many visceral effectors and rapidly produce widespread changes within our bodies. The middle column of Table 7-3 indicates many sympathetic responses. The heart beats faster. Most blood vessels constrict, causing blood pressure to increase. Blood vessels in skeletal muscles dilate, supplying the muscles with more blood. Sweat glands and adrenal glands secrete more abundantly. Salivary and other digestive glands secrete more sparingly. Digestive tract contractions (peristalsis) become sluggish, hampering digestion. Together, these sympathetic responses make us ready for strenuous muscular work, or they prepare us for *fight or flight.* The group of changes induced by sympathetic control is known as the **fight-or-flight response.**

PARASYMPATHETIC NERVOUS SYSTEM

Structure

The dendrites and cell bodies of parasympathetic preganglionic neurons are located in the gray matter of the brain stem and the sacral segments of the spinal cord. The parasympathetic system has also been referred to as the *craniosacral system.* The preganglionic parasympathetic axons extend some distance before terminating in the parasympathetic ganglia located in the head and in the thoracic and abdominal cavities close to the visceral effectors that they control. The dendrites and cell bodies of parasympathetic postganglionic neurons lie in these outlying parasympathetic ganglia, and their short axons extend into the nearby structures. Therefore each parasympathetic preganglionic neuron synapses only with postganglionic neurons to a single effector. For this reason, parasympathetic stimulation frequently involves response by only one organ. This is not true of sympathetic responses; as noted, sympathetic stimulation usually results in responses by numerous organs.

Functions of the Parasympathetic Nervous System

The parasympathetic system dominates control of many visceral effectors under normal, everyday conditions. Impulses over parasympathetic fibers, for example, tend to slow heartbeat, increase peristalsis, and increase secretion of digestive juices and insulin (Table 7-3).

TABLE 7–3   Autonomic Functions

VISCERAL EFFECTORS	SYMPATHETIC CONTROL	PARASYMPATHETIC CONTROL
Heart muscle	Accelerates heartbeat	Slows heartbeat
Smooth muscle		
Of most blood vessels	Constricts blood vessels	None
Of blood vessels in skeletal muscles	Dilates blood vessels	None
Of the digestive tract	Decreases peristalsis; inhibits defecation	Increases peristalsis
Of the anal sphincter	Stimulates—closes sphincter	Inhibits—opens sphincter for defecation
Of the urinary bladder	Inhibits—relaxes bladder	Stimulates—contracts bladder
Of the urinary sphincters	Stimulates—closes sphincter	Inhibits—opens sphincter for urination
Of the eye		
Iris	Stimulates radial fibers—dilation of pupil	Stimulates circular fibers—constriction of pupil
Ciliary	Inhibits—accomodation for far vision (flattening of lens)	Stimulates—accomodation for near vision (bulging of lens)
Of hairs (pilomotor muscles)	Stimulates—"goose pimples"	No parasympathetic fibers
Glands		
Adrenal medulla	Increases epinephrine secretion	None
Sweat glands	Increases sweat secretion	None
Digestive glands	Decreases secretion of digestive juices	Increases secretion of digestive juices

AUTONOMIC NEUROTRANSMITTERS

Turn your attention now to Figure 7-20. It reveals information about autonomic neurotransmitters, the chemical compounds released from the axon terminals of autonomic neurons. Observe that three of the axons shown in Figure 7-20—the sympathetic preganglionic axon, the parasympathetic preganglionic axon, and the parasympathetic postganglionic axon—release acetylcholine. These axons are therefore classified as **cholinergic fibers.** Only one type of autonomic axon releases the neurotransmitter norepinephrine (noradrenaline). This is the axon of a sympathetic postganglionic neuron, and such neurons are classified as **adrenergic fibers.** The fact that each division of the ANS signals its effectors with a different neurotransmitter explains how an organ can tell which division is stimulating it. The heart, for example, responds to acetylcholine from the parasympathetic division by slowing down. The presence of norepinephrine at the heart, on the other hand, is a signal from the sympathetic division, and the response is an increase in heart activity.

AUTONOMIC NERVOUS SYSTEM AS A WHOLE

The function of the **autonomic nervous system** is to regulate the body's automatic, involuntary functions in ways that maintain or quickly restore homeostasis. Many internal organs are doubly innervated by the ANS. In other words, they receive fibers from parasympathetic and sympathetic divisions. Parasympathetic and sympathetic impulses continually bombard them

FIGURE 7–20 *Autonomic Neurotransmitters.* Three of the four fiber types are cholinergic, secreting the neuro-transmitter acetylcholine (Ach) into a synapse. Only the sympathetic postganglionic fiber is adrenergic, secreting norepinephrine (NE) into a synapse.

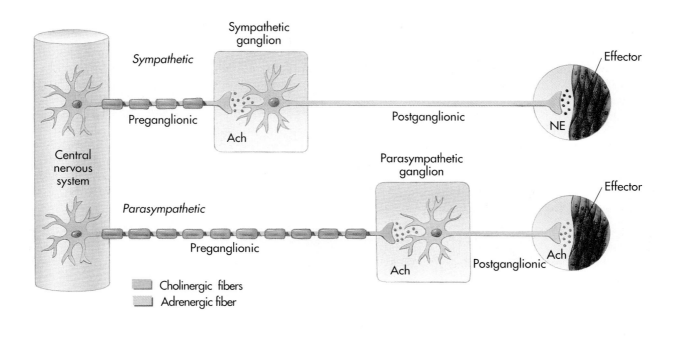

and, as Table 7-3 indicates, influence their function in opposite or antagonistic ways. For example, the heart continually receives sympathetic impulses that make it beat faster and parasympathetic impulses that slow it down. The ratio between these two antagonistic forces, determined by the ratio between the two different autonomic neurotransmitters, determines the actual heart rate.

The name *autonomic nervous system* is something of a misnomer. It seems to imply that this part of the nervous system is independent from other parts. This is not true. Dendrites and cell bodies of preganglionic neurons are located, as observed, in the spinal cord and brain stem. They

are continually influenced directly or indirectly by impulses from neurons located above them, notably by some in the hypothalamus and in the parts of the cerebral cortex called the *emotional brain*. Through conduction paths from these areas, emotions can produce widespread changes in the automatic functions of our bodies, in cardiac and smooth muscle contractions, and in secretion by glands. Anger and fear, for example, lead to increased sympathetic activity and the fight-or-flight response. According to some physiologists, the altered state of consciousness known as *meditation* leads to decreased sympathetic activity and a group of changes opposite to those of the fight-or-flight response.

Herpes Zoster or Shingles

Herpes zoster or **shingles** is a unique viral infection that almost always affects the skin of a single dermatome. It is caused by a varicella zoster virus of chickenpox. About 3% of the population will suffer from shingles at some time in their lives. In most cases the disease results from reactivation of the varicella virus. The virus probably traveled through a cutaneous nerve and remained dormant in a dorsal root ganglion for years after an episode of the chickenpox. If the body's immunological protective mechanism becomes diminished in the elderly after stress, or in individuals undergoing radiation therapy or taking immunosuppressive drugs, the virus may reacti-

vate. If this occurs, the virus travels over the sensory nerve to the skin of a single dermatome. The result is a painful eruption of red, swollen plaques or vesicles that eventually rupture and crust before clearing in 2 to 3 weeks. In severe cases, extensive inflammation, hemorrhagic blisters, and secondary bacterial infection may lead to permanent scarring. In most cases, the eruption of vesicles is preceded by 4 to 5 days of preeruptive pain, burning, and itching in the affected dermatome. Unfortunately, an attack of herpes zoster does not confer lasting immunity. Many individuals suffer three or four episodes in a lifetime.

ORGANS AND DIVISIONS OF THE NERVOUS SYSTEM (Figure 7-1)

A Central nervous system (CNS)—brain and spinal cord

B Peripheral nervous system (PNS)—all nerves

C Autonomic nervous system (ANS)

CELLS OF THE NERVOUS SYSTEM

A Neurons

  1 Consist of three main parts—dendrites: conduct impulses to cell body of neuron; cell body of neuron; and axon: conducts impulses away from cell body of neuron (Figure 7-2)

  2 Neurons classified according to function—sensory: conduct impulses to the spinal cord and brain; motor: conduct impulses away from brain and spinal cord to muscles and glands; and interneurons: conduct impulses from sensory neurons to motor neurons

B Glia—three main types of connective tissue cells of the CNS (Figure 7-3)

  1 Astrocytes—star-shaped cells that anchor small blood vessels to neurons

  2 Microglia—small cells that move in inflamed brain tissue carrying on phagocytosis

  3 Oligodendroglia—form myelin sheaths on axons in the CNS

NERVES (Figure 7-4)

A Nerve—bundle of peripheral axons

  1 Tract—bundle of central axons

  2 White matter—tissue composed primarily of myelinated axons (nerves or tracts)

  3 Gray matter—tissue composed primarily of cell bodies and unmyelinated fibers

B Nerve coverings—fibrous connective tissue

  1 Endoneurium—surrounds individual fibers within a nerve

  2 Perineurium—surrounds a group (fascicle) of nerve fibers

  3 Epineurium—surrounds the entire nerve

REFLEX ARCS

A Nerve impulses are conducted from receptors to effectors over neuron pathways or reflex arcs; conduction by a reflex arc results in a reflex (that is, contraction by a muscle or secretion by a gland)

B Simplest reflex arcs are two-neuron arcs—consist of sensory neurons synapsing in the spinal cord with motor neurons; three-neuron arcs consist of sensory neurons synapsing in the spinal cord with interneurons that synapse with motor neurons (Figure 7-5)

NERVE IMPULSES

A Definition—self-propagating wave of electrical disturbance that travels along the surface of a neuron membrane

B Mechanism

  1 A stimulus triggers the opening of Na^+ channels in the plasma membrane of the neuron

  2 Inward movement of positive sodium ions leaves a slight excess of negative ions outside at a stimulated point; marks the beginning of a nerve impulse

THE SYNAPSE

A Definition—chemical compounds released from axon terminals (of a presynaptic neuron) into a synaptic cleft

B Neurotransmitters bind to specific receptor molecules in the membrane of a postsynaptic neuron, opening ion channels and thereby stimulating impulse conduction by the membrane

C Names of neurotransmitters—acetylcholine, catecholamines (norepinephrine, dopamine, and serotonin), and other compounds

CENTRAL NERVOUS SYSTEM

A Divisions of the brain (Figure 7-9 and Table 7-1)

  1 Brain stem

    a Consists of three parts of brain; named in ascending order, they are the medulla oblongata, pons, and midbrain

    b Structure—white matter with bits of gray matter scattered through it

    c Function—gray matter in the brain stem functions as reflex centers (for example, for heartbeat, respirations, and blood vessel diameter); sensory tracts in the brain stem conduct impulses to the higher parts of the brain; motor tracts conduct from the higher parts of the brain to the spinal cord

  2 Diencephalon

    a Structure and function of the hypothalamus

      (1) Consists mainly of the posterior pituitary gland, pituitary stalk, and gray matter

      (2) Acts as the major center for controlling the ANS; therefore helps control the functioning of most internal organs

      (3) Controls hormone secretion by anterior and posterior pituitary glands; therefore indirectly helps control hormone secretion by most other endocrine glands

      (4) Contains centers for controlling appetite, wakefulness, pleasure, etc.

b Structure and function of the thalamus
  (1) Dumbbell-shaped mass of gray matter in each cerebral hemisphere
  (2) Relays sensory impulses to cerebral cortex sensory areas
  (3) In some way produces the emotions of pleasantness or unpleasantness associated with sensations
3 Cerebellum
  a Second largest part of the human brain
  b Helps control muscle contractions to produce coordinated movements so that we can maintain balance, move smoothly, and sustain normal postures
4 Cerebrum
  a Largest part of the human brain
  b Outer layer of gray matter is the cerebral cortex; made up of lobes; composed mainly of dendrites and cell bodies of neurons
  c Interior of the cerebrum composed mainly of white matter (that is nerve fibers arranged in bundles called *tracts*)
  d Functions of the cerebrum—mental processes of all types, including sensations, consciousness, memory, and voluntary control of movements
B Spinal cord (Figure 7-11)
  1 Outer part is composed of white matter made up of many bundles of axons called *tracts*; interior composed of gray matter made up mainly of neuron dendrites and cell bodies
  2 Functions as the center for all spinal cord reflexes; sensory tracts conduct impulses to the brain, and motor tracts conduct impulses from the brain
C Coverings and fluid spaces of the brain and spinal cord
  1 Coverings
    a Cranial bones and vertebrae
    b Cerebral and spinal meninges—the dura mater, the pia mater, and the arachnoid (Figure 7-13)
  2 Fluid spaces—subarachnoid spaces of meninges, central canal inside cord, and ventricles in brain (Figure 7-14)

PERIPHERAL NERVOUS SYSTEM

A Cranial nerves (Figure 7-16 and Table 7-2)
  1 Twelve pairs—attached to undersurface of the brain
  2 Connect brain with the neck and structures in the thorax and abdomen
B Spinal nerves
  1 Structure—contain dendrites of sensory neurons and axons of motor neurons
  2 Functions—conduct impulses necessary for sensations and voluntary movements

AUTONOMIC NERVOUS SYSTEM

A Autonomic nervous system—motor neurons that conduct impulses from the central nervous system to cardiac muscle, smooth muscle, and glandular epithelial tissue; regulates the body's automatic or involuntary functions (Figure 7-18)
B Autonomic neurons—preganglionic autonomic neurons conduct from spinal cord or brain stem to an autonomic ganglion; postganglionic neurons conduct from autonomic ganglia to cardiac muscle, smooth muscle, and glandular epithelial tissue
C Autonomic or visceral effectors—tissues to which autonomic neurons conduct impulses (that is, cardiac and smooth muscle and glandular epithelial tissue)
D Composed of two divisions—the sympathetic system and the parasympathetic system
E Autonomic conduction paths
  1 Consist of two-neuron relays (that is, preganglionic neurons from the central nervous system to autonomic ganglia, synapses, postganglionic neurons from ganglia to visceral effectors)
  2 In contrast, somatic motor neurons conduct all the way from the CNS to somatic effectors with no intervening synapses
F Sympathetic nervous system
  1 Structure
    a Dendrites and cell bodies of sympathetic preganglionic neurons are located in the gray matter of the thoracic and upper lumbar segments of the spinal cord
    b Axons leave the spinal cord in the anterior roots of spinal nerves, extend to sympathetic or collateral ganglia, and synapse with several postganglionic neurons whose axons extend to spinal or autonomic nerves to terminate in visceral effectors
    c A chain of sympathetic ganglia is in front of and at each side of the spinal column
  2 Functions
    a Serves as the emergency or stress system, controlling visceral effectors during strenuous exercise and strong emotions (anger, fear, hate, or anxiety)
    b Group of changes induced by sympathetic control is called the *fight-or-flight response*
G Parasympathetic nervous system
  1 Structure
    a Parasympathetic preganglionic neurons have dendrites and cell bodies in the gray matter of the brain stem and the sacral segments of the spinal cord

   b Parasympathetic preganglionic neurons terminate in parasympathetic ganglia located in the head and the thoracic and abdominal cavities close to visceral effectors
   c Each parasympathetic preganglionic neuron synapses with postganglionic neurons to only one effector
 2 Function—dominates control of many visceral effectors under normal, everyday conditions
H Autonomic neurotransmitters
 1 Cholinergic fibers—preganglionic axons of parasympathetic and sympathetic systems and parasympathetic postganglionic axons release acetylcholine
 2 Adrenergic fibers—axons of sympathetic postganglionic neurons release norepinephrine (noradrenaline)
I Autonomic nervous system as a whole
 1 Regulates the body's automatic functions in ways that maintain or quickly restore homeostasis
 2 Many visceral effectors are doubly innervated (that is, they receive fibers from parasympathetic and sympathetic divisions and are influenced in opposite ways by the two divisions)

N E W   W O R D S

acetylcholine	enkephalins	neurons	presynaptic neuron
anesthesia	fight-or-flight response	neurotransmitter	receptors
arachnoid	ganglion	node of Ranvier	reflex arc
astrocytes	glia	norepinephrine	saltatory conduction
axon	hydrocephalus	oligodendroglia	sensory neuron
catecholamines	interneuron	parasympathetic	serotonin
dendrite	meninges	system	sympathetic system
dopamine	microglia	pia mater	synapse
dura mater	motor neuron	postganglionic neurons	synaptic cleft
effectors	multiple sclerosis	postsynaptic neuron	tract
endorphins	myelin	preganglionic neuron	

CHAPTER   TEST

1. The nervous system as a whole is divided into two principal divisions: the _____ and _____ nervous systems.
2. The tough, fluid-containing membrane surrounding the brain and spinal cord is called the _____.
3. The medical name describing an accumulation of CSF in the ventricles of the brain is _____.
4. The two principal types of specialized cells found in the nervous system are _____ and _____.
5. Nervous system cells that form the myelin sheaths around nerve fibers in the brain and spinal cord are called _____.
6. Nervous impulses travel _____ from a nerve cell body in a single process called the _____.
7. The microscopic space that separates the axon endings of one neuron from the dendrites of another neuron is called a _____.
8. Chemical compounds released from axon terminals into synaptic clefts are called _____.
9. An area of skin supplied by a single spinal nerve is called a _____.
10. Two natural, morphinelike painkillers produced in the brain are called _____ and _____.
11. The two major structures that make up the diencephalon are called the _____ and _____.
12. There are _____ pairs of cranial nerves and _____ pairs of spinal nerves.
13. Motor neurons in the ANS conduct impulses to three kinds of tissues: _____, _____, and _____.
14. The ANS consists of two divisions called the _____ and _____ divisions.
15. Strong emotions and strenuous exercise activate the _____ division of the ANS.

Select the most correct answer from Column B for each statement in Column A. (Only one answer is correct.)

COLUMN A

16. _____Meninges
17. _____Transmit impulses to a neuron
18. _____Capable of phagocytosis
19. _____Produce myelin
20. _____Natural pain killer
21. _____Brain tumor
22. _____Nerve cell bodies
23. _____Largest part of the brain
24. _____Fight-or-flight response

COLUMN B

a. Microglia
b. Schwann cells
c. Pia mater
d. Response to stress
e. Endorphin
f. Gray matter
g. Cerebrum
h. Dendrites
i. Glioma

REVIEW QUESTIONS

1. What general function does the nervous system perform?
2. What other system performs the same general function as the nervous system?
3. What general functions does the spinal cord perform?
4. What does *CNS* mean? *PNS?*
5. What are the meninges?
6. What general functions does the cerebellum perform?
7. What general functions does the cerebrum perform?
8. What general functions do spinal nerves perform?
9. Define briefly each of the terms listed under "New Words."

10. Identify interneuron, motor neuron, reflex center, sensory neuron, somatic motor neuron, and synapse.
11. Compare parasympathetic and sympathetic functions.

CRITICAL THINKING

12. Why is the medulla considered the most vital part of the brain?
13. Explain the fight-or-flight response.
14. Contrast visceral and somatic effectors.

The Senses

OUTLINE

BOXED ESSAYS

OBJECTIVES

*After you have completed this chapter,
you should be able to:*

1. Classify sense organs as special or general and explain the basic differences between the two groups.
2. Discuss how a stimulus is converted into a sensation.
3. Discuss the general sense organs and their functions.
4. List the major senses.
5. Describe the structure of the eye and the functions of its components.
6. Discuss the anatomy of the ear and its sensory function in hearing and equilibrium.
7. Discuss the chemical receptors and their functions.

*I*f you were asked to name the sense organs, what organs would you name? Can you think of any besides the eyes, ears, nose, and taste buds? Actually there are millions of other sense organs throughout the body in our skin, internal organs, and muscles. They constitute the many **sensory receptors** that allow us to respond to stimuli such as touch, pressure, temperature, and pain. These microscopic receptors are located at the tips of dendrites of sensory neurons.

Our ability to "sense" changes in our external and internal environments is a requirement for maintaining homeostasis and for survival itself. We can initiate protective reflexes important to homeostasis only if we can sense a change or danger. External dangers may be detected by sight or hearing. If the danger is internal, such as overstretching a muscle, detecting an increase in body temperature (fever), or sensing the pain caused by an ulcer, other receptors make us aware of the problem and permit us to take appropriate action to maintain homeostasis.

Classification of Sense Organs

The sense organs are often classified as **special** sense organs and **general** sense organs. Special sense organs, such as the eye, are characterized by large and complex organs or by localized groupings of specialized receptors in areas such as the nasal mucosa or tongue. The general sense organs for detecting stimuli such as pain and touch are microscopic receptors widely distributed throughout the body.

Receptors are often classified according to the types of stimuli that activate them and whether or not they are *encapsulated* or *unencapsulated*, that is, whether or not they are covered by some sort of capsule or are "free" or "naked" of any such covering. Receptors in the retina of the eye are stimulated by light *(photoreceptors),* whereas taste and smell are activated by chemical stimuli *(chemoreceptors).* Still other receptors respond to physical damage or injury *(pain receptors),* changes in temperature *(thermoreceptors),* or to mechanical stimuli that change their position or in some way deform the capsule that surrounds them

(mechanoreceptors). Table 8-1 classifies the general sense organs and Table 8-2 on p. 176 classifies the special sense organs.

Converting a Stimulus Into a Sensation

All sense organs, regardless of size, type, or location, have in common some important functional characteristics. First, they must be able to sense or detect a stimulus in their environment. Of course, different sense organs detect different types of stimuli. Whether it is light, sound, temperature change, mechanical pressure, or the presence of chemicals ultimately identified as taste or smell, the stimulus must be changed into an electrical signal or nerve impulse. This signal is then transmitted over a nervous system "pathway" to the brain, where the sensation is actually perceived.

General Sense Organs

Groups of highly specialized and localized receptors are typically associated with the special senses. In the general sense organs, however, receptors are found in almost every part of the body, but are concentrated in the skin (Figure 8-1). To demonstrate this fact, try touching any point of your skin with the tip of a toothpick. You can hardly miss stimulating at least one receptor and almost instantaneously experiencing a sensation of touch. Stimulation of some receptors leads to the sensation of vibration; stimulation of others gives the sensation of pressure, and stimulation of still others gives the sensation of pain or temperature. General sense receptors are listed in Table 8-1 and illustrated in Figure 8-1.

Some specialized receptors found near the point of junction between tendons and muscles and others found deep within skeletal muscle tissue are called *proprioceptors.* When stimulated, they provide us with information concerning the position or movement of the different parts of the body as well as the length and the extent of contraction of our muscles. The Golgi tendon receptors and muscle spindles identified in Table 8-1 are important proprioceptors.

TABLE 8–1   General Sense Organs

TYPE	MAIN LOCATIONS	GENERAL SENSES	
Free nerve endings (naked nerve endings)	Skin and mucosa (epithelial layers)	Pain, crude touch, and possibly temperature	
Encapsulated nerve endings			
Meissner's corpuscles	Skin (in papillae of dermis) and fingertips and lips (numerous)	Fine touch and vibration	
Ruffini's corpuscles	Skin (dermal layer) and subcutaneous tissue of fingers	Touch and pressure	
Pacinian corpuscles	Subcutaneous, submucous, and subserous tissues; around joints; in mammary glands and external genitals of both sexes	Pressure and vibration	
Krause's end-bulbs	Skin (dermal layer), subcutaneous tissue, mucosa of lips and eyelids, and external genitals	Touch and possibly cold	

TABLE 8–1   General Sense Organs—cont'd

TYPE	MAIN LOCATIONS	GENERAL SENSES	
Encapsulated nerve endings—cont'd			
Golgi tendon receptors	Near junction of tendons and muscles	Proprioception	
Muscle spindles	Skeletal muscles	Proprioception	

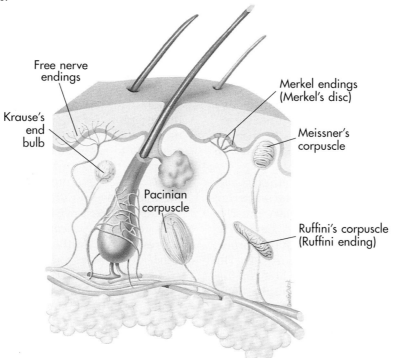

FIGURE 8–1   *General Sense Receptors.* This section of skin shows the placement of a number of receptors described in Table 8-1.

TABLE 8–2   Special Sense Organs

SENSE ORGAN	SPECIFIC RECEPTOR	TYPE OF RECEPTOR	SENSE
Eye	Rods and cones	Photoreceptor	Vision
Ear	Organ of Corti	Mechanoreceptor	Hearing
	Cristae ampullares	Mechanoreceptor	Balance
Nose	Olfactory cells	Chemoreceptor	Smell
Taste buds	Gustatory cells	Chemoreceptor	Taste

Special Sense Organs

THE EYE

When you look at a person's eye, you see only a small part of the whole eye. Three layers of tissue form the eyeball: the **sclera** (SKLE-rah), the **choroid** (KO-royd), and the **retina** (RET-i-nah) (Figure 8-2). The outer layer of sclera consists of tough fibrous tissue. The "white" of the eye is part of the front surface of the sclera. The other part of the front surface of the sclera is called the *cornea* and is sometimes spoken of as the window of the eye because of its transparency. At a casual glance, however, it does not look transparent but appears blue, brown, gray, or green because it lies over the **iris,** the colored part of the eye. A mucous membrane known as the **conjunctiva** (kon-junk-TEE-vah) lines the eyelids and covers the sclera in front. The conjunctiva is kept moist by tears formed in the **lacrimal gland** located in the upper lateral portion of the orbit.

The middle layer of the eyeball, the *choroid*, contains a dark pigment to prevent the scattering of incoming light rays. Two involuntary muscles make up the front part of the choroid. One is the *iris*, the colored structure seen through the cornea, and the other is the *ciliary muscle* (Figure 8-2). The black center of the iris is really a hole in this doughnut-shaped muscle; it is the **pupil** of the eye. Some of the fibers of the iris are arranged like spokes in a wheel. When they contract, the pupils dilate, letting in more light rays. Other fibers are circular. When they contract, the pupils constrict, letting in fewer light rays. Normally, the pupils constrict in bright light and dilate in dim light. When we look at distant objects, the ciliary muscle is relaxed, and the lens has only a slightly curved shape. To focus on near objects, however, the ciliary muscle must contract. As it contracts, it pulls the choroid coat forward toward the lens, thus causing the lens to bulge and curve even more. Most of us become more farsighted as we grow older and lose the ability to focus on close objects because our lenses lose their elasticity and can no longer bulge enough to bring near objects into focus. **Presbyopia** or "oldsightedness" is the name for this condition.

The *retina* or innermost layer of the eyeball contains microscopic receptor cells, called *rods* and *cones* because of their shapes. Dim light can stimulate the rods, but fairly bright light is necessary to stimulate the cones. In other words, **rods** are the receptors for night vision and **cones** for daytime vision. There are three kinds of cones; each is sensitive to a different color: red, green, or blue. Scattered throughout the central portion of the retina, these three types of cones allow us to distinguish between different colors.

There is a yellowish area near the center of the retina called the *macula lutea*. It surrounds a small depression, called the **fovea centralis,** which contains the greatest concentration of cones of any area of the retina (Figure 8-2). In good light, greater *visual acuity*, or sharpness of visual perception, can be obtained if we look directly at an object and focus the image on the fovea.

Fluids fill the hollow inside of the eyeball. They maintain the normal shape of the eyeball and help refract light rays; that is, the fluids bend light rays

FIGURE 8–2 *Horizontal Section Through the Left Eyeball.* The eye is viewed from above.

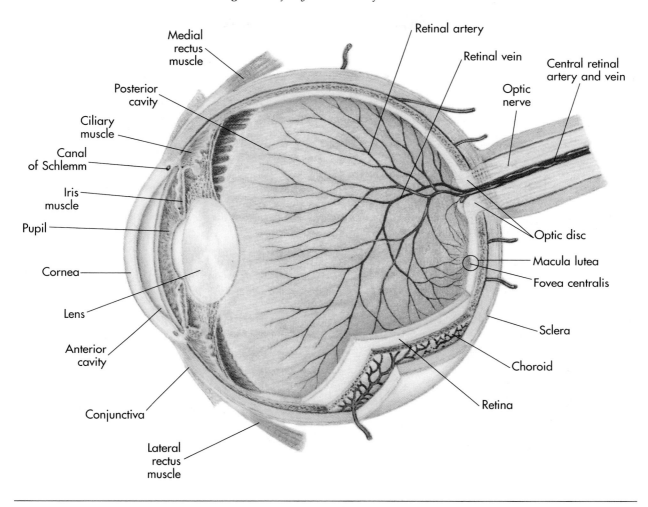

to bring them to focus on the retina. **Aqueous humor** is the name of the watery fluid in front of the lens (in the anterior cavity of the eye), and **vitreous humor** is the name of the jellylike fluid behind the lens (in the posterior cavity). Aqueous humor is constantly being formed, drained, and replaced in the anterior cavity. If drainage is blocked for any reason, the internal pressure within the eye will increase, and damage that could lead to blindness will occur. This condition is called **glaucoma** (glaw-KO-mah).

The **lens** of the eye lies directly behind the pupil. It is held in place by a ligament attached to the ciliary muscle. In most young people, the lens is transparent and somewhat elastic so that it is capable of changing shape. Exposure to ultraviolet (UV) radiation in sunlight causes the lens to become hard and milky. Over the years, repeated exposure to sunlight may cause **cataracts** or milky spots on the lens. Large or numerous cataracts may cause blindness. Cataracts can be removed surgically and replaced with artificial lenses. Many scientists believe that wearing sunglasses that filter UV radiation from an early age will help prevent the formation of cataracts.

Visual Pathway

Light is the stimulus that results in vision (that is, our ability to see objects as they exist in our environment). Light enters the eye through the pupil and is *refracted* or bent so that it is focused on the retina. Refraction occurs as light passes through the cornea, the aqueous humor, the lens, and the vitreous humor on its way to the retina.

The innermost layer of the retina contains the rods and cones, which are the *photoreceptor* cells of the eye (Figure 8-3). They respond to a light stimulus by producing a nervous impulse. The rod and cone photoreceptor cells synapse with neurons in the bipolar and ganglionic layers of the retina. Nervous signals eventually leave the retina and exit the eye through the optic nerve on the posterior surface of the eyeball. No rods or cones are present in the area of the retina where the optic nerve fibers exit. The result is a "blind spot" known as the *optic disc* (Figure 8-2).

FIGURE 8–3 *Cells of the Retina.* Photoreceptors called *rods* and *cones* (notice their shapes) detect changes in light and relay the information to bipolar neurons. The bipolar cells, in turn, conduct the information to ganglion cells. The information eventually leaves the eye by way of the optic nerve.

Focusing Problems

Focusing a clear image on the retina is essential for good vision. In the normal eye *(A)*, light rays enter the eye and are focused into a clear, upside-down image on the retina. The brain can easily right the upside-down image in our conscious perception but cannot fix an image that is not sharply focused. If our eyes are elongated *(B)*, the image focuses in front of the retina rather than on it. The retina receives only a fuzzy image. This condition, called **myopia** or *nearsightedness*, can be corrected by using contact lenses or glasses *(C)*. If our eyes are shorter than normal *(D)*, the image focuses behind the retina, also producing a fuzzy image. This condition, called **hyperopia** or *farsightedness*, can also be corrected by lenses *(E)*.

A

Normal

Uncorrected

B   Myopia

D   Hyperopia

Corrected

C

E

Color Blindness

Color blindness, usually an inherited condition, is caused by mistakes in producing three chemicals called *photopigments* in the cones. Each photopigment is sensitive to one of the three primary colors of light: green, blue, and red. In many cases, the green-sensitive photopigment is missing or deficient; other times, the red-sensitive photopigment is abnormal. (Deficiency of the blue-sensitive photopigment is very rare.) Color-blind individuals see colors, but they cannot distinguish between them normally.

Figures such as those shown here are often used to screen individuals for color blindness. A person with red-green blindness cannot see the *74* in Figure *A*, whereas a person with normal vision can. To determine which photopigment is deficient, a color-blind person may try a figure similar to *B*. Persons with a deficiency of red-sensitive photopigment can distinguish only the *2*; those deficient in green-sensitive photopigment can only see the *4*.

After leaving the eye, the optic nerves enter the brain and travel to the visual cortex of the occipital lobe. In this area, *visual interpretation* of the nervous impulses generated by light stimuli in the rods and cones of the retina result in "seeing."

THE EAR

In addition to its role in hearing, the ear also functions as the sense organ of equilibrium and balance. As we shall later see, the stimulation or "trigger" that activates receptors involved with hearing and equilibrium is mechanical, and the receptors themselves are called *mechanoreceptors* (mek-an-o-ree-SEP-tors). Physical forces that involve sound vibrations and fluid movements are responsible for initiating nervous impulses eventually perceived as sound and balance.

The ear is more than an appendage on the side of the head. A large part of the ear and its most important part lies hidden from view deep inside the temporal bone. It is divided into the following anatomical areas (Figure 8-4):

1. **External ear**
2. **Middle ear**
3. **Inner (internal) ear**

External Ear

The external ear has two parts: the **auricle** (AW-ri-kul) or pinna and the **external auditory canal.** The auricle is the appendage on the side of the head surrounding the opening of the external auditory canal. The canal itself is a curving tube about 2.5 cm (1 inch) in length. It extends into the temporal

FIGURE 8–4 *The Ear.* External, middle, and inner ears.

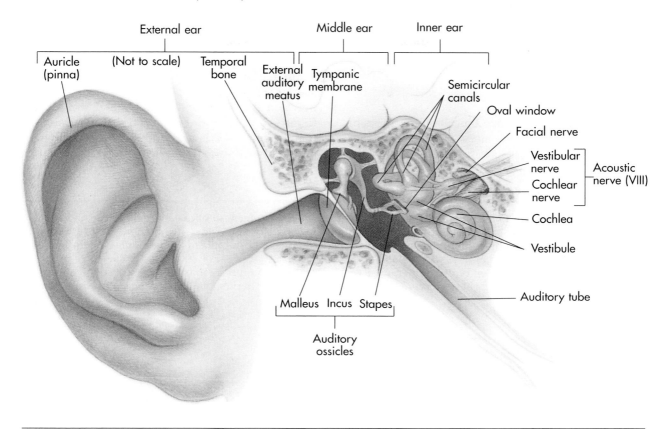

bone and ends at the **tympanic** (tim-PAN-ik) **membrane** or **eardrum,** which is a partition between the external and middle ear. The skin of the auditory canal, especially in its outer one third, contains many short hairs and **ceruminous** (se-ROO-mi-nus) **glands** that produce a waxy substance called *cerumen* that may collect in the canal and impair hearing by absorbing or blocking the passage of sound waves. Sound waves traveling through the external auditory canal strike the tympanic membrane and cause it to vibrate.

Middle Ear

The middle ear is a tiny and very thin epithelium-lined cavity hollowed out of the temporal bone. It houses three very small bones. The names of these ear bones, called **ossicles** (OS-si-kuls), describe their shapes—**malleus** (hammer), **incus** (anvil), and **stapes** (stirrup). The "handle" of the malleus attaches to the inside of the tympanic membrane, and the "head" attaches to the incus. The incus attaches to the stapes, and the stapes presses against a membrane that covers a small opening, the *oval window.* The oval window separates the middle ear from the inner ear. When sound waves cause the eardrum to vibrate, that movement is transmitted and amplified by the ear ossicles as it passes through the middle ear. Movement of the stapes against the oval window causes movement of fluid in the inner ear.

A point worth mentioning, because it explains the frequent spread of infection from the throat to the ear, is the fact that a tube—the **auditory** or

eustachian (yoo-STAY-shen) **tube**—connects the throat with the middle ear. The epithelial lining of the middle ears, auditory tubes, and throat are extensions of one continuous membrane. Consequently a sore throat may spread to produce a middle ear infection called *otitis* (o-TIE-tis) *media* (ME-dee-ah).

Inner Ear

The activation of specialized mechanoreceptors in the inner ear generates nervous impulses that result in hearing and equilibrium. Anatomically,

the inner ear consists of three spaces in the temporal bone, assembled in a complex maze called the **bony labyrinth** (LAB-i-rinth). This odd-shaped bony space is filled with a watery fluid called **perilymph** (PAIR-i-limf) and is divided into the following parts: **vestibule** (VES-ti-byool), **semicircular canals,** and **cochlea** (KOK-lee-ah). The vestibule is adjacent to the oval window between the semicircular canals and the cochlea (Figure 8-5). Note in Figure 8-5 that a balloonlike membranous sac is suspended in the perilymph and follows the shape of the bony labyrinth much

FIGURE 8–5   *The Inner Ear.* The bony labyrinth (orange) is the hard outer wall of the entire inner ear, and includes semicircular canals, vestibule, and cochlea. Within the bony labyrinth is the membranous labyrinth (purple), which is surrounded by perilymph and filled with endolymph. Each ampulla in the vestibule contains a crista ampullaris that detects changes in head position and sends sensory impulses through the vestibular nerve to the brain. The inset shows a section of the membranous cochlea. Hair cells in the organ of Corti detect sound and send the information through the cochlear nerve. The vestibular and cochlear nerves join to form the eighth cranial nerve.

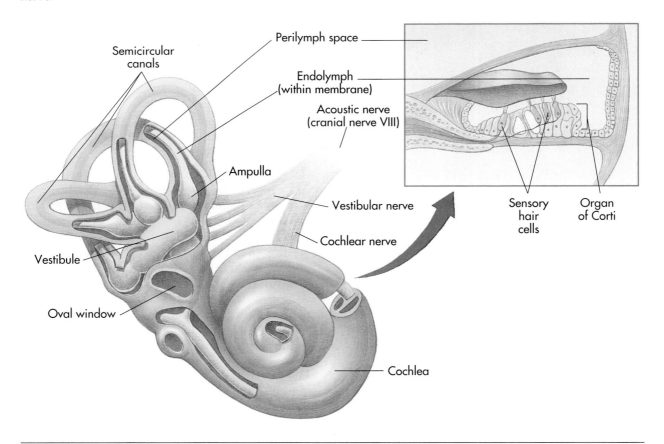

like a "tube within a tube." This is the **membranous labyrinth,** and it is filled with a thicker fluid called **endolymph** (EN-doe-limf).

The specialized mechanoreceptors for balance and equilibrium are located in the three semicircular canals and the vestibule. The three half-circle semicircular canals are oriented at right angles to one another (Figure 8-5). Within each canal is a dilated area called the *ampulla* that contains a specialized receptor called a **crista** (KRIS-tah) **ampullaris** (am-pyoo-LAIR-is), which generates a nerve impulse when you move your head. The sensory cells in the cristae ampullares have hairlike extensions that are suspended in the endolymph. The sensory cells are stimulated when movement of the head causes the endolymph to move, thus causing the hairs to bend. Nerves from other receptors in the vestibule join those from the semicircular canals to form the **vestibular nerve,** which joins with the cochlear nerve to form the acoustic nerve or cranial nerve VIII (Figure 8-5). Eventually, nervous impulses passing through this nerve reach the cerebellum and medulla. Other connections from these areas result in impulses reaching the cerebral cortex.

The organ of hearing, which lies in the snail-shaped cochlea, is the **organ of Corti** (KOR-tie). It is surrounded by endolymph, filling the membranous cochlea or **cochlear duct,** which is the membranous tube within the bony cochlea. Specialized hair cells on the organ of Corti generate nerve impulses when they are bent by the movement of endolymph set in motion by sound waves (Figures 8-5 and 8-6).

FIGURE 8–6 *Effect of Sound Waves on Cochlear Structures.* Sound waves strike the tympanic membrane and cause it to vibrate. This vibration causes the membrane of the oval window to vibrate. This vibration causes the perilymph in the bony labyrinth of the cochlea to move, which causes the endolymph in the membranous labyrinth of the cochlea or cochlear duct to move. This movement of endolymph stimulates hair cells on the organ of Corti to generate a nerve impulse. The nerve impulse travels over the cochlear nerve, which becomes a part of the eighth cranial nerve. Eventually, nerve impulses reach the auditory cortex and are interpreted as sound.

Swimmer's Ear

External otitis or *swimmer's ear* is a common infection of the external ear in athletes. It can be bacterial or fungal in origin and is usually associated with prolonged exposure to water. The infection generally involves, at least to some extent, the auditory canal and auricle. The ear as a whole is tender, red, and swollen. Treatment of swimmer's ear usually involves antibiotic therapy and prescription analgesics.

THE TASTE RECEPTORS

The chemical receptors, or chemoreceptors, that generate nervous impulses resulting in the sense of taste are called **taste buds.** About 10,000 of these microscopic receptors are found on the sides of much larger structures on the tongue called **papillae** (pah-PIL-ee) and also as portions of other tissues in the mouth and throat. Nervous impulses are generated by specialized cells in taste buds, called **gustatory** (GUS-tah-toe-ree) **cells.** They respond to dissolved chemicals in the saliva that bathe the tongue and mouth tissues (Figure 8-7).

Only four kinds of taste sensations—sweet, sour, bitter, and salty—result from stimulation of

FIGURE 8-7 *The Tongue.* **A,** Dorsal surface and regions sensitive to various tastes. **B,** Section through a papilla with taste buds on the side. **C,** Enlarged view of a section through a taste bud.

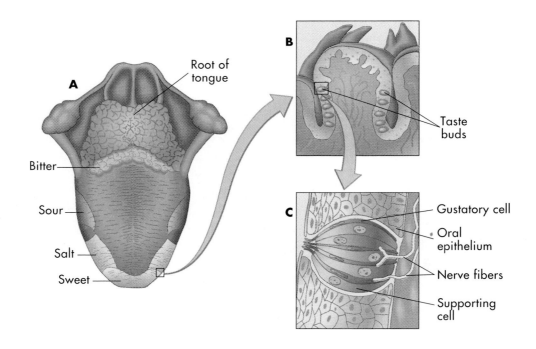

taste buds. All other flavors result from a combination of taste bud and olfactory receptor stimulation. In other words, the myriads of tastes recognized are not tastes alone but tastes plus odors. For this reason a cold that interferes with the stimulation of the olfactory receptors by odors from foods in the mouth markedly dulls taste sensations. Nervous impulses that are generated by stimulation of taste buds travel primarily through two cranial nerves (VII and IX) to end in the specialized taste area of the cerebral cortex.

THE SMELL RECEPTORS

The chemoreceptors responsible for the sense of smell are located in a small area of epithelial tissue in the upper part of the nasal cavity (Figure 8-8). The location of the **olfactory receptors** is somewhat hidden, and we are often forced to forcefully sniff air to smell delicate odors. Each olfactory cell has a number of specialized cilia that sense different chemicals and cause the cell to respond by generating a nervous impulse. To be detected by olfactory receptors, chemicals must be dissolved in the watery mucus that lines the nasal cavity.

Although the olfactory receptors are extremely sensitive (that is, stimulated by even very slight odors), they are also easily fatigued—a fact that explains why odors that are at first very noticeable are not sensed at all after a short time. After the olfactory cells are stimulated by odor-causing chemicals, the resulting nerve impulse travels through the olfactory nerves in the olfactory bulb and tract and then enters the thalamic and olfactory centers of the brain, where the nervous impulses are interpreted as specific odors. The pathways taken by olfactory nerve impulses and the areas where these impulses are interpreted are closely associated with areas of the brain important in memory and emotion. For this reason, we may retain vivid and long-lasting memories of particular smells and odors.

FIGURE 8–8 *Olfactory Structures.* Gas molecules stimulate olfactory cells in the nasal epithelium. Sensory information is then conducted along nerves in the olfactory bulb and olfactory tract to sensory processing centers in the brain.

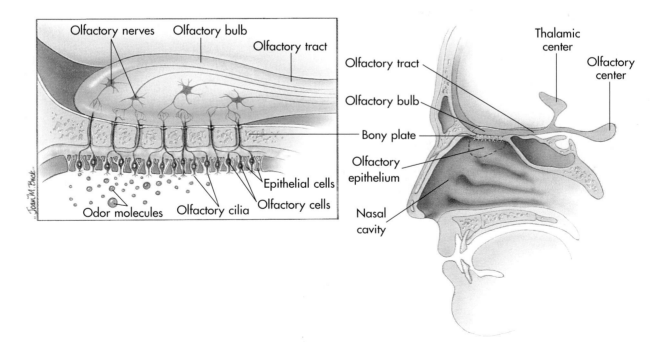

CLASSIFICATION OF SENSE ORGANS

A General sense organs (Table 8-1)
  1 Often exist as individual cells or receptor units
  2 Widely distributed throughout the body
B Special sense organs (Table 8-2)
  1 Large and complex organs
  2 Localized grouping of specialized receptors
C Classification by presence or absence of covering capsule
  1 Encapsulated
  2 Unencapsulated ("free" or "naked")
D Classification by type of stimuli required to activate receptors
  1 Photoreceptors (light)
  2 Chemoreceptors (chemicals)
  3 Pain receptors (injury)
  4 Thermoreceptors (temperature change)
  5 Mechanoreceptors (movement or deforming of capsule)
  6 Proprioceptors (position of body parts or changes in muscle length or tension)

CONVERTING A STIMULUS INTO A SENSATION

A All sense organs have common functional characteristics
  1 All are able to detect a particular stimulus
  2 A stimulus is converted into a nerve impulse
  3 A nerve impulse is perceived as a sensation in the CNS

GENERAL SENSE ORGANS (Table 8-1)

A Distribution is widespread; single-cell receptors are common
B Examples (Figure 8-1, Table 8-1)
  1 Free nerve endings—pain and crude touch
  2 Meissner's corpuscles—fine touch and vibration
  3 Ruffini's corpuscles—touch and pressure
  4 Pacinian corpuscles—pressure and vibration
  5 Krause's end-bulbs—touch
  6 Golgi tendon receptors—proprioception
  7 Muscle spindles—proprioception

SPECIAL SENSE ORGANS

A The eye (Figure 8-2)
  1 Layers of eyeball
    a Sclera—tough outer coat; "white" of eye; cornea is transparent part of sclera over iris
    b Choroid—pigmented vascular layer prevents scattering of light; front part of this layer made of ciliary muscle and iris, the colored part of the eye; the pupil is the hole in the center of the iris; contraction of iris muscle dilates or constricts pupil
    c Retina (Figure 8-3)—innermost layer of the eye; contains rods (receptors for night vision) and cones (receptors for day vision and color vision)
  2 Conjunctiva—mucous membrane covering the front surface of the sclera and lining the eyelid
  3 Lens—transparent body behind the pupil; focuses light rays on the retina
  4 Eye fluids
    a Aqueous humor—in the anterior cavity in front of the lens
    b Vitreous humor—in the posterior cavity behind the lens
  5 Visual pathway
    a Innermost layer of retina contains rods and cones
    b Impulse travels from the rods and cones through the bipolar and ganglionic layers of retina (Figure 8-3)
    c Nerve impulse leaves the eye through the optic nerve; the point of exit is free of receptors and is therefore called a *blind spot*
    d Visual interpretation occurs in the visual cortex of the cerebrum
B The ear
  1 The ear functions in hearing and in equilibrium and balance
    a Receptors for hearing and equilibrium are mechanoreceptors

2 Divisions of the ear (Figure 8-4)
  a External ear
    (1) Auricle (pinna)
    (2) External auditory canal
      (a) Curving canal 2.5 cm (1 inch) in length
      (b) Contains ceruminous glands
      (c) Ends at the tympanic membrane
  b Middle ear
    (1) Houses ear ossicles—malleus, incus, and stapes
    (2) Ends in the oval window
    (3) The auditory (eustachian) tube connects the middle ear to the throat
    (4) Inflammation called *otitis media*
  c Inner ear (Figure 8-5)
    (1) Bony labyrinth filled with perilymph
    (2) Subdivided into the vestibule, semicircular canals, and cochlea
    (3) Membranous labyrinth filled with endolymph

    (4) The receptors for balance in the semicircular canals are called *cristae ampullaris*
    (5) Specialized hair cells on the organ of Corti respond when bent by the movement of surrounding endolymph set in motion by sound waves (Figure 8-6)
C The taste receptors (Figure 8-7)
  1 Receptors are chemoreceptors called *taste buds*
  2 Cranial nerves VII and IX carry gustatory impulses
  3 Only four kinds of taste sensations—sweet, sour, bitter, salty
  4 Gustatory and olfactory senses work together
D The smell receptors (Figure 8-8)
  1 Receptors for fibers of olfactory or cranial nerve I lie in olfactory mucosa of nasal cavity
  2 Olfactory receptors are extremely sensitive but easily fatigued
  3 Odor-causing chemicals initiate a nervous signal that is interpreted as a specific odor by the brain

N E W   W O R D S

aqueous humor	crista ampullaris	organ of Corti	pupil
auricle	endolymph	ossicles	retina
cataract	glaucoma	papillae	sclera
chemoreceptor	gustatory cells	perilymph	semicircular canals
choroid	lacrimal gland	photoreceptor	tympanic membrane
cochlea	lens	presbyopia	vitreous humor
conjunctiva	mechanoreceptor	proprioception	

1. _____ sense organs are widely distributed throughout the body.
2. The receptors involved in vision are called _____ and _____.
3. The three layers of tissue that form the eyeball are the _____, the _____, and the _____.
4. The colored structure seen through the cornea of the eye is called the _____.
5. The fluid found in the anterior cavity of the eye is called _____.
6. The loss of lens transparency is a condition called _____.
7. The area of the retina that helps to reduce light scattering is called the _____ retina.
8. The type of receptors involved with hearing and equilibrium are called _____.
9. The major anatomical areas of the ear are called: _____ ear; _____ ear; and _____ ear.
10. The ear ossicle attached to the tympanic membrane is called the _____.
11. The bony labyrinth is filled with fluid called _____.

12. The hearing sense organ is called the organ of _____.
13. Sense receptors of the type used in taste and smell are called _____.
14. Receptor cells in the taste buds are called _____ cells.

Select the most correct answer from Column B for each statement in Column A. (Only one answer is correct.)

COLUMN A	COLUMN B
15. _____White portion of eyeball	a. Rods
16. _____Receptors for night vision	b. Auricle
17. _____Sense of position	c. Optic disk
18. _____Blind spot	d. Otitis media
19. _____External ear	e. Olfactory cell
20. _____Ear drum	f. Crista ampullaris
21. _____Middle ear infection	g. Sclera
22. _____Receptor for equilibrium	h. Tympanic membrane
23. _____Receptor for smell	i. Proprioception

REVIEW QUESTIONS

1. List, compare, and contrast special and general sense organs.
2. What are the functional characteristics required for a sense organ to convert a stimulus into a sensation?
3. How do rods and cones differ in terms of their ability to serve as receptors for vision?
4. Which pair or pairs of cranial nerves would you nickname "seeing nerves," "hearing nerves," "smelling nerves," and "tasting nerves"?
5. Explain the anatomical relationship between the bony and the membranous labyrinths.
6. Explain how photoreceptors, mechanoreceptors, and chemoreceptors differ in function.
7. What is the function of the ear ossicles?
8. Discuss the sense of balance or equilibrium, including the name, location, and mechanism of action of the specialized receptors involved.
9. Discuss the organ of Corti. Where is it located? How is it stimulated?
10. Discuss how odors are detected and perceived.
11. Identify and give the functions of five general sense organs.

CRITICAL THINKING

12. How does a cataract cause a problem with visual quality?
13. Trace and discuss the path of a sound wave as it influences the structures involved in hearing.
14. Briefly explain why most elderly people have difficulty in seeing objects close to their eyes.
15. Explain the difference between a tongue papilla, a taste bud, and a gustatory cell.

The Endocrine System

9

BOXED ESSAYS

Steroid Abuse
Growth Hormone Abnormalities
Thyroid Hormone Abnormalities
Adrenal Hormone Abnormalities
Exercise and Diabetes Mellitus

OBJECTIVES

*After you have completed this chapter,
you should be able to:*

1. Distinguish between endocrine and exocrine glands and define the terms *hormone* and *prostaglandin.*

2. Identify and locate the primary endocrine glands and list the major hormones produced by each gland.

3. Describe the mechanisms of steroid and protein hormone action.

4. Explain how negative and positive feedback mechanisms regulate the secretion of endocrine hormones.

5. Identify the principal functions of each major endocrine hormone and describe the conditions that may result from hyposecretion or hypersecretion.

6. Define *diabetes insipidus, diabetes mellitus, gigantism, goiter, cretinism, glycosuria.*

H ave you ever seen a giant or a dwarf? Have you ever known anyone who had "sugar diabetes" or a goiter? If so, you have had visible proof of the importance of the endocrine system for normal development and health.

The **endocrine system** performs the same general functions as the nervous system: communication and control. The nervous system provides rapid, brief control by fast-traveling nerve impulses. The endocrine system provides slower but longer-lasting control by **hormones** (chemicals) secreted into and circulated by the blood.

The organs of the endocrine system are located in widely separated parts of the body—in the neck; the cranial, thoracic, abdominal, and pelvic cavities; and outside of the body cavities. Note the names and locations of the endocrine glands shown in Figure 9-1.

All organs of the endocrine system are glands, but not all glands are organs of the endocrine system. Of the two types of glands in the body—**exocrine glands** and **endocrine glands**—only endocrine glands belong to this system. Exocrine glands secrete their products into ducts that empty onto a surface or into a cavity. For example, sweat glands produce a watery secretion that empties onto the surface of the skin. Salivary glands are also exocrine glands, secreting saliva that flows into the mouth. Endocrine glands are ductless glands. They secrete chemicals known as **hormones** into intercellular spaces. From there, the hormones diffuse directly into the blood and are carried throughout the body. Each hormone molecule may then bind to a cell that has specific receptors for that hormone, triggering a reaction in the cell. Such a cell is called a **target organ** cell. The list of endocrine glands and their target organs continues to grow. The names, locations, and functions of the well-known endocrine glands are given in Figure 9-1 and Table 9-1.

In this chapter you will read about the functions of the main endocrine glands and discover why their importance is almost impossible to exaggerate. Hormones are the main regulators of metabolism, growth and development, reproduction, and many other body activities. They play important roles in maintaining homeostasis—fluid and electrolyte, acid-base, and energy balances, for example. Hormones make the difference between normalcy and many kinds of abnormalities such as dwarfism, gigantism, and sterility. They are important not only for the healthy survival of each one of us but also for the survival of the human species.

Diseases of the endocrine glands are numerous, varied, and sometimes spectacular. Tumors or other abnormalities frequently cause a gland to secrete too much or too little hormone. Production of too much hormone by a diseased gland is called **hypersecretion.** If too little hormone is produced, the condition is called **hyposecretion.**

Mechanisms of Hormone Action

A hormone causes its target cells to respond in particular ways; this has been the subject of intense interest and research. The two major classes of hormones—**protein hormones** and **steroid hormones**—differ in the mechanisms by which they influence target organ cells.

PROTEIN HORMONES

The most widely accepted theory of protein hormone action is called the **second messenger hypothesis.** A *hypothesis* (hye-POTH-e-sis) is a proposed explanation that explains observed phenomena. The second messenger hypothesis attempts to explain why protein hormones cause specific effects in target organs but do not "recognize" or act on other organs of the body. According to this concept, a protein hormone, such as thyroid-stimulating hormone, acts as a "first messenger" (that is, it delivers its chemical message from the cells of an endocrine gland to highly specific membrane receptor sites on the cells of a target organ). This interaction between a hormone and its specific receptor site on the cell membrane of a target organ cell is often compared to the fitting of a unique key into a lock. After the hormone is attached to its specific receptor site, a number of chemical reactions occur. These reactions activate molecules within the cell called *second messengers.* One example of this mechanism occurs when the hormone-receptor interaction changes energy-rich ATP molecules inside the cell into **cyclic AMP** (adenosine monophosphate). Cyclic AMP serves as the second messenger,

FIGURE 9–1 *Location of the Endocrine Glands.* Thymus gland is shown at maximum size at puberty.

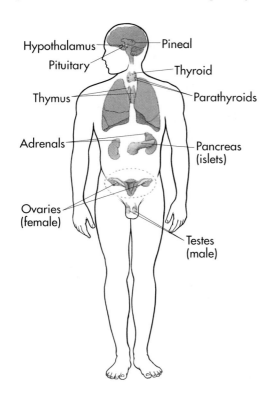

FIGURE 9–2 *Mechanism of Protein Hormone Action.* The hormone acts as "first messenger," delivering its message via the bloodstream to a membrane receptor in the target organ cell much like a key fits into a lock. The "second messenger" causes the cell to respond and perform its specialized function.

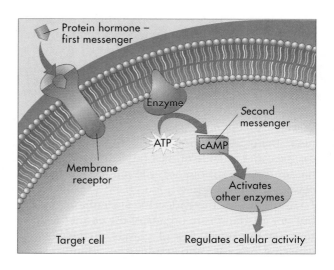

delivering information inside the cell that regulates the cell's activity. For example, cyclic AMP causes thyroid cells to respond to thyroid-stimulating hormone by secreting a thyroid hormone such as thyroxine. Cyclic AMP is only one of several second messengers that have been discovered.

In summary, protein hormones serve as first messengers, providing communication between endocrine glands and target organs. Another molecule, such as cyclic AMP, then acts as the second messenger, providing communication within a hormone's target cells. Figure 9-2 summarizes the mechanism of protein hormone action as explained by the second messenger hypothesis.

STEROID HORMONES

The action of small, lipid-soluble steroid hormones such as estrogen is not explained by the

Steroid Abuse

Some steroid hormones are called **anabolic steroids** because they stimulate the building of large molecules (anabolism). Specifically, they stimulate the building of proteins in muscle and bone. Steroids such as testosterone and its synthetic derivatives are often abused by athletes and others who want to increase their performance. The anabolic effects of the hormones increase the mass and strength of skeletal muscles.

Unfortunately, steroid abuse has other consequences. It disrupts the normal negative feedback control of hormones throughout the body and may result in tissue damage, sterility, mental imbalance, and many life-threatening metabolic problems.

TABLE 9–1 Endocrine Glands, Hormones, and Their Functions

GLAND/HORMONE	FUNCTION
Anterior Pituitary	
Thyroid-stimulating hormone (TSH)	Tropic hormone Stimulates secretion of thyroid hormones
Adrenocorticotropic hormone (ACTH)	Tropic hormone Stimulates secretion of adrenal cortex hormones
Follicle-stimulating hormone (FSH)	Tropic hormone *Female:* stimulates development of ovarian follicles and secretion of estrogens *Male:* stimulates seminiferous tubules of testes to grow and produce sperm
Luteinizing hormone (LH)	Tropic hormone *Female:* stimulates maturation of ovarian follicle and ovum; stimulates secretion of estrogen; triggers ovulation; stimulates development of corpus luteum (luteinization) *Male:* stimulates interstitial cells of the testes to secrete testosterone
Melanocyte-stimulating hormone (MSH)	Stimulates synthesis and dispersion of melanin pigment in the skin
Growth hormone (GH)	Stimulates growth in all organs; mobilizes food molecules, causing an increase in blood glucose concentration
Prolactin (lactogenic hormone)	Stimulates breast development during pregnancy and milk secretion after pregnancy
Posterior Pituitary*	
Antidiuretic hormone (ADH)	Stimulates retention of water by the kidneys
Oxytocin	Stimulates uterine contractions at the end of pregnancy; stimulates the release of milk into the breast ducts
Hypothalamus	
Releasing hormones (several)	Stimulate the anterior pituitary to release hormones
Inhibiting hormones (several)	Inhibit the anterior pituitary's secretion of hormones
Thyroid	
Thyroxine (T_4) and triiodothyronine (T_3)	Stimulate the energy metabolism of all cells
Calcitonin	Inhibits the breakdown of bone; causes a decrease in blood calcium concentration
Parathyroid	
Parathyroid hormone (PTH)	Stimulates the breakdown of bone; causes an increase in blood calcium concentration
Adrenal Cortex	
Mineralocorticoids: aldosterone	Regulate electrolyte and fluid homeostasis
Glucocorticoids: cortisol (hydrocortisone)	Stimulate gluconeogenesis, causing an increase in blood glucose concentration; also have antiinflammatory and antiimmunity, antiallergy effects
Sex hormones (androgens)	Stimulate sexual drive in the female but have negligible effects in the male
Adrenal/Medulla	
Epinephrine (adrenaline) and norepinephrine	Prolong and intensify the sympathetic nervous response during stress

Continued.

TABLE 9–1, cont'd   Endocrine Glands, Hormones, and Their Functions

GLAND/HORMONE	FUNCTION
Pancreatic Islets	
Glucagon	Stimulates liver glycogenolysis, causing an increase in blood glucose concentration
Insulin	Promotes glucose entry into all cells, causing a decrease in blood glucose concentration
Ovary	
Estrogens	Promotes development and maintenance of female sexual characteristics (see Chapter 19)
Progesterone	Promotes conditions required for pregnancy (see Chapter 19)
Testis	
Testosterone	Promotes development and maintenance of male sexual characteristics (see Chapter 19)
Thymus	
Thymosin	Promotes development of immune-system cells
Placenta	
Chorionic gonadotropin, estrogens, progesterone	Promote conditions required during early pregnancy
Pineal	
Melatonin	Inhibits tropic hormones that affect the ovaries; may be involved in the body's internal clock
Heart (atria)	
Atrial natriuretic hormone (ANH)	Regulates fluid and electrolyte homeostasis

*Posterior pituitary hormones are synthesized in the hypothalamus but released from axon terminals in the posterior pituitary.

second messenger hypothesis. Because they are lipid soluble, steroid hormones can pass intact directly through the cell membrane of the target organ cell. Once inside the cell, steroid hormones pass through the cytoplasm and enter the nucleus where they bind with a receptor to form a hormone-receptor complex. This complex acts on DNA, which utimately causes the formation of a new protein in the cytoplasm that then produces specific effects in the target cell. In the case of estrogen, that effect might be breast development in the female adolescent. Figure 9-3 summarizes the mechanism of steroid hormone action.

⬤ *Regulation of Hormone Secretion*

The regulation of hormone levels in the blood depends on a highly specialized homeostatic mechanism called *negative feedback*. The principle of **negative feedback** can be illustrated by using the hormone insulin as an example. When released from endocrine cells in the pancreas, insulin lowers blood sugar levels. Normally, elevated blood sugar levels occur after a meal, after the absorption of sugars from the digestive tract takes place. The elevated blood sugar stimulates the release of insulin from the pancreas. Insulin then assists in the transfer of sugar from the blood into cells, and blood sugar levels drop. Low blood sugar levels then cause endocrine cells in the pancreas to cease the production and release of insulin. These responses are *negative*. Therefore the homeostatic mechanism is called a negative feedback control mechanism because it reverses the change in blood sugar level (Figure 9-4).

Positive feedback mechanisms, which are uncommon, amplify changes rather than reverse them. Usually, such amplification threatens homeostasis, but in some situations it can help the body maintain its stability. For example, during labor,

FIGURE 9–3   *Mechanism of Steroid Hormone Action.* Steroid hormones pass through the plasma membrane and enter the nucleus to form a hormone receptor complex that acts on DNA. As a result, a new protein is formed in the cytoplasm that produces specific effects in the target cell.

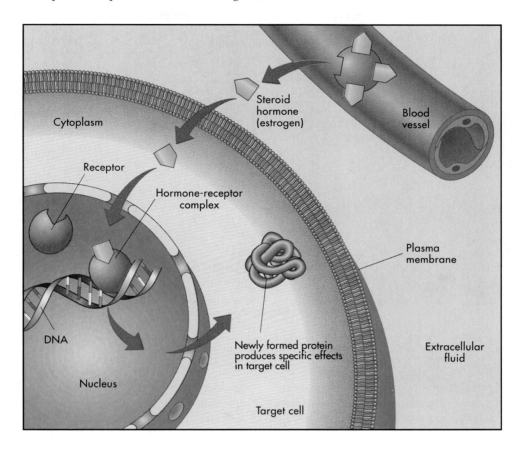

the muscle contractions that push the baby through the birth canal become stronger and stronger by means of a positive feedback mechanism that regulates secretion of the hormone oxytocin.

 Prostaglandins

Prostaglandins (PGs) or tissue hormones are important and extremely powerful substances found in a wide variety of tissues. They play an important role in communication and the control of many body functions but do not meet the defi-

nition of a typical hormone. The term *tissue hormone* is appropriate because in many instances a prostaglandin is produced in a tissue and diffuses only a short distance to act on cells within that tissue. Typical hormones influence and control activities of widely separated organs; typical prostaglandins influence activities of neighboring cells.

The prostaglandins in the body can be divided into several groups. Three classes of prostaglandins—prostaglandin A (PGA), prostaglandin E (PGE), and prostaglandin F (PGF)—are among the best known. Prostaglandins have profound effects on many body functions. They influence

FIGURE 9–4   *Negative Feedback.* The secretion of most hormones is regulated by negative feedback mechanisms that tend to reverse any deviations from normal. In this example, an increase in blood glucose triggers secretion of insulin. Since insulin promotes glucose uptake by cells, the blood glucose level is restored to its lower, normal level.

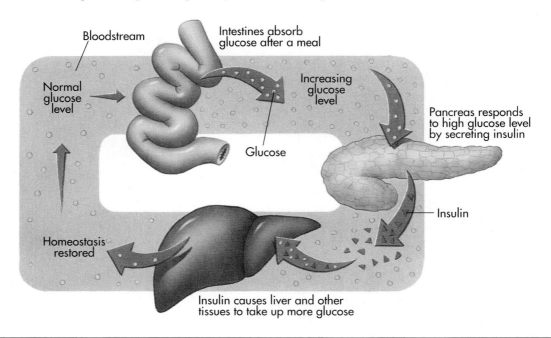

respiration, blood pressure, gastrointestinal secretions, and the reproductive system. Researchers believe that most prostaglandins regulate cells by influencing the production of cyclic AMP. Although much research is yet to be done, prostaglandins are already playing an important role in the treatment of conditions such as high blood pressure, asthma, and ulcers.

◑ *Pituitary Gland*

The **pituitary** (pi-TOO-i-tair-ee) **gland** is a small but mighty structure. Although no larger than a pea, it is really two endocrine glands. One is called the **anterior pituitary gland** or *adenohypophysis* (ad-e-no-hye-POF-i-sis), and the other is called the **posterior pituitary gland** or *neurohypophysis* (noo-ro-hye-POF-i-sis). Differences between the two glands are suggested by their names—*adeno* means "gland," and *neuro* means "nervous." The adenohypophysis has the structure of an endocrine gland, whereas the neurohypophysis has the structure of nervous tissue. Hormones secreted by the adenohypophysis serve very different functions from those released from the neurohypophysis.

The protected location of this dual gland suggests its importance. The pituitary gland lies buried deep in the cranial cavity, in the small depression of the sphenoid bone that is shaped like a saddle and called the *sella turcica* (Turkish saddle). A stemlike structure, the pituitary stalk, attaches the gland to the undersurface of the brain. More specifically, the stalk attaches the pituitary body to the hypothalamus.

ANTERIOR PITUITARY GLAND HORMONES

The anterior pituitary gland secretes several major hormones. Each of the four hormones listed as a **tropic** (TRO-pik) **hormone** in Table 9-1 stimulates another endocrine gland to grow and secrete its hormones. Because the anterior pitu-

Growth Hormone Abnormalities

Hypersecretion of growth hormone during the early years of life produces a condition called **gigantism** (jye-GAN-tizm). The name suggests the obvious characteristics of this condition. The child grows to giant size. Hyposecretion of the growth hormone produces pituitary **dwarfism** (DWARF-izm).

If the anterior pituitary gland secretes too much growth hormone after the normal growth years, then the disease called

acromegaly (ak-ro-MEG-ah-lee) develops. Characteristics of this disease are enlargement of the bones of the hands, feet, jaws, and cheeks. The facial appearance typical of acromegaly results from the combination of bone and soft tissue overgrowth. A prominent forehead and large nose are characteristic. In addition, the skin is characterized by large, widened pores, and the mandible grows in length so that separation of the lower teeth commonly occurs.

itary gland exerts this control over the structure and function of the thyroid gland, the adrenal cortex, the ovarian follicles, and the corpus luteum, it is sometimes referred to as the *master gland.*

Thyroid-stimulating hormone (TSH) acts on the thyroid gland. As its name suggests, it stimulates the thyroid gland to increase secretion of thyroid hormone.

The **adrenocorticotropic** (ad-re-no-kor-ti-ko-TRO-pik) **hormone (ACTH)** acts on the adrenal cortex. It stimulates the adrenal cortex to increase in size and to secrete larger amounts of its hormones, especially larger amounts of cortisol (hydrocortisone).

Follicle-stimulating hormone (FSH) stimulates the primary ovarian follicles in an ovary to start growing and to continue developing to maturity (that is, to the point of ovulation). FSH also stimulates follicle cells to secrete estrogens. In the male, FSH stimulates the seminiferous tubules to grow and form sperm.

Luteinizing (LOO-te-nye-zing) **hormone (LH)** acts with FSH to perform several functions. It stimulates a follicle and ovum to complete their growth to maturity, it stimulates follicle cells to secrete estrogens, and it causes ovulation (rupturing of the mature follicle with expulsion of its ripe ovum). Because of this function, LH is sometimes called the *ovulating hormone*. Finally, LH stimulates the formation of a golden body, the corpus luteum, in the ruptured follicle; the process is called *luteinization*. This function, of course, is the one that earned LH its title of *luteinizing hormone.* The male pituitary gland also secretes LH; it was formerly called *interstitial cell-stimulating hormone (ICSH)* because it stimulates interstitial cells in the testes to develop and secrete testosterone, the male sex hormone.

The **melanocyte-stimulating hormone (MSH)** causes a rapid increase in the synthesis and dispersion of melanin (pigment) granules in specialized skin cells.

Another important hormone secreted by the anterior pituitary gland is **growth hormone.** Growth hormone (GH) speeds up the movement of digested proteins (amino acids) out of the blood and into the cells, and this accelerates the cells' anabolism (build up) of amino acids to form tissue proteins; hence this action promotes normal growth. Growth hormone also affects fat and carbohydrate metabolism; it accelerates fat catabolism (breakdown) but slows glucose catabolism. This means that less glucose leaves the blood to enter cells, and therefore the amount of glucose in the blood increases. Thus growth hormone and insulin have opposite effects on blood glucose. Insulin decreases blood glucose, and growth hormone increases it. Too much insulin in the blood produces **hypoglycemia** (hye-po-glye-SEE-me-ah) (lower than normal blood glucose concentration). Too much growth hormone produces **hyperglycemia** (higher than normal blood glucose concentration).

The anterior pituitary gland also secretes **prolactin** (pro-LAK-tin) or lactogenic hormone. During pregnancy, prolactin stimulates the breast

development necessary for eventual lactation (milk secretion). Also, soon after delivery of a baby, prolactin stimulates the breasts to start secreting milk, a function suggested by prolactin's other name, lactogenic hormone.

For a brief summary of anterior pituitary hormone target organs and functions, see Figure 9-5.

POSTERIOR PITUITARY GLAND HORMONES

The posterior pituitary gland releases two hormones—**antidiuretic** (an-tie-dye-yoo-RET-ik) **hormone (ADH)** and **oxytocin** (ok-see-TOE-sin). ADH accelerates the reabsorption of water from urine in kidney tubules back into the blood. With more water moving out of the tubules into the blood, less water remains in the tubules, and therefore less urine leaves the body. The name *antidiuretic hormone* is appropriate because *anti-* means "against" and *diuretic* means "increasing the volume of urine excreted." Therefore antidiuretic means "acting against an increase in urine volume"; in other words, ADH acts to decrease urine volume. Hyposecretion of ADH results in **diabetes insipidus** (dye-ah-BEE-tes in-SIP-i-dus), a condition in which large volumes of urine are formed. Dehydration and electrolyte imbalances may cause serious problems unless the sufferer is treated with injections or nasal sprays containing ADH.

The posterior pituitary hormone oxytocin is secreted by a woman's body before and after she has a baby. Oxytocin stimulates contraction of the smooth muscle of the pregnant uterus and is believed to initiate and maintain labor. This is why physicians sometimes prescribe oxytocin injections to induce or increase labor. Oxytocin also performs a function important to a newborn baby. It causes the glandular cells of the breast to release milk into ducts from which a baby can obtain it by sucking. The right side of Figure 9-5 summarizes posterior pituitary functions.

○ *Hypothalamus*

In discussing ADH and oxytocin, we noted that these hormones were *released* from the posterior lobe of the pituitary. Actual production of these two hormones occurs in the hypothalamus. Two groups of specialized neurons in the hypothala-

FIGURE 9–5 *Pituitary Hormones.* Principal anterior and posterior pituitary hormones and their target organs.

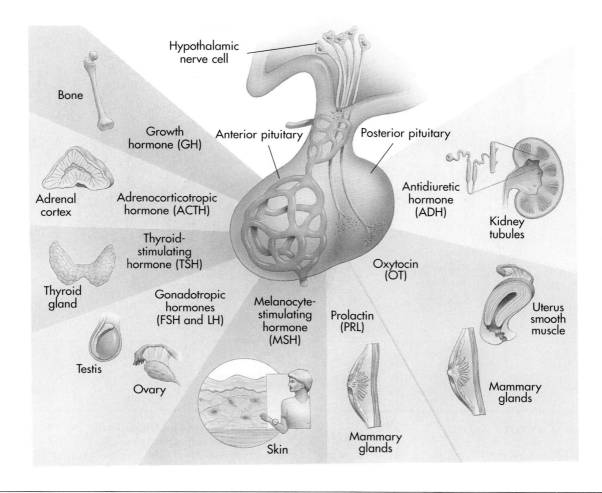

mus synthesize the posterior pituitary hormones, which then pass down along axons into the pituitary gland. Release of ADH and oxytocin into the blood is controlled by nervous stimulation.

In addition to oxytocin and ADH, the hypothalamus also produces substances called **releasing** and **inhibiting hormones.** These substances are produced in the hypothalamus and then travel directly through a specialized blood capillary system to the anterior pituitary gland, where

they cause the release of anterior pituitary hormones or, in a number of instances, inhibit their production and their release into the general circulation.

The combined nervous and endocrine functions of the hypothalamus allow it to play a dominant role in the regulation of many body functions related to homeostasis. Examples include the regulation of body temperature, appetite, and thirst.

FIGURE 9–6   *Thyroid and Parathyroid Glands.* Note their relationship to each other and to the larynx (voice box) and trachea.

Back of larynx

- Epiglottis
- Hyoid bone
- Larynx
- Superior parathyroid glands
- Thyroid gland
- Inferior parathyroid glands
- Trachea

Thyroid Gland

Earlier in this chapter, we mentioned that some endocrine glands are not located in a body cavity. The thyroid is one of these. It lies in the neck just below the larynx (Figure 9-6).

The thyroid gland secretes two thyroid hormones, **thyroxine** (thye-ROK-sin) or **T₄** and **triiodothyronine** (try-eye-o-doe-THY-ro-neen) or **T₃.** It also secretes the hormone **calcitonin** (kal-si-TOE-nin). Of the two thyroid hormones, T_4 is the more abundant; however, T3 is the more potent and is considered by physiologists to be the principal thyroid hormone. One molecule of T_4 contains four atoms of iodine, and one molecule of T_3, as its name suggests, contains three iodine atoms. For T_4 to be produced in adequate amounts, the diet must contain sufficient iodine.

Most endocrine glands do not store their hormones but secrete them directly into the blood as they are produced. The thyroid gland is different

FIGURE 9–7   *Thyroid Gland Tissue.* Note that each of the follicles is filled with colloid. The colloid serves as a storage medium for the thyroid hormones.

- Thyroid follicle
- Colloid in follicle

Thyroid Hormone Abnormalities

Hyperthyroidism (hye-per-THY-royd-izm) or oversecretion of the thyroid hormones dramatically increases the metabolic rate. Food material is burned by the cells at an excessive rate, and individuals who suffer from this condition lose weight, are irritable, have an increased appetite, and often show protrusion of the eyeballs due in part to edema of tissue at the back of the eye socket; see Figure *A*.

Hypothyroidism (hye-po-THY-royd-izm) or undersecretion of thyroid hormones can be caused by and result in a number of different conditions. Low dietary intake of iodine causes a painless enlargement of the thyroid gland called **simple goiter** (GOY-ter), shown in Figure *B*. This condition was once common in areas of the United States where the iodine content of the soil and water is inadequate. The use of iodized salt has dramatically reduced the incidence of simple goiter caused by low iodine intake. In simple goiter the gland enlarges to compensate for the lack of iodine in the diet necessary for the synthesis of thyroid hormones.

Hyposecretion of thyroid hormones during the formative years leads to a condition called **cretinism** (KREE-tin-izm). It is characterized by a low metabolic rate, retarded growth and sexual development, and, frequently, mental retardation. Later in life, deficient thyroid hormone secretion produces the disease called **myxedema** (mik-se-DEE-mah). The low metabolic rate that characterizes myxedema leads to lessened mental and physical vigor, weight gain, loss of hair, and swelling of tissues.

A

B

in that it stores considerable amounts of the thyroid hormones in the form of a colloid compound seen in Figure 9-7. The colloid material is stored in the follicles of the gland, and when the thyroid hormones are needed, they are released from the colloid and secreted into the blood.

T_4 and T_3 influence every one of the trillions of cells in our bodies. They make them speed up their release of energy from foods. In other words, these thyroid hormones stimulate cellular metabolism. This has far-reaching effects. Because all body functions depend on a normal supply of

FIGURE 9–8   *Regulation of Blood Calcium Levels.* Calcitonin and parathyroid hormones have antagonistic (opposite) effects on calcium concentration in the blood.

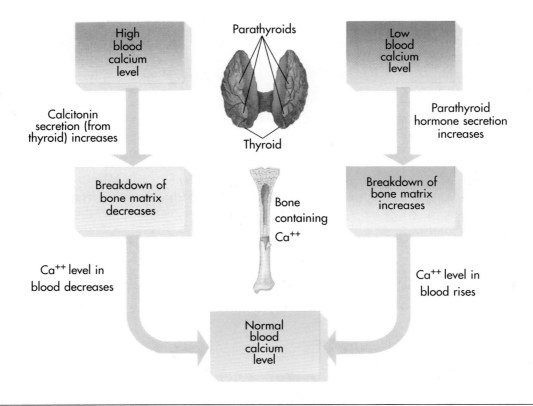

energy, they all depend on normal thyroid secretion. Even normal mental and physical growth and development depend on normal thyroid functioning.

Calcitonin decreases the concentration of calcium in the blood by first acting on bone to inhibit its breakdown. With less bone being resorbed, less calcium moves out of bone into blood, and, as a result, the concentration of calcium in blood decreases. An increase in calcitonin secretion quickly follows any increase in blood calcium concentration, even if it is a slight one. This causes blood calcium concentration to decrease to its normal level. Calcitonin thus helps maintain homeostasis of blood calcium. It prevents a harmful excess of calcium in the blood, a condition called **hypercalcemia** (hye-per-kal-SEE-me-ah), from developing.

Parathyroid Glands

The **parathyroid glands** are small glands. There are usually four of them, and they are found on the back of the thyroid gland (Figure 9-6). The parathyroid glands secrete **parathyroid hormone (PTH).**

Parathyroid hormone increases the concentration of calcium in the blood—the opposite effect of the thyroid gland's calcitonin. Whereas calcitonin acts to decrease the amount of calcium being resorbed from bone, parathyroid hormone acts to increase it. Parathyroid hormone stimulates bone-resorbing cells or osteoclasts to increase their breakdown of bone's hard matrix, a process that frees the calcium stored in the matrix. The released calcium then moves out of bone into

FIGURE 9–9 *The Adrenal Gland.* The three cell layers of the adrenal cortex are shown. The zona glomerulosa cells secrete mineralocorticoids (aldo-sterone). The zona fasciculata cells secrete glucocorticoids (hydorcortisone). The zona reticularis cells secrete sex hormones (androgens).

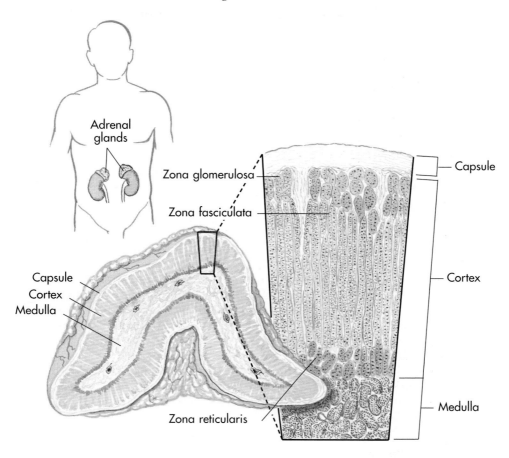

blood, and this in turn increases the blood's calcium concentration. For a summary of the antagonistic effects of calcitonin and parathyroid hormone, see Figure 9-8. This is a matter of life-and-death importance because our cells are extremely sensitive to changing amounts of blood calcium. They cannot function normally with too much or too little calcium. For example, with too much blood calcium, brain cells and heart cells soon do not function normally; a person becomes mentally disturbed, and the heart may stop. However, with too little blood calcium, nerve cells become overactive, sometimes to such a degree that they bombard muscles with so many impulses that the muscles go into spasms.

Adrenal Glands

As you can see in Figures 9-1 and 9-9, an adrenal gland curves over the top of each kidney. From the surface an adrenal gland appears to be only one organ, but it is actually two separate endocrine glands: the **adrenal cortex** and the **adrenal medulla.** Does this two-glands-in-one structure remind you of another endocrine organ? (See pp. 196-198). The adrenal cortex is the outer part of an adrenal gland, and the medulla is its inner part. Adrenal cortex hormones have different names and quite different actions from adrenal medulla hormones.

ADRENAL CORTEX

Three different zones or layers of cells make up the **adrenal cortex** (Figure 9-9). Starting with the zone or layer directly under the outer capsule of the gland and proceeding toward the center (that is, going from superficial to deep), their names are:

1. Zona (ZO-nah) glomerulosa (glo-mare-yoo-LO-sah)
2. Zona (ZO-nah) fasciculata (fas-ik-yoo-LAH-tah)
3. Zona (ZO-nah) reticularis (re-tic-yoo-LAIR-is)

Hormones secreted by the three cell layers or zona of the adrenal cortex are called **corticoids** (KOR-ti-koyds). The outer zone of adrenal cortex cells or zona glomerulosa secretes hormones called **mineralocorticoids** (min-er-al-o-KOR-ti-koyds) or **MCs** for short. The main mineralocorticoid is the hormone **aldosterone** (al-DOS-ste-rone). The middle zone or zona fasciculata secretes **glucocorticoids** (gloo-ko-KOR-ti-koyds) or **GCs. Cortisol** or **hydrocortisone** is the chief glucocorticoid. The innermost or deepest zone of the cortex or zona reticularis secretes small amounts of **sex hormones.** Sex hormones secreted by the adrenal cortex resemble testosterone. We shall now discuss briefly the functions of these three kinds of adrenal cortical hormones.

As their name suggests, **mineralocorticoids** help control the amount of certain mineral salts (mainly sodium chloride) in the blood. Aldosterone is the chief mineralocorticoid. Remember its main functions—to increase the amount of sodium and decrease the amount of potassium in the blood—because these changes lead to other profound changes. Aldosterone increases blood sodium and decreases blood potassium by influencing the kidney tubules. It causes them to speed up their reabsorption of sodium back into the blood so that less of it will be lost in the urine. At the same time, aldosterone causes the tubules to increase their secretion of potassium so that more of this mineral will be lost in the urine. The effects of aldosterone speed up kidney reabsorption of water.

One of the important functions of glucocorticoids is to help maintain normal blood glucose concentration. Glucocorticoids increase **gluconeogenesis** (gloo-ko-nee-o-JEN-e-sis), a process that converts amino acids or fatty acids to glucose and that is performed mainly by liver cells. Glucocorticoids act in several ways to increase gluconeogenesis. They promote the breakdown of tissue proteins to amino acids, especially in muscle cells. Amino acids thus formed move out of the tissue cells into blood and circulate to the liver. Liver cells then change them to glucose by the process of gluconeogenesis. The newly formed glucose leaves the liver cells and enters the blood. This action increases blood glucose concentration.

In addition to performing these functions that are necessary for maintaining normal blood glucose concentration, glucocorticoids also play an essential part in maintaining normal blood pressure. They act in a complicated way to make it possible for two other hormones secreted by the adrenal medulla to partially constrict blood vessels, a condition necessary for maintaining normal blood pressure. Also, glucocorticoids act with these hormones from the adrenal medulla to produce an antiinflammatory effect. They bring about a normal recovery from inflammations produced by many kinds of agents. The use of hydrocortisone to relieve skin rashes, for example, is based on the antiinflammatory effect of glucocorticoids.

Another effect produced by glucocorticoids is called their *antiimmunity, antiallergy effect.* Glucocorticoids bring about a decrease in the number of certain cells that produce antibodies, substances that make us immune to some factors and allergic to others.

When extreme stimuli act on the body, they produce an internal state or condition known as *stress.* Surgery, hemorrhage, infections, severe burns, and intense emotions are examples of extreme stimuli that bring on stress. The normal adrenal cortex responds to the condition of stress by quickly increasing its secretion of glucocorticoids. This fact is well established. What is still not known, however, is whether the increased amount of glucocorticoids helps the body cope successfully with stress. Increased glucocorticoid secretion is only one of many ways in which the body responds to stress, but it is one of the first stress responses, and it brings about many of the other stress responses. Examine Figure 9-10 to discover what stress responses are produced by a high concentration of glucocorticoids in the blood.

FIGURE 9–10 *Stress Responses Induced by High Concentrations of Glucocorticoids in Blood.*

The sex hormones that are secreted by the zona reticularis are weak male hormones **(androgens)** similar to testosterone. These hormones are secreted in small amounts in both males and females. In females, these androgens stimulate the female sexual drive. In males, so much testosterone is secreted by the testes that adrenal androgens are physiologically insignificant.

ADRENAL MEDULLA

The **adrenal medulla,** or inner portion of the adrenal gland shown in Figure 9-9, secretes the hormones **epinephrine** (ep-i-NEF-rin) and **norepinephrine** (nor-ep-i-NEF-rin).

Our bodies have many ways to defend themselves against enemies that threaten their well-being. A physiologist might say that the body resists stress by making many stress responses. We have just discussed increased glucocorticoid secretion. An even faster-acting stress response is increased secretion by the adrenal medulla. This occurs very rapidly because nerve impulses conducted by sympathetic nerve fibers stimulate the adrenal medulla. When stimulated, it literally squirts epinephrine and norepinephrine into the blood. Like glucocorticoids, these hormones may help the body resist stress. Unlike glucocorticoids, they are not essential for maintaining life. On the other hand, glucocorticoids, the hormones from the adrenal cortex, may help the body resist stress and are essential for life.

Suppose you suddenly faced some threatening situation. Imagine that a gunman threatened to kill you or that your doctor told you that you had to have a dangerous operation. Almost instantaneously, the medullas of your two adrenal glands would be galvanized into feverish activity. They would quickly secrete large amounts of epinephrine (adrenaline) into your blood. Many of your body functions would seem to be supercharged.

Adrenal Hormone Abnormalities

Injury, disease states, or malfunction of the adrenal glands can result in hypersecretion or hyposecretion of several different hormones.

Tumors of the adrenal cortex located in the zona fasciculata often result in the production of abnormally large amounts of glucocorticoids. The medical name for this is **Cushing's syndrome.** Figure *A* shows a boy just diagnosed with Cushing's syndrome. Figure *B* shows the same boy 4 months later after treatment. For some reason many more women than men develop Cushing's syndrome. Its most noticeable features are the so-called moon face and the buffalo hump on the upper back that develop because of redistribution of body fat. These individuals also have elevated blood sugar levels and suffer frequent infections. Surgical removal of a glucocorticoid-producing tumor may result in dramatic improvement of the moon-face symptom within 6 months.

Deficiency or hyposecretion of adrenal cortex hormones results in a condition called **Addison's disease.** Reduced cortical hormone levels result in muscle weakness, reduced blood sugar, nausea, loss of appetite, and weight loss.

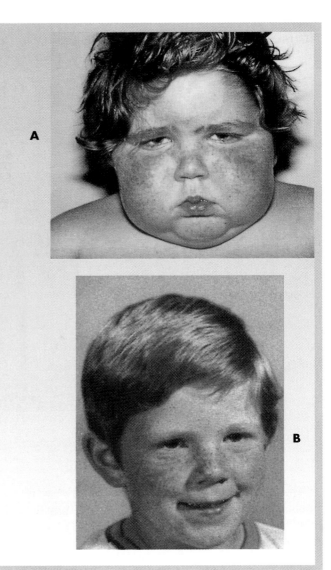

Your heart would beat faster; your blood pressure would rise; more blood would be pumped to your skeletal muscles; your blood would contain more glucose for more energy, and so on. In short, you would be geared for strenuous activity, for "fight or flight." Epinephrine prolongs and intensifies changes in body function brought about by the stimulation of the sympathetic subdivision of the autonomic nervous system. Recall from Chapter 7 that sympathetic or adrenergic nerve fibers release epinephrine and norepinephrine as neurotransmitter substances.

The close functional relationship between the nervous and the endocrine systems is perhaps most noticeable in the body's response to stress. The term **general-adaptation syndrome** or **GAS** is often used to describe how the body mobilizes different defense mechanisms when threatened by harmful stimuli. In generalized stress conditions the hypothalamus acts on the anterior pituitary gland to cause the release of ACTH, which stimulates the adrenal cortex to secrete glucocorticoids. In addition, the sympathetic subdivision of the autonomic nervous system is stimulated with

FIGURE 9–11 *Pancreas.* Two pancreatic islets (of Langerhans) or hormone-producing areas are evident among the pancreatic cells that produce the pancreatic digestive juice.

Common bile duct

Pancreatic duct (for digestive juice)

Pancreatic cells that produce digestive juice

Pancreatic islets (of Langerhans)

the adrenal medulla, so the release of epinephrine and norepinephrine occurs to assist the body in responding to the stressful stimulus. Unfortunately, during periods of prolonged stress, glucocorticoids may have harmful side effects because they are antiinflammatory and cause blood vessels to constrict. For example, decreased immune activity in the body may promote the spread of infections, and prolonged blood vessel constriction may lead to increased blood pressure.

Pancreatic Islets

All the endocrine glands discussed so far are big enough to be seen without a magnifying glass.

The **pancreatic islets** or **islets of Langerhans,** in contrast, are too tiny to be seen without a microscope. These glands are merely little clumps of cells scattered like islands in a sea among the pancreatic cells that secrete the pancreatic digestive juice (Figure 9-11).

There are two kinds of cells in the pancreatic islets: the *alpha cells* and *beta cells.* Alpha cells secrete a hormone called **glucagon,** whereas beta cells secrete one of the most famous of all hormones, **insulin.** Glucagon accelerates a process called **liver glycogenolysis** (glye-ko-jen-OL-i-sis). Glycogenolysis is a chemical process by which the glucose stored in the liver cells in the form of glycogen is converted to glucose. This glucose then leaves the liver cells and enters the blood.

Glucagon therefore increases blood glucose concentration.

Insulin and glucagon are antagonists. In other words, insulin decreases blood glucose concentration; glucagon increases it. Insulin is the only hormone that can decrease blood glucose concentration. Other hormones, however, increase its concentration, including glucocorticoids, growth hormone, and glucagon. Insulin decreases blood glucose by accelerating its movement out of the blood, through cell membranes, and into cells. As glucose enters the cells at a faster rate, the cells increase their metabolism of glucose. Briefly then, insulin decreases blood glucose and increases glucose metabolism.

If the pancreatic islets secrete a normal amount of insulin, a normal amount of glucose enters the cells, and a normal amount of glucose stays behind in the blood. ("Normal" blood glucose is about 80 to 120 mg of glucose in every 100 ml of blood.) If the pancreatic islets secrete too much

insulin, as they sometimes do when a person has a tumor of the pancreas, more glucose than usual leaves the blood to enter the cells, and blood glucose decreases. If the pancreatic islets secrete too little insulin, as they do in type I (insulin-dependent) **diabetes** (dye-ah-BEE-tes) **mellitus** (mell-EYE-tus), less glucose leaves the blood to enter the cells, so the blood glucose increases, sometimes to even three or more times the normal amount. Most cases of type II (insulin-independent) diabetes mellitus result from an abnormality of the insulin receptors, preventing the normal effects of insulin on its target cells and thus also raising blood glucose levels.

Screening tests for all types of diabetes mellitus rely on the fact that the blood glucose level is elevated in this condition. Today, most screening is done with a simple test with a drop of blood. Subjects with a high blood glucose level are suspected of having diabetes mellitus. Testing for sugar in the urine is another common screening procedure. In diabetes mellitus, excess glucose is filtered out of the blood by the kidneys and lost in the urine, producing the condition **glycosuria** (glye-ko-SOO-ree-ah).

Exercise and Diabetes Mellitus

Type I (insulin-dependent) diabetes mellitus is characterized by high blood glucose concentration because the lack of sufficient insulin prevents glucose from entering cells. However, exercise physiologists have found that aerobic training increases the number of insulin receptors in target cells and the insulin affinity (attraction) of the receptors. This condition allows a small amount of insulin to have a greater effect than it would have otherwise had. Thus exercise reduces the severity of the diabetic condition.

All forms of diabetes benefit from properly planned exercise therapy. Not only is this form of treatment natural and cost effective, but it also helps reduce or prevent other problems such as obesity and heart disease.

● Female Sex Glands

A woman's primary sex glands are her two ovaries. Each ovary contains two different kinds of glandular structures: the ovarian follicles and the corpus luteum. **Ovarian follicles** secrete estrogen, the "feminizing hormone." Estrogen is involved in the development and maturation of the breasts and external genitals. This hormone is also responsible for development of adult female body contours and initiation of the mentrual cycle. The **corpus luteum** chiefly secretes progesterone but also some estrogen. We shall save our discussion of the structure of these endocrine glands and the functions of their hormones for Chapter 19.

● Male Sex Glands

Some of the cells of the testes produce the male sex cells called **sperm.** Other cells in the testes, male reproductive ducts, and glands produce the liquid portion of the male reproductive fluid called *semen*. The interstitial cells in the testes secrete the male sex hormone called **testosterone** directly into the blood. These cells of the testes are therefore the male endocrine glands. Testosterone is the "masculinizing hormone." It is responsible for the maturation of the external genitals, beard growth, changes in voice at puberty, and for the muscular development and body contours typical of the male. Chapter 19 contains more information about the structure of the testes and the functions of testosterone.

Thymus

The thymus is located in the mediastinum (Figure 9-1), and in infants it may extend up into the neck as far as the lower edge of the thyroid gland. Like the adrenal gland, the thymus has a cortex and medulla. Both portions are composed largely of lymphocytes (white blood cells). As part of the body's immune system, the endocrine function of the thymus is not only important but essential. This small structure (it weighs at most less than a gram) plays a critical part in the body's defenses against infections—its vital immunity mechanism.

The hormone **thymosin** (THY-mo-sin) has been isolated from thymus tissue and is considered responsible for its endocrine activity. Thymosin is actually a group of several hormones that together play an important role in the development and function of the body's immune system.

Suppression of the immune system sometimes occurs in certain disease states and in patients who are undergoing massive chemotherapy or radiotherapy for the treatment of cancer. Such individuals are said to be "immunosuppressed" and are extremely susceptible to infections. Thymosin may prove useful as an activator of the immune system in such patients.

Placenta

The placenta functions as a temporary endocrine gland. During pregnancy, it produces **chorionic** (KO-ree-on-ik) **gonadotropins** (gon-ah-doe-TRO-pins), so called because they are tropic hormones secreted by cells of the **chorion** (KO-ree-on), the outermost membrane that surrounds the baby during development in the uterus. In addition to chorionic gonadotropins, the placenta also produces estrogen and progesterone. During pregnancy, the kidneys excrete large amounts of chorionic gonadotropins in the urine. This fact, discovered more than a half century ago, led to the development of the now-familiar pregnancy tests.

Pineal Gland

The pineal gland is a small, pine-cone-shaped gland near the roof of the third ventricle of the brain (see Figure 7-9). It is easily located in a child but becomes fibrous and encrusted with calcium deposits as a person ages. The pineal gland produces a number of hormones in very small quantities, with **melatonin** being the most significant. Melatonin inhibits the tropic hormones that affect the ovaries, and it is thought to be involved in regulating the onset of puberty and the menstrual cycle in women. Because the pineal gland receives and responds to sensory information from the optic nerves, it is sometimes called the *third eye*. The pineal gland uses information regarding changing light levels to adjust its output of melatonin; melatonin levels increase during the night and decrease during the day. This cyclic variation is thought to be an important timekeeping mechanism for the body's internal clock.

Other Endocrine Structures

Continuing research into the endocrine system has shown that nearly every organ and system has an endocrine function. Tissues in the kidneys, stomach, intestines, and other organs secrete hormones that regulate a variety of essential human functions. One of the most recently discovered hormones is **atrial natriuretic hormone (ANH).** Secreted by cells in the wall of the heart's atria (upper chambers), ANH is an important regulator of fluid and electrolyte homeostasis. ANH is an antagonist to aldosterone. Aldosterone stimulates the kidney to retain sodium ions and water, whereas ANH stimulates loss of sodium ions and water.

MECHANISMS OF HORMONE ACTION

A Endocrine glands secrete chemicals (hormones) into the blood (Figure 9-1)

B Hormones perform general functions of communication and control but a slower, longer-lasting type of control than that provided by nerve impulses

C Cells acted on by hormones are called *target organ cells*

D Protein hormones (first messengers) bind to receptors on the target cell membrane, triggering second messengers to affect the cell's activities (Figure 9-2)

E Steroid hormones bind to receptors within the target cell nucleus and influence cell activity by acting on DNA (Figure 9-3)

REGULATION OF HORMONE SECRETION

A Hormone secretion is controlled by homeostatic feedback

B Negative feedback—mechanisms that reverse the direction of a change in a physiological system (Figure 9-4)

C Positive feedback—(uncommon) mechanisms that amplify physiological changes

PROSTAGLANDINS

A Prostaglandins (PGs) are powerful substances found in a wide variety of body tissues

B PGs are often produced in a tissue and diffuse only a short distance to act on cells in that tissue

C Several classes of PGs include prostaglandin A (PGA), prostaglandin E (PGE), and prostaglandin F (PGF)

D PGs influence many body functions, including respiration, blood pressure, gastrointestinal secretions, and reproduction

PITUITARY GLAND (Figure 9-5)

A Anterior pituitary gland (adenohypophysis)

  1 Names of major hormones

    a Thyroid-stimulating hormone (TSH)

    b Adrenocorticotropic hormone (ACTH)

    c Follicle-stimulating hormone (FSH)

    d Luteinizing hormone (LH)

    e Melanocyte-stimulating hormone (MSH)

    f Growth hormone (GH)

    g Prolactin (lactogenic hormone)

  2 Functions of major hormones

    a TSH—stimulates growth of the thyroid gland; also stimulates it to secrete thyroid hormone

    b ACTH—stimulates growth of the adrenal cortex and stimulates it to secrete glucocorticoids (mainly cortisol)

    c FSH—initiates growth of ovarian follicles each month in the ovary and stimulates one or more follicles to develop to the stage of maturity and ovulation; FSH also stimulates estrogen secretion by developing follicles; stimulates sperm production in the male

    d LH—acts with FSH to stimulate estrogen secretion and follicle growth to maturity; causes ovulation; causes luteinization of the ruptured follicle and stimulates progesterone secretion by corpus luteum; causes interstitial cells in the testes to secrete testosterone in the male

    e MSH—causes a rapid increase in the synthesis and spread of melanin (pigment) in the skin

    f GH—stimulates growth by accelerating protein anabolism; also accelerates fat catabolism and slows glucose catabolism; by slowing glucose catabolism, tends to increase blood glucose to higher than normal level (hyperglycemia)

    g Prolactin or lactogenic hormone—stimulates breast development during pregnancy and secretion of milk after the delivery of the baby

B Posterior pituitary gland (neurohypophysis)

  1 Names of hormones

    a Antidiuretic hormone (ADH)

    b Oxytocin

  2 Functions of hormones

    a ADH—accelerates water reabsorption from urine in the kidney tubules into the blood, thereby decreasing urine secretion

    b Oxytocin—stimulates the pregnant uterus to contract; may initiate labor; causes glandular cells of the breast to release milk into ducts

HYPOTHALAMUS

A Actual production of ADH and oxytocin occurs in the hypothalamus

B After production in the hypothalamus, hormones pass along axons into the pituitary gland

C The secretion and release of posterior pituitary hormones is controlled by nervous stimulation

D The hypothalamus controls many body functions related to homeostasis (temperature, appetite, and thirst)

THYROID GLAND (Figure 9-6)

A Names of hormones

  1 Thyroid hormone—thyroxine (T_4) and triiodothyronine (T_3)

  2 Calcitonin

B Functions of hormones

  1 Thyroid hormones—accelerate catabolism (increase the body's metabolic rate)

  2 Calcitonin—decreases the blood calcium concentration by inhibiting breakdown of bone, which would release calcium into the blood

PARATHYROID GLANDS (Figure 9-6)

A Name of hormone—parathyroid hormone (PTH)

B Function of hormone—increases blood calcium concentration by increasing the breakdown of bone with the release of calcium into the blood

ADRENAL GLANDS (Figure 9-9)

A Adrenal cortex

  1 Names of hormones (corticoids)

    a Glucocorticoids (GCs)—chiefly cortisol (hydrocortisone)

    b Mineralocorticoids (MCs)—chiefly aldosterone

    c Sex hormones—small amounts of male hormones (androgens) secreted by adrenal cortex of both sexes

  2 Cell layers (zonae)

    a Zona glomerulosa—outermost layer, secretes mineralocorticoids

    b Zona fasciculata—middle layer, secretes glucocorticoids

    c Zona reticularis—deepest or innermost layer, secretes sex hormones

  3 Mineralocorticoids—increase blood sodium and decrease body potassium concentrations by accelerating kidney tubule reabsorption of sodium and excretion of potassium

  4 Functions of glucocorticoids

    a Help maintain normal blood glucose concentration by increasing gluconeogenesis—the formation of "new" glucose from amino acids produced by the breakdown of proteins, mainly those in muscle tissue cells; also the conversion to glucose of fatty acids produced by the breakdown of fats stored in adipose tissue cells

    b Play an essential part in maintaining normal blood pressure—make it possible for epinephrine and norepinephrine to maintain a normal degree of vasoconstriction, a condition necessary for maintaining normal blood pressure

    c Act with epinephrine and norepinephrine to produce an antiinflammatory effect, to bring about normal recovery from inflammations of various kinds

    d Produce antiimmunity, antiallergy effect; bring about a decrease in the number of lymphocytes and plasma cells and therefore a decrease in the amount of antibodies formed

    e Secretion of glucocorticoid quickly increases when the body is thrown into a condition of stress; high blood concentration of glucocorticoids, in turn, brings about many other stress responses (Figure 9-10)

B Adrenal medulla

  1 Names of hormones—epinephrine (adrenaline) and norepinephrine

  2 Functions of hormones—help the body resist stress by intensifying and prolonging the effects of sympathetic stimulation; increased epinephrine secretion is the first endocrine response to stress

PANCREATIC ISLETS (Figure 9-11)

A Names of hormones

  1 Glucagon—secreted by alpha cells

  2 Insulin—secreted by beta cells

B Functions of hormones

  1 Glucagon increases the blood glucose level by accelerating liver glycogenolysis (conversion of glycogen to glucose)

  2 Insulin decreases the blood glucose by accelerating the movement of glucose out of the blood into cells, which increases glucose metabolism by cells

FEMALE SEX GLANDS

The ovaries contain two structures that secrete hormones—the ovarian follicles and the corpus luteum; see Chapter 19

A Effects of estrogen (feminizing hormone)

  1 Development and maturation of breasts and external genitals

  2 Development of adult female body contours

  3 Initiation of menstrual cycle

MALE SEX GLANDS

The interstitial cells of testes secrete the male hormone testosterone; see Chapter 19

A Effects of testosterone (masculinizing hormone)

  1 Maturation of external genitals

  2 Beard growth

  3 Voice changes at puberty

  4 Development of musculature and body contours typical of the male

THYMUS

A Name of hormone—thymosin

B Function of hormone—plays an important role in the development and function of the body's immune system

PLACENTA

A Name of hormones—chorionic gonadotropins, estrogens, and progesterone
B Functions of hormones—maintain the corpus luteum during pregnancy

PINEAL GLAND

A A cone-shaped gland near the roof of the third ventricle of the brain
  1 Glandular tissue predominates in children and young adults
  2 Becomes fibrous and calcified with age
B Called *third eye* because its influence on secretory activity is related to the amount of light entering the eyes
C Secretes melatonin, which
  1 Inhibits ovarian activity
  2 Regulates the body's internal clock

OTHER ENDOCRINE STRUCTURES

A Many organs (for example, the stomach, intestines, and kidney) produce endocrine hormones
B The atrial wall of the heart secretes atrial natriuretic hormone (ANH), which stimulates sodium loss from the kidneys

---

N E W   W O R D S

corticoids	exocrine	hypercalcemia	negative feedback
cretinism	gigantism	hyperglycemia	prostaglandins
Cushing's syndrome	glucocorticoids	hypoglycemia	second messenger
diabetes insipidus	gluconeogenesis	luteinization	steroids
diabetes mellitus	glycogenolysis	mineralocorticoids	stress
diuresis	goiter	myxedema	target organ cell
endocrine	hormone		

C H A P T E R   T E S T

1. Chemicals secreted directly into the blood by endocrine glands are called _____.
2. Glands that discharge their secretions into ducts are called _____ glands.
3. Cyclic AMP is said to serve as a _____, providing communication within the cells acted on by protein hormones.
4. The regulation of hormone levels in the blood depends on a homeostatic mechanism that reverses physiological changes, called _____ feedback.
5. Another name for the group of powerful "tissue hormones" found in many body tissues is _____ .
6. Production of too much hormone by a diseased gland is called _____ .
7. The hormone that acts on the thyroid gland and is produced by the pituitary gland is called _____ hormone.
8. Growth hormone and prolactin are both secreted by the _____ gland.
9. The letters ADH stand for _____ hormone.
10. One of the primary thyroid hormones is called T_4 or _____.
11. Low dietary intake of iodine may cause an enlargement of the thyroid gland, called simple _____.
12. The layer of the adrenal cortex that secretes mineralocorticoids is called the zona _____ .
13. Epinephrine and norepinephrine are secreted by the adrenal _____.
14. The alpha cells of the pancreatic islets secrete _____.
15. Inadequate insulin secretion is associated with the disease _____.
16. The corpus luteum of the ovary secretes chiefly _____.
17. The male sex hormone is called _____.
18. The hormone of the thymus gland is called _____.
19. The hormone melatonin is secreted by the _____ gland.
20. In the female the ovarian follicles secrete the hormone _____.

Circle the T before each true statement and the F before each false statement.

T F 21. Endocrine and exocrine glands discharge their secretions directly into the blood.

T F 22. Hormones act on target organ cells.

T F 23. The placenta is considered an endocrine gland.

T F 24. The second messenger hypothesis is used to explain the action of steroid hormones.

T F 25. Prostaglandins are also called *tissue hormones*.

T F 26. The term *adenohypophysis* is used to describe the posterior pituitary gland.

T F 27. Hyposecretion of antidiuretic hormone causes diabetes insipidus.

T F 28. Oxytocin and ADH are produced in the anterior pituitary gland.

T F 29. Parathyroid hormone stimulates breakdown of bone with release of calcium into the blood.

T F 30. The zona fasciculata of the adrenal cortex secretes sex hormones.

T F 31. High blood glucocorticoid concentration inhibits the inflammatory response.

T F 32. Hyposecretion of glucocorticoids results in Addison's disease.

T F 33. Beta cells of the pancreatic islets secrete insulin.

T F 34. Glycogenolysis converts glycogen to glucose.

T F 35. The thymus does not function as an endocrine gland.

R E V I E W   Q U E S T I O N S

1. What endocrine glands are located in the abdominal cavity, cranial cavity, mediastinum, neck, and pelvic cavity?

2. How are the prostaglandins similar or dissimilar to regular hormones?

3. What endocrine gland secretes ACTH, aldosterone, calcitonin, chorionic gonadotropins, epinephrine, growth hormone, insulin, oxytocin, and progesterone?

4. What hormone is called the *water-retaining hormone* because it decreases the amount of urine formed?

5. What hormone causes potassium loss from the body through the urine?

6. What hormone is called the *salt-retaining hormone* because it makes the kidneys resorb sodium into the blood more rapidly so that less sodium is lost in the urine?

7. Name two or more hormones that increase blood glucose.

8. What hormone speeds up the rate of catabolism (that is, makes you burn up your foods faster)?

9. What is the main function of ACTH, ADH, calcitonin, epinephrine, insulin, parathyroid hormone, thyroid hormone, and testosterone?

10. Which hormone stimulates ovulation?

11. Cretinism results from abnormal secretion of what hormone?

12. Does diabetes insipidus result from abnormal secretion of ADH or insulin?

13. What endocrine disorder produces gigantism?

14. What hormone or hormones prepare the body for strenuous activity—for "fight or flight," in other words?

15. What hormone is important in the development and function of the immune system?

16. Name the hormone found at high levels only in the urine of pregnant women.

C R I T I C A L   T H I N K I N G

17. Many changes occur in the body when it is in a condition of stress (for example, after a person has had major surgery). Name two endocrine glands that greatly increase their secretion of hormones in times of stress. Name the hormones they secrete.

18. Metabolism changes when the body is in a condition of stress. How does the metabolism of proteins, of fats, and of carbohydrates change, and what hormones cause the changes?

19. What hormones, with a high concentration present in the blood, make us less immune to infectious diseases?

20. What hormone decreases blood calcium concentration? What effect does this have on bone breakdown?

Blood

10

OBJECTIVES

*After you have completed this chapter,
you should be able to:*

1. Describe the primary functions of blood.
2. Describe the characteristics of blood plasma.
3. List the formed elements of blood and identify the most important function of each.
4. Discuss anemia in terms of red blood cell numbers and hemoglobin content.
5. Explain the steps involved in blood clotting.
6. Describe ABO and Rh blood typing.
7. Define the following medical terms associated with blood: *hematocrit, leukocytosis, leukopenia, polycythemia, sickle cell, phagocytosis, acidosis, thrombosis, erythroblastosis fetalis, serum, fibrinogen, Rh factor, anemia.*

The next few chapters deal with **transportation** and **protection,** two of the body's most important functions. Have you ever thought of what would happen if the transportation ceased in your city or town? Or what would happen if the police, firefighters, and armed services stopped doing their jobs? Food would become scarce, garbage would pile up, and no one would protect you or your property. Stretch your imagination just a little, and you can imagine many disastrous results. Similarly, lack of transportation and protection for the cells—the "individuals" of the body—threatens the homeostasis of the body. The systems that provide these vital services for the body are the **circulatory system** and **lymphatic system.** In this chapter, we will discuss the primary transportation fluid—blood. Blood not only performs vital pickup and delivery services, but it also provides much of the protection necessary to withstand foreign "invaders." Blood vessels and the heart are discussed in Chapter 11. The lymphatic system is discussed in Chapter 12.

Blood Composition

Blood is a fluid tissue that has many kinds of chemicals dissolved in it and millions upon millions of cells floating in it (Figure 10-1). The liquid (extracellular) part is called **plasma.** Suspended in the plasma are many different types of cells and cell fragments, which make up the **formed elements** of blood.

BLOOD PLASMA

Blood plasma is the liquid part of the blood or blood minus its formed elements. It consists of water with many substances dissolved in it. All of the chemicals needed by cells to stay alive—food, oxygen, and salts, for example—have to be brought to them by the blood. Food and salts are dissolved in plasma; so, too, is a small amount of oxygen. (Most of the oxygen in the blood is carried in the RBCs as oxyhemoglobin.) Wastes that cells must get rid of are dissolved in plasma and transported to the excretory organs. The hormones and other regulatory chemicals that help control cells' activities are also dissolved in plasma. As Figure 10-1 shows, the most abundant type of solute in the plasma is a group of **plasma proteins.** These proteins include *albumins,* which help thicken the blood, *globulins,* which include the antibodies that help protect us from infections, and *fibrinogen,* which is necessary for blood clotting.

Blood **serum** is plasma minus its clotting factors such as fibrinogen. Serum is obtained from whole blood by allowing it to clot in the bottom of a tube and then pouring off the liquid serum. Serum still contains antibodies, so it can be used to treat patients that have a need for specific antibodies.

Many people seem curious about how much blood they have. The amount depends on how big they are and whether they are male or female. A big person has more blood than a small person, and a man has more blood than a woman. But as a general rule, most adults probably have between 4 and 6 L of blood. It normally accounts for about 7% to 9% of the total body weight.

The volume of the plasma part of blood is usually a little more than half the volume of whole blood. Examples of normal volumes are plasma—2.6 L; blood cell—2.4 L; total blood—5 L.

Blood is alkaline with a pH between 7.35 and 7.45; it rarely reaches even the neutral point (see Appendix A). If the alkalinity of your blood decreases toward neutral, you are a very sick person; in fact, you have **acidosis.** But even in this condition, blood almost never becomes the least bit acid; it just becomes less alkaline than normal.

FORMED ELEMENTS

There are three main types and several subtypes of formed elements:

1. Red blood cells (RBCs) or **erythrocytes** (e-RITH-ro-sites)
2. White blood cells (WBCs) or **leukocytes** (LOO-ko-sites)
   a. Granular leukocytes (have granules in their cytoplasm)
      (1) Neutrophils
      (2) Eosinophils
      (3) Basophils
   b. Nongranular leukocytes (do not have granules in their cytoplasm)
      (1) Lymphocytes
      (2) Monocytes
3. Platelets or **thrombocytes** (THROM-bo-sites)

FIGURE 10–1   *Components of Blood.* Approximate values for the components of blood in a normal adult.

Percentage by body weight

Other fluids and tissues 92%

Blood 8%

Percentage by volume

Plasma 55%

Formed elements 45%

Plasma (percentage by weight)

Proteins 7%

Water 91%

Other solutes 2%

Albumins 58%

Globulins 38%

Fibrinogen 4%

Ions

Nutrients

Waste products

Gases

Regulatory substances

Formed elements (number per cubic mm)

Platelets 250-400 thousand

Leukocytes 5-9 thousand

Erythrocytes 4.2-5.8 million

Leukocytes

Neutrophils 60-70%

Lymphocytes 20-25%

Monocytes 3-8%

Eosinophils 2-4%

Basophils 0.5-1%

Figure 10-1 shows the breakdown of numbers and percentages of the formed elements. Table 10-1 lists the functions of these different kinds of blood cells and shows what each looks like under the microscope.

It is difficult to believe how many blood cells and cell fragments there are in the body. For instance, 5,000,000 RBCs, 7500 WBCs, and 300,000 platelets in 1 cubic millimeter (mm^3) of blood (approximately 1 drop) would be considered normal RBC, WBC, and platelet counts. Because RBCs, WBCs, and platelets are continually being destroyed, the body must continually make new ones to take their place at a really staggering rate; a few million RBCs are manufactured *each second!*

Two kinds of connective tissue—**myeloid tissue** and **lymphatic tissue**—make blood cells for the body. Recall that formation of new blood cells is called *hemopoiesis.* Myeloid tissue is better known as *red bone marrow.* In the adult, it is chiefly in the sternum, ribs, and hipbones. A few other bones such as the vertebrae, clavicles, and cranial bones also contain small amounts of this valuable substance. Red bone marrow forms all types of blood cells except some lymphocytes and monocytes. Most of these others are formed by lymphatic tissue, which is located chiefly in the lymph nodes, thymus, and spleen.

As blood cells mature, they move into the circulatory vessels. Erythrocytes circulate up to 4

TABLE 10–1 Classes of Blood Cells

BODY CELL		FUNCTION	BODY CELL		FUNCTION
Erythrocyte		Oxygen and carbon dioxide transport	B-lymphocyte		Antibody production (precursor of plasma cells)
Neutrophil		Immune defense (phagocytosis)	T-lymphocyte		Cellular immune response
Eosinophil		Defense against parasites	Monocyte		Immune defenses (phagocytosis)
Basophil		Inflammatory response	Platelet		Blood clotting

months before they break apart and their components are removed from the bloodstream by the liver. Granular leukocytes often have a lifespan of only a few days, but nongranular leukocytes may live over 6 months.

RED BLOOD CELLS

As you can see in Figure 10-2, RBCs have an unusual shape. The cell is "caved in" on both sides so that each one has a thin center and thicker edges. Notice also that mature RBCs have no nucleus. Figure 10-2 shows RBCs photographed with a scanning electron microscope. With this instrument, extremely small objects can be enlarged far more than is possible with a standard light microscope, and, as you can see in the illustration, objects appear more three-dimensional. Because of the large numbers of RBCs and their unique shape, their total surface area is enormous. It provides an area larger than a football field for the exchange of oxygen and carbon dioxide between the blood and the body's cells.

RBCs perform several important functions. One essential function is to help transport carbon dioxide. Carbon dioxide (CO_2) is a harmful waste produced by the energy-producing processes of all living cells. It must be carried away from cells

FIGURE 10–2 *RBCs.* Color-enhanced scanning electron micrograph shows the detailed structure of normal RBCs.

Sickle Cell Anemia

Sickle cell anemia is a severe and sometimes fatal hereditary disease caused by an abnormal type of hemoglobin. A person who inherits only one defective gene develops a form of the disease called *sickle cell trait*. In this condition, RBCs contain a small amount of a type of hemoglobin that is less soluble than normal. It forms solid crystals when the blood oxygen level is low, causing distortion of the RBC. If two defective genes are inherited (one from each parent), more of the defective hemoglobin is produced, and the distortion of red blood cells becomes severe. The figure shows the characteristic sickle shape of many of the cells. Interestingly, this condition provides resistance to the parasite that causes malaria, an often fatal disease. Thus sickle cell conditions persist in areas in which malaria is prevalent.

and to the lungs for disposal into the external environment. RBCs also transport oxygen from the lungs to other cells in the body. A red pigment called **hemoglobin** (HEE-mo-glo-bin) in RBCs unites with oxygen to form **oxyhemoglobin** (ok-see-HEE-mo-glo-bin). This combined oxygen-hemoglobin complex makes possible the efficient transport of large quantities of oxygen to body cells. Hemoglobin can also carry a small proportion of the CO_2 carried by the blood.

The term **anemia** (ah-NEE-me-ah) is used to describe a number of different disease conditions caused by an inability of the blood to carry suffi-cient oxygen to the body cells. Anemias can result from inadequate numbers of RBCs or a deficiency of hemoglobin. Thus anemia can occur if the hemoglobin in RBCs is inadequate, even if adequate numbers of RBCs are present. Anemias caused by an actual decrease in the number of RBCs can occur if blood is lost by hemorrhage, as with accidents or bleeding ulcers, or if the blood-forming tissues cannot maintain normal numbers of blood cells. Such failures occur because of cancer, radiation (x-ray) damage, and certain types of infections. The term **pernicious** (per-NISH-us) **anemia** is used to describe a deficiency of RBCs caused by the lack of vitamin B_{12}. If bone marrow produces an *excess* of RBCs, the result is a condition called **polycythemia** (pol-ee-sye-THEE-me-ah). The blood in individuals suffering from this condition may contain so many RBCs that it may become too thick to flow properly.

Iron is a critical component of the hemoglobin molecule. Without adequate iron in the diet, the body cannot manufacture enough hemoglobin. The result is **iron deficiency anemia**—a world-wide medical problem. If hemoglobin falls below the normal level, as it does in this type of anemia, it starts an unhealthy chain reaction; less hemoglobin, less oxygen transported to cells, slower breakdown and use of nutrients by cells, less energy produced by cells, decreased cellular functions. If you understand this relationship between hemoglobin and energy, you can correctly guess that an anemic person's chief complaint will probably be that he or she feels "so tired all the time."

A common laboratory test called the **hematocrit** can tell a physician a great deal about the volume of RBCs in a blood sample. If whole blood is placed in a special hematocrit tube and then "spun down" in a centrifuge, the heavier formed elements will quickly settle to the bottom of the tube. During the hematocrit procedure, RBCs are forced to the bottom of the tube first. The WBCs and platelets then settle out in a layer called the **buffy coat.** In Figure 10-3 the buffy coat can be seen between the packed RBCs on the bottom of the hematocrit tube and the liquid layer of plasma above. Normally about 45% of the blood volume consists of RBCs. For a patient with anemia, the percentage of RBCs drops, and for a patient with polycythemia, it increases dramatically (Figure 10-3).

FIGURE 10–3 *Hematocrit Tubes Showing Normal Blood, Anemia, and Polycythemia.* Note the buffy coat located between the packed RBCs and the plasma. **A,** A normal percent of RBCs. **B,** Anemia (a low percent of RBCs). **C,** Polycythemia (a high percent of RBCs).

FIGURE 10–4 *Leukocytes in Human Blood Smears.* Each light micrograph shows a different type of stained WBC surrounded by several smaller RBCs.

WHITE BLOOD CELLS

WBCs have a function that is just as vital as that of RBCs. They defend the body from microorganisms that have succeeded in invading the tissues or bloodstream. For example, **neutrophils** (NOO-tro-fils) (Figure 10-4, *A*) and **monocytes** (MON-o-sites) (Figure 10-4, *E*) engulf microbes. They actually take them into their own cell bodies and digest them in the process of **phagocytosis** (see p. 27), and the cells that carry on this process are called **phagocytes** (FAG-o-sites) (Figure 10-5). The neutrophils are the most numerous of the phagocytes.

WBCs of the type called **lymphocytes** (LIM-fo-sites) (Figure 10-4, *D*) also help protect us against infections, but they do it by a process different from phagocytosis. Lymphocytes function in the immune mechanism, the complex process that makes us immune to infectious diseases.

The immune mechanism starts to operate, for example, when microbes invade the body. In some way, their presence stimulates lymphocytes to start multiplying and become active immune cells. Lymphocytes called *B-lymphocytes* begin to actively produce specific antibodies that inhibit the microbes. Other lymphocytes, called *T-lymphocytes*, may also become involved, directly attacking the microbes or aiding in the function of B-lymphocytes. Details of the immune system are discussed in Chapter 12.

FIGURE 10–5 *Phagocytosis.* This transmission electron micrograph shows a phagocytic cell consuming a foreign particle in a manner similar to that used by human WBCs. Extensions of the plasma membrane literally reach out and grab a particle, and then digest the particle within an intracellular vesicle. (Courtesy of *A. Arlan Hinchee*)

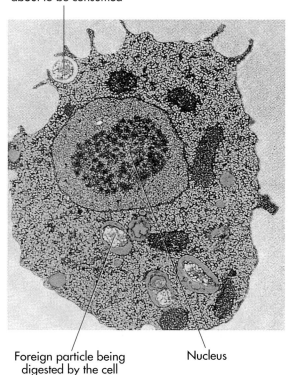

Foreign particle about to be consumed

Foreign particle being digested by the cell

Nucleus

Eosinophils (ee-o-SIN-o-fils) (Figure 10-4, *B*) are granulocytic WBCs that help protect the body from the numerous irritants that cause allergies. They are also capable of phagocytosis. **Basophils** (BAY-so-fils) (Figure 10-4, *C*) also function in allergic reactions. These leukocytes, which are less abundant than other types, also secrete a number of important substances. For example, they secrete the potent chemical **heparin,** which helps prevent the clotting of blood as it flows through the blood vessels of the body.

The term **leukopenia** (loo-ko-PEE-nee-ah) refers to an abnormally low WBC count (under 5000 WBCs/mm^3 of blood). A number of disease conditions may affect the immune system and decrease the amount of circulating WBCs. **Acquired immunodeficiency syndrome** or **AIDS,** which will be discussed in Chapter 12, is one example of a disease characterized by marked leukopenia. **Leukocytosis** (loo-ko-SYE-toe-sis) refers to an abnormally high WBC count (over 10,000 WBCs/mm^3 of blood). It is a much more common problem than leukopenia and almost always accompanies infections. There is also a malignant disease, **leukemia** (loo-KEE-mee-ah), in which the number of WBCs increases tremendously. The buffy coat is thicker and more noticeable in the hematocrit of blood from patients with leukemia because of the elevated WBC counts. You may have heard of this disease as "blood cancer." As in all cancers, the extra cells do not function properly.

PLATELETS AND BLOOD CLOTTING

Platelets, the third main type of formed element, play an essential part in blood clotting. Your life might someday be saved just because your blood can clot. A clot plugs up torn or cut vessels and stops bleeding that otherwise might prove fatal.

The story of how blood clots is the story of a chain of rapid-fire reactions. The first step in the chain is some kind of an injury to a blood vessel that makes a rough spot in its lining. (Normally the lining of blood vessels is extremely smooth.) Almost immediately, damaged tissue cells release certain clotting factors into the plasma. These factors rapidly react with other factors already present in the plasma to form **prothrombin activator** (pro-THROM-bin AK-tiv-ayt-or). At the same time this is happening, platelets become "sticky" at the point of injury and soon accumulate near the opening in the broken blood vessel, forming a soft, temporary *platelet plug*. As the platelets accumulate, they release additional clotting factors, forming even more prothrombin activator. If the normal amount of blood calcium is present, prothrombin activator triggers the next step of clotting by converting **prothrombin** (a protein in normal blood) to **thrombin** (THROM-bin). In the last step, thrombin reacts with **fibrinogen** (fi-BRIN-o-jen) (a normal plasma protein) to change it to a

FIGURE 10–6 *Blood Clotting.* The extremely complex clotting mechanism can be distilled into three basic steps: **1,** release of clotting factors from both injured tissue cells and sticky platelets at the injury site; **2,** formation of thrombin; and **3,** formation of fibrin and trapping of red blood cells to form a clot.

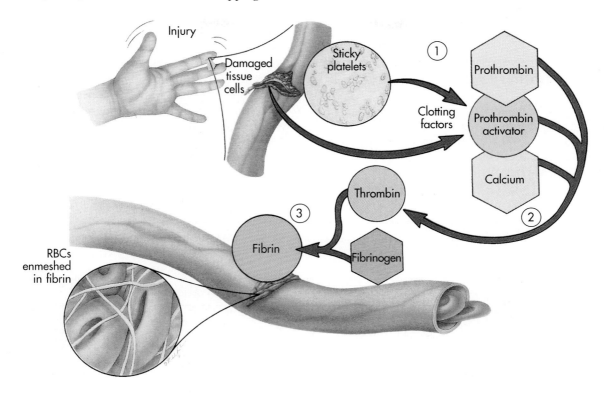

fibrous gel called **fibrin.** Under the microscope, fibrin looks like a tangle of fine threads with RBCs caught in the tangle. Figure 10-6 illustrates the steps in the blood-clotting mechanism.

The clotting mechanism contains clues for ways to stop bleeding by speeding up blood clotting. For example, you might simply apply gauze to a bleeding surface. Its slight roughness would cause more platelets to stick together and release more clotting factors. These additional factors would then make the blood clot more quickly.

Physicians sometimes prescribe vitamin K before surgery to make sure that the patient's blood will clot fast enough to prevent hemorrhage. Vitamin K stimulates liver cells to increase the synthesis of prothrombin. More prothrombin in blood allows faster production of thrombin during clotting and thus faster clot formation.

Unfortunately, clots sometimes form in unbroken blood vessels of the heart, brain, lungs, or some other organ—a dreaded thing because clots may produce sudden death by shutting off the blood supply to a vital organ. When a clot stays in the place where it formed, it is called a **thrombus** (THROM-bus) and the condition is spoken of as **thrombosis** (throm-BO-sus). If part of the clot dislodges and circulates through the bloodstream, the dislodged part is then called an **embolus** (EM-bo-lus), and the condition is called an **embolism** (EM-bo-lizm). Suppose that your doctor told you that you had a clot in one of your coronary arteries. Which diagnosis would he make—coronary thrombosis or coronary embolism—if he thought that the clot had formed originally in the coronary artery as a result of the accumulation of fatty material in the vessel wall? Physicians now have some drugs that they can use to help prevent thrombosis and embolism. Heparin, for example, can be used to prevent excessive blood clotting. Heparin inhibits the conversion of prothrombin to thrombin, thus preventing formation of a thrombus.

 ## *Blood Types*

ABO SYSTEM

Blood types are identified by certain antigens in RBCs (Figure 10-7). An **antigen** (AN-ti-jen) is a substance that can stimulate the body to make antibodies. Almost all substances that act as anti-gens are foreign proteins. That is, they are not the body's own natural proteins but are proteins that have entered the body from the outside by infec-tion, transfusion, or some other method.

The word *antibody* can be defined in terms of what causes its formation or in terms of how it functions. Defined the first way, an **antibody** (AN-ti-bod-ee) is a substance made by the body in response to stimulation by an antigen. Defined according to its functions, an antibody is a sub-stance that reacts with the antigen that stimulated its formation. Many antibodies react with their antigens to clump or **agglutinate** (ah-GLOO-tin-ate) them. In other words, they cause their anti-gens to stick together in little clusters.

FIGURE 10–7   *Results of Different Combinations of Donor and Recipient Blood.*   The left columns show the recip-ient's blood characteristics and the top row shows the donor's blood type.

Recipient's blood		Reactions with donor's blood			
RBC antigens	Plasma antibodies	Donor type O	Donor type A	Donor type B	Donor type AB
None (Type O)	Anti-A Anti-B				
A (Type A)	Anti-B				
B (Type B)	Anti-A				
AB (Type AB)	(none)				

 Normal blood   Agglutinated blood

Every person's blood is one of the following blood types in the ABO system of typing:

1. Type A
2. Type B
3. Type AB
4. Type O

Suppose that you have type A blood (as do about 41% of Americans). The letter *A* stands for a certain type of antigen (a protein) in the plasma membrane of your RBCs since birth. Because you were born with type A antigen, your body does not form antibodies to react with it. In other words, your blood plasma contains no anti-A antibodies. It does, however, contain anti-B antibodies. For some unknown reason, these antibodies are present naturally in type A blood plasma. The body did not form them in response to the presence of the B antigen. In summary, in type A blood the RBCs contain type A antigen and the plasma contains anti-B antibodies.

Similarly, in type B blood, the RBCs contain type B antigen, and the plasma contains anti-A antibodies. In type AB blood, as its name indicates, the RBCs contain both type A and type B antigens, and the plasma contains neither anti-A nor anti-B antibodies. The opposite is true of type O blood; its RBCs contain neither type A nor type B antigens, and its plasma contains both anti-A and anti-B antibodies.

Harmful effects or even death can result from a blood transfusion if the donor's RBCs become agglutinated by antibodies in the recipient's plasma. If a donor's RBCs do not contain any A or B antigen, they of course cannot be clumped by anti-A or anti-B antibodies. For this reason the type of blood that contains neither A nor B antigens—namely, type O blood—can be used in an emergency as donor blood without the danger of anti-A or anti-B antibodies clumping its RBCs. Type O blood has therefore been called **universal donor** blood. Similarly, blood type AB has been called **universal recipient** blood because it contains neither anti-A nor anti-B antibodies in its plasma. Therefore it does not clump any donor's RBCs containing A or B antigens. In a normal clinical setting, however, all blood intended for transfusion is matched carefully to the blood of the recipient for a variety of factors.

Figure 10-7 shows the results of different combinations of donor and recipient blood.

Blood Doping

A number of athletes have reportedly improved their performance by a practice called **blood doping.** A few weeks before an important event, an athlete has some blood drawn. The RBCs are separated and frozen. Just before competition, the RBCs are thawed and injected into the athlete. The increased hematocrit that results slightly improves the oxygen-carrying capacity of the blood, which theoretically improves performance. This method is judged to be an unfair and unwise practice in athletics.

RH SYSTEM

You may be familiar with the term **Rh-positive** blood. It means that the RBCs of this type of blood contain an antigen called the Rh factor. If, for example, a person has type AB, Rh-positive blood, his red blood cells contain type A antigen, type B antigen, and the Rh factor antigen. The term *Rh* is used because this important blood cell antigen was first discovered in the blood of Rhesus monkeys.

In **Rh-negative** blood the RBCs do not contain the Rh factor. Plasma never naturally contains anti-Rh antibodies. But if Rh-positive blood cells are introduced into an Rh-negative person's body, anti-Rh antibodies soon appear in the blood plasma. In this fact lies the danger for a baby born to an Rh-negative mother and an Rh-positive father. If the baby inherits the Rh-positive trait from his father, the Rh factor on his RBCs may stimulate the mother's body to form anti-Rh antibodies. Then, if she later carries another Rh-positive fetus, he may develop a disease called **erythroblastosis** (e-rith-ro-blas-TOE-sis) **fetalis** (fe-TAL-is), caused by the mother's Rh antibodies reacting with the baby's Rh-positive cells (Figure 10-8).

Some Rh-negative mothers who carry an Rh-positive baby are treated with a protein marketed as RhoGAM. RhoGAM stops the mother's body from forming anti-Rh antibodies and thus prevents the possibility of harm to the next Rh-positive baby.

FIGURE 10–8   *Erythroblastosis Fetalis.* **A,** Rh-positive blood cells enter the mother's blood stream during delivery of an Rh-positive baby. If not treated, the mother's body will produce anti-Rh antibodies. **B,** A later pregnancy involving an Rh-negative baby is normal because there are no Rh antigens in the baby's blood. **C,** A later pregnancy involving an Rh-positive baby may result in erythroblastosis fetalis. Anti-Rh antibodies enter the baby's blood supply and cause agglutination of RBCs with the Rh antigen.

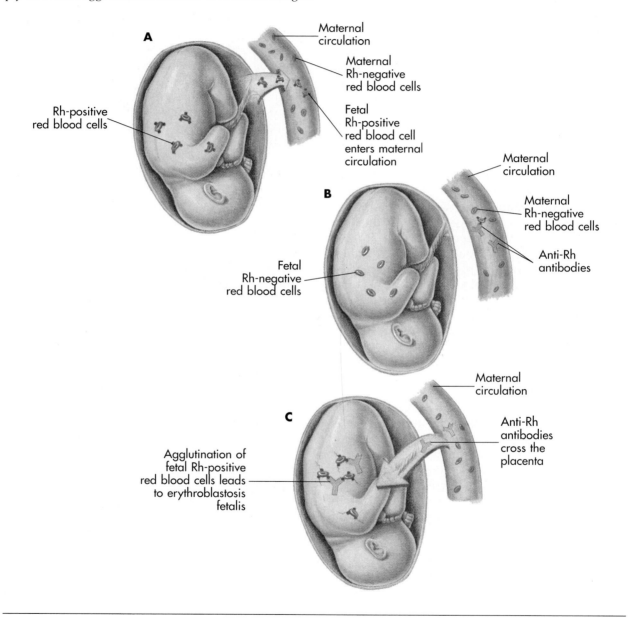

OUTLINE SUMMARY

BLOOD COMPOSITION (Table 10-1)

A Blood plasma
  1 Definition—blood minus its cells
  2 Composition—water containing many dissolved substances (for example, foods, salts, and hormones)
  3 Amount of blood—varies with size and sex; 4 to 6 L about average; about 7% to 9% of body weight
  4 Slightly alkaline
B Formed elements
  1 Kinds
    a RBCs (erythrocytes)
    b WBCs (leukocytes)
      (1) Granular leukocytes—neutrophils, eosinophils, and basophils
      (2) Nongranular leukocytes—lymphocytes and monocytes
    c Platelets or thrombocytes
  2 Numbers
    a RBCs—4½ to 5 million per mm^3 of blood
    b WBCs—5000 to 10,000 per mm^3 of blood
    c Platelets—300,000 per mm^3 of blood
  3 Formation—red bone marrow (myeloid tissue) forms all blood cells except some lymphocytes and monocytes, which are formed by lymphatic tissue in the lymph nodes, thymus, and spleen
C RBCs
  1 Structure—disk-shaped, without nuclei
  2 Functions—transport oxygen and carbon dioxide
  3 Anemia—inability of blood to carry adequate oxygen to tissues; caused, for example, by:
    a Inadequate RBC numbers
    b Deficiency of hemoglobin
    c Pernicious anemia—deficiency of vitamin B_{12}
  4 Hematocrit—medical test in which a centrifuge is used to separate whole blood into formed elements and liquid fraction (Figure 10-3)
    a Buffy coat is WBC and platelet fraction
    b Normal RBC level is about 45%
    c Polycythemia—abnormally high RBC count
D WBCs
  1 General function—defense
  2 Neutrophils and monocytes carry out phagocytosis

3 Lymphocytes produce antibodies (B-lymphocytes) or directly attack foreign cells (T-lymphocytes)
  4 Eosinophils protect against irritants that cause allergies
  5 Basophils produce heparin, which inhibits clotting
  6 Clinical conditions related to blood:
    a Leukopenia—abnormally low WBC count
    b Leukocytosis—abnormally high WBC count
    c Leukemia—cancer: elevated WBC count; cells do not function properly
E Platelets and blood clotting (Figure 10-6)
  1 Platelets play an essential role in blood clotting
  2 Blot clot formation
    a Clotting factors released at the injury site produce prothrombin activator
    b Prothrombin activator and calcium convert prothrombin to thrombin
    c Thrombin triggers formation of fibrin, which traps RBC to form a clot

BLOOD TYPES

A ABO system (Figure 10-7)
  1 Type A blood—type A antigens in RBCs; anti-B type antibodies in plasma
  2 Type B blood—type B antigens in RBCs; anti-A type antibodies in plasma
  3 Type AB blood—type A and type B antigens in RBCs; no anti-A or anti-B antibodies in plasma
  4 Type O blood—no type A or type B antigens in RBCs; both anti-A and anti-B antibodies in plasma
B Rh system
  1 Rh-positive blood—Rh factor antigen present in RBCs
  2 Rh-negative blood—no Rh factor present in RBCs; no anti-Rh antibodies present naturally in plasma; anti-Rh antibodies, however, appear in the plasma of Rh-negative persons if Rh-positive RBCs have been introduced into their bodies
  3 Erythroblastosis fetalis—may occur when Rh-negative mother carries a second Rh-positive fetus; caused by mother's Rh antibodies reacting with baby's Rh-positive cells

N E W   W O R D S

acidosis	erythroblastosis fetalis	leukocytosis	plasma protein
anemia	erythrocyte	leukopenia	polycythemia
antibodies	fibrin	lymphocyte	prothrombin
antigens	fibrinogen	monocyte	prothrombin activator
basophil	hematocrit	neutrophil	serum
buffy coat	hemoglobin	oxyhemoglobin	thrombin
embolism	heparin	pernicious anemia	thrombocyte
embolus	leukemia	phagocyte	thrombosis
eosinophil	leukocyte	plasma	thrombus

C H A P T E R   T E S T

1. Erythrocytes is another name for:
   a. WBCs
   b. RBCs
   c. Platelets
   d. Thrombocytes
2. Which of the following is classified as a granular leukocyte?
   a. Monocyte
   b. Neutrophil
   c. Lymphocyte
   d. None of the above are correct
3. Which of the following terms refers to an abnormally low WBC count?
   a. Anemia
   b. Leukemia
   c. Leukocytosis
   d. Leukopenia
4. The most numerous of the phagocytes are the:
   a. Neutrophils
   b. RBCs
   c. Basophils
   d. Lymphocytes
5. Heparin secretion is a function of:
   a. Monocytes
   b. Neutrophils
   c. Basophils
   d. Eosinophils

6. Vitamin K:
   a. Increases prothrombin levels in blood
   b. Is necessary for normal blood clotting
   c. Stimulates liver cells
   d. All of the above are correct
7. A thrombus:
   a. Is a stationary blood clot
   b. Is a moving blood clot
   c. Is seldom serious
   d. Is a normal component of plasma
8. Erythroblastosis fetalis is caused by:
   a. Anti-Rh antibodies in the father's blood
   b. Anti-Rh antibodies in the mother's blood
   c. Both of the above are correct
   d. None of the above are correct
9. The liquid part of blood that has not clotted is called _____.
10. A better-known name for myeloid tissue is

    _____.
11. An excess of RBCs is called _____.
12. A moving blood clot is called an _____.
13. A substance that can stimulate the body to produce antibodies is called an _____.
14. Type _____ blood contains both A and B antigens and neither anti-A nor anti-B antibodies.
15. The blood cell antigen first discovered in the blood of Rhesus monkeys is the _____ factor.

REVIEW QUESTIONS

1. What is the normal number of RBCs per mm^3 of blood? WBCs? Platelets?
2. Name the granular and nongranular leukocytes.
3. What two kinds of connective tissue make blood cells for the body?
4. What does the hematocrit blood test measure? What is the normal value for this test?
5. Your circulatory system is the transportation system of your body. Mention some of the substances it transports and tell whether each is carried in blood cells or in the blood plasma.
6. Briefly describe three ways in which blood cells defend the body against foreign cells.
7. Briefly explain what happens when blood clots, including what makes it start to clot.
8. You hear that a friend has a "coronary thrombosis." What does this mean to you?
9. Define *antibody* and *antigen* as they apply to blood typing.
10. Identify the four blood types.
11. What is the difference between ABO blood type and the Rh blood type?

CRITICAL THINKING

12. Suppose that your doctor told you that your "red count was 3 million." What does "red count" mean? Might the doctor say that you had any of the following conditions—anemia, leukocytosis, leukopenia, or polycythemia—with an RBC count of this amount? If so, which one?
13. If you had appendicitis or some other acute infection, would your WBC count be more likely to be 2000, 7000, or 15,000 per mm^3? Give a reason for your answer.
14. Why might a physician prescribe vitamin K before surgery for a patient with a history of bleeding problems?

The Circulatory System

BOXED ESSAYS

Changes in Blood Flow During Exercise
Blood Pressure Readings

OBJECTIVES

After you have completed this chapter,
you should be able to:

1. Discuss the location, size, and position of the heart in the thoracic cavity and identify the heart chambers, sounds, and valves.

2. Trace blood through the heart and compare the functions of the heart chambers on the right and left sides.

3. List the anatomical components of the heart conduction system and discuss the features of a normal electrocardiogram.

4. Explain the relationship between blood vessel structure and function.

5. Trace the path of blood through the systemic, pulmonary, hepatic portal, and fetal circulations.

6. Identify and discuss the primary factors involved in the generation and regulation of blood pressure and explain the relationships between these factors.

iffering amounts of nutrients and waste products enter and leave the fluid surrounding each body cell continually. In addition, requirements for hormones, body salts, water, and other critical substances constantly change. However, homeostasis or constancy of the body fluid contents surrounding the billions of cells that make up our bodies is required for survival. The system that supplies our cells' transportation needs is the **circulatory system.** The levels of dozens of substances in the blood can remain constant even though the absolute amounts that are needed or produced may change because we have this extremely effective system that transports these substances to or from each cell as circumstances change.

We will begin the study of the circulatory system with the heart—the pump that keeps blood moving through a closed circuit of blood vessels. Details related to heart structure will be followed by a discussion of how the heart functions. This chapter concludes with a study of the vessels through which blood flows as a result of the pumping action of the heart. As a group, these vessels are multipurpose structures. Some allow for rapid movement of blood from one body area to another. Others, such as the microscopic capillaries, permit the movement or exchange of many substances between the blood and fluid surrounding body cells. Chapter 12 will cover the lymphatic system and immunity topics that relate in many ways to the structure and functions of the circulatory system.

 Heart

LOCATION, SIZE, AND POSITION

No one needs to be told where the heart is or what it does. Everyone knows that the heart is in the chest, that it beats night and day to keep the blood flowing, and that if it stops, life stops.

Most of us probably think of the heart as located on the left side of the body. As you can see in Figure 11-1, the heart is located between the lungs in the lower portion of the mediastinum. Draw an imaginary line through the middle of the trachea in Figure 11-1 and continue the line down through the thoracic cavity to divide it into right

and left halves. Note that about two thirds of the mass of the heart is to the left of this line and one third to the right.

The heart is often described as a triangular organ, shaped and sized roughly like a closed fist. In Figure 11-1 you can see that the **apex** or blunt point of the lower edge of the heart lies on the diaphragm, pointing toward the left. Doctors and nurses often listen to the heart sounds by placing a stethoscope on the chest wall directly over the apex of the heart. Sounds of the so-called apical beat are easily heard in this area (that is, in the space between the fifth and sixth ribs on a line even with the midpoint of the left clavicle).

The heart is positioned in the thoracic cavity between the sternum in front and the bodies of the thoracic vertebrae behind. Because of this placement, it can be compressed or squeezed by application of pressure to the lower portion of the body of the sternum using the heel of the hand. Rhythmic compression of the heart in this way can maintain blood flow in cases of cardiac arrest and, if combined with effective artificial respiration, the resulting procedure, called **cardiopulmonary resuscitation (CPR),** can be lifesaving.

ANATOMY
Heart Chambers

If you cut open a heart, you can see many of its main structural features (Figure 11-2). This organ is hollow, not solid. A partition divides it into right and left sides. The heart contains four cavities or hollow chambers. The two upper chambers are called **atria** (AY-tree-ah), and the two lower chambers are called **ventricles** (VEN-tri-kuls). The atria are smaller than the ventricles, and their walls are thinner and less muscular. Atria are often called *receiving chambers* because blood enters the heart through veins that open into these upper cavities. Eventually, blood is pumped from the heart into arteries that exit from the ventricles; therefore the ventricles are sometimes referred to as the *discharging chambers* of the heart. Each heart chamber is named according to its location. Thus there is a right and left atrial chamber above and a right and left ventricular chamber below. The wall of each heart chamber is composed of cardiac muscle tissue usually referred to as the **myocardium** (my-o-KAR-dee-um). The septum

FIGURE 11–1 *The Heart.* The heart and major blood vessels viewed from the front (anterior). Inset shows the relationship of the heart to other structures in the thoracic cavity.

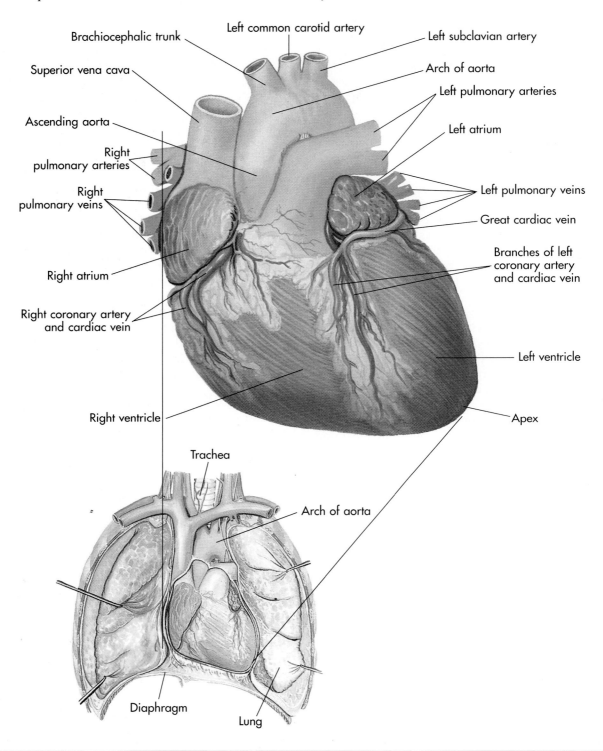

Brachiocephalic trunk

Left common carotid artery

Left subclavian artery

Superior vena cava

Arch of aorta

Left pulmonary arteries

Ascending aorta

Left atrium

Right pulmonary arteries

Right pulmonary veins

Left pulmonary veins

Great cardiac vein

Right atrium

Branches of left coronary artery and cardiac vein

Right coronary artery and cardiac vein

Left ventricle

Right ventricle

Apex

Trachea

Arch of aorta

Diaphragm

Lung

FIGURE 11–2 *An Internal View of the Heart.* The inset shows a cross section of the heart wall, including the pericardium.

Parietal pericardium
Visceral pericardium (epicardium)
Fatty connective tissue
OUTSIDE OF HEART
Coronary vessels
INSIDE OF HEART
Pericardial space
Myocardium
Endocardium

Superior vena cava
Aorta
Pulmonary semilunar valve
Pulmonary arteries
Left pulmonary veins
Right atrium
Tricuspid valve
Left atrium
Aortic semilunar valve
Bicuspid valve
Chordae tendineae
Right ventricle
Interventricular septum
Left ventricle

between the atrial chambers is called the *interatrial septum;* the *interventricular septum* separates the ventricles.

Each chamber of the heart is lined by a thin layer of very smooth tissue called the **endocardium** (en-doe-KAR-dee-um) (Figure 11-2). Inflammation of this lining is referred to as **endocarditis** (en-doe-kar-DYE-tis). If inflamed, the endocardial lining can become rough and abrasive to RBCs passing over its surface. Blood flowing over a rough surface is subject to clotting, and a **thrombosis** (throm-BO-sis) or clot may form (see Chapter 10). Unfortunately, rough spots caused by endocarditis or injuries to blood vessel walls often cause the release of platelet factors. The result is often the formation of a fatal blood clot.

Covering Sac or Pericardium

The heart has a covering and a lining. Its covering, called the **pericardium** (pair-i-KAR-dee-um), consists of two layers of fibrous tissue with a small space in between. The inner layer of the pericardium is called the **visceral pericardium** or **epicardium** (ep-i-KAR-dee-um). It covers the heart the way an apple skin covers an apple. The outer layer of pericardium is called the **parietal pericardium.** It fits around the heart like a loose-fitting sack, allowing enough room for the heart to beat. It is easy to remember the difference between the *endocardium,* which lines the heart chambers, and the *epicardium,* which covers the surface of the heart (Figure 11-2), if you understand the meaning of the prefixes *endo-* and *epi-*. *Endo-* comes from the Greek word meaning "inside" or "within," and *epi-* comes from the Greek word meaning "upon" or "on."

The two pericardial layers slip against each other without friction when the heart beats because these are serous membranes with moist, not dry, surfaces. A thin film of pericardial fluid furnishes the lubricating moistness between the heart and its enveloping pericardial sac. If the pericardium becomes inflamed, a condition called **pericarditis** (pair-i-kar-DYE-tis) results.

Heart Action

The heart serves as a muscular pumping device for distributing blood to all parts of the body. Contraction of the heart is called **systole** (SIS-toe-lee), and relaxation is called **diastole** (dye-AS-toe-lee). When the heart beats (that is, when it contracts), the atria contract first (atrial systole), forcing blood into the ventricles. Once filled, the two ventricles contract (ventricular systole) and force blood out of the heart (Figure 11-3). For the heart to be efficient in its pumping action, more than just the rhythmic contraction of its muscular fibers is required. The direction of blood flow must be directed and controlled. This is accomplished by four sets of valves located at the entrance and near the exit of the ventricles.

Heart Valves

The two valves that separate the atrial chambers above from the ventricles below are called **AV** or **atrioventricular** (ay-tree-o-ven-TRIK-yoo-lar) **valves.** The two AV valves are called the **bicuspid** or **mitral** (MY-tral) **valve,** located between the left atrium and ventricle, and the **tricuspid valve,** located between the right atrium and ventricle. The AV valves prevent backflow of blood into the atria when the ventricles contract. Locate the AV valves in Figures 11-2 and 11-3. Note that a number of stringlike structures called **chordae tendineae** (KOR-dee ten-DIN-ee) attach the AV valves to the wall of the heart.

The **SL** or **semilunar** (sem-i-LOO-nar) **valves** are located between the two ventricular chambers and the large arteries that carry blood away from the heart when contraction occurs (Figure 11-3). The ventricles, like the atria, contract together. Therefore the two semilunar valves open and close at the same time. The **pulmonary semilunar valve** is located at the beginning of the pulmonary artery and allows blood going to the lungs to flow out of the right ventricle but prevents it from flowing back into the ventricle. The **aortic semilunar valve** is located at the beginning of the aorta and allows blood to flow out of the left ventricle up into the aorta but prevents backflow into this ventricle.

HEART SOUNDS

If a stethoscope is placed on the anterior chest wall, two distinct sounds can be heard. They are rhythmical and repetitive sounds that are often described as **lub dup.**

The first or *lub* sound is caused by the vibration and abrupt closure of the AV valves as the ventricles contract. Closure of the AV valves prevents blood from rushing back up into the atria during

FIGURE 11–3 *Heart Action.* **A,** During atrial systole (contraction) cardiac muscle in the atrial wall contracts, forcing blood through the AV valves and into the ventricles. Bottom illustration shows superior view of all four valves, with SL valves closed and AV valves open. **B,** During ventricular systole that follows, the AV valves close, and blood is forced out of the ventricles through the semilunar valves and into the arteries. Bottom illustration shows superior view of SL valves open and AV valves closed.

contraction of the ventricles. This first sound is of longer duration and lower pitch than the second. The pause between this first sound and the *dup* or second sound is shorter than that after the second sound and the *lub dup* of the next systole. The second heart sound is caused by the closing of both the semilunar valves when the ventricles undergo diastole (relax).

BLOOD FLOW THROUGH THE HEART

The heart acts as two separate pumps. The right atrium and the right ventricle perform a task quite different from the left atrium and the left ventricle. When the heart "beats," first the atria contract simultaneously. This is atrial systole. Then the ventricles fill with blood, and they, too, contract together during ventricular systole. Although the atria contract as a unit followed by the ventricles below, the right and left sides of the heart act as separate pumps. As we study the blood flow through the heart, the separate functions of the two pumps will become clearer.

Note in Figure 11-3 that blood enters the right atrium through two large veins called the **superior vena** (VEE-nah) **cava** (KAY-vah) and **inferior vena cava.** The right heart pump receives oxygen-poor blood from the veins. After entering the right atrium, it is pumped through the right AV or tricuspid valve and enters the right ventricle. When the ventricles contract, blood in the right ventricle is pumped through the pulmonary semilunar valve into the **pulmonary artery** and eventually to the lungs, where oxygen is added and carbon dioxide is lost.

As you can see in Figure 11-3, blood rich in oxygen returns to the left atrium of the heart through four **pulmonary veins.** It then passes through the left AV or bicuspid valve into the left ventricle. When the left ventricle contracts, blood is forced through the aortic semilunar valve into the **aorta** (ay-OR-tah) and is distributed to the body as a whole.

As you can tell from Figure 11-4, the two sides of the heart actually pump blood through two separate "circulations" and function as two separate pumps. The **pulmonary circulation** involves movement of blood from the right ventricle to the lungs, and the **systemic circulation** involves movement of blood from the left ventricle throughout the body as a whole. The pulmonary and systemic circulations are discussed later in the chapter.

BLOOD SUPPLY TO THE HEART MUSCLE

To sustain life, the heart must pump blood throughout the body on a regular and ongoing basis. As a result, the heart muscle or myocardium requires a constant supply of blood containing nutrients and oxygen to function effectively. The delivery of oxygen and nutrient-rich arterial blood to cardiac muscle tissue and the return of oxygen-poor blood from this active tissue to the venous system is called the **coronary circulation.**

Blood flows into the heart muscle by way of two small vessels that are surely the most famous of all the blood vessels—the **right** and **left coronary arteries**—famous because coronary heart disease kills many thousands of people every year. The coronary arteries are the aorta's first branches. The openings into these small vessels lie behind the flaps of the aortic semilunar valves. In both coronary thrombosis and coronary **embolism** (EM-bo-lizm), a blood clot occludes or plugs up some part of a coronary artery. Blood cannot pass through the occluded vessel and so cannot reach the heart muscle cells it normally supplies. Deprived of oxygen, these cells soon die or are damaged. In medical terms, **myocardial** (my-o-KAR-dee-al) **infarction** (in-FARK-shun) or tissue death occurs. Myocardial infarction or "heart attack" is a common cause of death during middle and late adulthood. Recovery from a myocardial infarction is possible if the amount of heart tissue damaged was small enough so that the remaining undamaged heart muscle can pump blood effectively enough to supply the needs of the rest of the heart and the body. The term **angina** (an-JYE-nah) **pectoris** (PEK-tor-is) is used to describe the severe chest pain that occurs when the myocardium is deprived of adequate oxygen. It is often a warning that the coronary arteries are no longer able to supply enough blood and oxygen to the heart muscle. **Coronary bypass surgery** is a frequent treatment for those who suffer from severely restricted coronary artery blood flow. In this procedure, veins are "harvested" or removed from other areas of the body and used to bypass partial blockages in coronary arteries.

FIGURE 11–4 *Blood Flow Through the Circulatory System.* In the pulmonary circulatory route, blood is pumped from the right side of the heart to the gas-exchange tissues of the lungs. In the systemic circulation, blood is pumped from the left side of the heart to all other tissues of the body.

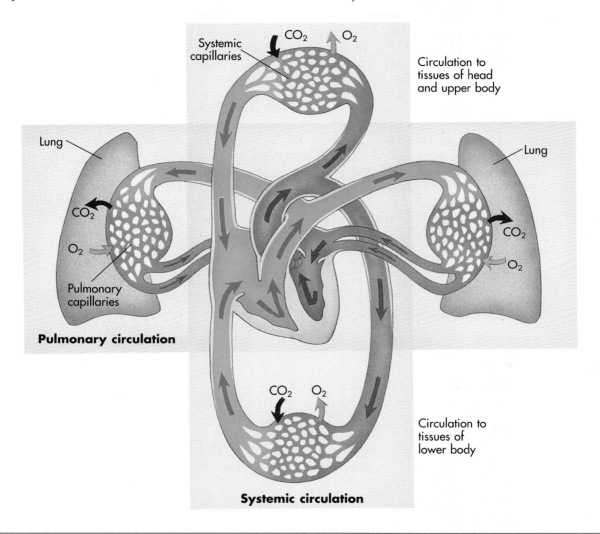

After blood has passed through the capillary beds in the myocardium, it flows into **cardiac veins,** which empty into the **coronary sinus** and finally into the right atrium.

CARDIAC CYCLE

The beating of the heart is a regular and rhythmic process. Each complete heart beat is called a **cardiac cycle** and includes the contraction (sys-tole) and relaxation (diastole) of atria and ventricles. Each cycle takes about 0.8 seconds to complete if the heart is beating at an average rate of about 72 beats per minute. The term **stroke volume** refers to the volume of blood ejected from the ventricles during each beat. **Cardiac output,** or the volume of blood pumped by one ventricle per minute, averages about 5 L in a normal, resting adult.

Changes in Blood Flow During Exercise

Not only does the overall rate of blood flow increase during exercise, but the relative blood flow through the different organs of the body also changes. During exercise, blood is routed away from the kidneys and digestive organs and toward the skeletal muscles, cardiac muscle, and skin. Rerouting of blood is accomplished by contracting precapillary sphincters in some tissues (thus reducing blood flow) while relaxing precapillary sphincters in other tissues (thus increasing blood flow). How can homeostasis be better maintained by these changes? One reason is because glucose and oxygen levels are dropping rapidly in muscles as they use up these substances to produce energy. Increased blood flow restores normal levels of glucose and oxygen rapidly. Blood warmed up in active muscles flows to the skin for cooling. This helps keep the body temperature from getting too high. Can you think of other ways this situation helps maintain homeostasis? Typical changes in organ blood flow with exercise are shown in the illustration. The green bar in each pair shows the resting blood flow; the blue bar shows the flow during exercise.

CONDUCTION SYSTEM OF THE HEART

Cardiac muscle fibers can contract rhythmically on their own. However, they must be coordinated by electrical signals (impulses) if the heart is to pump effectively. Although the rate of the cardiac muscle's rhythm is controlled by autonomic nerve signals, the heart has its own built-in conduction system for coordinating contractions during the cardiac cycle. The most important thing to realize about this conduction system is that all of the cardiac muscle fibers in each region of the heart are electrically linked together. The *intercalated disks* that were first introduced in Chapter 2 (see Figure 2-20, p. 39) are actually electrical connectors that join muscle fibers into a single unit that can conduct an impulse through the entire wall of a heart chamber without stopping. Thus both atrial walls will contract at about the same time because all their fibers are electrically linked. Likewise, both ventricular walls will contract at about the same time.

FIGURE 11–5 *Conduction System of the Heart.* Specialized cardiac muscle cells in the wall of the heart rapidly conduct an electrical impulse throughout the myocardium. The signal is initiated by the SA node (pacemaker) and spreads to the rest of the atrial myocardium and to the AV node. The AV node then initiates a signal that is conducted through the ventricular myocardium by way of the AV bundle (of His) and Purkinje fibers.

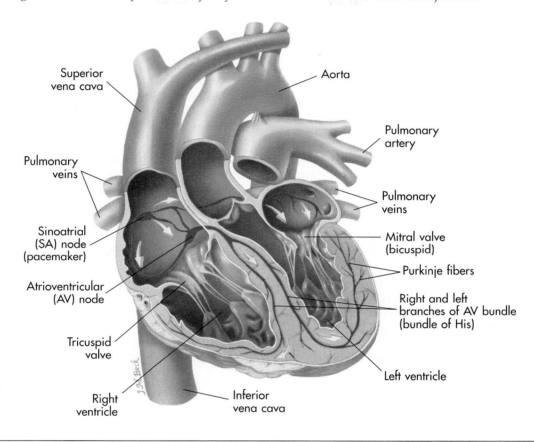

Four structures embedded in the wall of the heart are specialized to generate strong impulses and conduct them rapidly to certain regions of the heart wall. Thus they make sure that the atria contract and then the ventricles contract in an efficient manner. The names of the structures that make up this conduction system of the heart follow:

1. **Sinoatrial** (sye-no-AY-tree-al) **node,** which is sometimes called the SA node or the *pacemaker*
2. **Atrioventricular** (ay-tree-o-ven-TRIK-yoo-lar) **node** or **AV node**

3. **AV bundle** or **bundle of His**
4. **Purkinje** (pur-KIN-jee) **fibers**

Impulse conduction normally starts in the heart's pacemaker, namely, the SA node. From there, it spreads, as you can see in Figure 11-5, in all directions through the atria. This causes the atrial fibers to contract. When impulses reach the AV node, it relays them by way of the bundle of His and Purkinje fibers to the ventricles, causing them to contract. Normally, therefore, a ventricular beat follows each atrial beat. Various conditions such as endocarditis or myocardial infarction, however, can damage the heart's conduction system and thereby disturb its rhythmical beating. One such disturbance is the condition commonly called *heart block.* Impulses are blocked from getting through to the ventricles, resulting in the heart beating at a much slower rate than normal. A physician may treat heart block by implanting in the heart an **artificial pacemaker,** an electrical device that causes ventricular contractions at a rate fast enough to maintain an adequate circulation of blood.

ELECTROCARDIOGRAM

The specialized structures of the heart's conduction system generate tiny electrical currents that spread through surrounding tissues to the surface of the body. This fact is of great clinical significance because these electrical signals can be picked up from the body surface and transformed into visible tracings by an instrument called an **electrocardiograph** (e-lek-tro-KAR-dee-o-graf).

The **electrocardiogram** (e-lek-tro-KAR-dee-o-gram) or **ECG** is the graphic record of the heart's electrical activity. Skilled interpretation of these ECG records may sometimes make the difference between life and death. A normal ECG tracing is shown in Figure 11-6.

A normal ECG tracing has three very characteristic deflections or waves called the **P wave,** the **QRS complex,** and the **T wave.** These deflections represent the electrical activity that regulates the contraction or relaxation of the atria or ventricles. The term *depolarization* describes the electrical activity that triggers contraction of the heart muscle. *Repolarization* begins just before the relaxation phase of cardiac muscle activity. In the normal ECG shown in Figure 11-6, the small P wave occurs with depolarization of the atria. The QRS complex occurs as a result of depolarization of the ventricles, and the T wave results from electrical activity generated by repolarization of the ventricles. You may wonder why no visible record of atrial repolarization is noted in a normal ECG. The reason is simply that the deflection is very small and is hidden by the large QRS complex that occurs at the same time.

Damage to cardiac muscle tissue that is caused by a myocardial infarction or disease affecting the heart's conduction system results in distinctive changes in the ECG. Therefore ECG tracings are extremely valuable in the diagnosis and treatment of heart disease.

Blood Vessels

KINDS

Arterial blood is pumped from the heart through a series of large distribution vessels—the **arteries.** The largest artery in the body is the aorta. Arteries subdivide into vessels that become progressively smaller and finally become tiny **arterioles** (ar-TEER-ee-ols) that control the flow into microscopic exchange vessels called **capillaries** (KAP-i-lair-ees). In the so-called **capillary beds,** the exchange of nutrients and respiratory gases occurs between the blood and tissue fluid around the cells. Blood exits or is drained from the capillary beds and then enters the small **venules** (VEN-yools), which join with other venules and increase in size, becoming **veins.** The largest veins are the superior vena cava and the inferior vena cava.

As noted before (Figure 11-4), arteries carry blood away from the heart toward capillaries. Veins carry blood toward the heart away from capillaries, and capillaries carry blood from the tiny arterioles into tiny venules. The aorta carries blood out of the left ventricle of the heart, and the venae cavae return blood to the right atrium after the blood has circulated through the body.

STRUCTURE

Arteries, veins, and capillaries differ in structure. Three coats or layers are found in both arteries and veins (Figure 11-7). The outermost layer is called the **tunica adventitia.** Note that smooth muscle tissue is found in the middle layer or **tunica media** of arteries and veins. However, the muscle layer is much thicker in arteries than in veins. Why is this important? Because the thicker muscle layer in the artery wall is able to resist great pressures generated by ventricular systole. In arteries, the tunica media plays a critical role in maintaining blood pressure and controlling blood distribution. This is a smooth muscle, so it is controlled by the autonomic nervous system. The tunica media also includes a thin layer of elastic fibrous tissue.

An inner layer of endothelial cells called the **tunica intima** lines arteries and veins. The tunica intima is actually a single layer of squamous epithelial cells called **endothelium** (en-doe-THEE-lee-um) that lines the inner surface of the entire circulatory system.

FIGURE 11–6 *Events Represented by the Electrocardiogram (ECG).* **A** through **C,** The P wave represents the depolarization of cardiac muscle tissue in the SA node and atrial walls. **C** and **D,** Before the QRS complex is observed, the AV node and AV bundle depolarize. **E** and **F,** The QRS complex occurs as the atrial walls repolarize and the ventricular walls depolarize. **G,** The T wave is observed as the ventricular walls repolarize. Depolarization triggers contraction in the affected muscle tissue. Thus cardiac muscle contraction occurs *after* depolarization begins.

FIGURE 11–7 *Artery and Vein.* **A,** Schematic drawings of an artery and a vein show comparative thicknesses of the three layers: the outer layer or tunica adventitia, the muscle layer or tunica media, and the tunica intima made of endothelium. Note that the muscle and outer layer are much thinner in veins than in arteries and that veins have valves. **B,** Photomicrograph of artery and vein.

ARTERY

A

Thinner than
tunica media

Tunica adventitia
(connective tissue)

(thickest layer)

Thicker in arteries

Tunica media
(smooth muscle layer
and elastic tissue)

(thinner in veins)

Tunica intima
(endothelium)

VEIN

Semilunar valve

B

Vein

Artery

As you can see in Figure 11-7, veins have a unique structural feature not present in arteries. They are equipped with one-way valves that prevent the backflow of blood. When a surgeon cuts into the body, only arteries, arterioles, veins, and venules can be seen. Capillaries cannot be seen because they are microscopic. The most important structural feature of capillaries is their extreme thinness—only one layer of flat, endothelial cells composes the capillary membrane. Instead of three layers or coats, the capillary wall is composed of only one—the tunica intima. Substances such as glucose, oxygen, and wastes can quickly pass through it on their way to or from cells. Smooth muscle cells called **precapillary sphincters** guard the entrance to the capillary and determine into which capillary blood will flow.

FUNCTIONS

Arteries, veins, and capillaries have different functions. Arteries and arterioles distribute blood from the heart to capillaries in all parts of the body. In addition, by constricting or dilating, arterioles help maintain arterial blood pressure at a normal level. Venules and veins collect blood from capillaries and return it to the heart. They also serve as blood reservoirs because they can expand to hold a larger volume of blood or constrict to hold a much smaller amount. Capillaries function as exchange vessels. For example, glucose and oxygen move out of the blood in capillaries into interstitial fluid and on into cells. Carbon dioxide and other substances move in the opposite direction (that is, into the capillary blood from the cells). Fluid is also exchanged between capillary blood and interstitial fluid (see Chapter 18).

Study Figure 11-8 and Table 11-1 to learn the names of the main arteries of the body and Figures 11-9 and 11-10 and Table 11-2 for the names of the main veins.

Circulation

SYSTEMIC AND PULMONARY CIRCULATION

The term **circulation of blood** is self-explanatory, meaning that blood flows through vessels that are arranged to form a circuit or circular pattern. Blood flow from the left ventricle of the heart

TABLE 11–1 The Major Arteries

ARTERY	TISSUES SUPPLIED
Head and Neck	
Occipital	Posterior head and neck
Facial	Mouth, pharynx, and face
Internal carotid	Anterior brain and meninges
External carotid	Superficial neck, face, eyes, and larynx
Right common carotid	Right side of the head and neck
Left common carotid	Left side of the head and neck
Thorax	
Left subclavian	Left upper extremity
Brachiocephalic	Head and arm
Arch of aorta	Branches to head, neck, and upper extremities
Right and left coronary	Heart muscle
Abdomen	
Celiac	Stomach, spleen, and liver
Splenic	Spleen
Renal	Kidneys
Superior mesenteric	Small intestine; first half of the large intestine
Inferior mesenteric	Lower half of the large intestine
Upper Extremity	
Axillary	Axilla (armpit)
Brachial	Arm
Radial	Lateral side of the hand
Ulnar	Medial side of the hand
Lower Extremity	
Internal iliac	Pelvic viscera and rectum
External iliac	Genitalia and lower trunk muscles
Deep femoral	Deep thigh muscles
Femoral	Thigh
Popliteal	Leg and foot
Anterior tibial	Leg

FIGURE 11–8 *Principal Arteries of the Body.*

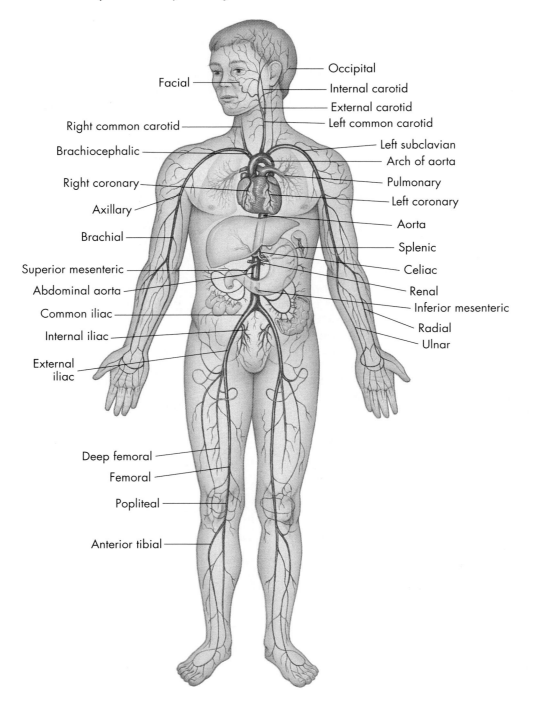

FIGURE 11–9   *Principal Veins of the Body.*

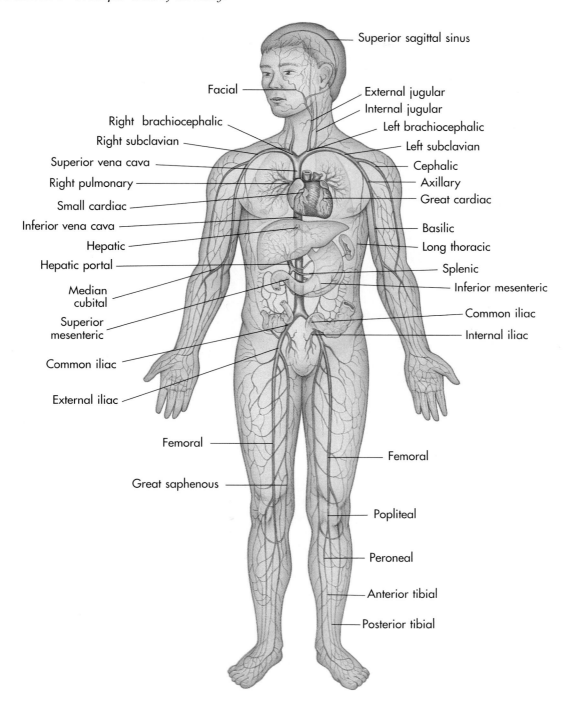

Superior sagittal sinus

Facial

External jugular
Internal jugular

Right brachiocephalic
Right subclavian

Left brachiocephalic

Left subclavian

Superior vena cava

Cephalic

Right pulmonary

Axillary

Small cardiac

Great cardiac

Inferior vena cava

Basilic

Hepatic

Long thoracic

Hepatic portal

Splenic

Median cubital

Inferior mesenteric

Superior mesenteric

Common iliac
Internal iliac

Common iliac

External iliac

Femoral

Femoral

Great saphenous

Popliteal

Peroneal

Anterior tibial

Posterior tibial

TABLE 11–2 The Major Veins

VEIN	TISSUES DRAINED
Head and Neck	
Superior sagittal sinus	Brain
Anterior facial	Anterior and superficial face
External jugular	Superficial tissues of the head and neck
Internal jugular	Sinuses of the brain
Thorax	
Brachiocephalic	Viscera of the thorax
Subclavian	Upper extremities
Superior vena cava	Head, neck, and upper extremities
Pulmonary	Lungs
Right and left coronary	Heart
Inferior vena cava	Lower body
Abdomen	
Hepatic	Liver
Long thoracic	Abdominal and thoracic muscles
Hepatic portal	Liver and gallbladder
Splenic	Spleen
Superior mesenteric	Small intestine and most of the colon
Inferior mesenteric	Descending colon and rectum
Upper Extremity	
Cephalic	Lateral arm
Axillary	Axilla and arm
Basilic	Medial arm
Median cubital	Cephalic vein (to basilic vein)
Lower Extremity	
External iliac	Lower limb
Internal iliac	Pelvic viscera
Femoral	Thigh
Great saphenous	Leg
Popliteal	Lower leg
Peroneal	Foot
Anterior tibial	Deep anterior leg and dorsal foot
Posterior tibial	Deep posterior leg and plantar aspect of foot

FIGURE 11–10 *Main Superficial Veins of the Arm.*

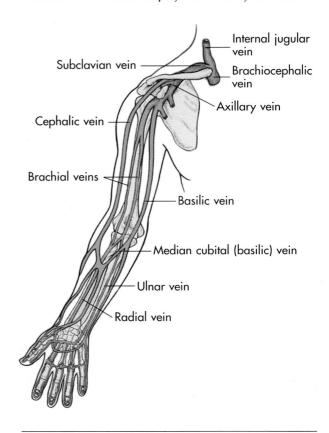

through blood vessels to all parts of the body and back to the right atrium of the heart has already been described as the **systemic circulation.** The left ventricle pumps blood into the aorta. From there, it flows into arteries that carry it into the tissues and organs of the body. As indicated in Figure 11-11, within each structure, blood moves from arteries to arterioles to capillaries. There, the vital two-way exchange of substances occurs between blood and cells. Next, blood flows out of each organ by way of its venules and then its veins to drain eventually into the inferior or superior vena cava. These two great veins return venous blood to the right atrium of the heart to complete the systemic circulation. But the blood has not quite come full circle back to its starting point in the left ventricle. To do this and start on its way again, it must first flow through another circuit, referred to earlier as the **pulmonary circulation.** Observe in Figure 11-11 that venous blood

FIGURE 11–11 *Diagram of Blood Flow in the Circulatory System.* Blood leaves the heart through arteries, then travels through arterioles, capillaries, venules, and veins before returning to the opposite side of the heart. Compare this figure with Figure 11-4.

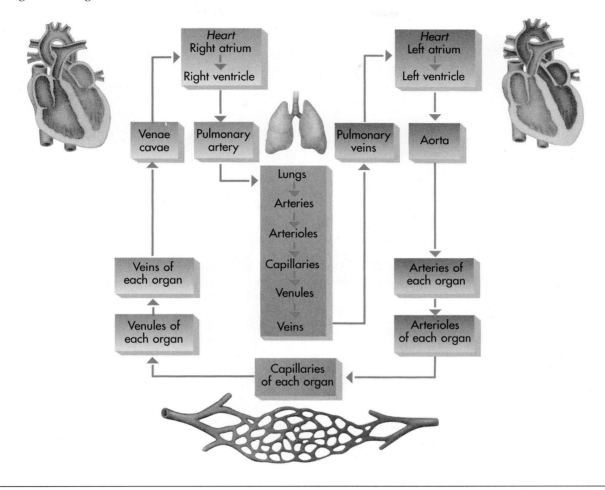

moves from the right atrium to the right ventricle to the pulmonary artery to lung arterioles and capillaries. There, the exchange of gases between the blood and air takes place, converting the deep crimson typical of venous blood to the scarlet of arterial blood. This oxygenated blood then flows through lung venules into four pulmonary veins and returns to the left atrium of the heart. From the left atrium, it enters the left ventricle to be pumped again through the systemic circulation.

HEPATIC PORTAL CIRCULATION

The term **hepatic portal circulation** refers to the route of blood flow through the liver. Veins from the spleen, stomach, pancreas, gallbladder, and intestines do not pour their blood directly into the inferior vena cava as do the veins from other abdominal organs. Instead, they send their blood to the liver by means of the hepatic portal vein (Figure 11-12). The blood then must pass through the liver before it reenters the regular venous

FIGURE 11–12 *Hepatic Portal Circulation.* In this very unusual circulation, a vein is located between two capillary beds. The hepatic portal vein collects blood from capillaries in visceral structures located in the abdomen and empties it into the liver. Hepatic veins return blood to the inferior vena cava. (Organs are not drawn to scale.)

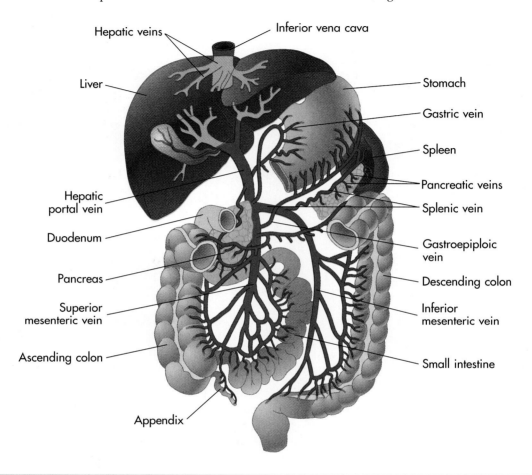

return to the heart. Blood leaves the liver by way of the hepatic veins, which drain into the inferior vena cava. As noted in Figure 11-11, blood normally flows from arteries to arterioles to capillaries to venules to veins and back to the heart. Blood flow in the hepatic portal circulation, however, does not follow this typical route. Venous blood, which would ordinarily return directly to the heart, is sent instead through a second capillary bed in the liver. The hepatic portal vein shown in Figure 11-12 is located between two capillary beds—one set in the digestive organs and the other in the liver. From the liver capillary beds, the path of blood returns to its normal route.

The detour of venous blood through a second capillary bed in the liver before its return to the heart serves some valuable purposes. For example, when a meal is being absorbed, the blood in the portal vein contains a higher-than-normal concentration of glucose. Liver cells remove the excess glucose and store it as glycogen; therefore blood leaving the liver usually has a normal blood glucose concentration. Liver cells also

FIGURE 11–13 *The Fetal Circulation.*

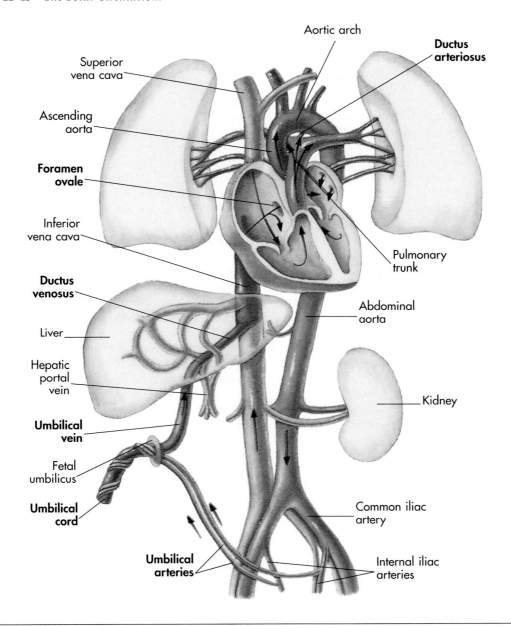

remove and detoxify various poisonous substances that may be present in the blood. The hepatic portal system is an excellent example of how "structure follows function" in helping the body maintain homeostasis.

FETAL CIRCULATION

Circulation in the body before birth differs from circulation after birth because the fetus must secure oxygen and food from maternal blood instead of from its own lungs and digestive organs. For the exchange of nutrients and oxygen to occur between fetal and maternal blood, specialized blood vessels must carry the fetal blood to the **placenta** (plah-SEN-tah), where the exchange occurs, and then return it to the fetal body. Three vessels (shown in Figure 11-13 as part of the **umbilical cord**) accomplish this purpose.

They are the two small **umbilical arteries** and a single, much larger **umbilical vein.** The movement of blood in the umbilical vessels may seem unusual at first in that the umbilical vein carries oxygenated blood, and the umbilical artery carries oxygen-poor blood. Remember that arteries are vessels that carry blood away from the heart, whereas veins carry blood toward the heart, regardless of the oxygen supply they may have.

Another structure unique to fetal circulation is called the **ductus venosus** (DUK-tus ve-NO-sus). As you can see in Figure 11-13, it is actually a continuation of the umbilical vein. It serves as a shunt, allowing most of the blood returning from the placenta to bypass the immature liver of the developing baby and empty directly into the inferior vena cava. Two other structures in the developing fetus allow most of the blood to bypass the developing lungs, which remain collapsed until birth. The **foramen ovale** (fo-RAY-men o-VAL-ee) shunts blood from the right atrium directly into the left atrium, and the **ductus arteriosus** (DUK-tus ar-teer-ee-O-sus) connects the aorta and the pulmonary artery.

At birth, the specialized fetal blood vessels and shunts must be rendered nonfunctional. When the newborn infant takes its first deep breaths, the circulatory system is subjected to increased pressure. The result is closure of the foramen ovale and rapid collapse of the umbilical blood vessels, the ductus venosus, and ductus arteriosus.

● *Blood Pressure*

UNDERSTANDING BLOOD PRESSURE

A good way to understand blood pressure might be to try to answer a few questions about it. What is blood pressure? Just what the words say— blood pressure is the pressure or push of blood.

Where does blood pressure exist? It exists in all blood vessels, but it is highest in the arteries and lowest in the veins. In fact, if we list blood vessels in order according to the amount of blood pressure in them and draw a graph, as in Figure 11-14, the graph looks like a hill, with aortic blood pressure at the top and vena caval pressure at the bottom. This blood pressure "hill" is spoken of as the *blood pressure gradient.* More precisely, the blood pressure gradient is the difference between two blood pressures. The blood pressure gradient for the entire systemic circulation is the difference between the average or mean blood pressure in the aorta and the blood pressure at the termination of the venae cavae where they join the right atrium of the heart. The mean blood pressure in the aorta, given in Figure 11-14, is 100 mm of mercury (mm Hg), and the pressure at the termination of the venae cavae is 0. Therefore, with these typical normal figures, the systemic blood pressure gradient is 100 mm Hg (100 minus 0).

Why is it important to understand blood pressure? What is its function? The blood pressure gradient is vitally involved in keeping the blood flowing. When a blood pressure gradient is present, blood circulates; conversely, when a blood pressure gradient is not present, blood does not circulate. For example, suppose that the blood pressure in the arteries were to decrease so that it became equal to the average pressure in arterioles. There would no longer be a blood pressure gradient between arteries and arterioles, and therefore there would no longer be a force to move blood out of arteries into arterioles. Circulation would stop, in other words, and very soon life itself would cease. This is why when arterial blood pressure is observed to be falling rapidly, whether in surgery or elsewhere, emergency measures must be quickly started to try to reverse this fatal trend.

What we have just said may start you wondering about why high blood pressure (meaning, of course, high arterial blood pressure) and low blood pressure are bad for circulation. High blood pressure is bad for several reasons. For one thing, if it becomes too high, it may cause the rupture of one or more blood vessels (for example, in the brain, as happens in a stroke). But low blood pressure also can be dangerous. If arterial pressure falls low enough, circulation and life cease. Massive hemorrhage, which dramatically reduces blood pressure, kills in this way.

FACTORS THAT INFLUENCE BLOOD PRESSURE

What causes blood pressure, and what makes blood pressure change from time to time? Factors such as blood volume, the strength of each heart contraction, heart rate, and the thickness of blood are all discussed in the following paragraphs.

FIGURE 11–14 *Pressure Gradients in Blood Flow.* Blood flows down a "blood pressure hill" from arteries, where blood pressure is highest, into arterioles, where it is somewhat lower, into capillaries, where it is still lower, and so on. All numbers on the graph indicate blood pressure measured in millimeters of mercury. The broken line, starting at 100 mm, represents the average pressure in each part of the circulatory system.

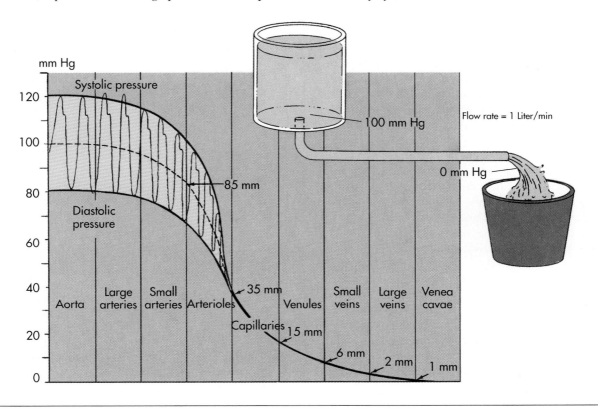

Blood Volume

The direct cause of blood pressure is the volume of blood in the vessels. The larger the volume of blood in the arteries, for example, the more pressure the blood exerts on the walls of the arteries, or the higher the arterial blood pressure.

Conversely, the less blood in the arteries, the lower the blood pressure tends to be. Hemorrhage demonstrates this relation between blood volume and blood pressure. In hemorrhage a pronounced loss of blood occurs, and this decrease in the volume of blood causes blood pressure to drop. In fact, the major sign of hemorrhage is a rapidly falling blood pressure.

The volume of blood in the arteries is determined by how much blood the heart pumps into the arteries and how much blood the arterioles drain out of them. The diameter of the arterioles plays an important role in determining how much blood drains out of arteries into arterioles.

Strength of Heart Contractions

The strength and the rate of the heartbeat affect cardiac output and therefore blood pressure. Each time the left ventricle contracts, it squeezes a certain volume of blood (the stroke volume) into the aorta and on into other arteries. The stronger that

Blood Pressure Readings

A device called a **sphygmomanometer** (sfig-mo-ma-NAH-me-ter) is often used to measure blood pressures in both clinical and home health care situations. The traditional sphygmomanometer is an inverted tube of mercury (Hg) with a balloonlike air cuff attached via an air hose. The air cuff is placed around a limb, usually the subject's upper arm as shown in the figure. A stethoscope sensor is placed over a major artery (the *brachial artery* in the figure) to listen for the arterial pulse. A hand-operated pump fills the air cuff, increasing the air pressure and pushing the column of mercury higher. While listening through the stethoscope, the operator opens the air cuff's outlet valve and slowly reduces the air pressure around the limb. Loud, tapping *Korotkoff sounds* suddenly begin when the cuff pressure measured by the mercury column equals the systolic pressure—usually about 120 mm. As the air pressure surrounding the arm continues to decrease, the Korotkoff sounds disappear. The pressure measurement at which the sounds disappear is equal to the diastolic pressure—usually about 80 mm. The subject's blood pressure is then expressed as **systolic pressure** (the maximum arterial pressure during each contraction) over the **diastolic pressure** (the minimum arterial pressure), such as 120/80. Mercury sphygmomanometers have been replaced in many clinical settings by nonmercury devices that similarly measure the maximum and minimum arterial blood pressures. In home health care settings, patients can often learn to monitor their own blood pressure.

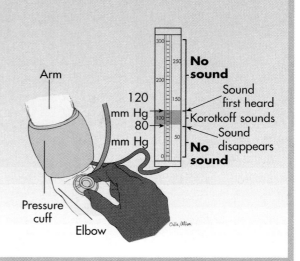

each contraction is, the more blood it pumps into the aorta and arteries. Conversely, the weaker that each contraction is, the less blood it pumps. Suppose that one contraction of the left ventricle pumps 70 ml of blood into the aorta, and suppose that the heart beats 70 times a minute; 70 ml × 70 equals 4900 ml. Almost 5 L of blood would enter the aorta and arteries every minute (the cardiac output). Now suppose that the heartbeat were to become weaker and that each contraction of the left ventricle pumps only 50 ml instead of 70 ml of blood into the aorta. If the heart still contracts 70 times a minute, it will obviously pump much less blood into the aorta—only 3500 ml instead of the more normal 4900 ml per minute. This decrease in the heart's output decreases the volume of blood in the arteries, and the decreased arterial blood volume decreases arterial blood pressure. In summary, the strength of the heartbeat affects blood pressure in this way: a stronger heartbeat increases blood pressure, and a weaker beat decreases it.

Heart Rate

The rate of the heartbeat may also affect arterial blood pressure. You might reason that when the heart beats faster, more blood enters the aorta, and therefore the arterial blood volume and blood pressure increase. This is true only if the stroke volume does not decrease sharply when the heart rate increases. Often, however, when the heart beats faster, each contraction of the left ventricle takes place so rapidly that it has little time to fill, and it squeezes out much less blood than usual into the aorta. For example, suppose that the heart rate speeded up from 70 to 100 times a minute and that at the same time its stroke volume decreased from 70 ml to 40 ml. Instead of a cardiac output of 70×70 or 4900 ml per minute, the cardiac output would have changed to 100×40 or 4000 ml per minute. Arterial blood volume decreases under these conditions, and therefore blood pressure also decreases, even though the heart rate has increased. What generalization, then, can we make? We can only say that an increase in the rate of the heartbeat increases blood pressure, and a decrease in the rate decreases blood pressure. But whether a change in the heart rate actually produces a similar change in blood pressure depends on whether the stroke volume also changes and in which direction.

Blood Viscosity

Another factor that we ought to mention in connection with blood pressure is the viscosity of blood, or in plainer language, its thickness. If blood becomes less viscous than normal, blood pressure decreases. For example, if a person suffers a hemorrhage, fluid moves into the blood from the interstitial fluid. This dilutes the blood and decreases its viscosity, and blood pressure then falls because of the decreased viscosity. After hemorrhage, whole blood or plasma is preferred to saline solution for transfusions. The reason is that saline solution is not a viscous liquid and so cannot keep blood pressure at a normal level.

FLUCTUATIONS IN BLOOD PRESSURE

No one's blood pressure stays the same all the time. It fluctuates, even in a perfectly healthy individual. For example, it goes up when a person exercises strenuously. Not only is this normal, but the increased blood pressure serves a good purpose. It increases circulation to bring more blood to muscles each minute and thus supplies them with more oxygen and food for more energy.

A normal average arterial blood pressure is about 120/80, or 120 mm Hg systolic pressure (as the ventricles contract) and 80 mm Hg diastolic pressure (as the ventricles relax). Remember, however, that what is "normal" varies somewhat among individuals and also with age.

The venous blood pressure, as you can see in Figure 11-14, is very low in the large veins and falls almost to 0 by the time blood leaves the venae cavae and enters the right atrium. The venous blood pressure within the right atrium is called the **central venous pressure.** This pressure level is important because it influences the pressure that exists in the large peripheral veins. If the heart beats strongly, the central venous pressure is low as blood enters and leaves the heart chambers efficiently. However, if the heart is weakened, central venous pressure increases, and the flow of blood into the right atrium is slowed. As a result, a person suffering heart failure, who is sitting at rest in a chair, often has distended external jugular veins as blood "backs up" in the venous network.

Five mechanisms help to keep venous blood moving back through the circulatory system and into the right atrium. They include a strongly beating heart, an adequate arterial blood pressure, valves in veins, the "milking action" of skeletal muscles as they contract, and changing pressures in the chest cavity caused by breathing.

Pulse

What you feel when you take a pulse is an artery expanding and then recoiling alternately. To feel a pulse, you must place your fingertips over an artery that lies near the surface of the body and over a bone or other firm base. The pulse is a valuable clinical sign. It can provide information, for example, about the rate, strength, and rhythmicity of the heart beat. It is also easily determined with little or no danger or discomfort. There are nine major "pulse points" named after the arteries where they are felt. Locate each pulse point on Figure 11-15 and on your own body.

FIGURE 11–15   *Pulse Points.* Each pulse point is named after the artery with which it is associated.

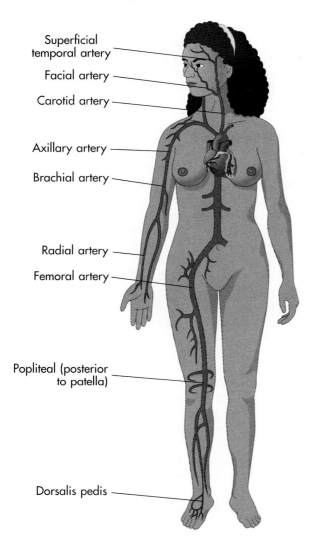

Superficial temporal artery

Facial artery

Carotid artery

Axillary artery

Brachial artery

Radial artery

Femoral artery

Popliteal (posterior to patella)

Dorsalis pedis

The following pulse points are located on each side of the head and neck: (1) over the superficial temporal artery in front of the ear, (2) the common carotid artery in the neck along the front edge of the sternocleidomastoid muscle, and (3) over the facial artery at the lower margin of the mandible at a point below the corner of the mouth.

A pulse is also detected at three points in the upper limb: (1) in the axilla over the axillary artery; (2) over the brachial artery at the bend of the elbow along the inner or medial margin of the biceps brachii muscle, and (3) at the radial artery at the wrist. The so-called radial pulse is the most frequently monitored and easily accessible in the body.

The pulse can also be felt at three locations in the lower extremity: (1) over the femoral artery in the groin, (2) at the popliteal artery behind and just proximal to the knee, and (3) at the dorsalis pedis artery on the front surface of the foot, just below the bend of the ankle joint.

HEART

A Location, size, and position

  1 Triangular organ located in mediastinum with two thirds of the mass to the left of the body midline and one third to the right; the apex on the diaphragm; shape and size of a closed fist (Figure 11-1)

  2 Cardiopulmonary resuscitation (CPR)—heart lies between the sternum in front and the bodies of the thoracic vertebrae behind; rhythmic compression of the heart between the sternum and vertebrae can maintain blood flow during cardiac arrest; if combined with artificial respiration procedure, it can be life saving

B Anatomy

  1 Heart chambers (Figure 11-2)

    a Two upper chambers are called *atria* (receiving chambers)—right and left atria

    b Two lower chambers called *ventricles* (discharging chambers)—right and left ventricles

    c Wall of each heart chamber is composed of cardiac muscle tissue called *myocardium*

    d Endocardium—smooth lining of heart chambers—inflammation of endocardium called *endocarditis*

  2 Covering sac or pericardium

    a Pericardium is a two-layered fibrous sac with a lubricated space between the two layers

    b Inner layer is called *visceral pericardium* or *epicardium*

    c Outer layer called *parietal pericardium*

  3 Heart action

    Contraction of the heart is called *systole;* relaxation is called *diastole.*

  4 Heart valves (Figure 11-3)

    Four valves keep blood flowing through the heart; prevent backflow (two atrioventricular or AV and two semilunar valves)

    a Tricuspid—at the opening of the right atrium into the ventricle

    b Bicuspid (mitral)—at the opening of the left atrium into the ventricle

    c Pulmonary semilunar—at the beginning of the pulmonary artery

    d Aortic semilunar—at the beginning of the aorta

C Heart sounds

  1 Two distinct heart sounds in every heartbeat or cycle—"lub-dup"

  2 First (lub) sound is caused by the vibration and closure of AV valves during contraction of the ventricles

  3 Second (dup) sound is caused by the closure of the semilunar valves during relaxation of the ventricles

D Blood flow through the heart (Figure 11-4)

  1 Heart acts as two separate pumps—the right atrium and ventricle performing different functions from the left atrium and ventricle

  2 Sequence of blood flow: venous blood enters the right atrium through the superior and inferior venae cavae—passes from the right atrium through the tricuspid valve to the right ventricle; from the right ventricle it passes through the pulmonary semilunar valve to the pulmonary artery to the lungs—blood moves from the lungs to the left atrium, passing through the bicuspid (mitral) valve to the left ventricle; blood in the left ventricle is pumped through the aortic semilunar valve into the aorta and is distributed to the body as a whole

E Blood supply to the heart muscle

  1 Blood, which supplies oxygen and nutrients to the myocardium of the heart, flows through the right and left coronary arteries

  2 Blockage of blood flow through the coronary arteries is called *myocardial infarction (heart attack)*

  3 Angina pectoris—chest pain caused by inadequate oxygen to the heart

F Cardiac cycle

  1 Heart beat is regular and rhythmic—each complete beat called a *cardiac cycle*—average is about 72 beats per minute

  2 Each cycle, about 0.8 seconds long, subdivided into systole (contraction phase) and diastole (relaxation phase)

  3 Stroke volume—volume of blood ejected from one ventricle with each beat

  4 Cardiac output—amount of blood that one ventricle can pump each minute; average is about 5 L per minute at rest

G Conduction system (Figure 11-5)

1 Intercalated disks are electrical connectors that join all the cardiac muscle fibers in a region together so that they receive their impulse, and thus contract, at about the same time

2 SA (sinoatrial) node, the pacemaker—located in the wall of the right atrium near the opening of the superior vena cava

3 AV (atrioventricular) node—located in the right atrium along the lower part of the interatrial septum

4 AV bundle (bundle of His)—located in the septum of the ventricle

5 Purkinje fibers—located in the walls of the ventricles

H Electrocardiogram (Figure 11-6)

1 Specialized conduction system structures generate and transmit the electrical impulses that result in contraction of the heart

2 These tiny electrical impulses can be picked up on the surface of the body and transformed into visible tracings by a machine called an electrocardiograph

3 The visible tracing of these electrical signals is called an *electrocardiogram* or *ECG*

4 The normal ECG has three deflections or waves called the *P wave,* the *QRS complex,* and the *T wave*

   a P wave—associated with depolarization of the atria

   b QRS complex—associated with depolarization of the ventricles

   c T wave—associated with repolarization of the ventricles

BLOOD VESSELS

A Kinds

1 Arteries—carry blood away from the heart

2 Veins—carry blood toward the heart

3 Capillaries—carry blood from the arterioles to the venules

B Structure (Figure 11-7)

1 Arteries

   a Tunica intima—inner layer of endothelial cells

   b Tunica media—smooth muscle with some elastic tissue, thick in arteries; important in blood pressure regulation

   c Tunica adventitia—thin layer of elastic tissue

2 Capillaries—microscopic vessels

   a Only layer is the tunica intima

3 Veins

   a Tunica intima—inner layer; valves prevent retrograde movement of blood

   b Tunica media—smooth muscle; thin in veins

   c Tunica adventitia—heavy layer in many veins

C Functions

1 Arteries—distribution of nutrients, gases, etc., with movement of blood under high pressure; assist in maintaining the arterial blood pressure

2 Capillaries—serve as exchange vessels for nutrients, wastes, and fluids

3 Veins—collect blood for return to the heart; low pressure vessels

D Names of main arteries—see Figure 11-8 and Table 11-1

E Names of main veins—see Figure 11-9 and Table 11-2

CIRCULATION

A Plan of circulation—refers to the blood flow through the vessels arranged to form a circuit or circular pattern (Figure 11-11)

B Types of circulation

1 Systemic circulation

   a Carries blood throughout the body

   b Path goes from left ventricle through aorta, smaller arteries, arterioles, capillaries, venules, venae cavae, to right atrium

2 Pulmonary circulation

   a Carries blood to and from the lungs; arteries deliver deoxygenated blood to the lungs for gas exchange

   b Path goes from right ventricle through pulmonary arteries, lungs, pulmonary veins, to left atrium

3 Hepatic portal circulation (Figure 11-12)

   a Unique blood route through the liver

   b Vein (hepatic portal vein) exists between two capillary beds

   c Assists with homeostasis of blood glucose levels

4 Fetal circulation (Figure 11-13)
  a Refers to circulation before birth
  b Modifications required for fetus to efficiently secure oxygen and nutrients from the maternal blood
  c Unique structures include the placenta, umbilical arteries and vein, ductus venosus, ductus arteriosus, and foramen ovale

BLOOD PRESSURE

A Blood pressure is push or force of blood in the blood vessels
B Highest in arteries, lowest in veins (Figure 11-14)
C Blood pressure gradient causes blood to circulate—liquids can flow only from the area where pressure is higher to where it is lower
D Blood volume, heartbeat, and blood viscosity are main factors that produce blood pressure
E Blood pressure varies within normal range from time to time

PULSE

A Definition—alternate expansion and recoil of the blood vessel wall
B Places where you can count the pulse easily (Figure 11-15)

NEW WORDS

angina pectoris
arteriole
artery
atrioventricular (AV) valve
atrium
bicuspid valve
capillary
cardiac output
cardiopulmonary resuscitation (CPR)
coronary circulation

diastolic pressure
ductus arteriosus
ductus venosus
electrocardiogram (ECG)
endocarditis
endocardium
epicardium
foramen ovale
hepatic portal circulation

mitral valve
myocardial infarction
myocardium
P wave
pacemaker
pericardium
pulmonary circulation
pulse
Purkinje fibers
QRS complex

semilunar valve
sinoatrial node
systemic circulation
T wave
tricuspid valve
umbilical
vein
ventricle
venule

CHAPTER TEST

1. The two upper chambers of the heart are called _____, and the two lower chambers are called _____.
2. Cardiac muscle tissue, which forms the wall of each heart chamber, is called the _____.
3. The covering sac around the heart is called the _____.
4. The two valves that separate the upper from the lower chambers of the heart are called _____ valves.
5. The valve located between the right atrium and ventricle is called the _____ valve.
6. Blood enters the right atrium through two large veins called the superior and inferior _____.
7. Blood passing through the pulmonary semilunar valve enters the _____ artery.
8. Severe chest pain caused by inadequate blood flow to the myocardium is called _____.
9. The sinoatrial or SA node is often called the _____ of the heart.
10. A graphic recording of the heart's electrical activity is called an _____.
11. In a normal electrocardiogram the QRS complex occurs as a result of depolarization of the _____.
12. Blood is carried toward the heart by large vessels called _____.
13. The muscular middle layer of both arteries and veins is called the _____.
14. Blood flow from the left ventricle of the heart to all parts of the body and back to the right atrium of the heart is called the _____ circulation.
15. In portal circulation a portal vein is located between two _____ beds.
16. In fetal circulation, blood is shunted from the right atrium directly into the left atrium by passing through the _____ _____ of the heart.
17. The volume of blood pumped into the arteries from one ventricle in 1 minute is called the _____ _____.
18. The _____ _____ is a continuation of the umbilical vein.
19. In the fetus the _____ _____ connects the aorta and pulmonary artery.
20. Contraction of the heart is called _____.
21. Each complete heart beat is called a _____.
22. The venous blood pressure within the right atrium is called the _____ _____ _____.

Select the most correct answer from Column B for each statement in Column A. (Only one answer is correct.)

COLUMN A		COLUMN B
23. _____ Receiving chambers of heart		a. P wave
		b. Coronary arteries
24. _____ Attach valves to heart wall		c. Portal circulation
		d. Heart attack
25. _____ Supply blood to heart muscle		e. Ductus arteriosus
		f. Chordae tendinea
26. _____ Myocardial infarction		g. Systole
		h. Atria
27. _____ Depolarization of atria		
28. _____ Vein between two capillary beds		
29. _____ Fetal circulation		
30. _____ Heart contraction		

R E V I E W   Q U E S T I O N S

1. Describe the position of the heart in the mediastinum.
2. What is cardiopulmonary resuscitation (CPR)?
3. What is meant by the term *cardiac cycle?*
4. A patient has had an operation to repair the bicuspid valve. Where is the valve, and what is its function?
5. What are some differences between an artery, a vein, and a capillary?
6. The right ventricle of the heart pumps blood to and through only one organ. Which one?
7. What part of the heart pumps blood through the systemic circulation (that is, to and through all organs other than the lungs)?
8. All blood returns from the systemic circulation to what part of the heart?
9. Blood returning from the pulmonary circulation (from the lungs, in other words) enters what part of the heart?
10. How do arterial blood and venous blood differ with regard to their oxygen and carbon dioxide contents?
11. How does cardiac output influence blood pressure?
12. Explain what causes a pulse in an arterial vessel.

C R I T I C A L   T H I N K I N G

13. Outline the path of blood flow through the heart. Explain the function of the valves as you proceed.
14. Considering that the function of the circulatory system is to transport substances to and from the cells, do you think it is true that in one sense capillaries are our most important blood vessels? Give a reason for your answer.
15. Does every artery carry oxygenated blood and every vein carry deoxygenated blood? If not, what exceptions are there?

The Lymphatic System and Immunity

OBJECTIVES

*After you have completed this chapter,
you should be able to:*

1. Describe the generalized functions of the lymphatic system and list the primary lymphatic structures.

2. Define and compare nonspecific and specific immunity, inherited and acquired immunity, and active and passive immunity.

3. Discuss the major types of immune system molecules and indicate how antibodies and complements function.

4. Discuss and contrast the development and functions of B and T cells.

5. Compare and contrast humoral and cell-mediated immunity.

All of us live in a hostile and dangerous environment. Each day we are faced with potentially harmful toxins, disease-causing bacteria, viruses, and even cells from our own bodies that have been transformed into cancerous invaders. Fortunately, we are protected from this staggering variety of differing biological enemies by a remarkable set of defense mechanisms. We refer to this protective "safety net" as the **immune system.**

This chapter deals with the immune system. This system is characterized by structural components, many of them lymphatic organs, and by a functional group of specialized cells and molecules that protect us from infection and disease. This chapter begins with an overview of the lymphatic system, discussing vessels that help maintain fluid balance and lymphoid tissues that help protect the internal environment. We will then discuss the concept of immunity and the ways that highly specialized cells and molecules provide us with effective and very specific resistance to disease.

The Lymphatic System

LYMPH AND LYMPH VESSELS

Maintaining the constancy of the fluid around each body cell is possible only if numerous homeostatic mechanisms function effectively together in a controlled and integrated response to changing conditions. We know from Chapter 11 that the circulatory system provides a key role in bringing many needed substances to cells and then removing the waste products that accumulate as a result of metabolism. This exchange of substances between blood and tissue fluid occurs in capillary beds. Many additional substances that cannot enter or return through the capillary walls, including excess fluid and protein molecules, are returned to the blood as **lymph.** Lymph is a specialized fluid formed in the tissue spaces that is transported by way of specialized **lymphatic vessels** to eventually reenter the circulatory system. In addition to lymph and the lymphatic vessels, the lymphatic system includes lymph nodes and specialized lymphatic organs such as the thymus and spleen (Figure 12-1).

Lymph forms in this way: blood plasma filters out of the capillaries into the microscopic spaces between tissue cells because of the pressure generated by the pumping action of the heart. There, the liquid is called **interstitial fluid** or tissue fluid. Much of the interstitial fluid goes back into the blood by the same route it came out (that is, through the capillary membrane). The remainder of the interstitial fluid enters the lymphatic system before it returns to the blood. The fluid, now called *lymph,* enters a network of tiny blind-ended tubes distributed in the tissue spaces. These tiny vessels, called *lymphatic capillaries,* permit excess tissue fluid and some other substances such as dissolved protein molecules to leave the tissue spaces. Figure 12-2 shows the role of the lymphatic system in fluid homeostasis.

Lymphatic and blood capillaries are similar in many ways. Both types of vessels are microscopic and both are formed from sheets consisting of a cell layer of simple squamous epithelium called *endothelium* (en-doe-THEE-lee-um). The flattened endothelial cells that form blood capillaries, however, fit tightly together so that large molecules cannot enter or exit from the vessel. The "fit" between endothelial cells forming the lymphatic capillaries is not as tight. As a result, they are more porous and allow larger molecules, including proteins and other substances, as well as the fluid itself, to enter the vessel and eventually return to the general circulation. The movement of lymph in the lymphatic vessels is one way. Unlike blood, lymph does not flow over and over again through vessels that form a circular route.

Lymph flowing through the lymphatic capillaries next moves into successively larger and larger vessels called *lymphatic venules* and *veins* and eventually empties into two terminal vessels called the **right lymphatic duct** and the **thoracic duct,** which empty their lymph into the blood in veins in the neck region. Lymph from about three fourths of the body eventually drains into the thoracic duct, which is the largest lymphatic vessel in the body. Lymph from the right upper extremity and from the right side of the head, neck, and upper torso flows into the right lymphatic duct (Figure 12-3). The lymphatic vessels have a "beaded" appearance caused by the presence of valves that assist in maintaining a one-way flow of lymph. Note in Fig-

FIGURE 12–1   *Principal Organs of the Lymphatic System.*

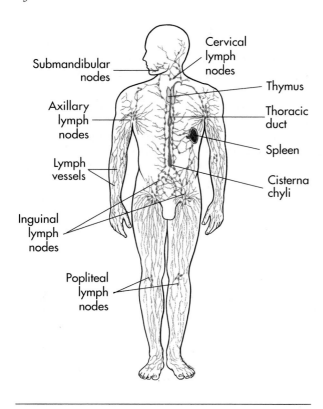

FIGURE 12–2   *Role of Lymphatic System in Fluid Homeostasis.* Fluid from blood plasma that is not reabsorbed by blood vessels drains into lymphatic vessels. Lymphatic drainage prevents accumulation of too much tissue fluid.

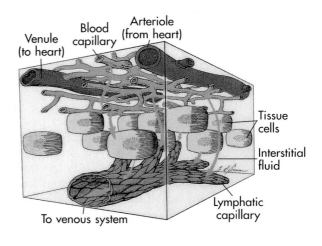

FIGURE 12–3   *Lymph Drainage.* The right lymphatic duct drains lymph from the upper right quarter of the body into the right subclavian vein. The thoracic duct drains lymph from the rest of the body into the left subclavian vein.

Effects of Exercise on Immunity

Exercise physiologists have found that moderate exercise increases the number of WBCs, specifically granular leukocytes and lymphocytes. Not only is the number of circulating immune cells higher after exercise, but the activity of sensitized T cells is also increased. But at the same time, research also shows that strenuous exercise may actually inhibit immune function. Nevertheless, immediate moderate exercise such as walking after a trauma such as surgery is often encouraged because of its immunity-strengthening effects.

ure 12-1 that the thoracic duct in the abdomen has an enlarged pouchlike structure called the **cisterna chyli** (sis-TER-nah KI-li) that serves as a storage area for lymph moving toward its point of entry into the venous system.

Lymphatic capillaries in the wall of the small intestine are given the special name of **lacteals** (LAK-tee-als). They transport fats obtained from food to the bloodstream and are discussed in Chapter 14.

LYMPH NODES

As lymph moves from its origin in the tissue spaces toward the thoracic or right lymphatic ducts and then into the venous blood, it is filtered by moving through **lymph nodes,** which are located in clusters along the pathway of lymphatic vessels. Some of these nodes may be as small as a pinhead, and others may be as large as a lima bean. With the exception of a compara-

tively few single nodes, most lymph nodes occur in groups or clusters in certain areas. Figure 12-1 shows the locations of the clusters of greatest clinical importance. You may have experienced swelling of the submandibular nodes during a "sore throat" or as a result of an infected tooth. The structure of the lymph nodes makes it possible for them to perform two important functions: defense and white blood cell formation.

Defense Function: Biological Filtration

Figure 12-4 shows the structure of a typical lymph node. In this example, a small node located next to an infected hair follicle is shown filtering bacteria from lymph. Lymph nodes perform biological filtration, a process in which cells (phagocytic cells in this case) alter the contents of the filtered fluid. Biological filtration of bacteria and other abnormal cells by phagocytosis prevents local infections from spreading. Note that lymph enters

Figure 12–4 *Infection of Hair Follicle.* Diagrammatic representation of a skin section in which an infection surrounds a hair follicle. The yellow areas represent dead and dying cells (pus). The black dots around the yellow areas represent bacteria. Bacteria entering the node via the afferent lymphatics are filtered out.

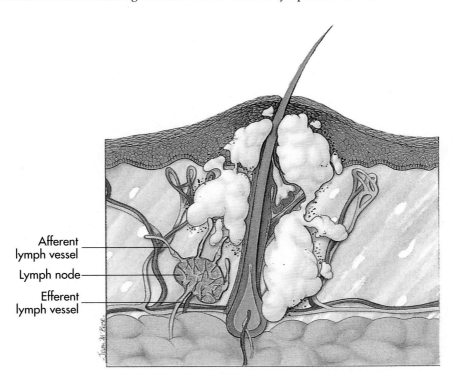

Afferent lymph vessel

Lymph node

Efferent lymph vessel

the node through four **afferent** (from the Latin "to carry toward") **lymph vessels.** These vessels deliver lymph to the node. When passing through the node, lymph is filtered so that injurious particles such as bacteria, soot, and cancer cells are removed and prevented from entering the blood and circulating all over the body. Lymph exits from the node through a single **efferent** (from the Latin "to carry away from") **lymph vessel.** Clusters of lymph nodes allow a very effective biological filtration of lymph flowing from specific body areas. A knowledge of lymph node location and function is important in clinical medicine. For example, a school nurse monitoring the progress of a child with an infected finger will watch the elbow and axillary regions for swelling and tenderness of the lymph nodes. These nodes filter lymph returning from the hand and may become infected by the bacteria they trap. A surgeon uses knowledge of lymph node function when removing lymph nodes under the arms (axillary nodes) and in other areas during an operation for breast cancer. These nodes may contain cancer cells filtered out of the lymph drained from the breast. Cancer of the breast is one of the most common forms of this disease in women. Unfortunately, cancer cells from a single tumorous growth in the breast often spread to other areas of the body through the lymphatic system. Figure 12-5 shows how lymph from the breast drains into many different and widely placed nodes.

THYMUS

The **thymus** (THYE-mus) (Figure 12-1) is a small lymphoid tissue organ located in the mediastinum, extending upward in the midline of the neck. It is composed of lymphocytes in a unique epithelial tissue framework. The thymus is largest at puberty and even then weighs only about 35 or 40 g—less than an ounce. Although small in size the thymus plays a central and critical role in the body's vital immunity mechanism. First, it is a source of lymphocytes before birth and is then especially important in the "maturation" or development of specialized lymphocytes that then leave the thymus and circulate to the spleen, tonsils, lymph nodes, and other lymphatic tissues. These **T-lymphocytes** or T cells are critical to the functioning of the immune system and are discussed later. They develop under the influence of

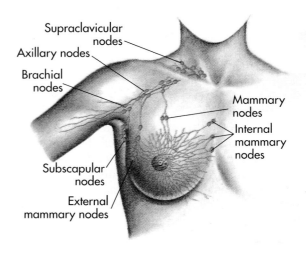

FIGURE 12–5 *Lymphatic Drainage of the Breast.* Note the extensive network of nodes that receive lymph from the breast.

Allergy

The term **allergy** is used to describe hypersensitivity of the immune system to relatively harmless environmental antigens. One in six Americans have a genetic predisposition to an allergy. Immediate allergic responses involve antigen-antibody reactions that trigger the release of histamine, kinins, and other inflammatory substances. These responses usually cause symptoms such as a runny nose, conjunctivitis, and hives. In some cases, these substances may cause constriction of airways, relaxation of blood vessels, and irregular heart rhythms that can lead to a life-threatening condition called *anaphylactic shock*. Delayed allergic responses, on the other hand, involve cell-mediated immunity. In contact dermatitis, for example, T-lymphocytes trigger events that lead to local skin inflammation a few hours or days after initial exposure.

a hormone secreted by the thymus, **thymosin** (THYE-mo-sin). The thymus essentially completes its work early in childhood and is then replaced by fat and connective tissue, a process called *involution*.

TONSILS

Masses of lymphoid tissue called tonsils are located in a protective ring under the mucous membranes in the mouth and back of the throat (Figure 12-6). They help protect us against bacteria that may invade tissues in the area around the openings between the nasal and oral cavities. The **palatine tonsils** are located on each side of the throat. The **pharyngeal tonsils,** known as **adenoids** (AD-e-noyds) when they become swollen, are near the posterior opening of the nasal cavity. A third type of tonsil, the **lingual tonsils,** are near the base of the tongue. The tonsils serve as the first line of defense from the exterior and as such are subject to chronic infection. They may have to be removed surgically if antibiotic therapy is not successful or if swelling impairs breathing.

FIGURE 12–6 *Location of the Tonsils.* Small segments of the roof and floor of the mouth have been removed to show the protective ring of tonsils (lymphoid tissue) around the internal opening of the nose and throat.

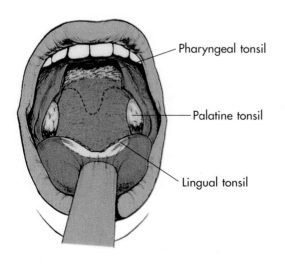

Pharyngeal tonsil

Palatine tonsil

Lingual tonsil

SPLEEN

The spleen is the largest lymphoid organ in the body. It is located high in the upper left quadrant of the abdomen lateral to the stomach (Figures 12-1 and 1-7). Although the spleen is protected by the lower ribs, it can be injured by abdominal trauma. The spleen has a very rich blood supply and may contain over 1 pint of blood. If damaged and bleeding, surgical removal, called a **splenectomy** (splen-NEK-toe-mee), may be required to stop the loss of blood.

After entering the spleen, blood flows through dense, pulplike accumulations of lymphocytes. As blood flows through the pulp, the spleen removes by filtration and phagocytosis many bacteria and other foreign substances, destroys worn out RBCs and salvages the iron found in hemoglobin for future use, and serves as a reservoir for blood that can be returned to the circulatory system when needed.

The Immune System

FUNCTION OF THE IMMUNE SYSTEM

The body's defense mechanisms protect us from disease-causing microorganisms that invade our bodies, foreign tissue cells that may have been transplanted into our bodies, and our own cells that have turned malignant or cancerous. The body's specific defense system is called the **immune system.** The immune system makes us immune (that is, able to resist these enemies). Unlike other systems of the body, which are made up of groups of organs, the immune system is made up of billions of cells and trillions of molecules.

NONSPECIFIC IMMUNITY

Nonspecific immunity is maintained by mechanisms that attack any irritant or abnormal substance that threatens the internal environment. In other words, nonspecific immunity confers general protection rather than protection from certain kinds of invading cells or chemicals. The skin and mucous membranes, for example, are mechanical barriers to prevent entry into the body of bacteria and many other substances such as toxins and harmful chemicals. Tears and mucus also contribute to nonspecific immunity. Tears wash harmful substances from the eyes, and mucus traps foreign material that may enter the respiratory tract. Phagocytosis of bacteria by WBCs is a nonspecific form of immunity.

The **inflammatory response** is a set of nonspecific responses that often occurs in the body. In the example shown in Figure 12-7, bacteria cause tissue damage that, in turn, triggers the release of mediators from any of a variety of immune cells. Some of the mediators attract white blood cells to the area. Many of these factors produce the characteristic signs of inflammation: heat, redness, pain, and swelling. These signs are caused by increased blood flow (resulting in heat and redness) and vascular permeability (resulting in tissue swelling and the pain it causes) in the affected region. Such changes help phagocytic white blood cells reach the general area and enter the affected tissue.

SPECIFIC IMMUNITY

Specific immunity includes protective mechanisms that confer very specific protection against certain types of invading bacteria or other toxic materials. Specific immunity involves memory and the ability to recognize and respond to certain particular harmful substances or bacteria. For example, when the body is first attacked by particular bacteria or viruses, disease symptoms may occur as the body fights to destroy the invading organism. However, if the body is exposed a second time, no symptoms occur because the organism is destroyed quickly—the person is said to be **immune.** Immunity to one type of disease-causing bacteria or virus does not protect the body against others. Immunity can be very selective.

Immunity to disease is classified as **inherited** or **acquired.** If inherited, it is described as **inborn immunity** (Table 12-1). Humans are immune from birth to certain diseases that affect other animals. Distemper, for example, is an often fatal viral disease in dogs. The virus does not produce symptoms in humans; we have an inborn or inherited immunity to the disease.

Acquired immunity may be further classified as "natural" or "artificial" depending on how the body is exposed to the harmful agent. Natural exposure is not deliberate and occurs in the

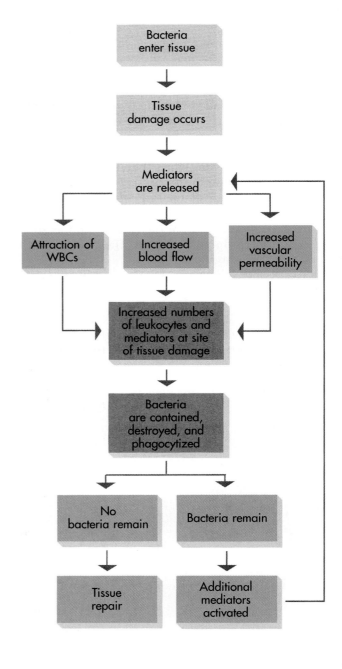

FIGURE 12–7 *Inflammatory Response.* In this example, bacterial infection triggers a set of responses that tend to inhibit or destroy the bacteria.

TABLE 12–1 Specific Immunity

TYPE	EXAMPLE
Inborn immunity	Immunity to certain diseases (for example, canine distemper) is inherited.
Acquired immunity	
Natural immunity	Exposure to the causative agent is not deliberate.
Active (exposure)	A child develops measles and acquires an immunity to a subsequent infection.
Passive (exposure)	A fetus receives protection from the mother through the placenta, or an infant receives protection via the mother's milk.
Artificial immunity	Exposure to the causative agent is deliberate.
Active (exposure)	Injection of the causative agent, such as a vaccination against polio, confers immunity.
Passive (exposure)	Injection of protective material (antibodies) that was developed by another individual's immune system is given.

course of everyday living. We are naturally exposed to many disease-causing agents on a regular basis. Artificial exposure is called *immunization* and is the deliberate exposure of the body to a potentially harmful agent.

Natural and artificial immunity may be "active" or "passive." Active immunity occurs when an individual's own immune system responds to a harmful agent, regardless of whether that agent was naturally or artificially encountered. Passive immunity results when immunity to a disease that has developed in another individual or animal is transferred to an individual who was not previously immune. For example, antibodies in a mother's milk confer passive immunity to her nursing infant. Active immunity generally lasts longer than passive immunity. Passive immunity, although temporary, provides immediate protection. Table 12-1 lists the various forms of specific immunity and gives examples of each.

🔵 *Immune System Molecules*

The immune system functions because of adequate amounts of highly specialized protein molecules and unique cells. The protein molecules critical to immune system functioning are called **antibodies** (AN-ti-bod-ees) and **complements**

(KOM-ple-ments). Antibody molecules outnumber the cells of the immune system by about 100 million to one.

ANTIBODIES
Definition

Antibodies are protein compounds that are normally present in the body. A defining characteristic of an antibody molecule is the uniquely shaped concave regions called combining sites on its surface. Another defining characteristic is the ability of an antibody molecule to combine with a specific compound called an **antigen** (AN-ti-jen). All antigens are compounds whose molecules have small regions on their surfaces that are uniquely shaped to fit into the combining sites of a specific antibody molecule as precisely as a key fits into a specific lock. Antigens are usually foreign proteins, most often the molecules in the surface membranes of invading or diseased cells such as microorganisms or cancer cells.

Functions

In general, antibodies produce **humoral** or **antibody-mediated immunity** by changing the antigens so that they cannot harm the body (Figure 12-8). To do this, an antibody must first bind to its specific antigen. This forms an antigen-antibody complex. The antigen-antibody complex then acts in one or more ways to make the antigen, or the

FIGURE 12–8 *Antibody Function.* Antibodies produce humoral immunity by binding to specific antigens to form antigen-antibody complexes. These complexes produce a variety of changes that inactivate or kill invading cells.

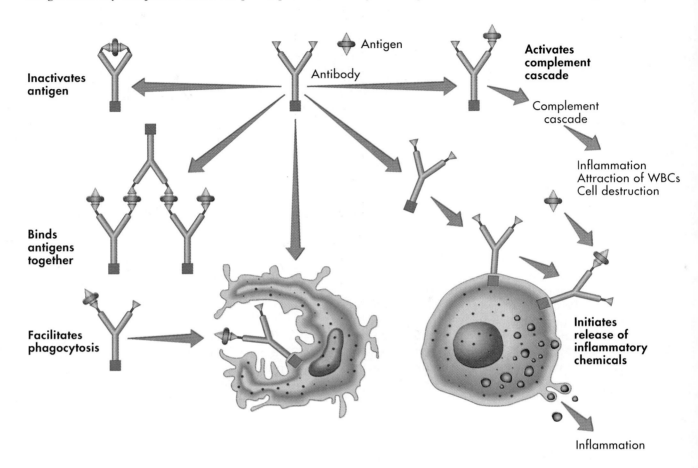

cell on which it is present, harmless. For example, if the antigen is a toxin, a substance poisonous to body cells, the toxin is neutralized or made non-poisonous by becoming part of an antigen-antibody complex. Or if antigens are molecules in the surface membranes of invading cells, when antibodies combine with them, the resulting antigen-antibody complexes may agglutinate the enemy cells (that is, make them stick together in clumps). Then macrophages or the other phagocytes can rapidly destroy them by ingesting and digesting large numbers of them at one time.

Another important function of antibodies is promotion and enhancement of phagocytosis.

Certain antibody fractions help promote the attachment of phagocytic cells to the object they will engulf. As a result, the contact between the phagocytic cell and its victim is enhanced, and the object is more easily ingested. This process contributes to the efficiency of immune system phagocytic cells, which is described on p. 268.

Probably the most important way in which antibodies act is one we will consider last. It is a process called **complement fixation.** In many instances, when antigens that are molecules on an antigenic or foreign cell's surface combine with antibody molecules, they change the shape of the antibody molecule slightly but just enough to

Monoclonal Antibodies

Techniques that have permitted biologists to produce large quantities of pure and very specific antibodies have resulted in dramatic advances in medicine. As a new medical technology, the development of **monoclonal antibodies** has been compared in importance with advances in recombinant DNA or genetic engineering.

Monoclonal antibodies are specific antibodies produced or derived from a population or culture of identical, or **monoclonal,** cells. In the past, antibodies produced by the immune system against a specific antigen had to be "harvested" from serum containing literally hundreds of other antibodies. The total amount of a specific antibody that could be recovered was very limited, so the cost of recovery was high. Monoclonal antibody techniques are based on the ability of immune system cells to produce individual antibodies that bind to and react with very specific antigens. We know, for example, that if the body is exposed to the varicella virus of chickenpox, WBCs will produce an antibody that will react very specifically with that virus and no other.

With monoclonal antibody techniques, lymphocytes that are produced by the body after the injection of a specific antigen are "harvested" and then "fused" with other cells that have been transformed to grow and divide indefinitely in a tissue culture medium. These fused or hybrid cells, called *hybridomas* (HYE-brid-o-mahs) continue to produce the same antibody produced by the original lymphocyte. The result is a rapidly growing population of identical or monoclonal cells that produce large quantities of a very specific antibody. Monoclonal antibodies have now been produced against a wide array of different antigens, including disease-producing organisms and various types of cancer cells.

The availability of very pure antibodies against specific disease-producing agents is the first step in the commercial preparation of diagnostic tests that can be used to identify viruses, bacteria, and even specific cancer cells in the blood or other body fluids. The use of monoclonal antibodies may serve as the basis for specific treatment of many human diseases.

---

expose two previously hidden regions. These are called **complement-binding sites.** Their exposure initiates a series of events that kill the cell on whose surface they take place. The next section describes these events.

COMPLEMENT PROTEINS

Complement is the name used to describe a group of 14 proteins normally present in an inactive state in blood. These proteins are activated by exposure of complement-binding sites. The result is formation of highly specialized antigen-antibody complexes that target foreign cells for

destruction. The process is a rapid-fire cascade or sequence of events collectively called **complement fixation.** In this process a doughnut-shaped assemblage complete with a hole in the middle is formed when antibodies, antigens in the invading cell's plasma membrane, and complement molecules combine.

Complement fixation kills invading cells of various types. How? In effect, by drilling a hole in their plasma membranes! The tiny holes allow sodium to rapidly diffuse into the cell; then water follows through osmosis. The cell literally bursts as the internal osmotic pressure increases (Figure 12-9).

FIGURE 12–9 *Complement Fixation.* **A,** Complement molecules activated by antibodies form doughnut-shaped complexes in a bacterium's plasma membrane. **B,** Holes in the complement complex allow sodium (Na^+) and then water (H_2O) to diffuse into the bacterium. **C,** After enough water has entered, the swollen bacterium bursts.

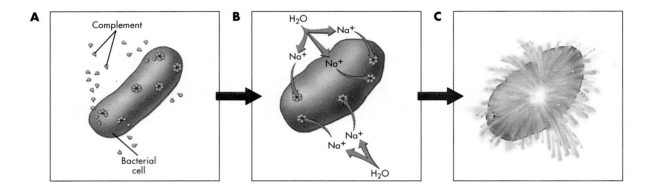

Immune System Cells

The primary cells of the immune system include:

1. Phagocytes
   a. Neutrophils
   b. Monocytes
   c. Macrophages
2. Lymphocytes
   a. T-lymphocytes
   b. B-lymphocytes

PHAGOCYTES

Phagocytic WBCs are an important part of the immune system. In Chapter 10, phagocytes were described as cells derived from the bone marrow that carry on phagocytosis or ingestion and digestion of foreign cells or particles. The most important phagocytes are neutrophils and monocytes (see Figure 10-4, p. 219). These blood phagocytes migrate out of the blood and into the tissues in response to an infection. The neutrophils are functional but short lived in the tissues. Once in the tissues, monocytes develop into phagocytic cells called **macrophages** (MAK-ro-fay-jes). Some macrophages "wander" through the tissues to engulf bacteria wherever they find them. Other macrophages become permanent residents of other organs. Macrophages found in spaces between liver cells, for example, are called *Kupffer's cells,* whereas those that ingest particulate matter in the small air sacs of the lungs are called *dust cells.* Macrophages can also be found in the spleen and lymph nodes and on the lining membranes of the abdominal and thoracic cavities. Specialized antibodies that bind to and coat certain foreign particles help macrophages function effectively. They serve as "flags" that alert the macrophage to the presence of foreign material, infectious bacteria, or cellular debris. They also help bind the phagocyte to the foreign material so that it can be engulfed more effectively (Figure 12-10).

LYMPHOCYTES

The most numerous cells of the immune system are the lymphocytes; they are ultimately responsible for antibody production. Several million strong, lymphocytes continually patrol the body, searching out any enemy cells that may have entered. Lymphocytes circulate in the body's fluids. Huge numbers of them wander vigilantly through most of its tissues. Lymphocytes densely populate the body's widely scattered lymph nodes and its other lymphatic tissues, especially the thymus gland in the chest and the spleen and liver in the abdomen. There are two major types of lymphocytes, designated as *B-* and *T-lymphocytes* but usually called *B cells* and *T cells.*

FIGURE 12–10 *Phagocytosis.* This series of scanning electron micrographs shows the progressive steps in phagocytosis of damaged RBCs by a macrophage. **A,** RBCs *(R)* attach to the macrophage *(M)*. **B,** Plasma membrane of the macrophage begins to enclose the RBC. **C,** The RBCs are almost totally ingested by the macrophage.

 Interferon

Interferon (in-ter-FEER-on) is a small protein compound that plays a very significant role in producing immunity from viral infections. It is produced by T cells within hours after they have been infected by a virus. The interferon released from the T cells protects other cells by interfering with the ability of the virus to reproduce as it moves from cell to cell. In the past, thousands of pints of blood had to be processed to harvest tiny quantities of leukocyte (T cell) interferon for study. Synthetic human interferon is now being "manufactured" in bacteria as a result of gene-splicing techniques and is available in quantities sufficient for clinical use. Synthetic interferon decreases the severity of many virus-related diseases including chickenpox and measles. Interferon also shows promise as an anticancer agent. It has been shown to be effective in treating breast, skin, and other forms of cancer.

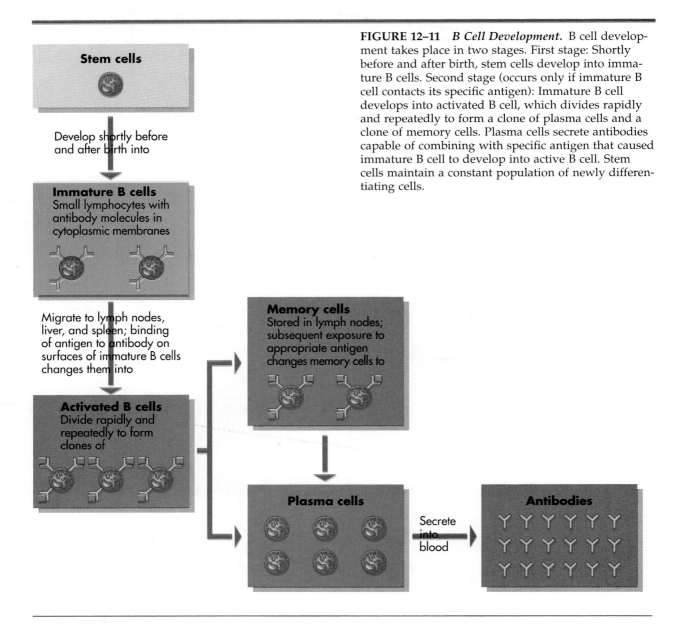

FIGURE 12–11 *B Cell Development.* B cell development takes place in two stages. First stage: Shortly before and after birth, stem cells develop into immature B cells. Second stage (occurs only if immature B cell contacts its specific antigen): Immature B cell develops into activated B cell, which divides rapidly and repeatedly to form a clone of plasma cells and a clone of memory cells. Plasma cells secrete antibodies capable of combining with specific antigen that caused immature B cell to develop into active B cell. Stem cells maintain a constant population of newly differentiating cells.

Development of B Cells

All lymphocytes that circulate in the tissues arise from primitive cells in the bone marrow called *stem cells* and go through two stages of development. The first stage of B-cell development, transformation of stem cells into immature B cells, occurs in the liver and bone marrow before birth but only in the bone marrow in adults. Because

this process was first discovered in a bird organ called the *bursa,* these cells were named B cells.

Immature B cells are small lymphocytes that have synthesized and inserted into their cytoplasmic membranes numerous molecules of one specific kind of antibody (Figure 12-11). These antibody-bearing immature B cells leave the tissue

AIDS

AIDS or **acquired immunodeficiency syndrome** was first recognized as a new disease by the Centers for Disease Control (CDC) in 1981. This syndrome (collection of symptoms) is caused by the human immunodeficiency virus or HIV. HIV, a retrovirus, contains RNA that undergoes reverse transcription inside infected cells to form its own DNA. The viral DNA often becomes part of the cell's DNA. When the viral DNA is activated, it directs the synthesis of its own RNA and protein coat, thus "stealing" raw materials from the cell. When this occurs in certain T cells, the cell is destroyed, and immunity is impaired. As the T cell dies, it releases new retroviruses that can spread the HIV infection.

Although HIV can invade several types of cells, it has its most obvious effects in certain types of T cells. When T cell function is impaired, infectious organisms and cancer cells can grow and spread much more easily than normal. Unusual conditions, such as pneumocystosis (a protozoan infection) and Kaposi sarcoma (a type of skin cancer), may also appear. Because their immune systems are deficient, AIDS patients usually die from one of these infections or cancers.

After infection with HIV, a person may not show signs of AIDS for months or years. This is because the immune system can hold the infection at bay for a long time before finally succumbing to it.

There are several strategies for controlling AIDS. Many agencies are trying to slow the spread of AIDS by educating people about how to avoid contact with the HIV retrovirus. HIV is spread by means of direct contact of body fluids, so preventing such contact reduces HIV transmission. Sexual relations, contaminated blood transfusions, and intravenous use of contaminated needles are the usual modes of HIV transmission. Several teams of researchers are working on HIV vaccines. Like many viruses, such as those that cause the common cold, HIV changes rapidly enough to make development of a vaccine difficult at best.

Another way to inhibit the disease is by means of chemicals such as azidothymidine (AZT) that block HIV's ability to reproduce within infected cells. At least 80 such compounds are being evaluated for use in halting the progress of HIV infections.

---

where they were formed, enter the blood, and are transported to their new place of residence, chiefly the lymph nodes. There they act as seed cells. Each immature B cell undergoes repeated mitosis (cell division) and forms a clone of immature B cells. A **clone** is a family of many identical cells all descended from one cell. Because all the cells in a clone of immature B cells have descended from one immature B cell, all of them bear the same surface antibody molecules as did their single ancestor cell.

The second stage of B cell development changes an immature B cell into an activated B cell. Not all immature B cells undergo this change. They do so only if an immature B cell comes in contact with certain protein molecules—antigens—whose shape fits the shape of the immature B cell's surface antibody molecules. If this happens, the antigens lock onto the antibodies and by so doing change the immature B cell into an activated B cell. Then the activated B cell, by dividing rapidly and repeatedly, develops into clones of two kinds of cells, **plasma cells** and **memory cells** (Figure 12-11). Plasma cells secrete copious amounts of antibody into the blood—reportedly, 2000 antibody molecules per second by each plasma cell for every second of the few days that it lives. Antibodies circulating in the blood constitute an enormous, mobile, ever-on-duty army.

FIGURE 12–12   *T Cell Development.* The first stage takes place in the thymus gland shortly before and after birth. Stem cells maintain a constant population of newly differentiating cells as they are needed. The second stage occurs only if a T cell contacts antigen, which combines with certain proteins on the T cell's surface.

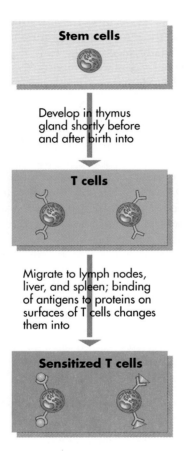

FIGURE 12–13   *T Cells.* The blue spheres seen in this scanning electron microscope view are T cells attacking a much larger cancer cell. The cells are a significant part of our defense against cancer and other types of foreign cells.

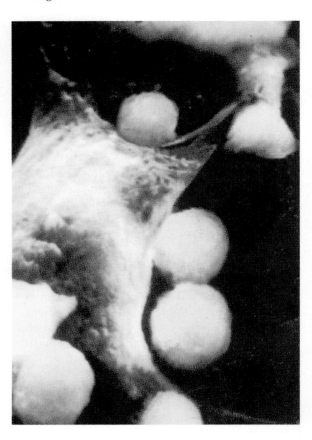

Memory cells can secrete antibodies but do not immediately do so. They remain in reserve in the lymph nodes until they are contracted by the same antigen that led to their formation. Then, very quickly, the memory cells develop into plasma cells and secrete large amounts of antibody. Memory cells, in effect, seem to remember their ancestor-activated B cell's encounter with its appropriate antigen. They stand ready, at a moment's notice, to produce antibody that will combine with this antigen.

Function of B Cells

B cells function indirectly to produce humoral immunity. Recall that humoral immunity is resistance to disease organisms produced by the actions of antibodies binding to specific antigens while circulating in body fluids. Activated B cells develop into plasma cells. Plasma cells secrete antibodies into the blood; they are the "antibody factories" of the body. These antibodies, like other proteins manufactured for extracellular use, are formed on the endoplasmic reticulum of the cell.

FIGURE 12–14 *T Cell Function.* Sensitized T cells produce cell-mediated immunity by releasing various compounds in the vicinity of invading cells. Some act directly, and some act indirectly to kill invading cells.

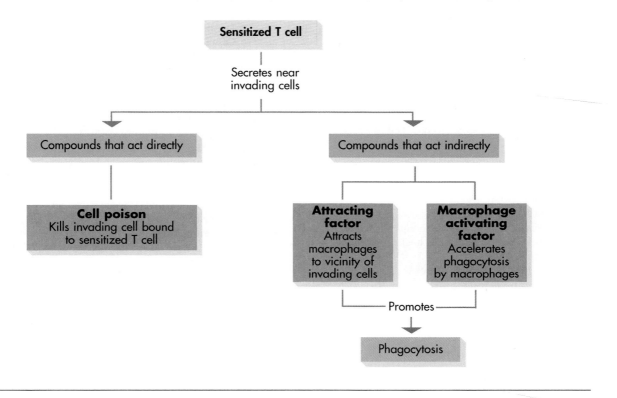

Development of T Cells

T cells are lymphocytes that have undergone their first stage of development in the thymus gland. Stem cells from the bone marrow seed the thymus, and shortly before and after birth, they develop into T cells. The newly formed T cells stream out of the thymus into the blood and migrate chiefly to the lymph nodes, where they take up residence. Embedded in each T cell's cytoplasmic membrane are protein molecules shaped so that they can fit only one specific kind of antigen molecule. The second stage of T-cell development takes place when and if a T cell comes into contact with its specific antigen. If this happens, the antigen binds to the protein on the T cell's surface, thereby changing the T cell into a sensitized T cell (Figure 12-12).

Functions of T Cells

Sensitized T cells produce cell-mediated immunity. As the name suggests, **cell-mediated immunity** is resistance to disease organisms resulting from the actions of cells—chiefly sensitized T cells. Some sensitized T cells kill invading cells directly (Figure 12-13). When bound to antigens on an invading cell's surface, they release a substance that acts as a specific and lethal poison against the bound cell. Many sensitized cells produce their deadly effects indirectly by means of compounds that they release into the area around enemy cells. Among these are a substance that attracts macrophages into the neighborhood of the enemy cells. The assembled macrophages then destroy the cells by phagocytosing (ingesting and digesting) them (Figure 12-14).

O U T L I N E   S U M M A R Y

THE LYMPHATIC SYSTEM (Figure 12-1)

A Lymph—fluid in the tissue spaces that carries protein molecules and other substances back to the blood

B Lymphatic vessels—permit only one-way movement of lymph
  1 Lymphatic capillaries—tiny blind-ended tubes distributed in tissue spaces (Figure 12-2)
    a Microscopic in size
    b Sheets consisting of one cell layer of simple squamous epithelium
    c Poor "fit" between adjacent cells results in porous walls
    d Called *lacteals* in the intestinal wall (for fat transportation)
  2 Right lymphatic duct
    a Drains lymph from the right upper extremity and right side of head, neck, and upper torso
  3 Thoracic duct
    a Largest lymphatic vessel
    b Has an enlarged pouch along its course, called *cisterna chyli*
    c Drains lymph from about three fourths of the body (Figure 12-3)

C Lymph nodes
  1 Filter lymph (Figure 12-4)
  2 Located in clusters along the pathway of lymphatic vessels (Figure 12-5)
  3 Functions include defense and WBC formation
  4 Flow of lymph: to node via several afferent lymph vessels and drained from node by a single efferent lymph vessel

D Thymus
  1 Lymphoid tissue organ located in mediastinum
  2 Total weight of 35 to 40 g—less than an ounce
  3 Plays a vital and central role in immunity
  4 Produces T-lymphocytes or T cells
  5 Secretes hormone called *thymosin*
  6 Lymphoid tissue is replaced by fat in the process called *involution*

E Tonsils (Figure 12-6)
  1 Composed of three masses of lymphoid tissue around the openings of the mouth and throat
    a Palatine tonsils ("the tonsils")
    b Pharyngeal tonsils (adenoids)
    c Lingual tonsils
  2 Subject to chronic infection
  3 Enlargement of pharyngeal tonsils may impair breathing

F Spleen
  1 Largest lymphoid organ in body
  2 Located in upper left quadrant of abdomen
  3 Often injured by trauma to abdomen
  4 Surgical removal called *splenectomy*
  5 Functions include phagocytosis of bacteria and old RBCs; acts as a blood reservoir

THE IMMUNE SYSTEM (Table 12-1)

A Protects body from pathological bacteria, foreign tissue cells, and cancerous cells

B Made up of specialized cells and molecules

C Nonspecific immunity
  1 Skin—mechanical barrier to bacteria and other harmful agents
  2 Tears and mucus—wash eyes and trap and kill bacteria
  3 Inflammation—attracts immune cells to site of injury, increases local blood flow, increases vascular permeability; promotes movement of WBCs to site of injury or infection (Figure 12-7)

D Specific immunity—ability of body to recognize, respond to, and remember harmful substances or bacteria

E Inherited or inborn immunity—inherited immunity to certain diseases from birth

F Acquired immunity
  1 Natural immunity—exposure to causative agent is not deliberate
    a Active—active disease produces immunity
    b Passive—immunity passes from mother to fetus through placenta or from mother to child through mother's milk
  2 Artificial immunity—exposure to causative agent is deliberate
    a Active—vaccination results in immunity
    b Passive—protective material developed in another individual's immune system and given to previously nonimmune individual

IMMUNE SYSTEM MOLECULES

A Antibodies
  1 Protein compounds with specific combining sites
  2 Combining sites attach antibodies to specific antigens (foreign proteins), forming an antigen-antibody complex—called *humoral* or *antibody-mediated immunity* (Figure 12-8)
  3 Antigen-antibody complexes may:
    a Neutralize toxins
    b Clump or agglutinate enemy cells
    c Promote phagocytosis

B Complement proteins
   1 Group of 14 proteins normally present in blood in inactive state
   2 Complement fixation
      a Important mechanism of action for antibodies
      b Causes cell lysis by permitting entry of water through a defect created in the plasma membrane (Figure 12-9)

IMMUNE SYSTEM CELLS

A Phagocytes—ingest and destroy foreign cells or other harmful substances via phagocytosis (Figure 12-10)
   1 Types
      a Neutrophils
      b Monocytes
      c Macrophages (Figure 12-10)
         (1) Kupffer's cells (liver)
         (2) Dust cells (lung)
B Lymphocytes
   1 Most numerous of immune system cells
   2 Development of B cells—primitive stem cells migrate from bone marrow and go through two stages of development (Figure 12-11)
      a First stage—stem cells develop into immature B cells; takes place in the liver and bone marrow before birth and in the bone marrow only in adults; immature B cells are small lymphocytes with antibody molecules (which they have synthesized) in their plasma membranes; migrate chiefly to lymph nodes
      b Second stage—immature B cell develops into activated B cell; initiated by immature B cell's contact with antigens, which bind to its surface antibodies; activated B cell, by dividing repeatedly, forms two clones of cells; plasma cells and memory cells; plasma cells secrete antibodies into blood; memory cells stored in lymph nodes; if subsequent exposure to antigen that activated B cell occurs, memory cells become plasma cells and secrete antibodies
   3 Function of B cells—indirectly, B cells produce humoral immunity; activated B cells develop into plasma cells; plasma cells secrete antibodies into the blood; circulating antibodies produce humoral immunity (Figure 12-11)
   4 Development of T cells—stem cells from bone marrow migrate to thymus gland (Figure 12-12)
      a Stage 1—stem cells develop into T cells; occurs in thymus during few months before and after birth; T cells migrate chiefly to lymph nodes
      b Stage 2—T cells develop into sensitized T cells; occurs when, and if, antigen binds to T cell's surface proteins
   5 Functions of T cells—produce cell-mediated immunity; kill invading cells by releasing a substance that poisons cells and also by releasing chemicals that attract and activate macrophages to kill cells by phagocytosis (Figures 12-13 and 12-14)

NEW WORDS

afferent	cell-mediated immunity	complement fixation	macrophage
AIDS		efferent	memory cells
antibodies	clone	humoral immunity	monoclonal antibodies
antigen	combining sites	inflammatory response	plasma cells
B cells (lymphocytes)	complement	interferon	T cells (lymphocytes)
		lymph	

C H A P T E R   T E S T

1. The largest lymphatic vessel in the body is called the _____ _____.
2. Phagocytic cells located in the _____ _____ perform biological filtration.
3. Deliberate exposure to a causative agent produces _____ immunity.
4. The two protein molecules most critical to immune system functioning are called _____ and _____.
5. Antibodies are characterized by their ability to combine with very specific compounds called _____.
6. Antibodies make antigens unable to harm the body—a process called _____ immunity.
7. Plasma cells produce _____.
8. The most numerous cells of the immune system are the _____.
9. The most important phagocytes of the immune system are the _____ and _____.
10. B cells and T cells are the two major types of _____ in the immune system.
11. B cells indirectly produce _____ immunity in the body.
12. Sensitized T cells produce _____ immunity in the body.

Circle the T before each true statement and the F before each false statement.

T F 13. Lymphatic vessels permit only one-way movement of lymph.
T F 14. The right lymphatic duct is the largest lymph vessel in the body.
T F 15. Vaccination is an example of active artificial immunity.
T F 16. Cells of the immune system outnumber antibody molecules by about 100 million to one.
T F 17. Antibodies function to produce cell-mediated immunity.
T F 18. After leaving the blood and entering the tissue space, monocytes are called macrophages.
T F 19. Macrophages are capable of phagocytosis.
T F 20. Complement fixation is an important way in which antibodies act.
T F 21. Antigen-antibody complexes may neutralize toxins.
T F 22. Activated B cells develop into Kupffer's cells.
T F 23. Memory cells develop from sensitized T cells.
T F 24. T cells begin as stem cells that develop in the thymus gland.

REVIEW QUESTIONS

1. Discuss two functions of the lymph nodes.
2. Explain nonspecific immunity, specific immunity, inherited immunity, acquired immunity, natural immunity, artificial immunity, active immunity, and passive immunity.
3. Explain the terms *antigen* and *antibody*.
4. What are combining sites? Complement binding sites?
5. Name four kinds of cells that constitute part of the immune system. Which are most numerous? How numerous are they estimated to be?
6. Name the two major immune system molecules.
7. What are the two major types of lymphocytes called? What makes these names appropriate?
8. Describe briefly the two stages of development that B cells undergo.
9. Describe briefly the two stages of development that T cells undergo.
10. What cells secrete antibodies?
11. Explain the function of memory cells.
12. Describe the function of activated B cells.
13. Describe the function of sensitized T cells.

CRITICAL THINKING

14. Sometimes a woman's arm becomes very swollen for a while after removal of a breast and the nearby lymph nodes and lymphatic vessels, including some of those in the upper arm. Can you think of any reason why swelling occurs?
15. How does the "circulation" of lymph differ from that of blood?
16. Outline the inflammatory response and explain its advantages in an infection.

The Respiratory System

13

BOXED ESSAYS

Heimlich Maneuver
Infant Respiratory Distress Syndrome
Hiccup
Maximum Oxygen Consumption
Oxygen Therapy

OBJECTIVES

*After you have completed this chapter,
you should be able to:*

1. Discuss the generalized functions of the respiratory system.
2. List the major organs of the respiratory system and describe the function of each.
3. Compare, contrast, and explain the mechanism responsible for the exchange of gases that occurs during internal and external respiration.
4. List and discuss the volumes of air exchanged during pulmonary ventilation.
5. Identify and discuss the mechanisms that regulate respiration.

No one needs to be told how important the **respiratory system** is. The respiratory system serves the body much as a lifeline to an oxygen tank serves a deep-sea diver. Think how panicked you would feel if suddenly your lifeline became blocked—if you could not breathe for a few seconds! Of all the substances that cells and therefore the body as a whole must have to survive, oxygen is by far the most crucial. A person can live a few weeks without food, a few days without water, but only a few minutes without oxygen. Constant removal of carbon dioxide from the body is just as important for survival as a constant supply of oxygen.

The organs of the respiratory system are designed to perform two basic functions; they serve as an **air distributor** and as a **gas exchanger** for the body. The respiratory system ensures that oxygen is supplied to and carbon dioxide is removed from the body's cells. The process of respiration therefore is an important **homeostatic mechanism.** By constantly supplying adequate oxygen and by removing carbon dioxide as it forms, the respiratory system helps maintain a constant environment that enables our body cells to function effectively.

In addition to air distribution and gas exchange, the respiratory system effectively **filters, warms,** and **humidifies** the air we breathe. Respiratory organs or organs closely associated with the respiratory system, such as the **sinuses,** also influence speech or sound production and make possible the sense of smell or **olfaction** (ol-FAK-shun). In this chapter the structural plan of the respiratory system will be considered first, then the respiratory organs will be discussed individually, and finally some facts about gas exchange and the nervous system's control of respiration will be discussed.

Structural Plan

Respiratory organs include the **nose, pharynx** (FAIR-inks), **larynx** (LAIR-inks), **trachea** (TRAY-kee-ah), **bronchi** (BRONG-ki), and **lungs.** The basic structural design of this organ system is that of a tube with many branches ending in millions of extremely tiny, very thin-walled sacs called **alveoli** (al-VEE-o-li). Figure 13-1 shows the exten-sive branching of the "respiratory tree" in both lungs. Think of this air distribution system as an "upside-down tree." The trachea or windpipe then becomes the trunk and the bronchial tubes the branches. This idea will be developed when the types of bronchi and the alveoli are studied in more detail later in the chapter. A network of capillaries fits like a hairnet around each microscopic alveolus. Incidentally, this is a good place for us to think again about a principle already mentioned several times, namely, that structure and function are intimately related. The function of alveoli—in fact, the function of the entire respiratory system—is to distribute air close enough to blood for a gas exchange to take place between air and blood. The passive transport process of **diffusion,** which was described in Chapter 2, is responsible for the exchange of gases that occurs in the respiratory system. You may want to review the discussion of diffusion on pp. 23 to 25 before you study the mechanism of gas exchange that occurs in the lungs and body tissues.

Two characteristics about the structure of alveoli assist in diffusion and make them able to perform this function admirably. First, the wall of each alveolus is made up of a single layer of cells and so are the walls of the capillaries around it. This means that, between the blood in the capillaries and the air in the alveolus, there is a barrier probably less than 1 micron thick! This extremely thin barrier is called the **respiratory membrane** (Figure 13-2). Second, there are millions of alveoli. This means that together they make an enormous surface (approximately 100 square meters, an area many times larger than the surface of the entire body) where larger amounts of oxygen and carbon dioxide can rapidly be exchanged.

Respiratory Tracts

The respiratory system is often divided into upper and lower tracts or divisions to assist in the description of symptoms associated with common respiratory problems such as a cold. The organs of the upper respiratory tract are located outside of the thorax or chest cavity, whereas those in the lower tract or division are located almost entirely within it. The **upper respiratory** tract is composed of the nose, pharynx, and larynx. The **lower res-**

FIGURE 13–1   *Structural Plan of the Respiratory Organs Showing the Pharynx, Trachea, Bronchi, and Lungs.* The inset shows the alveolar sacs where the interchange of oxygen and carbon dioxide takes place through the walls of the grapelike alveoli. Capillaries surround the alveoli.

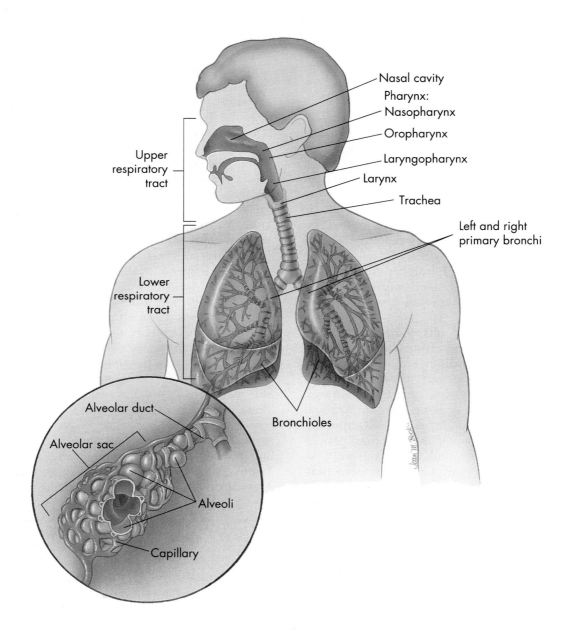

FIGURE 13–2   *The Gas-Exchange Structures of the Lung.* Each alveolus is continually ventilated with fresh air. The inset shows a magnified view of the respiratory membrane composed of the alveolar wall (surfactant, epithelial cells, and basement membrane), interstitial fluid, and the wall of a pulmonary capillary (basement membrane and endothelial cells). The gases, CO_2 (carbon dioxide) and O_2 (oxygen), diffuse across the respiratory membrane.

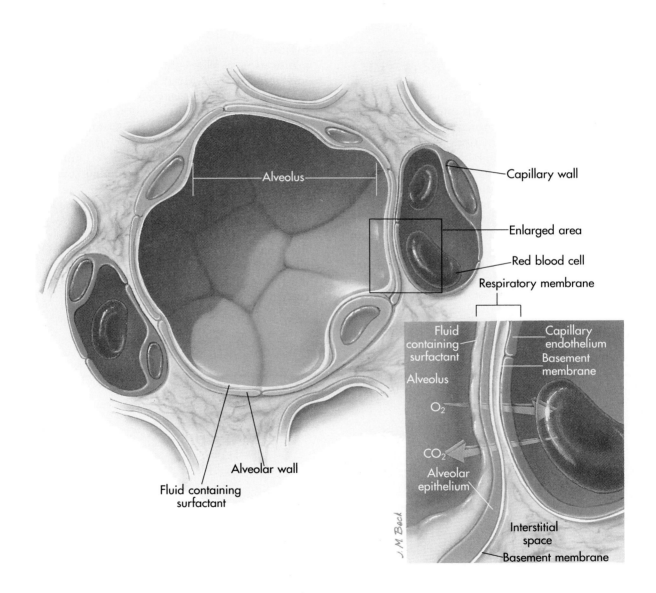

piratory tract or division consists of the trachea, all segments of the bronchial tree, and the lungs. The designation "upper respiratory infection" or URI is often used to describe a head cold. Typically the symptoms of an upper respiratory infection involve the sinuses, nasal cavity, pharynx, and larynx, whereas the symptoms of a chest cold are similar to pneumonia and involve the organs of the lower respiratory tract.

Respiratory Mucosa

Before beginning the study of individual organs in the respiratory system, it is important to review the histology or microscopic anatomy of the **respiratory mucosa**—the membrane that lines most of the air distribution tubes in the system. Do not confuse the respiratory membrane with the respiratory mucosa! The **respiratory membrane** (Figure 13-2) separates the air in the alveoli from the blood in surrounding capillaries. The respiratory mucosa (Figure 13-3) is covered with mucus and lines the tubes of the respiratory tree.

Recall that in addition to serving as air distribution passageways or gas exchange surfaces, the anatomical components of the respiratory tract and lungs cleanse, warm, and humidify inspired air. Air entering the nose is generally contaminated with one or more common irritants; examples include insects, dust, pollen, and bacterial organisms. A remarkably effective air purification mechanism removes almost every form of contaminant before inspired air reaches the alveoli or terminal air sacs in the lungs.

The layer of protective mucus that covers a large portion of the membrane that lines the respiratory tree serves as the most important air purification mechanism. Over 125 ml of respiratory mucus is produced daily. It forms a continuous sheet called a *mucous blanket* that covers the lining of the air distribution tubes in the respiratory tree. This layer of cleansing mucus moves upward to the pharynx from the lower portions of the bronchial tree on millions of hairlike cilia that cover the epithelial cells in the respiratory mucosa (Figure 13-3). The microscopic cilia that cover epithelial cells in the respiratory mucosa beat or move only in one direction. The result is movement of mucus toward the pharynx. Cigarette smoke paralyzes these cilia and results in accumulations of mucus and the typical smoker's cough, which is an effort to clear the secretions.

FIGURE 13–3 *Respiratory Mucosa Lining the Trachea.* A layer of mucus covers the hairlike cilia.

⬤ *Nose*

Air enters the respiratory tract through the **external nares** (NA-rees) or nostrils. It then flows into the right and left **nasal cavities,** which are lined by respiratory mucosa. A partition called the *nasal septum* separates these two cavities.

The surface of the nasal cavities is moist from mucus and warm from blood flowing just under it. Nerve endings responsible for the sense of smell (olfactory receptors) are located in the nasal mucosa. Four **paranasal sinuses**—frontal, maxillary, sphenoidal, and ethmoidal—drain into the nasal cavities (Figure 13-4). Because the mucosa that lines the sinuses is continuous with the mucosa that lines the nose, sinus infections, called **sinusitis** (sye-nyoo-SYE-tis), often develop from colds in which the nasal mucosa is inflamed. The paranasal sinuses are lined with a mucous membrane that assists in the production of mucus for the respiratory tract. In addition, these hollow spaces help to lighten the skull bones and serve as resonant chambers for the production of sound.

FIGURE 13–4 *The Paranasal Sinuses.* The anterior view shows the anatomical relationship of the paranasal sinuses to each other and to the nasal cavity. The inset is a lateral view of the position of the sinuses.

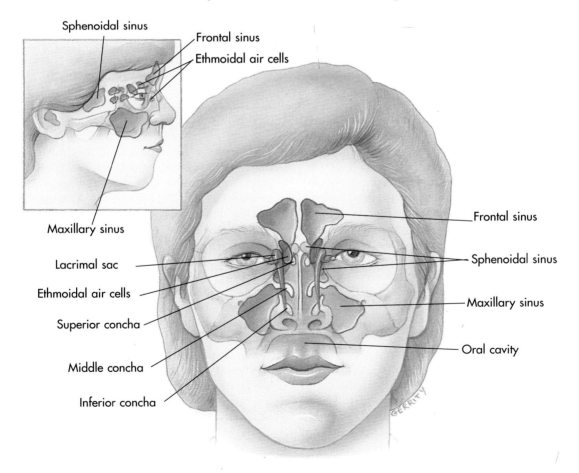

Two ducts from the **lacrimal sacs** (LAK-rim-al saks) also drain into the nasal cavity, as Figure 13-4 shows. The lacrimal sacs collect tears from the corner of the each eyelid and drain them into the nasal cavity.

Note in Figure 13-5 that three shelflike structures called **conchae** (KONG-kee) protrude into the nasal cavity on each side. The mucosa-covered conchae greatly increase the surface over which air must flow as it passes through the naval cavity. As air moves over the conchae and through the nasal cavities, it is warmed and humidified. This helps explain why breathing through the nose is more effective in humidifying inspired air than is breathing through the mouth. If an individual who is ill requires supplemental oxygen, it is first bubbled through water to reduce the amount of moisture that would otherwise have to be removed from the lining of the respiratory tree to humidify it. Administration of "dry" oxygen pulls water from the mucosa and results in respiratory discomfort and irritation.

Pharynx

The **pharynx** is the structure that many of us call the throat. It is about 12.5 cm (5 inches) long and can be divided into three portions (Figure 13-5). The uppermost part of the tube just behind the nasal cavities is called the **nasopharynx** (nay-zo-FAIR-inks). The portion behind the mouth is called the **oropharynx** (o-ro-FAIR-inks). The last or lowest segment is called the **laryngopharynx** (lah-ring-go-FAIR-inks). The pharynx as a whole serves the same purpose for the respiratory and digestive tracts as a hallway serves for a house. Air and food pass through the pharynx on their way to the lungs and the stomach respectively. Air enters the pharynx from the two nasal cavities and leaves it by way of the larynx; food enters it from the mouth and leaves it by way of the esophagus. The right and left **auditory** or *eustachian* (yoo-STAY-she-an) **tubes** open into the nasopharynx; they connect the middle ears with the nasopharynx (Figure 13-5). This connection permits equalization of air pressure between the middle and the exterior ear. The lining of the auditory tubes is continuous with the lining of the nasopharynx and middle ear. Thus just as sinus infections can develop from colds in

which the nasal mucosa is inflamed, middle ear infections can develop from inflammation of the nasopharynx.

Masses of lymphatic tissue called **tonsils** are embedded in the mucous membrane of the pharynx (see p. 263). The **pharyngeal** (fa-RIN-jee-al) **tonsils** or **adenoids** (AD-e-noyds) are in the nasopharynx. The **palatine tonsils** are located in the oropharynx (Figure 13-5). Both tonsils are generally removed in a **tonsillectomy** (ton-si-LEK-toe-mee). Although this surgical procedure is still rather common, the number of tonsillectomies reported each year continues to decrease as new and more effective antibiotics become available. Physicians now recognize the value of lymphatic tissue in the body defense mechanism and delay removal of the tonsils—even in cases of inflammation or **tonsillitis**—unless antibiotic treatment is ineffective. When the pharyngeal tonsils become swollen, they are referred to as adenoids. Such swelling caused by infections may make it difficult or impossible for air to travel from the nose into the throat. In these cases the individual is forced to breathe through the mouth.

Larynx

The **larynx** or voice box is located just below the pharynx. It is composed of several pieces of cartilage. You know the largest of these (the *thyroid cartilage*) as the "Adam's apple" (Figure 13-6).

Two short fibrous bands, the **vocal cords,** stretch across the interior of the larynx. Muscles that attach to the larynx cartilages can pull on these cords in such a way that they become tense or relaxed. When they are tense, the voice is high pitched; when they are relaxed, it is low pitched. The space between the vocal cords is the **glottis.** Another cartilage, the **epiglottis** (ep-i-GLOT-is) partially covers the opening of the larynx (Figure 13-6). The epiglottis acts like a trapdoor, closing off the larynx during swallowing and preventing food from entering the trachea.

Trachea

The **trachea** or windpipe is a tube about 11 cm (4½ inches) long that extends from the larynx in

FIGURE 13–5 *Sagittal Section of the Head and Neck.* The nasal septum has been removed, exposing the right lateral wall of the nasal cavity so that the nasal conchae can be seen. Note also the divisions of the pharynx and the position of the tonsils.

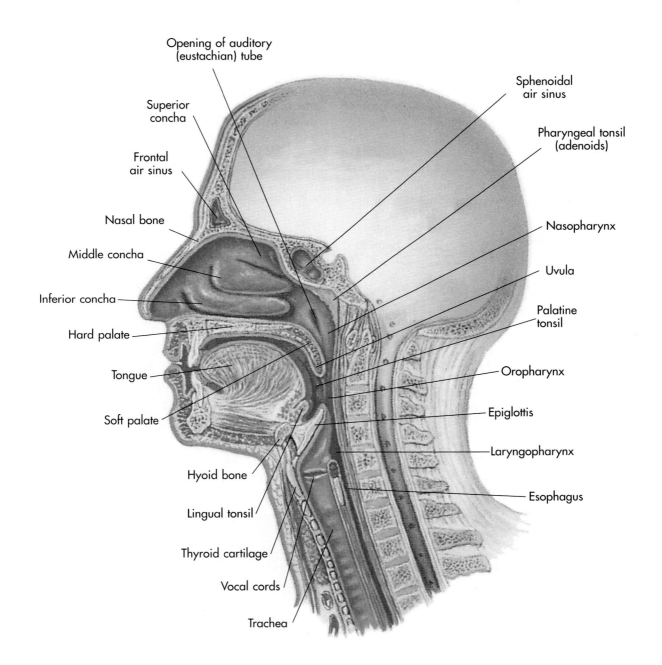

the neck to the bronchi in the chest cavity (Figures 13-1 and 13-7). The trachea performs a simple but vital function; it furnishes part of the open passageway through which air can reach the lungs from the outside.

By pushing with your fingers against your throat about an inch above the sternum, you can feel the shape of the trachea or windpipe. Only if you use considerable force can you squeeze it closed. Nature has taken precautions to keep this lifeline open. Its framework is made of an almost noncollapsible material—15 or 20 C-shaped rings of cartilage placed one above the other with only a little soft tissue between them (Figure 13-7). The trachea is lined by the typical respiratory mucosa. Glands below the ciliated epithelium help produce the blanket of mucus that continually moves upward toward the pharynx.

Despite the structural safeguard of cartilage rings, closing of the trachea sometimes occurs. A tumor or an infection may enlarge the lymph nodes of the neck so much that they squeeze the trachea shut, or a person may aspirate (breathe in) a piece of food or something else that blocks the windpipe. Because air has no other way to get to the lungs, complete tracheal obstruction causes death in a matter of minutes. Choking on food and other substances caught in the trachea kills

FIGURE 13–6   *The Larynx.* **A,** Sagittal section of the larynx. **B,** Superior view of the larynx. **C,** Photograph of the larynx taken with an endoscope (optical device) inserted through the mouth and pharynx to the epiglottis.

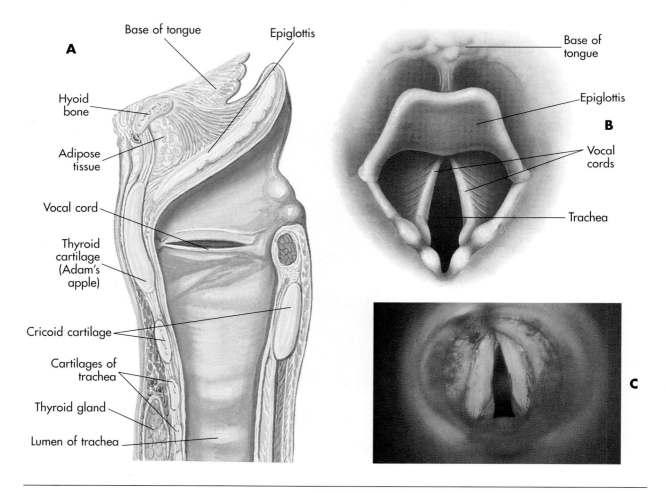

over 4000 people each year and is the fifth major cause of accidental deaths in the United States. A lifesaving technique developed by Dr. Henry Heimlich (see p. 288), is now widely used to free the trachea of ingested food or other foreign objects that would otherwise block the airway and cause death in choking victims.

⬭ *Bronchi, Bronchioles, and Alveoli*

Recall that one way to picture the thousands of air tubes that make up the lungs is to think of an

upside-down tree. The trachea is the main trunk of this tree; the right bronchus (the tube leading into the right lung) and the left bronchus (the tube leading into the left lung) are the trachea's first branches or **primary bronchi.** In each lung, they branch into smaller or **secondary bronchi** whose walls, like those of the trachea and bronchi, are kept open by rings of cartilage for air passage. These bronchi divide into smaller and smaller tubes, ultimately branching into tiny tubes whose walls contain only smooth muscle. These very small passageways are called **bronchioles.** The bronchioles subdivide into microscopic tubes called **alveolar ducts,** which resemble the main

FIGURE 13–7 *Cross Section of the Trachea.* Inset at top shows where the section was cut. The scanning electron micrograph shows the tip of one of the C-shaped cartilage rings.

Heimlich Maneuver

The **Heimlich maneuver** is an effective and often lifesaving technique that can be used to open a windpipe that is suddenly obstructed. The maneuver (see figures) uses air already present in the lungs to expel the object obstructing the trachea. Most accidental airway obstructions result from pieces of food aspirated (breathed in) during a meal; the condition is sometimes referred to as a *cafe coronary*. Other objects such as chewing gum or balloons are frequently the cause of obstructions in children.

Individuals trained in emergency procedures must be able to tell the difference between airway obstruction and other conditions such as heart attacks that produce similar symptoms. The key question they must ask the person who appears to be choking is, "Are you choking?" A person with an obstructed airway will not be able to speak, even while conscious. The Heimlich maneuver, if the victim is standing, consists of the rescuer's grasping the victim with both arms around the victim's waist just below the rib cage and above the navel. The rescuer makes a fist with one hand, grasps it with the other, and then delivers an upward thrust against the diaphragm just below the xiphoid process of the sternum. Air trapped in the lungs is compressed, forcing the object that is choking the victim out of the airway.

Technique if victim can be lifted (see *A*)

1. The rescuer stands behind the victim and wraps both arms around the victim's chest slightly below the rib cage and above the navel. The victim is allowed to fall forward with the head, arms, and chest over the rescuer's arms.

2. The rescuer makes a fist with one hand and grasps it with the other hand, pressing the thumb side of the fist against the victim's abdomen just below the end of the xiphoid process and above the navel.

3. The hands only are used to deliver the upward subdiaphragmatic thrusts. It is performed with sharp flexion of the elbows, in an upward rather than inward direction, and is usually repeated four times. It is very important not to compress the rib cage or actually press on the sternum during the Heimlich maneuver.

Technique if victim has collapsed or cannot be lifted (see *B*)

1. Rescuer places victim on floor face up.

2. Facing victim, rescuer straddles the victim's hips.

3. Rescuer places one hand on top of the other, with the bottom hand on the victim's abdomen slightly above the navel and below the rib cage.

4. Rescuer performs a forceful upward thrust with the heel of the bottom hand, repeating several times if necessary.

A

B

FIGURE 13–8 *Alveoli.* Bronchioles subdivide to form tiny tubes called *alveolar ducts,* which end in clusters of alveoli called *alveolar sacs.*

Infant Respiratory Distress Syndrome

Infant respiratory distress syndrome or **IRDS** is a very serious, life-threatening condition that often affects prematurely born infants of less than 37 weeks' gestation or those who weigh less than 2.2 kg (5 lbs) at birth. IRDS is the leading cause of death among premature infants in the United States, claiming over 5000 premature babies each year. The disease, characterized by a lack of **surfactant** in the alveolar air sacs, affects 50,000 babies annually.

Surfactant is manufactured by specialized cells in the walls of the alveoli. Surfactant reduces the surface tension of the fluid on the free surface of the alveolar walls and permits easy movement of air into and out of the lungs. The ability of the body to manufacture this important substance is not fully developed until shortly before birth—normally about 40 weeks after conception.

In newborn infants who are unable to manufacture surfactant, many air sacs collapse during expiration because of the increased surface tension. The effort required to reinflate these collapsed alveoli is much greater than that needed to reinflate normal alveoli with adequate surfactant. The baby soon develops labored breathing, and symptoms of respiratory distress appear shortly after birth.

In the past, treatment of IRDS was limited to keeping the alveoli open so that delivery and exchange of oxygen and carbon dioxide could occur. To accomplish this, a tube is inserted into the respiratory tract, and oxygen-rich air is delivered under sufficient pressure to keep the alveoli from collapsing at the end of expiration. A newer treatment involves delivering air under pressure and applying prepared surfactant directly into the baby's airways by means of a tube.

stem of a bunch of grapes (Figure 13-8). Each alveolar duct ends in several **alveolar sacs,** each of which resembles a cluster of grapes, and the wall of each alveolar sac is made up of numerous **alveoli,** each of which resembles a single grape.

Alveoli are very effective in gas exchange, mainly because they are extremely thin walled; each alveolus lies in contact with a blood capillary, and there are millions of alveoli in each lung. The surface of the respiratory membrane inside the alveolus is covered by a substance called **surfactant** (sur-FAK-tant). This important substance helps reduce surface tension in the alveoli and keeps them from collapsing as air moves in and out during respiration.

Lungs and Pleura

The **lungs** are fairly large organs. Note in Figure 13-9 that the right lung has three lobes and the left lung has two. Figure 13-9 shows the relationship of the lungs to the rib cage at the end of a normal expiration. The narrow, superior position of each lung, up under the collarbone, is the apex; the broad, inferior portion resting on the diaphragm, is the base.

The **pleura** covers the outer surface of the lungs and lines the inner surface of the rib cage. The pleura resembles other serous membranes in structure and function. Like the peritoneum or pericardium, the pleura is an extensive, thin, moist, slippery membrane. It lines a large, closed cavity of the body and covers the organs located within it. The parietal pleura lines the walls of the thoracic cavity; the visceral pleura covers the lungs, and the intrapleural space lies between the two pleural membranes (Figure 13-10). Pleurisy is an inflammation of the pleura that causes pain when the pleural membranes rub together.

Normally the intrapleural space contains just enough fluid to make both portions of the pleura moist and slippery and able to glide easily against each other as the lungs expand and deflate with each breath. **Pneumothorax** (noo-mo-THO-raks) is the presence of air in the intrapleural space on one side of the chest. The additional air increases the pressure on the lung on that side and causes it to collapse. While collapsed, the lung does not function in breathing.

Respiration

Respiration means exchange of gases (oxygen and carbon dioxide) between a living organism and its environment. If the organism consists of only one cell, gases can move directly between it and the environment. If, however, the organism consists of billions of cells, as do our bodies, most of its cells are too far from the air for a direct exchange of gases. To overcome this difficulty, a pair of organs—the lungs—provides a place where air and a circulating fluid (blood) can come close enough to each other for oxygen to move out of the air into blood while carbon dioxide moves out of the blood into air. Breathing or **pulmonary ven-**

tilation is the process that moves air into and out of the lungs. It makes possible the exchange of gases between air in the lungs and in the blood. This exchange is often called **external respiration.** In addition, exchange of gases occurs between the blood and the cells of the body—a process called **internal respiration.** *Cellular respiration* refers to the actual use of oxygen by cells in the process of metabolism, which is discussed in Chapter 15.

MECHANICS OF BREATHING

Pulmonary ventilation or breathing has two phases. **Inspiration** or inhalation moves air into the lungs, and **expiration** or exhalation moves air out of the lungs. The lungs are enclosed within the thoracic cavity. Thus changes in the shape and size of the thoracic cavity result in changes in the air pressure within that cavity and in the lungs. This difference in air pressure causes the movement of air into and out of the lungs. Air moves from an area where pressure is high to an area where pressure is lower. Respiratory muscles are responsible for the changes in the shape of the thoracic cavity that cause the air movements involved in breathing.

Hiccup

The term **hiccup** (HIK-up) is used to describe an involuntary, spasmodic contraction of the diaphragm. When such a contraction occurs, generally at the beginning of an inspiration, the glottis suddenly closes, producing the characteristic sound.

Hiccups lasting for extended periods of time can be disabling. They may be produced by irritation of the phrenic nerve or the sensory nerves in the stomach or by direct injury or pressure on certain areas of the brain. Fortunately, most cases of hiccups last only a few minutes and are harmless.

FIGURE 13–9 *Lungs.* The trachea is an airway that branches to form a treelike formation of bronchi and bronchioles. Note that the right lung has three lobes and that the left lung two lobes.

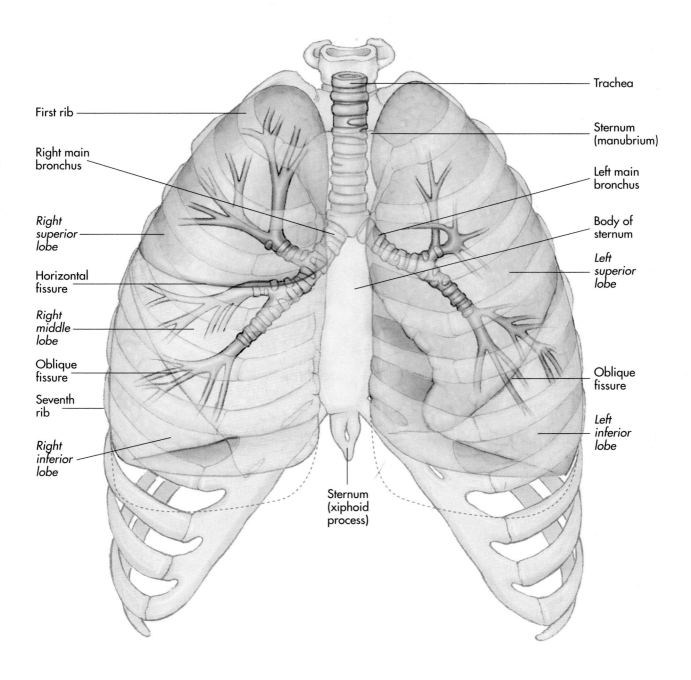

First rib

Right main
bronchus

*Right
superior
lobe*

Horizontal
fissure

*Right
middle
lobe*

Oblique
fissure

Seventh
rib

*Right
inferior
lobe*

Trachea

Sternum
(manubrium)

Left main
bronchus

Body of
sternum

*Left
superior
lobe*

Oblique
fissure

*Left
inferior
lobe*

Sternum
(xiphoid
process)

FIGURE 13–10 *Lungs and Pleura.* The inset shows where the body was cut to show this transverse section of the thorax. A serous membrane lines the thoracic wall (parietal pleura) and then folds inward near the bronchi to cover the lung (visceral pleura). The intrapleural space contains a small amount of serous pleural fluid.

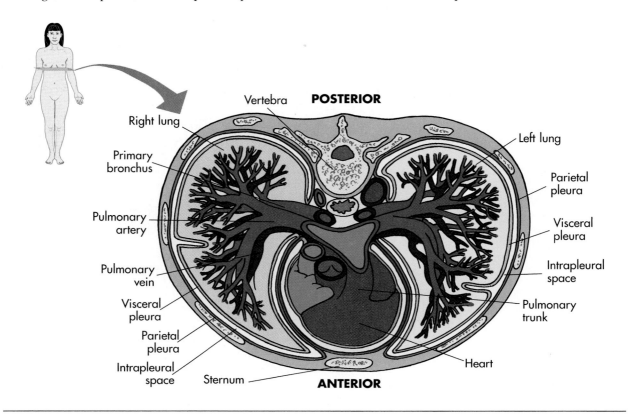

Inspiration

Inspiration occurs when the chest cavity enlarges. As the thorax enlarges, the lungs expand along with it, and air rushes into them and down into the alveoli. Muscles of respiration that are classified as **inspiratory muscles** include the **diaphragm** (DYE-a-fram) and the external intercostals. The diaphragm is the dome-shaped muscle separating the abdominal cavity from the thoracic cavity. The diaphragm flattens out when it contracts during inspiration. Instead of protruding up into the chest cavity, it moves down toward the abdominal cavity. Thus the contraction or flattening of the diaphragm makes the chest cavity longer from top to bottom. The diaphragm is the most important muscle of inspiration. Nerve impulses passing through the *phrenic nerve* stimulate the diaphragm to contract. The external intercostal muscles are located between the ribs. When they contract, they enlarge the thorax by increasing the size of the cavity from front to back and from side to side. Contraction of the inspiratory muscles increases the volume of the thoracic cavity and reduces the air pressure within it, drawing air into the lungs (Figure 13-11).

Expiration

Quiet expiration is ordinarily a passive process that begins when the inspiratory muscles relax. The thoracic cavity then returns to its smaller size.

FIGURE 13–11 *Mechanics of Breathing.* During *inspiration,* the diaphragm contracts, increasing the volume of the thoracic cavity. This increase in volume results in a decrease in pressure, which causes air to rush into the lungs. During *expiration,* the diaphragm returns to an upward position, reducing the volume in the thoracic cavity. Air pressure increases then, forcing air out of the lungs. The insets show the classic model in which a jar represents the rib cage, a rubber sheet represents the diaphragm, and a balloon represents the lungs.

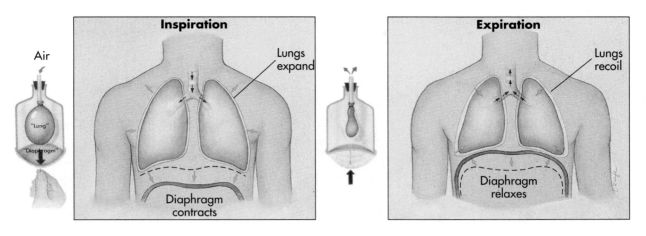

The elastic nature of lung tissue also causes these organs to "recoil" and decrease in size as air leaves the alveoli and flows outward through the respiratory passageways. When we speak, sing, or do heavy work, we may need more forceful expiration to increase the rate and depth of ventilation. During more forceful expiration, the **expiratory muscles** (internal intercostals and abdominal muscles) contract. When contracted, the internal intercostal muscles depress the rib cage and decrease the front-to-back size of the thorax. Contraction of the abdominal muscles pushes the abdominal organs against the underside of the diaphragm, thus elevating it and making it more "domeshaped." The result is to further shorten or decrease the top to bottom size of the thoracic cavity. As the thoracic cavity decreases in size, the air pressure within it increases and air flows out of the lungs (Figure 13-11).

EXCHANGE OF GASES IN LUNGS

Blood pumped from the right ventricle of the heart enters the pulmonary artery and eventually enters the lungs. It then flows through the thousands of tiny lung capillaries that are in close proximity to the air-filled alveoli (Figure 13-1). External respiration or the exchange of gases between the blood and alveolar air occurs by diffusion.

Diffusion is a passive process that results in movement down a concentration gradient; that is, substances move from an area of high concentration to an area of low concentration of the diffusing substance. Blood flowing into lung capillaries is low in oxygen. Oxygen is continually removed from the blood and used by the cells of the body. By the time it enters the lung capillaries, it is low in oxygen content. Because alveolar air is rich in oxygen, diffusion causes movement of oxygen from the area of high concentration (alveolar air) to the area of low concentration (capillary blood). Note in Figure 13-12 that most of the oxygen (O_2) entering the blood combines with hemoglobin (Hb) in the RBCs to form **oxyhemoglobin** (ok-see-HEE-mo-glo-bin) (HbO_2) so that it can be carried to the tissues and used by the body cells.

Diffusion of carbon dioxide (CO_2) also occurs between blood in lung capillaries and alveolar air.

FIGURE 13–12 *Exchange of Gases in Lung and Tissue Capillaries.* The left insets show O_2 diffusing out of alveolar air into blood and associating with hemoglobin (Hb) in lung capillaries to form oxyhemoglobin. In tissue capillaries, oxyhemoglobin dissociates, releasing O_2, which diffuses from the RBC and then crosses the capillary wall to reach the tissue cells. As the left insets show, CO_2 diffuses in the opposite direction (into RBCs) and some of it associates with Hb to form carbaminohemoglobin. However, most CO_2 combines with water to form carbonic acid (H_2CO_3), which dissociates to form H^+ and HCO_3^- (bicarbonate) ions. Back in the lung capillaries, CO_2 dissociates from the bicarbonate and carbaminohemoglobin molecules and diffuses out of blood into alveolar air.

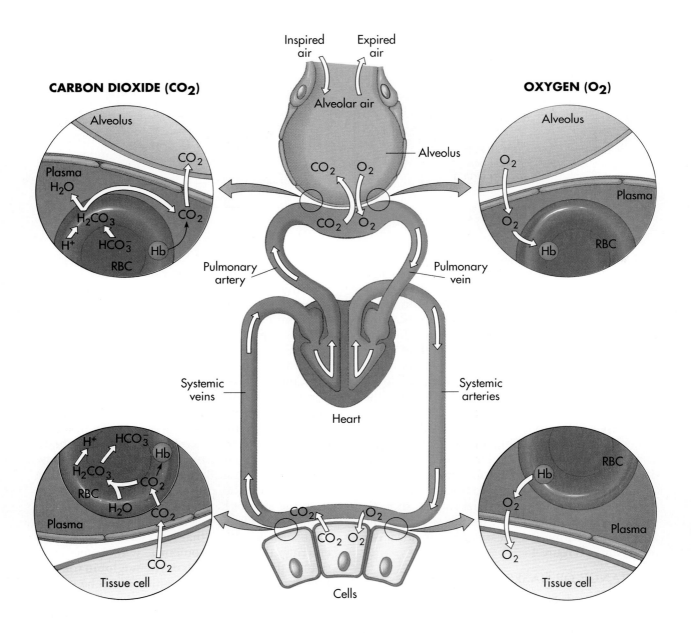

Blood flowing through the lung capillaries is high in carbon dioxide. Most carbon dioxide is carried as bicarbonate ion (HCO_3^-) in the blood. Some, as in Figure 13-12, combines with the hemoglobin in RBCs to form **carbaminohemoglobin** (kar-bam-i-no-HEE-mo-glo-bin) (HbNHCOOH). As cells remove oxygen from circulating blood, they add the waste product carbon dioxide to it. As a result, the blood in pulmonary capillaries eventually becomes low in oxygen and high in carbon dioxide. Diffusion of carbon dioxide results in its movement from an area of high concentration in the pulmonary capillaries to an area of low concentration in alveolar air. Then from the alveoli, carbon dioxide leaves the body in expired air.

EXCHANGE OF GASES IN TISSUES

The exchange of gases that occurs between blood in tissue capillaries and the body cells is called *internal respiration.* As you would expect, the direction of movement of oxygen and carbon dioxide during internal respiration is just the opposite of that noted in the exchange that occurs during external respiration when gases are exchanged between the blood in the lung capillaries and the air in alveoli. As shown in Figure 13-12, oxyhemoglobin breaks down into oxygen and hemoglobin in the tissue capillaries. Oxygen molecules move rapidly out of the blood through the tissue capillary membrane into the interstitial fluid and on into the cells that compose the tissues. The oxygen is used by the cells in their metabolic activities. Diffusion results in the movement of oxygen from an area of high concentration to an area of low concentration in the cells where it is needed. While this is happening, carbon dioxide molecules leave the cells, entering the tissue capillaries where bicarbonate ions are formed and where hemoglobin molecules unite with carbon dioxide to form carbaminohemoglobin. Once again, diffusion is responsible for the movement of carbon dioxide from an area of high concentration in the cells to an area of lower concentration in the capillary blood. In other words, oxygenated blood enters tissue capillaries and is changed into deoxygenated blood as it flows through them. In the process of losing oxygen, the waste product carbon dioxide is picked up and transported to the lungs for removal from the body.

Maximum Oxygen Consumption

Exercise physiologists use **maximum oxygen consumption** ($VO_{2\ max}$) as a predictor of a person's capacity to do aerobic exercise. An individual's $VO_{2\ max}$ represents the amount of oxygen taken up by the lungs, transported to the tissues, and used to do work. $VO_{2\ max}$ is determined largely by hereditary factors, but aerobic (endurance) training can increase it by as much as 35%. Many endurance athletes are now using $VO_{2\ max}$ measurements to help them determine and then maintain their peak condition.

VOLUMES OF AIR EXCHANGED IN PULMONARY VENTILATION

A special device called a **spirometer** is used to measure the amount of air exchanged in breathing. Figure 13-13 illustrates the various pulmonary volumes, which can be measured as a subject breathes into a spirometer. We take 500 ml (about a pint) of air into our lungs with each normal inspiration and expel it with each normal expiration. Because this amount comes and goes regularly like the tides of the sea, it is referred to as the **tidal volume (TV).** The largest amount of air that we can breathe out in one expiration is known as the **vital capacity (VC).** In normal young men, this is about 4800 ml. Tidal volume and vital capacity are frequently measured in patients with lung or heart disease, conditions that often lead to abnormal volumes of air being moved in and out of the lungs.

Observe the area in Figure 13-13 that represents the **expiratory reserve volume (ERV).** This is the amount of air that can be forcibly exhaled after expiring the tidal volume. Compare this with the area in Figure 13-13 that represents the **inspiratory reserve volume (IRV).** The IRV is the amount of air that can be forcibly inspired over and above a normal inspiration. As the tidal volume

FIGURE 13–13 *Pulmonary Ventilation Volumes.* The chart in *A* shows a tracing like that produced with a spirometer. The diagram in *B* shows the pulmonary volumes as relative proportions of an inflated balloon (see Figure 13-11). During normal, quiet breathing, about 500 ml of air is moved into and out of the respiratory tract, an amount called the *tidal volume*. During forceful breathing (like that during and after heavy exercise), an extra 3300 ml can be inspired (the inspiratory reserve volume), and an extra 1000 ml or so can be expired (the expiratory reserve volume). The largest volume of air that can be moved in and out during ventilation is called the *vital capacity*. Air that remains in the respiratory tract after a forceful expiration is called the residual volume.

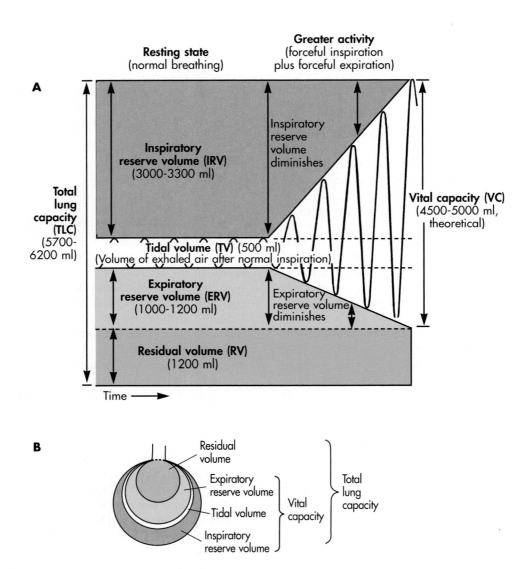

increases, the ERV and IRV decrease. Note in Figure 13-13 that vital capacity (VC) is the total of tidal volume, inspiratory reserve volume, and expiratory reserve volume—or expressed in another way: VC = TV + IRV + ERV. **Residual volume (RV)** is simply the air that remains in the lungs after the most forceful expiration.

Regulation of Respiration

We know that the body uses oxygen to obtain energy for the work it has to do. The more work the body does, the more oxygen that must be delivered to its millions of cells. One way this is accomplished is by increasing the rate and depth of respirations. Although we may take only 12 to 18 breaths a minute when we are not moving about, we take considerably more than this when we are exercising. Not only do we take more breaths, but our tidal volume also increases.

To help supply cells with more oxygen when they are doing more work, automatic adjustments occur not only in respirations but also in circulation. Most notably, the heart beats faster and harder and therefore pumps more blood through the body each minute. This means that the millions of RBCs make more round trips between the lungs and tissues each minute and so deliver more oxygen per minute to tissue cells.

Working cells not only require more oxygen, but they also produce more waste products such as carbon dioxide and certain metabolic acids. The increase in respirations during exercise shows us how the body automatically regulates its vital functions. By increasing the rate and depth of respiration, we can adjust to the varying demands for increased oxygen while increasing the elimination of metabolic waste products in expired air to maintain homeostasis.

Normal respiration depends on proper functioning of the muscles of respiration. These muscles are stimulated by nervous impulses that originate in **respiratory control centers** located in the medulla and pons of the brain. These centers are in turn regulated by a number of inputs from receptors located in varying areas of the body. These receptors can sense the need for changing the rate or depth of respirations to maintain

Oxygen Therapy

Oxygen therapy is the administration of oxygen, often by a registered *respiratory therapist*, to patients suffering from **hypoxia** (hi-POK-see-a)—an insufficient oxygen supply to the tissues. Respiratory problems that involve a decrease in ventilation or a lack of effecient gas exchange in the lungs often require home treatment with oxygen therapy.

Oxygen (O_2) gas is commonly stored in and dispensed from small, green storage tanks that hold the gas under high pressure until it is used. Because the oxygen dispensed from such tanks is cold and very dry, it must first be warmed and moistened to prevent damage to the respiratory tract. This is usually done by simply bubbling the oxygen through warm water as it leaves the tank. The oxygen gas may then pass through a mask or through tubes that lead into the nasal passage (nasal prongs).

Oxygen therapy is used in a variety of home health situations. For example, the hypoxia that often accompanies chronic pulmonary disease and heart problems such as myocardial infarctions may be treated with oxygen therapy.

homeostasis. Certain receptors sense carbon dioxide or oxygen levels, whereas others sense blood acid levels or the amount of stretch in lung tissues. The two most important control centers are in the medulla and are called the **inspiratory center** and the **expiratory center.** Centers in the pons have a modifying function. Under rest conditions, neurons in the inspiratory and expiratory centers "fire" at a rate that will produce a normal breathing rate of about 12 to 18 breaths a minute.

The depth and rate of respiration can be influenced by many "inputs" to the respiratory control centers from other areas of the brain or from specialized receptors located outside of the central nervous system (Figure 13-14).

FIGURE 13–14 *Regulation of Respiration.* Respiratory control centers in the brain stem control the basic rate and depth of breathing. The brain stem also receives input from other parts of the body; information from chemoreceptors and stretch receptors can alter the basic breathing pattern, as can emotional and sensory input. Despite these controls, the cerebral cortex can override the "automatic" control of breathing to some extent to accomplish activities such as singing or blowing up a balloon. Green arrows show the flow of regulatory information flows into the respiratory control centers. The purple arrow shows the flow of regulatory information from the control centers to the respiratory muscles that drive breathing.

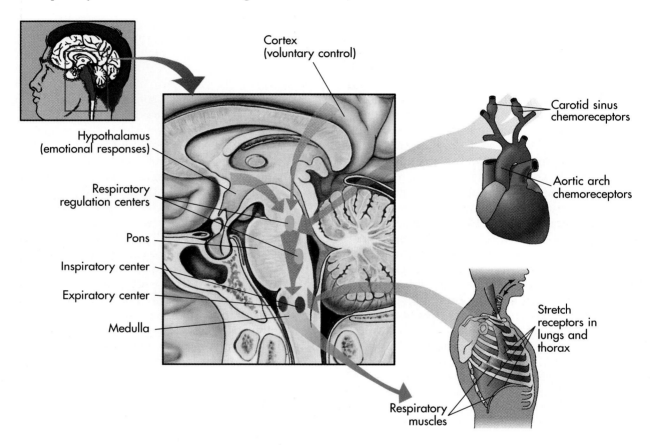

CEREBRAL CORTEX

The cerebral cortex can influence respiration by modifying the rate at which neurons "fire" in the inspiratory and expiratory centers of the medulla. In other words, an individual may voluntarily speed up or slow down the breathing rate or greatly change the pattern of respiration during activities. This ability permits us to change respiratory patterns and even to hold our breath for short periods to accommodate activities such as underwater swimming, speaking, or eating. This voluntary control of respiration, however, has limits. As indicated in a later section, other factors such as blood carbon dioxide levels are much more powerful in controlling respiration than conscious control. Regardless of cerebral intent to the contrary, we resume breathing when our bodies sense the need for more oxygen or if carbon dioxide levels increase to certain levels.

RECEPTORS INFLUENCING RESPIRATION

Chemoreceptors

Chemoreceptors (KEE-mo-ree-SEP-tors) located in the **carotid** and **aortic bodies** are specialized receptors that are sensitive to increases in blood carbon dioxide level and decreases in blood oxygen level. They also can sense and respond to increasing blood acid levels. The carotid body receptors are found at the point where the common carotid arteries divide, and the aortic bodies are small clusters of chemosensitive cells that lie adjacent to the aortic arch near the heart (Figure 13-14). When stimulated by increasing levels of blood carbon dioxide, decreasing oxygen levels, or increasing blood acidity, these receptors send nerve impulses to the respiratory regulatory centers that in turn modify respiratory rates.

Pulmonary Stretch Receptors

Specialized stretch receptors in the lungs are located throughout the pulmonary airways and in the alveoli (Figure 13-14). Nervous impulses generated by these receptors influence the normal pattern of breathing and protect the respiratory system from excess stretching caused by harmful overinflation. When the tidal volume of air has been inspired, the lungs are expanded enough to stimulate stretch receptors that then send inhibitory impulses to the inspiratory center. Relaxation of inspiratory muscles occurs, and expiration follows. After expiration, the lungs are sufficiently deflated to inhibit the stretch receptors, and inspiration is then allowed to start again.

Types of Breathing

A number of terms are used to describe breathing patterns. **Eupnea** (YOOP-nee-ah), for example, refers to a normal respiratory rate. During eupnea, the need for oxygen and carbon dioxide exchange is being met, and the individual is usually not aware of the breathing pattern. The terms **hyperventilation** and **hypoventilation** describe very rapid and deep or slow and shallow respirations, respectively. Hyperventilation sometimes results from a conscious voluntary effort preceding exertion or from psychological factors—"hysterical hyperventilation." **Dyspnea** (DISP-nee-ah) refers to labored or difficult breathing and is often associated with hypoventilation. If breathing stops completely for a brief period, regardless of cause, it is called **apnea** (AP-nee-ah). Failure to resume breathing after a period of apnea is called **respiratory arrest**.

STRUCTURAL PLAN (Figure 13-1)

Basic plan of respiratory system would be similar to an inverted tree if it were hollow; leaves of the tree would be comparable to alveoli, with the microscopic sacs enclosed by networks of capillaries

RESPIRATORY TRACTS

A Upper respiratory tract—nose, pharynx, and larynx
B Lower respiratory tract—trachea, bronchial tree, and lungs

RESPIRATORY MUCOSA (Figure 13-3)

A Specialized membrane that lines the air distribution tubes in the respiratory tree
B Over 125 ml of mucus produced each day forms a "mucous blanket" over much of the respiratory mucosa
C Mucus serves as an air purification mechanism by trapping inspired irritants such as dust and pollen
D Cilia on mucosal cells beat in only one direction, moving mucus upward to pharynx for removal

NOSE

A Structure
  1 Nasal septum separates interior of nose into two cavities
  2 Mucous membrane lines nose
  3 Frontal, maxillary, sphenoidal, and ethmoidal sinuses drain into nose (Figure 13-4)
B Functions
  1 Warms and moistens inhaled air
  2 Contains sense organs of smell

PHARYNX

A Structure (Figure 13-5)
  1 Pharynx (throat) about 12.5 cm (5 inches) long
  2 Divided into nasopharynx, oropharynx, and laryngopharynx
  3 Two nasal cavities, mouth, esophagus, larynx, and auditory tubes all have openings into pharynx
  4 Pharyngeal tonsils and openings of auditory tubes open into nasopharynx; tonsils found in oropharynx
  5 Mucous membrane lines pharynx
B Functions
  1 Passageway for food and liquids
  2 Air distribution; passageway for air

LARYNX

A Structure (Figure 13-6)
  1 Several pieces of cartilage form framework
    a Thyroid cartilage (Adam's apple) is largest
    b Epiglottis partially covers opening into larynx

  2 Mucous lining
  3 Vocal cords stretch across interior of larynx
B Functions
  1 Air distribution; passageway for air to move to and from lungs
  2 Voice production

TRACHEA

A Structure (Figure 13-7)
  1 Tube about 11 cm (4½ inches) long that extends from larynx into the thoracic cavity
  2 Mucous lining
  3 C-shaped rings of cartilage hold trachea open
B Function—passageway for air to move to and from lungs
C Obstruction
  1 Blockage of trachea occludes the airway and if complete causes death in minutes
  2 Tracheal obstruction causes over 4000 deaths annually in the United States
  3 Heimlich maneuver (p. 288) is a lifesaving technique used to free the trachea of obstructions

BRONCHI, BRONCHIOLES, AND ALVEOLI

A Structure
  1 Trachea branches into right and left bronchi
  2 Each bronchus branches into smaller and smaller tubes eventually leading to bronchioles
  3 Bronchioles end in clusters of microscopic alveolar sacs, the walls of which are made up of alveoli (Figure 13-8)
B Function
  1 Bronchi and bronchioles—air distribution; passageway for air to move to and from alveoli
  2 Alveoli—exchange of gases between air and blood

LUNGS AND PLEURA

A Structure (Figure 13-9)
  1 Size—large enough to fill the chest cavity, except for middle space occupied by heart and large blood vessels
  2 Apex—narrow upper part of each lung, under collarbone
  3 Base—broad lower part of each lung; rests on diaphragm
  4 Pleura—moist, smooth, slippery membrane that lines chest cavity and covers outer surface of lungs; reduces friction between the lungs and chest wall during breathing (Figure 13-10)
B Function—breathing (pulmonary ventilation)

RESPIRATION

A Mechanics of breathing (Figure 13-11)
  1 Pulmonary ventilation includes two phases called *inspiration* (movement of air into lungs) and *expiration* (movement of air out of lungs)
  2 Changes in size and shape of thorax cause changes in air pressure within that cavity and in the lungs
  3 Air pressure differences actually cause air to move into and out of the lungs

B Inspiration
  1 Active process—air moves into lungs
  2 Inspiratory muscles include diaphragm and external intercostals
    a Diaphragm flattens during inspiration—increases top to bottom length of thorax
    b External intercostals contraction elevates the ribs and increases the size of the thorax from the front to the back and from side to side
  3 The increase in the size of the chest cavity reduces pressure within it, and air enters the lungs

C Expiration
  1 Quiet expiration is ordinarily a passive process
  2 During expiration, thorax returns to its resting size and shape
  3 Elastic recoil of lung tissues aids in expiration
  4 Expiratory muscles used in forceful expiration are internal intercostals and abdominal muscles
    a Internal intercostals—contraction depresses the rib cage and decreases the size of the thorax from the front to back
    b Contraction of abdominal muscles elevates the diaphragm, thus decreasing size of the thoracic cavity from the top to bottom
  5 Reduction in the size of the thoracic cavity increases its pressure and air leaves the lungs

D Exchange of gases in lungs (Figure 13-12)
  1 Carbaminohemoglobin breaks down into carbon dioxide and hemoglobin
  2 Carbon dioxide moves out of lung capillary blood into alveolar air and out of body in expired air
  3 Oxygen moves from alveoli into lung capillaries
  4 Hemoglobin combines with oxygen, producing oxyhemoglobin

E Exchange of gases in tissues
  1 Oxyhemoglobin breaks down into oxygen and hemoglobin
  2 Oxygen moves out of tissue capillary blood into tissue cells
  3 Carbon dioxide moves from tissue cells into tissue capillary blood

  4 Hemoglobin combines with carbon dioxide, forming carbaminohemoglobin

F Volumes of air exchanged in pulmonary ventilation (Figure 13-13)
  1 Volumes of air exchanged in breathing can be measured with a spirometer
  2 Tidal volume (TV)—amount normally breathed in or out with each breath
  3 Vital capacity (VC)—largest amount of air that one can breathe out in one expiration
  4 Expiratory reserve volume (ERV)—amount of air that can be forcibly exhaled after expiring the tidal volume
  5 Inspiratory reserve volume (IRV)—amount of air that can be forcibly inhaled after a normal inspiration
  6 Residual volume (RV)—air that remains in the lungs after the most forceful expiration
  7 Rate—usually about 12 to 18 breaths a minute; much faster during exercise

G Regulation of respiration (Figure 13-14)
  1 Regulation of respiration permits the body to adjust to varying demands for oxygen supply and carbon dioxide removal
  2 Most important central regulatory centers in medulla are called *respiratory control centers* (inspiratory and expiratory centers)
    a Under resting conditions nervous activity in the respiratory control centers produces a normal rate and depth of respirations (12 to 18 per minute)
  3 Respiratory control centers in the medulla are influenced by "inputs" from receptors located in other body areas:
    a Cerebral cortex—voluntary (but limited) control of respiratory activity
    b Chemoreceptors respond to changes in carbon dioxide, oxygen, and blood acid levels—located in carotid and aortic bodies
    c Pulmonary stretch receptors—respond to the stretch in lungs, thus protecting respiratory organs from overinflation

TYPES OF BREATHING

A Eupnea—normal breathing
B Hyperventilation—rapid and deep respirations
C Hypoventilation—slow and shallow respirations
D Dyspnea—labored or difficult respirations
E Apnea—stopped respiration
F Respiratory arrest—failure to resume breathing after a period of apnea

---

N E W   W O R D S

alveoli	eupnea	larynx	respiratory arrest
aortic body	expiratory reserve	oxyhemoglobin	respiratory membrane
apnea	volume (ERV)	paranasal sinuses	spirometer
bronchi	Heimlich maneuver	pharynx	surfactant
carbaminohemoglobin	hyperventilation	pleurisy	tidal volume (TV)
carotid body	hypoventilation	pulmonary ventilation	tonsillectomy
conchae	inspiratory reserve	residual volume (RV)	trachea
dyspnea	volume (IRV)	respiration	vital capacity (VC)

C H A P T E R   T E S T

1. The exchange of gases in the respiratory system depends on the passive transport process of _____.

2. Surface tension in the alveoli is reduced by a substance called _____.

3. The trachea is located in the _____ respiratory tract.

4. The portion of the pharynx located behind the mouth is called the _____.

5. The adenoids or pharyngeal tonsils are located in the _____.

6. The voice box is also known as the _____.

7. During swallowing, food is prevented from entering the trachea by the _____.

8. The presence of air in the pleural space is called _____.

9. The exchange of gases between air in the lungs and the blood is called _____ respiration.

10. Most oxygen entering the blood combines with hemoglobin in the RBCs to form _____.

11. The largest amount of air we can breathe out in one expiration is known as the _____.

12. The most important respiratory control centers are located in the _____.

13. Gas exchange in the lung takes places across the thin, moist membranes of the _____, which are surrounded by a dense network of capillaries.

Select the most correct answer from Column B for each statement in Column A. (Only one answer is correct.)

COLUMN A	COLUMN B
14. _____ Maximum expiration	a. Diaphragm
15. _____ "Adam's apple"	b. Thyroid cartilage
16. _____ Lines thoracic cavity	c. Chemoreceptor
17. _____ Pulmonary ventilation	d. Breathing
18. _____ Labored breathing	e. Vital capacity
19. _____ Carotid body	f. Dyspnea
20. _____ Major muscle of inspiration	g. Parietal pleura

REVIEW QUESTIONS

1. Discuss the location, microscopic structure, and functions of the respiratory mucosa.
2. List the paranasal air sinuses.
3. Discuss the functions of the nose in respiration.
4. What and where are the pharynx and larynx?
5. What structures open into the pharynx?
6. What are the anatomical subdivisions of the pharynx?
7. Where are the tonsils and adenoids located?
8. What function do the C-shaped rings of cartilage serve in the trachea?
9. What and where is the "Adam's apple"?
10. Define parietal pleura, pleural space, pleurisy, pneumothorax, and visceral pleura.
11. Do breathing and respiration mean the same thing? Define each term.
12. What does *residual volume* mean?
13. Discuss the anatomy of the "respiratory membrane."
14. Compare the structure, location, and functions of the respiratory membrane and the respiratory mucosa.
15. Discuss the Heimlich maneuver.
16. Discuss the respiratory control mechanisms.

CRITICAL THINKING

17. Why does sinusitis or middle ear infection occur so frequently after the common cold?
18. Briefly explain how oxygen and carbon dioxide can move between alveolar air, blood, and tissue cells.
19. Explain the following equation:

$$VC = TV + IRV + ERV$$

20. How can a brain hemorrhage affect the respiratory system?

The Digestive System

14

BOXED ESSAYS

Dental Diseases
Heartburn
Fractal Geometry of the Body
Ulcers
Exercise and Fluid Uptake
Colostomy

OBJECTIVES

After you have completed this chapter, you should be able to:

1. List in sequence each of the component parts or segments of the alimentary canal from the mouth to the anus and identify the accessory organs of digestion.

2. List and describe the four layers of the wall of the alimentary canal. Compare the lining layer in the esophagus, stomach, small intestine, and large intestine.

3. Discuss the basics of protein, fat, and carbohydrate digestion and give the end products of each process.

4. Define and contrast mechanical and chemical digestion.

5. Define: *peristalsis, bolus, chyme, jaundice, ulcer,* and *diarrhea.*

The principal structure of the **digestive system** is an irregular tube, open at both ends, called the **alimentary** (al-i-MEN-tar-ee) **canal** or the **gastrointestinal** (gas-tro-in-TES-ti-nal) **(GI) tract.** In the adult, this hollow tube is about 9 m (29 feet) long. Although this may seem strange, food or other material that enters the digestive tube is not really inside the body. Most parents of young children quickly learn that a button or pebble swallowed by their child will almost always pass unchanged and with little difficulty through the tract. Think of the tube as a passageway that extends through the body like a hallway through a building. Food must be broken down or **digested** and then absorbed through the walls of the digestive tube before it can actually enter the body and be used by cells. The breakdown or digestion of food material is both mechanical and chemical in nature. The teeth are used to physically break down food material before it is swallowed. The churning of food in the stomach then continues the mechanical breakdown process. Chemical breakdown results from the action of digestive enzymes and other chemicals acting on food as it passes through the GI tract. In chemical digestion, large food molecules are reduced to smaller molecules that can be absorbed through the lining of the intestinal wall and then distributed to body cells for use. This process of altering the chemical and physical composition of food so that it can be absorbed and used by body cells is known as *digestion*, and it is the function of the digestive system. Part of the digestive system, the large intestine, serves also as an organ of elimination, ridding the body of the waste material or **feces** resulting from the digestive process. Table 14-1 names both main and accessory digestive organs. Note that the accessory organs include the teeth, tongue, gallbladder, and appendix, as well as a number of glands that secrete their products into the digestive tube.

Foods undergo three kinds of processing in the body: **digestion, absorption,** and **metabolism.** Digestion and absorption are performed by the organs of the digestive system (Figure 14-1). Metabolism, on the other hand, is performed by all body cells. In this chapter, we shall begin by describing digestive organs and then discuss digestion and absorption. Later, in Chapter 15, we will discuss the metabolism of food after it has been absorbed.

Wall of the Digestive Tract

The digestive tract has been described as a tube that extends from the mouth to the anus. The wall of this digestive tube is fashioned of four layers of tissue (Figure 14-2). The inside or hollow space within the tube is called the **lumen.** The four layers, named from the inside coat to the outside of the tube, follow:

1. Mucosa or mucous membrane
2. Submucosa
3. Muscularis
4. Serosa

Although the same four tissue coats form the organs of the alimentary tract, their structures vary in different organs. The **mucosa** of the esophagus, for example, is composed of tough and stratified abrasion-resistant epithelium. The mucosa of the remainder of the tract is a delicate layer of simple columnar epithelium designed for absorption and secretion. The mucus produced by either type of epithelium coats the lining of the alimentary canal.

The **submucosa,** as the name implies, is a connective tissue layer that lies just below the mucosa. It contains many blood vessels and nerves. The two layers of muscle tissue called the **muscularis** have an important function in the digestive process. By a wavelike, rhythmic contraction of the muscular coat, called **peristalsis** (pair-i-STAL-sis), food material is moved through the digestive tube. In addition, the contraction of the muscularis also assists in the mixing of food with digestive juice and in the further mechanical breakdown of larger food particles.

The **serosa** is the outermost covering or coat of the digestive tube. In the abdominal cavity it is composed of the parietal peritoneum. The loops of the digestive tract are anchored to the posterior wall of the abdominal cavity by a large double fold of peritoneal tissue called the **mesentery** (MEZ-en-tair-ee).

TABLE 14–1 Organs of the Digestive System

MAIN ORGAN	ACCESSORY ORGAN	MAIN ORGAN	ACCESSORY ORGAN
Mouth	Teeth and tongue	Large intestine	Vermiform appendix
Pharynx (throat)	Salivary glands	Cecum	
Esophagus (foodpipe)	Parotid	Colon	
Stomach	Submandibular	Ascending colon	
Small intestine	Sublingual	Transverse colon	
Duodenum		Descending colon	
Jejunum	Liver	Sigmoid colon	
Ileum	Gallbladder	Rectum	
	Pancreas	Anal canal	

FIGURE 14–1 *Location of Digestive Organs.*

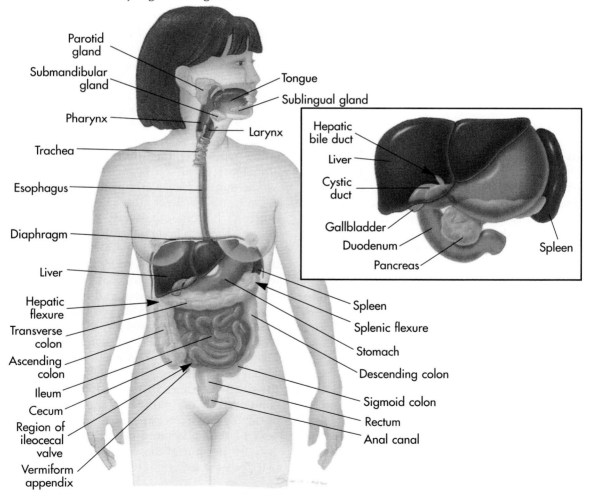

FIGURE 14–2 *Section of the Small Intestine.* The four layers typical of walls of the gastrointestinal tract are shown. Circular folds of mucous membrane called *plicae* increase the surface area of the lining coat.

Mouth

The **mouth** or **oral cavity** is a hollow chamber with a roof, a floor, and walls. Food enters or is ingested into the digestive tract through the mouth, and the process of digestion begins immediately. Like the remainder of the digestive tract, the mouth is lined with mucous membrane. It may be helpful if you review the structure and function of mucous membranes in Chapter 4. Typically, mucous membranes line hollow organs, such as the digestive tube, that open to the exterior of the body. Mucus produced by the lining of the GI tract protects the epithelium from digestive juices and lubricates food passing through the lumen.

FIGURE 14–3   *The Mouth Cavity.*

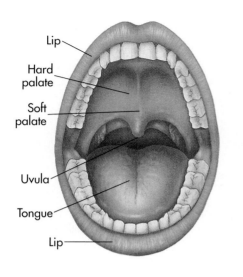

The roof of the mouth is formed by the **hard** and **soft palates** (Figure 14-3). The hard palate is a bony structure in the anterior or front portion of the mouth, formed by parts of the palatine and maxillary bones. The soft palate is located above the posterior or rear portion of the mouth. It is soft because it consists chiefly of muscle. Hanging down from the center of the soft palate is a cone-shaped process, the **uvula** (YOO-vyoo-lah). If you look in the mirror, open your mouth wide, and say "Ah," you can see your uvula. The uvula and the soft palate prevent any food and liquid from entering the nasal cavities above the mouth.

The floor of the mouth consists of the tongue and its muscles. The tongue is made of skeletal muscle covered with mucous membrane. It is anchored to bones in the skull and to the hyoid bone in the neck. A thin membrane called the **frenulum** (FREN-yoo-lum) attaches the tongue to the floor of the mouth. Occasionally the frenulum is too short to allow free movements of the tongue. Individuals with this condition cannot enunciate words normally and are said to be tongue-tied. Note in Figure 14-4 that the tongue

FIGURE 14–4   *The Tongue.* A, Surface. **B,** Mouth cavity showing the undersurface of the tongue.

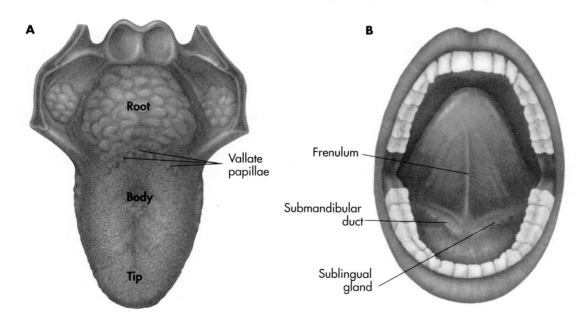

can be divided into a blunt rear portion called the *root*, a pointed *tip*, and a central *body*.

Have you ever noticed the many small elevations on the surface of your tongue? They are **papillae.** The largest are the **vallate** type; as you can see in Figure 14-4, they form an inverted V-shaped row of about 10 to 12 mushroomlike elevations. The taste buds, which contain sensory receptors for salty, sour, sweet, and bitter compounds, are located on the papillae (see Chapter 8).

○ *Teeth*

The shape and placement of the teeth assist in their functions. The four major types of teeth follow:

1. Incisors
2. Canines
3. Premolars
4. Molars

Note in Figure 14-5 that the incisors have a sharp cutting edge. They have a cutting function during **mastication** (mas-ti-KAY-shun) or chewing of food. The canine teeth are sometimes called **cuspids.** They pierce or tear the food that is being eaten. This tooth type is particularly apparent in meat-eating mammals such as dogs. Premolars or **bicuspids** and molars or **tricuspids** have rather large, flat surfaces with two or three grinding or crushing "cusps" on their surface. They provide extensive breakdown of food in the mouth. After food has been chewed, it is formed into a small rounded mass called a **bolus** (BO-lus) so that it can be swallowed.

By the time a baby is 2 years old, the child probably has his full set of 20 baby teeth. When a young adult is somewhere between 17 and 24 years old, a full set of 32 permanent teeth is generally present. The average age for cutting the first tooth is about 6 months, and the average age for losing the first baby tooth and starting to cut

FIGURE 14–5 *The Deciduous (Baby) Teeth and Adult Teeth.* In the deciduous set, there are no premolars and only two pairs of molars in each jaw. Generally the lower teeth erupt before the corresponding upper teeth.

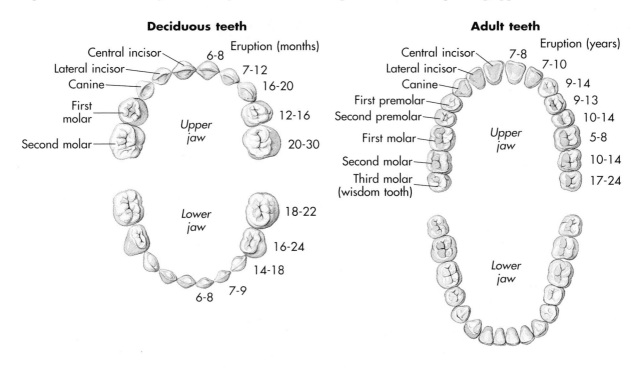

FIGURE 14–6 *Longitudinal Section of a Tooth.* A molar is sectioned to show its bony socket and details of its three main parts: crown, neck, and root. Enamel (over the crown) and cementum (over the neck and root) surround the dentin layer. The pulp contains nerves and blood vessels.

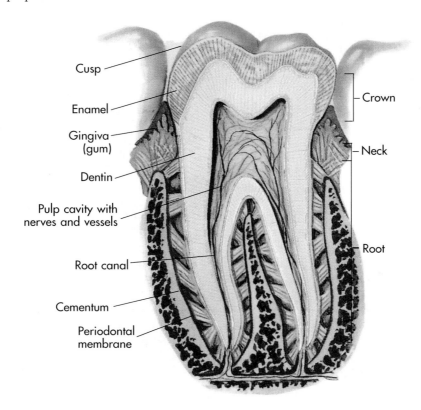

the permanent teeth is about 6 years. Figure 14-5 gives the names of the teeth and shows which ones are lacking in the deciduous or baby set.

TYPICAL TOOTH

A typical tooth can be divided into three main parts: crown, neck, and root. The **crown** is the portion that is exposed and visible in the mouth. It is covered by enamel—the hardest tissue in the body. Enamel is ideally suited to withstand the grinding that occurs during the chewing of hard and brittle foods. In addition to enamel, the outer shell of each tooth is covered by two other dental

tissues—dentin and cementum (Figure 14-6). Dentin makes up the greatest proportion of the tooth shell. It is covered by enamel in the crown and by cementum in the neck and root areas. The center of the tooth contains a pulp cavity consisting of connective tissue, blood and lymphatic vessels, and sensory nerves.

The **neck** of a tooth is the narrow portion, shown in Figure 14-6, surrounded by the pink gingiva or gum tissue. It joins the crown of the tooth to the root. The **root** fits into the socket of the upper or lower jaw. A fibrous **periodontal membrane** lines each tooth socket.

⬤ *Salivary Glands*

Three pairs of salivary glands—the parotids, submandibulars, and sublinguals—secrete most (about 1 L) of the saliva produced each day in the adult. The salivary glands (Figure 14-7) are typical of the accessory glands associated with the digestive system. They are located outside of the digestive tube itself and must convey their secretions by way of ducts into the tract.

The **parotid glands,** largest of the salivary glands, lie just below and in front of each ear at the angle of the jaw—an interesting anatomical position because it explains why people who have mumps (an infection of the parotid gland) often complain that it hurts when they open their mouths or chew; these movements squeeze the tender, inflamed gland. To see the openings of the parotid ducts, look in a mirror at the insides of your cheeks opposite the second molar tooth on either side of the upper jaw.

The ducts of the **submandibular glands** open into the mouth on either side of the lingual frenulum (Figure 14-4). The ducts of the **sublingual glands** open into the floor of the mouth.

FIGURE 14–7 *Location of the Salivary Glands.*

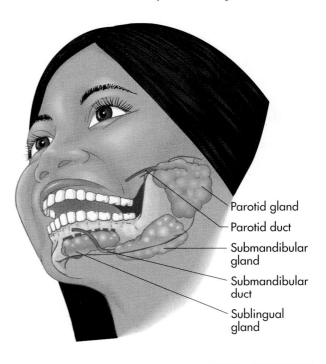

Parotid gland
Parotid duct
Submandibular gland
Submandibular duct
Sublingual gland

Dental Diseases

Tooth decay or dental caries is one of the most common diseases in the civilized world. It is a disease of the enamel, dentin, and cementum of teeth that results in the formation of a permanent defect called a **cavity.** Most people living in the United States, Canada, and Europe are significantly affected by the disease. Decay occurs on tooth surfaces where food debris, acid-secreting bacteria, and plaque accumulate.

Regular and thorough brushing of the teeth is important in preventing decay. Reducing dietary intake of refined sugars, which promote growth of decay bacteria, also reduces the rate of cavity formation. The introduction of fluoride to water supplies has proved to be the most effective and practical method of reducing the rate of tooth decay in children, and it is considered an important public health measure in many areas of the United States.

If the disease is untreated, tooth decay results in infection, loss of teeth, and inflammation of the soft tissues in the mouth. Bacteria may also invade the paranasal sinuses or extend to the surface of the face and neck, causing serious complications.

Saliva contains mucus and a digestive enzyme that is called **salivary amylase** (AM-i-lase). Mucus moistens the food and allows it to pass with less friction through the esophagus and into the stomach. Salivary amylase begins the chemical digestion of carbohydrates.

Pharynx

The **pharynx** is a tubelike structure made of muscle and lined with mucous membrane. Observe its location in Figure 14-1. Because of its location behind the nasal cavities and mouth, it functions as part of the respiratory and digestive systems. Air must pass through the pharynx on its way to the lungs, and food must pass through it on its way to the stomach. The pharynx as a whole is subdivided into three anatomical components, as described in Chapter 13.

Esophagus

The **esophagus** (e-SOF-ah-gus) or foodpipe is the muscular, mucus-lined tube that connects the pharynx with the stomach. It is about 25 centimeters (10 inches) long. The esophagus serves as a dynamic passageway for food, pushing the food toward the stomach.

Stomach

The **stomach** (Figure 14-8) lies in the upper part of the abdominal cavity just under the diaphragm. It serves as a pouch that food enters after it has been chewed, swallowed, and passed through the esophagus. The stomach looks small after it is emptied, not much bigger than a large sausage, but it expands considerably after a large meal. Have you ever felt so uncomfortably full after eating that you could not take a deep breath? If so, it probably meant that your stomach was so full of food that it occupied more space than usual and pushed up against the diaphragm. This made it hard for the diaphragm to contract and move downward as much as necessary for a deep breath.

Heartburn

Heartburn is often described as a burning sensation characterized by pain and a feeling of fullness beneath the sternum. It is a common problem caused by irritation of esophageal mucosa by acidic stomach contents that reenter the esophagus—a process called **reflux** (REE-fluhks). Even very small quantities of acidic stomach contents regurgitating back into the esophagus through the cardiac sphincter can cause discomfort and even inflammation. Total emptying of stomach contents back through the cardiac sphincter and ultimately up the esophagus and out of the mouth is called **vomiting** or **emesis** (EM-e-sis).

After food has entered the stomach by passing through the muscular **cardiac sphincter** (SFINGK-ter) at the end of the esophagus, the digestive process continues. Sphincters are rings of muscle tissue. The cardiac sphincter keeps food from reentering the esophagus when the stomach contracts.

Contraction of the stomach's muscular walls mixes the food thoroughly with the gastric juice and breaks it down into a semisolid mixture called **chyme** (KIME). Gastric juice contains hydrochloric acid and enzymes that function in the digestive process. Chyme formation is a continuation of the mechanical digestive process that begins in the mouth.

Note in Figure 14-8 that three layers of smooth muscle are in the stomach wall. The muscle fibers that run lengthwise, around, and obliquely make the stomach one of the strongest internal organs—well able to break up food into tiny particles and to mix them thoroughly with gastric juice to form chyme. Stomach muscle contractions result in **peristalsis,** which propels food down the digestive tract. Mucous membrane lines the stomach; it contains thousands of microscopic **gastric glands** that secrete gastric juice and hydrochloric acid into the stomach. When the stomach is empty, its lining lies in folds called **rugae.**

FIGURE 14–8 *Stomach.* A portion of the anterior wall has been cut away to reveal the muscle layers of the stomach wall. Notice that the mucosa lining the stomach forms folds called *rugae*.

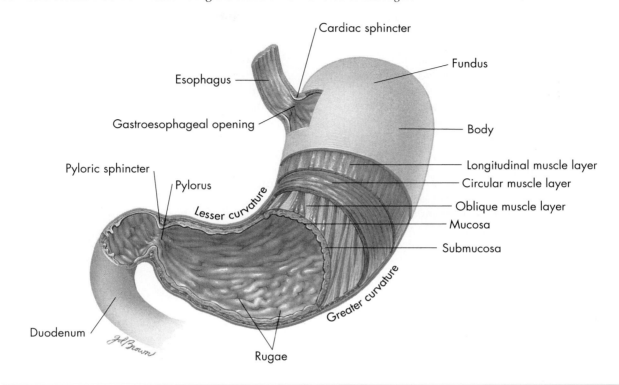

The three divisions of the stomach shown in Figure 14-8 are the **fundus, body,** and **pylorus** (pye-LOR-us). The fundus is the enlarged portion to the left of and above the opening of the esophagus into the stomach. The body is the central part of the stomach, and the pylorus is its lower narrow section, which joins the first part of the small intestine. Partial digestion occurs after food is held in the stomach by the **pyloric** (pi-LOR-ik) **sphincter** muscle. The smooth muscle fibers of the sphincter stay contracted most of the time and thereby close off the opening of the pylorus into the small intestine. Note also in Figure 14-8 that the upper right border of the stomach is known as the **lesser curvature,** and the lower left border is called the **greater curvature.** After food has been mixed in the stomach, chyme begins its passage through the pyloric sphincter into the first part of the small intestine.

Fractal Geometry of the Body

Biologists have just begun applying the principles of the new field of **fractal geometry** to human anatomy. Specialists in fractal geometry study surfaces with a seemingly infinite area, such as the lining of the small intestine. Fractal surfaces have bumps that have bumps that have bumps, and so on. The fractal nature of the intestinal lining is represented in Figure 14-9. The plicae (folds) have villi, the villi have microvilli, and even the microvilli have bumps that cannot be seen in the figure. Thus, the absorptive surface area of the small intestine is almost limitless.

Ulcers

An **ulcer** is a craterlike wound in a membrane caused by tissue destruction. The two most common sites for GI ulcers are the stomach (gastric ulcers) and the duodenum (duodenal ulcers). Both sites are regularly exposed to high levels of hydrochloric acid from the stomach. Excessive secretion of acid or **hyperacidity** is an important factor in the formation of ulcers. If the acid is not neutralized by other substances in chyme or stopped by the protective mucus, it destroys enough tissue to perforate the wall of the stomach or duodenum. This can lead to massive hemorrhage and widespread inflammation of the abdominal cavity and its contents. Even if the ulcer does not perforate the wall, episodes of moderate bleeding can eventually cause anemia. Until recently, drugs such as cimetidine (Tagamet) that reduce secretion of hydrochloric acid were a primary treatment. However, recent research shows that in many cases hyperacidity is only partly to blame. The basic cause seems to be a spiral-shaped bacterium called *Helicobacter pylori.* This bacterium burrows through the protective mucus lining the GI tract and impairs the lining's ability to produce more mucus. This opens the way for stomach acid to begin its ulcer-producing attack. Many ulcer treatments now center around antibiotics that inhibit the *H. pylori* bacterium.

 ## Small Intestine

The **small intestine** seems to be misnamed if you look at its length—it is roughly 7 meters (20 feet) long. However, it is noticeably smaller in diameter than the large intestine, so in this respect its name is appropriate. Different names identify different sections of the small intestine. In the order in which food passes through them, they are the **duodenum** (doo-o-DEE-num), **jejunum**, (je-JOO-num), and **ileum** (IL-ee-um).

The mucous lining of the small intestine, like that of the stomach, contains thousands of microscopic glands. These **intestinal glands** secrete the intestinal digestive juice. Another structural feature of the lining of the small intestine makes it especially well suited to absorption of food and water; it is not perfectly smooth, as it appears to the naked eye. Instead, the intestinal lining is arranged into multiple circular folds called **plicae** (PLYE-kee) (Figures 14-2 and 14-9). These folds are themselves covered with thousands of tiny "fingers" called **villi** (VILL-eye). Under the microscope, the villi can be seen projecting into the hollow interior of the intestine. Inside each villus lies a rich network of blood capillaries that absorb the products of carbohydrate and protein digestion (sugars and amino acids). Millions and millions of villi jut inward from the mucous lining. Imagine the lining as perfectly smooth without any villi; think how much less surface area there would be for contact between capillaries and intestinal lining. Consider what an advantage a large contact area offers for faster absorption of food from the intestine into the blood and lymph—one more illustration that structure and function are intimately related.

Note also in Figure 14-9 that each villus in the intestine contains a lymphatic vessel or **lacteal** that absorbs lipid or fat materials from the chyme passing through the small intestine. In addition to the thousands of villi that increase surface area in the small intestine, each villus is itself covered by epithelial cells, which have a brushlike border composed of **microvilli.** The microvilli further increase the surface area of each villus for absorption of nutrients.

FIGURE 14–9 *The Small Intestine.* Note that the folds of mucosa are covered with villi and that each villus is covered with epithelium, which increases the surface area for absorption of food.

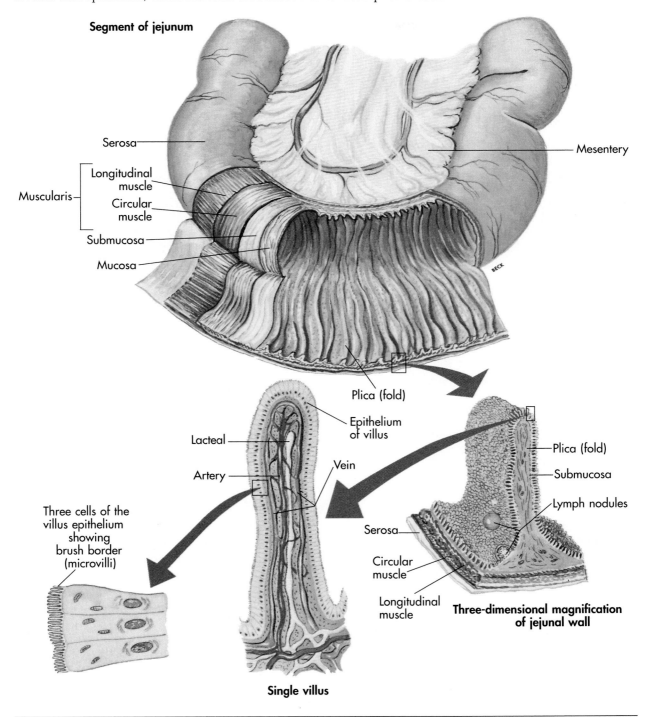

Segment of jejunum

Serosa

Mesentery

Muscularis — Longitudinal muscle

Circular muscle

Submucosa

Mucosa

Plica (fold)

Epithelium of villus

Lacteal

Artery

Vein

Three cells of the villus epithelium showing brush border (microvilli)

Serosa

Circular muscle

Longitudinal muscle

Plica (fold)

Submucosa

Lymph nodules

Three-dimensional magnification of jejunal wall

Single villus

FIGURE 14–10   *The Gallbladder and Bile Ducts.* Obstruction of the hepatic or common bile duct by stone or spasm blocks the exit of bile from the liver, where it is formed, and prevents bile from being ejected into the duodenum.

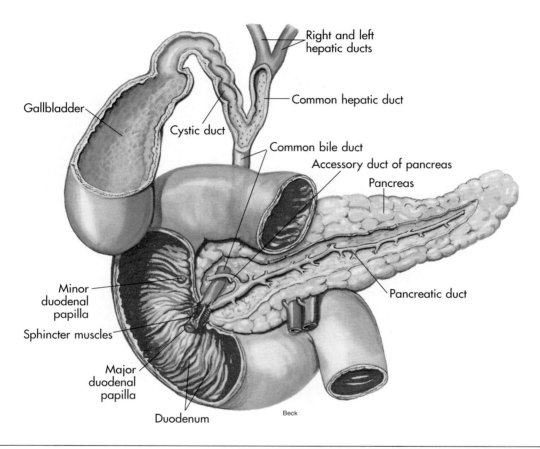

Right and left hepatic ducts

Common hepatic duct

Common bile duct

Accessory duct of pancreas

Pancreas

Gallbladder

Cystic duct

Pancreatic duct

Minor duodenal papilla

Sphincter muscles

Major duodenal papilla

Duodenum

Beck

Most of the chemical digestion occurs in the first subdivision of the small intestine or duodenum. The duodenum is C shaped (Figure 14-10) and curves around the head of the pancreas. The acid chyme enters the duodenum from the stomach. This area is the site of frequent ulceration (duodenal ulcers.) The middle third of the duodenum contains the openings of ducts that empty pancreatic digestive juice and bile from the liver into the small intestine. As you can see in Figure 14-10, the two openings are called the **minor** and **major duodenal papillae.** Occasionally a gallstone blocks the major duodenal papilla, causing symptoms such as severe pain, jaundice, and digestive problems. Smooth muscle in the wall of the small intestine contracts to produce peristalsis, the wavelike contraction that moves food through the tract.

◯ *Liver and Gallbladder*

The liver is so large that it fills the entire upper right section of the abdominal cavity and even extends partway into the left side. Because its cells secrete a substance called **bile** into ducts, the liver is classified as an exocrine gland; in fact, it is the largest gland in the body.

Exercise and Fluid Uptake

Replacement of fluids lost during exercise, primarily through sweating, is essential for maintaining homeostasis. Nearly everyone increases his or her intake of fluids during and after exercise. The main limitation to efficient fluid replacement is how quickly fluid can be absorbed, rather than how much a person drinks. Very little water is absorbed until it reaches the intestines, where it is absorbed almost immediately. Thus the rate of **gastric emptying** into the intestine is critical.

Large volumes of fluid leave the stomach and enter the intestines more rapidly than small volumes. However, large volumes may be uncomfortable during exercise. Cool fluids (8° to 13° C) empty more quickly than warm fluids. Fluids with a high solute concentration empty slowly and may cause nausea or stomach cramps. Thus large amounts of cool, dilute, or isotonic fluids are best for replacing fluids quickly during exercise.

The duration of exercise does not affect gastric emptying, but the intensity can. Strenuous exercise practically shuts down gastric emptying. Thus the harder you work, the harder it is to replace lost fluids.

Look again at Figure 14-10. First, identify the hepatic ducts. They drain bile out of the liver, a fact suggested by the name "hepatic," which comes from the Greek word for liver *(hepar).* Next, notice the duct that drains bile into the small intestine (duodenum), the common bile duct. It is formed by the union of the common hepatic duct with the cystic duct.

Because fats form large globules, they must be broken down into smaller particles to increase the surface area for digestion. This is the function of bile. It mechanically breaks up or **emulsifies** (e-MUL-se-fi) fats. When chyme containing lipid or fat enters the duodenum, it initiates a mechanism that contracts the gallbladder and forces bile into the small intestine. Fats in chyme stimulate or "trigger" the secretion of the hormone **cholecystokinin** (ko-le-sis-toe-KYE-nin) or **CCK** from the intestinal mucosa of the duodenum. This hormone then stimulates the contraction of the gallbladder, and bile flows into the duodenum. Between meals, a lot of bile moves up the cystic duct into the gallbladder on the undersurface of the liver. The gallbladder thus concentrates and stores bile produced in the liver.

Visualize a gallstone blocking the common bile duct shown in Figure 14-10. Bile could not then drain into the duodenum. Feces would then appear gray-white because the pigments from bile give feces its characteristic color. Furthermore, excessive amounts of bile would be absorbed into the blood. A yellowish skin discoloration called **jaundice** (JAWN-dis) would result. Obstruction of the common hepatic duct also leads to jaundice. Because bile cannot then drain out of the liver, excessive amounts of it are absorbed. Because bile is not resorbed from the gallbladder, no jaundice occurs if the cystic duct is blocked.

Pancreas

The pancreas lies behind the stomach in the concavity produced by the C shape of the duodenum. It is an exocrine gland that secretes pancreatic juice into ducts and an endocrine gland that secretes hormones into the blood. Pancreatic juice is the most important digestive juice. It contains enzymes that digest all three major kinds of foods. It also contains sodium bicarbonate, an alkaline substance that neutralizes the hydrochloric acid in the gastric juice that enters the intestines. Pancreatic juice enters the duodenum of the small intestine at the same place that bile enters. As you can see in Figure 14-10, the common bile and pancreatic ducts open into the duodenum at the major duodenal papilla.

FIGURE 14–11 *Horizontal (Transverse) Section of the Abdomen.*   The photograph of a cadaver section shows the relative position of some of the major digestive organs of the abdomen. Such a view is typical in newer imaging methods such as computed tomography (CT) scanning and magnetic resonance imaging (MRI).

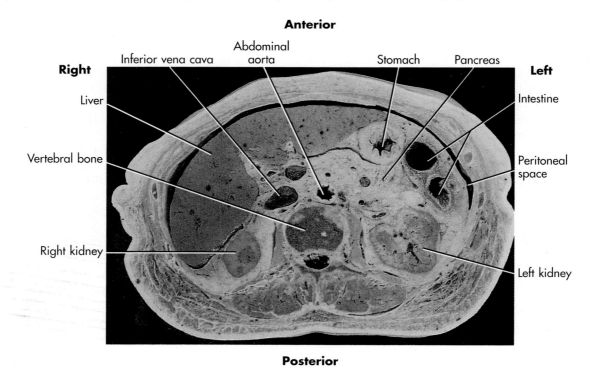

Between the cells that secrete pancreatic juice into ducts lie clusters of cells that have no contact with any ducts. These are the pancreatic islets (of Langerhans), which secrete the hormones of the pancreas described in Chapter 9. Locate the pancreas and nearby structures in Figure 14-11, which shows a transverse section of a human cadaver.

⬤ *Large Intestine*

The **large intestine** is only about 1.5 meters (5 feet) in length. As the name implies, it has a much larger diameter than the small intestine. It forms the lower or terminal portion of the digestive tract. Undigested and unabsorbed food material enters the large intestine after passing through a sphincterlike structure (Figure 14-12) called the **ileocecal** (il-ee-o-SEE-kal) **valve.** The

word *chyme* is no longer appropriate in describing the contents of the large intestine. Chyme, which has the consistency of soup and is found in the small intestine, changes to the consistency of fecal matter as water and salts are reabsorbed during its passage through the small intestine. During its movement through the large intestine, material that escaped digestion in the small intestine is acted upon by bacteria. As a result of this bacterial action, additional nutrients may be released from cellulose and other fibers and absorbed. In addition to their digestive role, bacteria in the large intestine have other important functions. They are responsible for the synthesis of vitamin K needed for normal blood clotting and for the production of some of the B-complex vitamins. Once formed, these vitamins are absorbed from the large intestine and enter the blood.

FIGURE 14–12 *Divisions of the Large Intestine.*

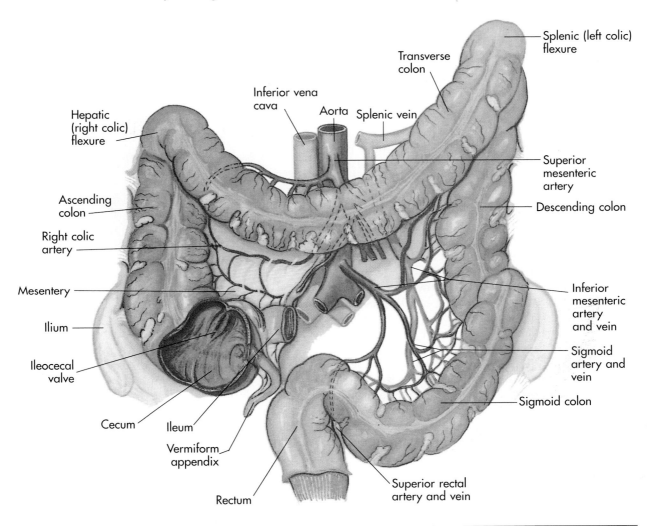

Although some absorption of water, salts, and vitamins occurs in the large intestine, this segment of the digestive tube is not as well suited for absorption as is the small intestine. Salts, especially sodium, are absorbed by active transport, and water is moved into the blood by osmosis. No villi are present in the mucosa of the large intestine. As a result, much less surface area is available for absorption, and the efficiency and speed of movement of substances through the wall of the large intestine is lower than in the small intestine. Normal passage of material through the large intestine takes about 3 to 5 days. If the rate of passage of material quickens, the consistency of the stools or fecal material becomes more and more fluid, and **diarrhea** (dye-ah-REE-ah) results. If the time of passage through the large intestine is prolonged beyond 5 days, the feces lose volume and becomes more solid because of excessive water absorption. This reduces stimulation of the bowel emptying reflex, resulting in retention of feces, a condition called **constipation.**

The subdivisions of the large intestine are listed below in the order in which food passes through them.

1. Cecum
2. Ascending colon
3. Transverse colon
4. Descending colon
5. Sigmoid colon
6. Rectum
7. Anal canal

These areas can be studied and identified by tracing the passage of material from its point of entry into the large intestine at the ileocecal valve to its elimination from the body through the external opening called the **anus.**

Note in Figure 14-12 that the ileocecal valve opens into a pouchlike area called the **cecum** (SEE-kum). The opening itself is about 5 or 6 cm (2 inches) above the beginning of the large intestine. Food residue in the cecum flows upward on the right side of the body in the **ascending colon.** The **hepatic** or **right colic flexure** is the bend between the ascending colon and the **transverse colon,** which extends across the front of the abdomen from right to left. The **splenic** or **left colic flexure** marks the point where the **descending colon** turns downward on the left side of the abdomen. The **sigmoid colon** is the S-shaped segment that terminates in the **rectum.** The terminal portion of the rectum is called the **anal canal,** which ends at the external opening or anus.

Two sphincter muscles stay contracted to keep the anus closed except during defecation. Smooth or involuntary muscle composes the **inner anal sphincter,** but striated, or voluntary, muscle composes the outer one. This anatomical fact sometimes becomes highly important from a practical standpoint. For example, often after a person has had a stroke, the voluntary anal sphincter at first becomes paralyzed. This means, of course, that the individual has no control at this time over bowel movements.

Appendix

The **vermiform appendix** (Latin *vermiformis* from *vermis* "worm" and *forma* "shape") is, as the name implies, a wormlike, tubular structure. Although

FIGURE 14–13 *The Large Intestine.* A special x-ray technique produces a clear image of the large intestine and its position relative to the skeleton.

it serves no important digestive function in humans, it contains lymphatic tissue and may play a minor role in the immunologic defense mechanisms of the body described in Chapter 12. Note in Figure 14-12 that the appendix is directly attached to the cecum. The appendix contains a blind, tubelike interior lumen that communicates with the lumen of the large intestine 3 cm (1 inch) below the opening of the ileocecal valve into the cecum. If the mucous lining of the appendix becomes inflamed, the resulting condition is the well-known affliction, **appendicitis.** As you can see in Figures 14-12 and 14-13, the appendix is very close to the rectal wall. For patients with suspected appendicitis, a physician often evaluates the appendix by a digital rectal examination.

Peritoneum

The **peritoneum** is a large, moist, slippery sheet of serous membrane that lines the abdominal cavity and covers the organs located in it, including most of the digestive organs. The parietal layer of the peritoneum lines the abdominal cavity. The

FIGURE 14–14   *The Peritoneum.* The parietal layer of the peritoneum lines the abdominopelvic cavity, and then extends as a series of mesenteries to form the visceral layer that covers abdominal organs.

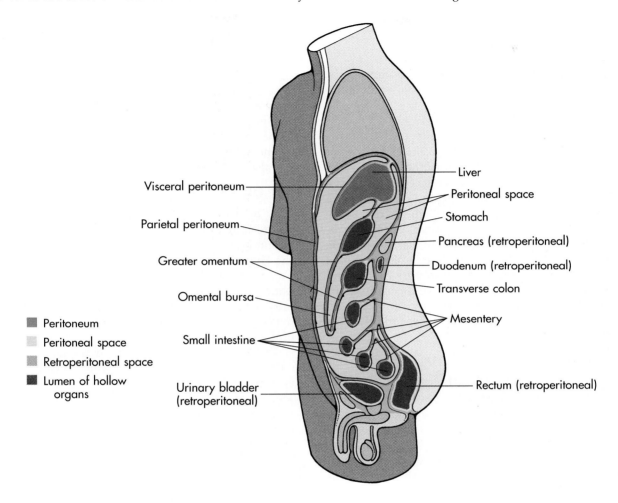

visceral layer of the peritoneum forms the outer or covering layer of each abdominal organ. The small space between the parietal and visceral layers is called the *peritoneal space*. It contains just enough peritoneal fluid to keep both layers of the peritoneum moist and able to slide freely against each other during breathing and digestive movements (Figure 14-14). Organs outside of the peritoneum are said to be retroperitoneal.

EXTENSIONS

The two most prominent extensions of the peritoneum are the mesentery and the greater omentum. The **mesentery,** an extension between the parietal and visceral layers of the peritoneum, is shaped like a giant, pleated fan. Its smaller edge attaches to the lumbar region of the posterior abdominal wall, and its long, loose outer edge encloses most of the small intestine, anchoring it to the posterior abdominal wall. The **greater omentum** is a pouchlike extension of the visceral peritoneum from the lower edge of the stomach, part of the duodenum, and the transverse colon. Shaped like a large apron, it hangs down over the intestines, and because spotty deposits of fat give it a lacy appearance, it has been nicknamed the *lace apron.* It usually envelops a badly inflamed appendix, walling it off from the rest of the abdominal organs.

Digestion

Digestion, a complex process that occurs in the alimentary canal, consists of physical and chemical changes that prepare food for absorption. **Mechanical digestion** breaks food into tiny particles, mixes them with digestive juices, moves them along the alimentary canal, and finally eliminates the digestive wastes from the body. Chewing or mastication, swallowing or **deglutition** (deg-loo-TISH-un), peristalsis, and defecation are the main processes of mechanical digestion. **Chemical digestion** breaks down large, nonabsorbable food molecules into smaller, absorbable molecules—molecules that are able to pass through the intestinal mucosa into blood and lymph. Chemical digestion consists of numerous chemical reactions catalyzed by enzymes in saliva, gastric juice, pancreatic juice, and intestinal juice.

CARBOHYDRATE DIGESTION

Very little digestion of carbohydrates (starches and sugars) occurs before food reaches the small intestine. Salivary amylase usually has little time to do its work because so many of us swallow our food so fast. Gastric juice contains no carbohydrate-digesting enzymes. But after the food reaches the small intestine, pancreatic and intestinal juice enzymes digest the starches and sugars. A pancreatic enzyme (amylase) starts the process by changing starches into a double sugar, namely, maltose. Three intestinal enzymes—maltase, sucrase, and lactase—digest double sugars by changing them into simple sugars, chiefly glucose (dextrose). Maltase digests maltose (malt sugar), sucrase digest sucrose (ordinary cane sugar), and lactase digests lactose (milk sugar). The end products of carbohydrate digestion are the so-called simple sugars; the most abundant is glucose.

PROTEIN DIGESTION

Protein digestion starts in the stomach. Two enzymes (rennin and pepsin) in the gastric juice cause the giant protein molecules to break up into somewhat simpler compounds. Pepsinogen, a component of gastric juice, is converted into active pepsin enzyme by hydrochloric acid (also in gastric juice). In the intestine, other enzymes (trypsin in the pancreatic juice and peptidases in the intestinal juice) finish the job of protein digestion. Every protein molecule is made up of many amino acids joined together. When enzymes have split up the large protein molecule into its separate amino acids, protein digestion is completed. Hence the end product of protein digestion is amino acids. For obvious reasons, the amino acids are also referred to as *protein building blocks.*

FAT DIGESTION

Very little carbohydrate and fat digestion occurs before food reaches the small intestine. Most fats are undigested until after emulsification by bile in the duodenum (that is, fat droplets are broken into very small droplets). After this takes place, pancreatic lipase splits up the fat molecules into fatty acids and glycerol (glycerin). The end products of fat digestion, then, are fatty acids and glycerol.

Table 14-2 summarizes the main facts about chemical digestion. Enzyme names indicate the type of food digested by the enzyme. For example, the name *amylase* indicates that the enzyme digests carbohydrates (starches and sugars), *protease* indicates a protein-digesting enzyme, and *lipase* means a fat-digesting enzyme. When carbohydrate digestion has been completed, starches (polysaccharides) and double sugars (disaccharides) have been changed mainly to glucose, a simple sugar (monosaccharide). The end products of protein digestion, on the other hand, are amino acids. Fatty acid and glycerol are the end products of fat digestion.

Absorption

After food is digested, it is absorbed; that is, it moves through the mucous membrane lining of the small intestine into the blood and lymph. In other words, food absorption is the process by which molecules of amino acids, glucose, fatty acids, and glycerol go from the inside of the

TABLE 14–2 Chemical Digestion

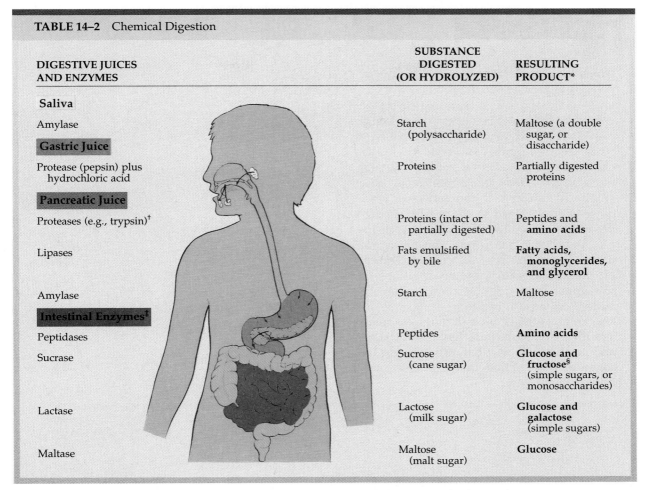

DIGESTIVE JUICES AND ENZYMES	SUBSTANCE DIGESTED (OR HYDROLYZED)	RESULTING PRODUCT*
Saliva		
Amylase	Starch (polysaccharide)	Maltose (a double sugar, or disaccharide)
Gastric Juice		
Protease (pepsin) plus hydrochloric acid	Proteins	Partially digested proteins
Pancreatic Juice		
Proteases (e.g., trypsin)†	Proteins (intact or partially digested)	Peptides and **amino acids**
Lipases	Fats emulsified by bile	**Fatty acids, monoglycerides, and glycerol**
Amylase	Starch	Maltose
Intestinal Enzymes‡		
Peptidases	Peptides	**Amino acids**
Sucrase	Sucrose (cane sugar)	**Glucose and fructose§** (simple sugars, or monosaccharides)
Lactase	Lactose (milk sugar)	**Glucose and galactose** (simple sugars)
Maltase	Maltose (malt sugar)	**Glucose**

*Substances in boldface type are end products of digestion (that is, completely digested nutrients ready for absorption).
†Secreted in inactive form (trypsinogen); activated by enterokinase, an enzyme in the intestinal brush order.
‡Brush-border enzymes.
§Glucose is also called *dextrose;* fructose is also called *levulose.*

intestines into the circulating fluids of the body. Absorption of foods is just as essential as digestion of foods. The reason is fairly obvious. As long as food stays in the intestines, it cannot nourish the millions of cells that compose all other parts of the body. Their lives depend on the absorption of digested food and its transportation to them by the circulating blood.

Structural adaptations of the digestive tube, including folds in the lining mucosa, villi, and microvilli, increase the absorptive surface and the efficiency and speed of transfer of materials from the intestinal lumen to body fluids. Many salts such as sodium are actively transported through the intestinal mucosa. Water follows by osmosis. Other nutrients are also actively transported into the blood of capillaries in the intestinal villi. Fats enter the lymphatic vessels or lacteals found in intestinal villi.

Colostomy

Colostomy (ko-LAH-sto-me) is a surgical procedure in which an artificial anus is created on the abdominal wall by cutting the colon and bringing the cut end or ends out to the surface to form an opening called a *stoma* (see the figure). In the home health care situation, patients are helped to accept the change in body image that may cause emotional discomfort. The patient or caregiver is also trained in the regular changing of the disposable bag, including cleaning of the stoma and preventing of irritation, chapping, or infection. Irrigation of the colon with isotonic solutions is sometimes necessary. Deodorants may be added to the fresh bag to prevent unpleasant odors. Patients are also taught to manage their diet to include low-residue food and to avoid foods that produce gas or cause diarrhea. Fluid intake is also carefully managed.

Stoma

External collection bag

WALL OF THE DIGESTIVE TRACT (Figure 14-2)

The wall of the digestive tube is formed by four layers of tissue:

A Mucosa—mucous epithelium

B Submucosa—connective tissue

C Muscularis—two or three layers of smooth muscle

D Serosa—serous membrane that covers the outside of abdominal organs; it attaches the digestive tract to the wall of the abdominopelvic cavity by forming folds called *mesenteries*

MOUTH

A Roof—formed by hard palate (parts of maxillary and palatine bones) and soft palate, an arch-shaped muscle separating mouth from pharynx; uvula, a downward projection of soft palate (Figure 14-3)

B Floor—formed by tongue and its muscles; papillae, small elevations on mucosa of tongue; taste buds, found in many papillae; lingual frenulum, fold of mucous membrane that helps anchor tongue to floor of mouth (Figure 14-4)

TEETH

A Names of teeth—incisors, cuspids, bicuspids, and tricuspids

B Twenty teeth in temporary set; average age for cutting first tooth about 6 months; set complete at about 2 years of age

C Thirty-two teeth in permanent set; 6 years about average age for starting to cut first permanent tooth; set complete usually between ages of 17 and 24 years (Figure 14-5)

D Structures of a typical tooth—crown, neck, and root (Figure 14-6)

SALIVARY GLANDS (Figure 14-7)

A Parotid glands

B Submandibular glands

C Sublingual glands

PHARYNX

ESOPHAGUS

STOMACH (Figure 14-8)

A Size—expands after large meal; about size of large sausage when empty

B Pylorus—lower part of stomach; pyloric sphincter muscle closes opening of pylorus into duodenum

C Wall—many smooth muscle fibers; contractions produce churning movements (peristalsis)

D Lining—mucous membrane; many microscopic glands that secrete gastric juice and hydrochloric acid into stomach; mucous membrane lies in folds (rugae) when stomach is empty

SMALL INTESTINE (Figure 14-9)

A Size—about 7 meters (20 feet) long but only 2 cm or so in diameter

B Divisions

1 Duodenum

2 Jejunum

3 Ileum

C Wall—contains smooth muscle fibers that contract to produce peristalsis

D Lining—mucous membrane; many microscopic glands (intestinal glands) secrete intestinal juice; villi (microscopic finger-shaped projections from surface of mucosa into intestinal cavity) contain blood and lymph capillaries

LIVER AND GALLBLADDER

A Size and location—liver is largest gland; fills upper right section of abdominal cavity and extends over into left side

B Liver secretes bile

C Ducts (Figure 14-10)

1 Hepatic—drains bile from liver

2 Cystic—duct by which bile enters and leaves gallbladder

3 Common bile—formed by union of hepatic and cystic ducts; drains bile from hepatic or cystic ducts into duodenum

D Gallbladder

1 Location—undersurface of the liver

2 Function—concentrates and stores bile produced in the liver

PANCREAS

A Location—behind stomach

B Functions

1 Pancreatic cells secrete pancreatic juice into pancreatic ducts; main duct empties into duodenum

2 Pancreatic islets (of Langerhans)—cells not connected with pancreatic ducts; secrete hormones glucagon and insulin into the blood

LARGE INTESTINE (Figure 14-12)

A Divisions

1 Cecum

2 Colon—ascending, transverse, descending, and sigmoid

3 Rectum

B Opening to exterior—anus

C Wall—contains smooth muscle fibers that contract to produce churning, peristalsis, and defecation

D Lining—mucous membrane

APPENDIX

Blind tube off cecum; no important digestive functions in humans

PERITONEUM (Figure 14-14)

A Definitions—peritoneum, serous membrane lining abdominal cavity and covering abdominal organs; parietal layer of peritoneum lines abdominal cavity; visceral layer of peritoneum covers abdominal organs; peritoneal space lies between parietal and visceral layers

B Extensions—largest ones are the mesentery and greater omentum; mesentery is extension of parietal peritoneum, which attaches most of small intestine to posterior abdominal wall; greater omentum, or "lace apron," hangs down from lower edge of stomach and transverse colon over intestines

DIGESTION (Table 14-2)

Meaning—changing foods so that they can be absorbed and used by cells

A Mechanical digestion—chewing, swallowing, and peristalsis break food into tiny particles, mix them well with digestive juices, and move them along the digestive tract

B Chemical digestion—breaks up large food molecules into compounds having smaller molecules; brought about by digestive enzymes

C Carbohydrate digestion—mainly in small intestine
  1 Pancreatic amylase—changes starches to maltose
  2 Intestinal juice enzymes
    a Maltase—changes maltose to glucose
    b Sucrase—changes sucrose to glucose
    c Lactase—changes lactose to glucose

D Protein digestion—starts in stomach; completed in small intestine
  1 Gastric juice enzymes, rennin and pepsin, partially digest proteins
  2 Pancreatic enzyme, trypsin, completes digestion of proteins to amino acids
  3 Intestinal enzymes, peptidases, complete digestion of partially digested proteins to amino acids

E Fat digestion
  1 Bile contains no enzymes but emulsifies fats (breaks fat droplets into very small droplets)
  2 Pancreatic lipase changes emulsified fats to fatty acids and glycerol in small intestine

ABSORPTION

A Meaning—digested food moves from intestine into blood or lymph

B Where absorption occurs—foods and most water from small intestine; some water also absorbed from large intestine

---

N E W   W O R D S

absorption	diarrhea	mastication	plica
alimentary canal	digestion	mesentery	rugae
appendicitis	emesis	papilla	ulcer
bolus	feces	peristalsis	uvula
cavity	frenulum	peritoneum	villus
chyme	jaundice		
	lumen		

C H A P T E R   T E S T

1. In the digestive system the breakdown of food material is _____ and _____ in nature.
2. Undigested waste material resulting from the digestive process is called _____.
3. The liver and gallbladder are classified as _____ organs of digestion.
4. The hollow space within the digestive tube is called the _____.
5. The inside or lining coat of the digestive tube is called the _____.
6. The roof of the mouth is formed by the hard and soft _____.
7. The portion of a tooth that is exposed and visible is called the _____.
8. The teeth serve cutting and grinding functions during the chewing of food—a process called _____.
9. Saliva contains the enzyme called salivary _____.
10. Food moves from the pharynx to the stomach by passing through the _____.
11. The semisolid mixture of food and gastric juice in the stomach is called _____.
12. The movement of food through the digestive tract results from contractions called _____.
13. The middle segment of the small intestine is called the _____.

Select the most correct answer from Column B for each statement in Column A. (Only one answer is correct.)

COLUMN A

14. _____ Waste material
15. _____ Accessory organ of digestion
16. _____ Double fold of peritoneum
17. _____ Hardest tissue in body
18. _____ Semisolid mixture
19. _____ Division of stomach
20. _____ Yellowish skin condition
21. _____ Digestive enzyme

COLUMN B

a. Liver
b. Enamel
c. Fundus
d. Feces
e. Jaundice
f. Mesentery
g. Lipase
h. Chyme

Circle the T before each true statement and the F before each false statement.

T F 22. The salivary glands are considered accessory organs of digestion.

T F 23. The ileum is the portion of the small intestine found between the duodenum and jejunum.

T F 24. The large intestine is classified as an accessory organ of digestion.

T F 25. The serosa is the outermost coat of the digestive tube.

T F 26. The uvula is attached to the soft palate.

T F 27. In humans there are 32 deciduous (baby) teeth and 20 permanent teeth.

T F 28. No chemical digestion can occur in the mouth.

T F 29. The pyloric sphincter is located between the esophagus and stomach.

T F 30. Plica, villi, and microvilli all increase the surface area of the small intestine.

REVIEW QUESTIONS

1. What organs form the gastrointestinal tract?
2. Identify the jejunum, cecum, colon, duodenum, and ileum.
3. How many teeth should an adult have?
4. Identify the pancreatic islets, parotid glands, pylorus, rugae, and villi.
5. In what organ does the digestion of starches begin?
6. What digestive juice contains no enzymes?
7. Only one digestive juice contains enzymes for digesting all three kinds of food. Which juice is this? In what organ does it do its work?
8. Which digestive juice emulsifies fats?
9. What three digestive juices act on foods in the small intestine?
10. What juices digest carbohydrates? Proteins? Fats?
11. Where are simple sugars and amino acids absorbed into blood capillaries? Where are lipids absorbed into lacteals?
12. Where is most of the water absorbed from the lumen of the digestive tract?

CRITICAL THINKING

13. If you inserted 9 inches of an enema tube through the anus, the tip of the tube would probably be in what structure?
14. How many teeth should a child 2½ years old have? Would he have some of each of the following teeth: incisors, canines, premolars, and molars? If not, which ones would he not have?
15. What kinds of food are not digested in the stomach?

Nutrition and Metabolism

15

OBJECTIVES

*After you have completed this chapter,
you should be able to:*

1. Define and contrast *catabolism* and *anabolism*.
2. Describe the metabolic roles of carbohydrates, fats, proteins, vitamins, and minerals.
3. Define basal metabolic rate and list some factors that affect it.
4. Discuss the physiological mechanisms that regulate body temperature.

*N*utrition and *metabolism* are words that are often used together—but what do they mean? *Nutrition* is a term that refers to the food (nutrients) that we eat. Proper nutrition requires a balance of the three basic food types, *carbohydrates*, *fats*, and *proteins*, plus essential *vitamins* and *minerals*. Malnutrition is a deficiency or imbalance in the consumption of food, vitamins, and minerals.

A good phrase to remember in connection with the word metabolism is "use of foods" because basically this is what metabolism is—the use the body makes of foods after they have been digested, absorbed, and circulated to cells. It uses them in two ways: as an energy source and as building blocks for making complex chemical compounds. Before they can be used in these two ways, foods have to be *assimilated*. Assimilation occurs when food molecules enter cells and undergo many chemical changes there. All the chemical reactions that release energy from food molecules make up the process of catabolism, a vital process because it is the only way that the body has of supplying itself with energy for doing any work. The many chemical reactions that build food molecules into more complex chemical compounds constitute the process of anabolism. Catabolism and anabolism make up the process of metabolism.

This chapter explores many of the basic ideas about why certain nutrients are necessary for survival and how they are used by the body.

Role of the Liver

As we discussed in Chapter 14, the liver plays an important role in the mechanical digestion of lipids because it secretes *bile*. As you recall, bile breaks large fat globules into smaller droplets of fat that are more easily broken down. In addition, liver cells perform other functions necessary for healthy survival. They play a major role in the metabolism of all three kinds of foods. They help maintain a normal blood glucose concentration by carrying on complex and essential chemical reactions. Liver cells also carry on the first steps of protein and fat metabolism and synthesize several kinds of protein compounds. They release them into the blood, where they are called the *blood pro-*

teins or *plasma proteins*. Prothrombin and fibrinogen, two of the plasma proteins formed by liver cells, play essential parts in blood clotting (see pp. 220 and 221). Another protein made by liver cells, albumin, helps maintain normal blood volume. Liver cells detoxify various poisonous substances such as bacterial products and certain drugs. Liver cells store several substances, notably iron and vitamins A and D.

The liver is assisted by an interesting structural feature of the blood vessels that supply it. As you may recall from Chapter 11, the hepatic portal vein delivers blood directly from the gastrointestinal tract to the liver (see Figure 11-12). This arrangement allows blood that has just absorbed nutrients and other substances to be processed by the liver before being distributed throughout the body. Thus excess nutrients and vitamins can be stored and toxins can be removed from the bloodstream.

Nutrient Metabolism

CARBOHYDRATE METABOLISM

Carbohydrates are the preferred energy food of the body. They are composed of smaller "building blocks"—primarily *glucose*. (See Appendix A). Human cells catabolize (break down) glucose rather than other substances as long as enough glucose enters them to supply their energy needs. Three series of chemical reactions, occurring in a precise sequence, make up the process of glucose catabolism. **Glycolysis** (glye-KOL-i-sis) is the name given the first series of reactions; **citric acid cycle** is the name of the second series, and **electron transfer system** is the third. Glycolysis, as Figure 15-1 shows, changes glucose to pyruvic acid. The citric acid cycle changes the pyruvic acid to carbon dioxide. Glycolysis takes place in the cytoplasm of a cell, whereas the citric acid cycle goes on in the mitochondria, the cell's miniature power plants. Glycolysis uses no oxygen; it is an **anaerobic** (an-er-O-bik) process. The citric acid cycle, in contrast, is an oxygen-using or **aerobic** (aer-O-bik) process.

While the chemical reactions of glycolysis and the citric acid cycle occur, energy stored in the glucose molecule is being released. Over half the released energy is in the form of high-energy elec-

FIGURE 15–1   *Catabolism of Glucose.* Glycolysis splits one molecule of glucose (six carbon atoms) into two molecules of pyruvic acid (three carbon atoms each). The citric acid cycle converts each pyruvic acid molecule into three carbon dioxide molecules (one carbon atom each).

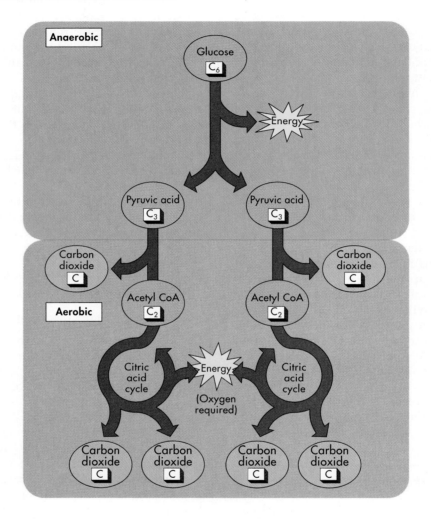

trons. The electron transport system, located in the mitochondria, almost immediately transfers the energy to molecules of adenosine triphosphate (ATP). The rest of the energy originally stored in the glucose molecule is released as heat. ATP serves as the direct source of energy for doing cellular work in all kinds of living organisms from one-cell plants to billion-cell animals, including man. Among biological compounds, therefore, ATP ranks as one of the most important. The energy transferred to ATP molecules differs in two ways from the energy stored in food molecules; the energy in ATP molecules is not stored but is released almost instantaneously, and it can be used directly to do cellular work. Release of energy from food molecules occurs much more slowly because it accompanies the long series of chemical reactions that make up the process of catabolism. Energy released from food molecules cannot be used directly for doing cellular work. It must first be transferred to ATP molecules and be released explosively from them.

FIGURE 15–2 *ATP.* **A,** The structure of ATP. A single adenosine group *(A)* has three attached phosphate groups *(P)*. The high-energy bonds between the phosphate groups can release chemical energy to do cellular work. **B,** ATP energy cycle. ATP stores energy in its last high-energy phosphate bond. When that bond is later broken, energy is released to do cellular work. The ADP and phosphate groups that result can be resynthesized into ATP capturing additional energy from nutrient catabolism.

As Figure 15-2 shows, ATP comprises an adenosine group and three phosphate groups. The capacity of ATP to store large amounts of energy is found in the high-energy bonds that hold the phosphate groups together, illustrated as curvy lines. When a phosphate group breaks off the molecule, an adenosine diphosphate (ADP) molecule and free phosphate group result. Energy that had been holding the phosphate bond together is freed to do cellular work (muscle fiber contractions, for example). As you can see in Figure 15-2, the ADP and phosphate are reunited by the energy produced by carbohydrate catabolism, making ATP a reusable energy-storage molecule. Only enough ATP for immediate cellular requirements is made at any one time. Glucose that is not needed is anabolized into larger molecules that are stored for later use.

Glucose anabolism is called **glycogenesis** (glye-ko-JEN-e-sis). Carried on chiefly by liver and muscle cells, glycogenesis consists of a series of reactions that join glucose molecules together, like many beads in a necklace, to form *glycogen,* a compound sometimes called *animal starch.*

Something worth noticing is that the amount of nutrients in the blood normally does not change very much, not even when we go without food for many hours, when we exercise and use a lot of food for energy, or when we sleep and use little food for energy. The amount of glucose in our blood, for example, usually stays at about 80 to 120 mg in 100 ml of blood.

Several hormones help regulate carbohydrate metabolism to keep blood glucose normal. **Insulin** is one of the most important of these. It acts in some way not yet definitely known to make glucose leave the blood and enter the cells at a more rapid rate. As insulin secretion increases, more glucose leaves the blood and enters the cells. The amount of glucose in the blood therefore decreases as the rate of glucose metabolism in cells increases (see p. 207). Too little insulin secretion, such as that which occurs with diabetes mellitus, produces the opposite effects. Less glucose leaves the blood and enters the cells; more glucose therefore remains in the blood, and less glucose is metabolized by cells. In other words, high blood glucose (hyperglycemia)

Carbohydrate Loading

A number of athletes and others who must occasionally sustain endurance exercise for a significant period practice **carbohydrate loading** or **glycogen loading.** Like liver cells, some skeletal muscle fibers can take up and store glucose in the form of glycogen. By ceasing intense exercise and switching to a diet high in carbohydrates 2 or 3 days before an endurance event, an athlete can cause the skeletal muscles to store almost twice as much glycogen as usual. This allows the muscles to sustain aerobic exercise for up to 50% longer than usual.

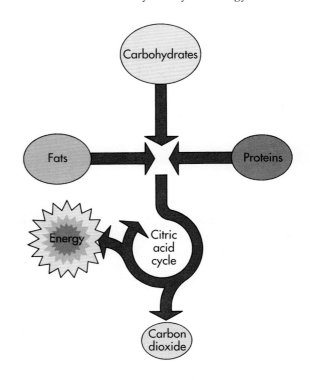

FIGURE 15–3 *Catabolism of Nutrients.* Fats, carbohydrates, and proteins can be converted to products that enter the citric acid cycle to yield energy.

and a low rate of glucose metabolism characterize insulin deficiency. Insulin is the only hormone that lowers the blood glucose level. Several other hormones, on the other hand, increase it. Growth hormone secreted by the anterior pituitary gland, hydrocortisone secreted by the adrenal cortex, epinephrine secreted by the adrenal medulla, and glucagon secreted by the pancreatic islets are four of the most important hormones that increase blood glucose. More information about these hormones appears in Chapter 9.

FAT METABOLISM

Fats, like carbohydrates, are primarily energy foods. If cells have inadequate amounts of glucose to catabolize, they immediately shift to the catabolism of fats for energy. They are simply converted into a form of glucose that can enter the citric acid cycle. This happens normally when a person goes without food for many hours. It happens abnormally in diabetic individuals. Because of an insulin deficiency, too little glucose enters the cells of a diabetic person to supply all energy needs. Result? The cells catabolize fats to make up the difference (Figure 15-3). In all persons, fats not needed for catabolism are anabolized to form tryglycerides and stored in adipose tissue.

PROTEIN METABOLISM

In a healthy person, proteins are catabolized to release energy to a very small extent. When fat reserves are low, as they are in the starvation that accompanies certain eating disorders such as anorexia nervosa, the body can start to use its protein molecules as an energy source. Specifically, the amino acids that make up proteins are each broken apart to yield an amine group that is converted to a form of glucose that can enter the citric acid cycle. After a shift to reliance on protein catabolism as a major energy source occurs, death may quickly follow because vital proteins in the muscles and nerves are catabolized (Figure 15-3).

A more common situation in normal bodies is protein anabolism, the process by which the body builds amino acids into complex protein compounds (for example, enzymes and proteins that form the structure of the cell). Proteins are assembled from a pool of 20 different kinds of amino acids. If any one type of amino acid is deficient, vital proteins cannot be synthesized—a serious health threat. One way your body maintains a constant supply of amino acids is by making them from other compounds already present in the body. Only about half of the required 20 types of amino acids can be made by the body, however. The remaining types of amino acids must be supplied in the diet. **Essential amino acids** are those that must be in the diet. **Nonessential amino acids** can be missing from the diet because they can be made by the body. See Table 15-1.

Vitamins and Minerals

One glance at the label of any packaged food product reveals the importance we place on vitamins and minerals. We know that carbohydrates, fats, and proteins are used by our bodies to build important molecules and to provide energy. So why do we need vitamins and minerals?

First, let's discuss the importance of vitamins. Vitamins are organic molecules needed in small quantities for normal metabolism throughout the body. Vitamin molecules attach to enzymes and help them work properly. Many enzymes are totally useless without the appropriate vitamins to activate them. Most vitamins cannot be made by the body, so we must eat them in our food. The body can store fat-soluble vitamins—A, D, E, and K—in the liver for later use. Because the body cannot store water-soluble vitamins such as B vitamins and vitamin C, they must be continually supplied in the diet. Vitamin deficiencies can lead to severe metabolic problems. Table 15-2 lists some of the more well-known vitamins, their sources, functions, and symptoms of deficiency.

Minerals are just as important as vitamins. Minerals are inorganic elements or salts found naturally in the earth. Like vitamins, mineral ions can attach to enzymes and help them work. Minerals also function in a variety of other vital chemical reactions. For example, sodium, calcium, and other minerals are required for nerve conduction and for contraction in muscle fibers. Without these minerals, the brain, heart, and respiratory tract would cease to function. Information about some of the more important minerals is summarized in Table 15-3.

TABLE 15–1 Amino Acids	
ESSENTIAL (INDISPENSABLE)	**NONESSENTIAL (DISPENSABLE)**
Histidine*	Alanine
Isoleucine	Arginine
Leucine	Asparagine
Lysine	Aspartic acid
Methionine	Cysteine
Phenylalanine	Glutamic acid
Threonine	Glutamine
Tryptophan	Glycine
Valine	Proline
	Serine
	Tyrosine†

*Essential in infants and, perhaps, adult males.
†Can be synthesized from phenylalanine; therefore is nonessential as long as phenylalanine is in the diet.

Vitamin Supplements for Athletes

Because a deficiency of vitamins (**avitaminosis**) can cause poor athletic performance, many athletes regularly consume vitamin supplements. However, research suggests that vitamin supplementation has little or no effect on athletic performance. A reasonably well-balanced diet supplies more than enough vitamins for even the elite athlete. The use of vitamin supplements therefore has fueled somewhat of a controversy among exercise experts. Opponents of vitamin supplements cite the cost and the possibility of liver damage associated with some forms of **hypervitaminosis,** whereas supporters cite the benefit of protecting against vitamin deficiency.

TABLE 15–2   Major Vitamins

VITAMIN	DIETARY SOURCE	FUNCTIONS	SYMPTOMS OF DEFICIENCY
Vitamin A	Green and yellow vegetables, dairy products, and liver	Maintains epithelial tissue and produces visual pigments	Night blindness and flaking skin
B-complex vitamins			
B_1 (thiamine)	Grains, meat, and legumes	Helps enzymes in the citric acid cycle	Nerve problems (beriberi), heart muscle weakness, and edema
B_2 (riboflavin)	Green vegetables, organ meats, eggs, and dairy products	Aids enzymes in the citric acid cycle	Inflammation of skin and eyes
B_3 (niacin)	Meat and grains	Helps enzymes in the citric acid cycle	Pellagra (scaly dermatitis and mental disturbances) and nervous disorders
B_5 (pantothenic acid)	Organ meat, eggs, and liver	Aids enzymes that connect fat and carbohydrate metabolism	Loss of coordination (rare)
B_6 (pyridoxine)	Vegetable, meats, and grains	Helps enzymes that catabolize amino acids	Convulsions, irritability, and anemia
B_{12} (cyanocobalamin)	Meat and dairy products	Involved in blood production and other processes	Pernicious anemia
Biotin	Vegetables, meat, and eggs	Helps enzymes in amino acid catabolism and fat and glycogen synthesis	Mental and muscle problems (rare)
Folic acid	Vegetables	Aids enzymes in amino acid catabolism and blood production	Digestive disorders and anemia
Vitamin C (ascorbic acid)	Fruits and green vegetables	Helps in manufacture of collagen fibers	Scurvy and degeneration of skin, bone, and blood vessels
Vitamin D (calciferol)	Dairy products and fish liver oil	Aids in calcium absorption	Rickets and skeletal deformity
Vitamin E (tocopherol)	Green vegetable and seeds	Protects cell membranes from being catabolized	Muscle and reproductive disorders (rare)

Metabolic Rates

The **basal metabolic rate (BMR)** is the rate at which food is catabolized under basal conditions (that is, when the individual is resting but awake, is not digesting food, and is not adjusting to a cold external temperature). Or, stated differently, the BMR is the number of calories of heat that must be produced per hour by catabolism just to keep the body alive, awake, and comfortably warm. To provide energy for muscular work and digestion and absorption of food, an additional amount of food must be catabolized. The amount of additional food depends mainly on how much

work the individual does. The more active he or she is, the more food the body must catabolize and the higher the total metabolic rate will be. The **total metabolic rate (TMR)** is the total amount of energy used by the body per day (Figure 15-4).

When the number of calories in your food intake equals your TMR, your weight remains constant (except for possible variations resulting from water retention or water loss). When your food intake provides more calories than your TMR, you gain weight; when your food intake

TABLE 15–3   Major Minerals

MINERAL	DIETARY SOURCE	FUNCTIONS	SYMPTOMS OF DEFICIENCY
Calcium (Ca)	Dairy products, legumes, and vegetables	Helps blood clotting, bone formation, and nerve and muscle function	Bone degeneration and nerve and muscle malfunction
Chlorine (Cl)	Salty foods	Aids in stomach acid production and acid-base balance	Acid-base imbalance
Cobalt (Co)	Meat	Helps vitamin B_{12} in blood cell production	Pernicious anemia
Copper (Cu)	Seafood, organ meats, and legumes	Involved in extracting energy from the citric acid cycle and in blood production	Fatigue and anemia
Iodine (I)	Seafood and iodized salt	Aids in thyroid hormone synthesis	Goiter (thyroid enlargement) and decrease of metabolic rate
Iron (Fe)	Meat, eggs, vegetables, and legumes	Involved in extracting energy from the citric acid cycle and in blood production	Fatigue and anemia
Magnesium (Mg)	Vegetables and grains	Helps many enzymes	Nerve disorders, blood vessel dilation, and heart rhythm problems
Manganese (Mn)	Vegetables, legumes, and grains	Helps many enzymes	Muscle and nerve disorders
Phosphorus (P)	Dairy products and meat	Aids in bone formation and is used to make ATP, DNA, RNA, and phospholipids	Bone degeneration and metabolic problems
Potassium (K)	Seafood, milk, fruit, and meats	Helps muscle and nerve function	Muscle weakness, heart problems, and nerve problems
Sodium (Na)	Salty foods	Aids in muscle and nerve function and fluid balance	Weakness and digestive upset
Zinc (Zn)	Many foods	Helps many enzymes	Metabolic problems

FIGURE 15–4   *Factors That Determine the Basal and Total Metabolic Rates.*

Measuring Energy

Physiologists studying metabolism must be able to express a quantity of energy in mathematical terms. The unit of energy measurement most often used is the calorie (cal). A calorie is the amount of energy needed to raise the temperature of 1 g of water 1° C. Because physiologists often deal with very large amounts of energy, the larger unit, *kilocalorie* (kcal) or *Calorie* (notice the upper-case C), is used. There are 1000 cal in 1 kcal or Calorie. Nutritionists prefer to use *Calorie* when they express the amount of energy stored in a food.

provides fewer calories than your TMR, you lose weight. These weight control principles rarely fail to operate. Nature does not forget to count calories. Reducing diets make use of this knowledge. They contain fewer calories than the TMR of the individual eating the diet.

 Body Temperature

Considering the fact that over 60% of the energy released from food molecules during catabolism is converted to heat rather than being transferred to ATP, it is no wonder that maintaining a constant body temperature is a challenge. Maintaining homeostasis of body temperature or thermoregulation is the function of the hypothalamus. The hypothalamus operates a variety of negative-feedback mechanisms that keep body temperature in its normal range (36.2° to 37.6° C or 97° to 100° F).

The skin is often involved in negative-feedback loops that maintain body temperature. When the body is overheated, blood flow to the skin increases (Figure 15-5). Warm blood from the body's core can then be cooled by the skin, which acts as a radiator. At the skin, heat can be lost from blood by the following mechanisms:

1. Radiation—flow of heat waves away from the blood
2. Conduction—transfer of heat energy to the skin and then the external environment
3. Convection—transfer of heat energy to air that is continually flowing away from the skin
4. Evaporation—absorption of heat by water (sweat) vaporization

When necessary, heat can be conserved by reducing blood flow in the skin (Figure 15-5).

A number of other mechanisms can be called on to help maintain the homeostasis of body temperature. Heat-generating muscle activity such as shivering and secretion of metabolism-regulating hormones are two of the body's processes that can be altered to adjust the body's temperature. The concept of using feedback control loops in homeostatic mechanisms was introduced in Chapter 1.

FIGURE 15–5 *The Skin as a Thermoregulatory Organ.* When homeostasis requires that the body conserve heat, blood flow in the warm organs of the body's core increases *(left)*. When heat must be lost to maintain the stability of the internal environment, flow of warm blood to the skin increases *(right)*. Heat can be lost from the blood and skin by means of radiation, conduction, convection, and evaporation.

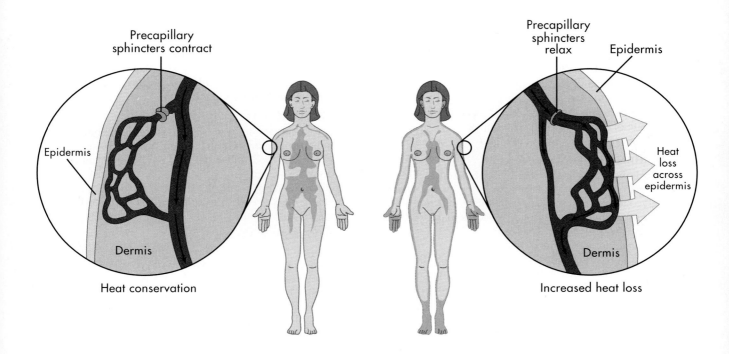

OUTLINE SUMMARY

DEFINITIONS

A Nutrition—food, vitamins, and minerals that are ingested and assimilated into the body

B Metabolism—process of using food molecules as energy sources and as building blocks for our own molecules

C Catabolism—breaks food molecules down, releasing their stored energy; oxygen used in catabolism

D Anabolism—builds food molecules into complex substances

ROLE OF THE LIVER

A Processes blood immediately after it leaves the gastrointestinal tract

  1 Helps maintain normal blood glucose level

  2 Site of protein and fat metabolism

  3 Removes toxins from the blood

NUTRIENT METABOLISM

A Carbohydrates are primarily catabolized for energy (Figure 15-1), but small amounts are anabolized by glycogenesis (a series of chemical reactions that changes glucose to glycogen—occurs mainly in liver cells where glycogen is stored)

B Blood glucose (imprecisely, blood sugar)—normally stays between about 80 and 120 mg per 100 ml of blood; insulin accelerates the movement of glucose out of the blood into cells, therefore decreases blood glucose and increases glucose catabolism

C Adenosine triphosphate (ATP)—molecule in which energy obtained from breakdown of foods is stored; serves as a direct source of energy for cellular work (Figure 15-2)

D Fats catabolized to yield energy and anabolized to form adipose tissue (Figure 15-3)

E Proteins primarily anabolized and secondarily catabolized

VITAMINS AND MINERALS

A Vitamins—organic molecules that are needed in small amounts for normal metabolism (Table 15-2)

B Minerals—inorganic molecules required by the body for normal function (Table 15-3)

METABOLIC RATES

A Basal metabolic rate (BMR)—rate of metabolism when a person is lying down but awake and not digesting food and when the environment is comfortably warm

B Total metabolic rate (TMR)—the total amounts of energy, expressed in calories, used by the body per day (Figure 15-4)

BODY TEMPERATURE

A Hypothalamus—regulates the homeostasis of body temperature through a variety of processes

B Skin—can cool the body by losing heat from the blood through four processes: radiation, conduction, convection, evaporation (Figure 15-5)

NEW WORDS

anabolism	catabolism	glycogenesis	total metabolic rate
basal metabolic rate (BMR)	citric acid cycle	glycolysis	(TMR)
calorie	electron transport system	kilocalorie	vitamin
		thermoregulation	

CHAPTER TEST

1. The type of metabolism that involves the breakdown of food molecules is called _____.
2. The type of metabolism that involves the synthesis of large molecules is called _____.
3. *Nutrition* refers to the food that we eat, whereas _____ refers to the use of foods after they have entered the cells.
4. The _____ vein delivers blood from the gastrointestinal tract directly to the liver.
5. The chemical process that changes glucose into pyruvic acid, producing energy, is called _____.
6. The series of reactions in liver cells that joins glucose molecules together to form glycogen is called _____.
7. When the body runs low on carbohydrates several hours after a meal, it begins catabolizing _____ instead.
8. _____ are organic molecules needed in small quantities for normal metabolism throughout the body.
9. The molecule in which energy from catabolism is stored until it is used by the cell is called _____.
10. The rate at which food is catabolized in a resting individual who is not digesting a meal is called the _____.
11. Thermoregulation is controlled mainly by the _____ in the brain.

Circle the T before each true statement and the F before each false statement.

T F 12. The citric acid cycle changes glucose into pyruvic acid, releasing energy for the cell's use.
T F 13. Insulin promotes the entry of glucose into body cells.
T F 14. Anorexia nervosa can result in death.
T F 15. Minerals can also be called vitamins.
T F 16. Sodium is a mineral that is found in only a few types of food.
T F 17. Measuring oxygen consumption or blood concentration of protein-bound iodine can be used to determine a person's rate of metabolism.
T F 18. The skin is an important organ for maintaining the body's temperature homeostasis.

REVIEW QUESTIONS

1. Briefly and clearly explain anabolism, catabolism, metabolism, and nutrition.
2. In words or in a sketched diagram, describe the metabolic pathway taken by a glucose molecule when energy is extracted from it.
3. Liver cells perform a process that prevents the blood glucose level from getting too high just after a large meal. What is it?
4. How are fats used by the body? Proteins?
5. Explain the metabolic roles of vitamins and minerals.
6. Explain what is meant by the term *metabolic rate*.
7. What is the difference between basal metabolic rate and total metabolic rate?
8. How does the body maintain the body temperature within a normal range?
9. Name the four mechanisms by which the skin removes heat from the body.

CRITICAL THINKING

10. What adaptive advantage is gained by detouring blood from the gastrointestinal tract to the liver before returning it to the heart?
11. How does the body get energy during fasting?
12. Explain why you think the following statement is true or false: "If you do not want to gain or lose weight but just stay the same, you must eat just enough food to supply the kilocalories of your BMR. If you eat more than this, you will gain; if you eat less than this, you will lose."

The Urinary System

16

OBJECTIVES

*After you have completed this chapter,
you should be able to:*

1. Identify the major organs of the urinary system and give the generalized function of each.
2. Name the parts of a nephron and describe the role each component plays in the formation of urine.
3. Explain the importance of filtration, tubular reabsorption, and tubular secretion in urine formation.
4. Discuss the mechanisms that control urine volume.
5. Explain how the kidneys act as vital organs in maintaining homeostasis.

As you might guess from its name, the urinary system performs the functions of producing and excreting urine from the body. What you might not guess so easily is how essential these functions are for the maintenance of homeostasis and healthy survival. The constancy of body fluid volumes and the levels of many important chemicals depend on normal urinary system function. Unless the urinary system operates normally, the normal composition of blood cannot be maintained long, and serious consequences soon follow. The kidneys "clear" or clean the blood of the many waste products continually produced as a result of metabolism of foodstuffs in the body cells. As nutrients are burned for energy, the waste products produced must be removed from the blood, or they quickly accumulate to toxic levels—a condition called **uremia** (yoo-REE-mee-ah) or **uremic poisoning.** The kidneys also play a vital role in maintaining electrolyte, water, and acid-base balances in the body. In this chapter, we will discuss the structure and function of each organ of the urinary system. There are two kidneys, two ureters, one bladder, and one urethra (Figure 16-1). We will also briefly mention disease conditions produced by abnormal functioning of the urinary system.

Kidneys

LOCATION

To locate the kidneys on your own body, stand erect and put your hands on your hips with your thumbs meeting over your backbone. When you are in this position, your kidneys lie just above your thumb; in short, the kidneys lie just above the waistline. Usually the right kidney is a little lower than the left. They are located under the muscles of the back and behind the parietal peritoneum (the membrane that lines the abdominal cavity). Because of this retroperitoneal location, a surgeon can operate on a kidney without cutting through the peritoneum. A heavy cushion of fat normally encases each kidney and helps hold it in place.

Note the relatively large diameter of the renal arteries in Figure 16-1. Normally a little over 20% of the total blood pumped by the heart each minute enters the kidneys. The rate of blood flow through this organ is among the highest in the body. This is understandable because one of the main functions of the kidney is to remove waste products from the blood. Maintenance of a high rate of blood flow and normal blood pressure in the kidneys is essential for the formation of urine.

INTERNAL STRUCTURE

If you were to slice through a kidney from side to side and open it like the pages of a book, you would see the structures shown in Figure 16-2. Identify each of the following parts:

1. **Cortex** (KOR-teks)—the outer part of the kidney (The word *cortex* comes from the Latin word for "bark" or "rind," so the cortex of an organ is its outer layer; each kidney and adrenal gland, as well as the brain, has a cortex.)
2. **Medulla** (me-DUL-ah)—the inner portion of the kidney
3. **Pyramids** (PIR-ah-mids)—the triangular divisions of the medulla of the kidney
4. **Papilla** (pah-PIL-ah) (pl. *papillae*)—narrow, innermost end of a pyramid
5. **Pelvis**—(the kidney or renal pelvis) an expansion of the upper end of a ureter (the tube that drains urine into the bladder)
6. **Calyx** (KAY-liks) (pl. *calyces*)—a division of the renal pelvis (The papilla of a pyramid opens into each calyx.)

MICROSCOPIC STRUCTURE

More than a million microscopic units called **nephrons** (NEF-rons) make up each kidney's interior. The shape of a nephron is unique, unmistakable, and admirably suited to its function of producing urine. It looks a little like a tiny funnel with a very long stem, but it is an unusual stem in that it is highly convoluted (that is, it has many bends in it). The nephron is composed of two principle components: the **renal corpuscle** and the **renal tubule.** The renal corpuscle can be subdivided still further into two parts and the renal tubule into four regions or segments. Identify each part of the renal corpuscle and renal tubule described below in Figures 16-3 and 16-4.

FIGURE 16–1 *Urinary System.* **A,** Location of urinary system organs. **B,** X-ray film of the urinary organs.

A

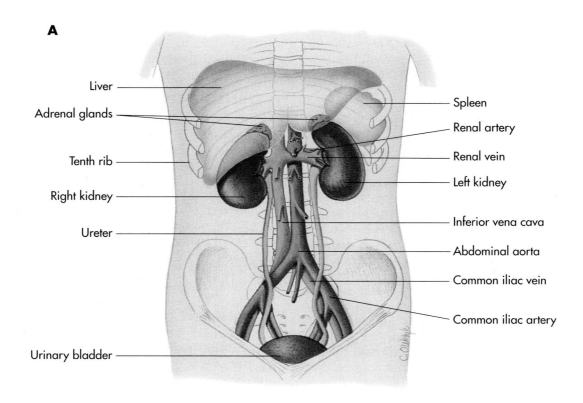

Liver
Adrenal glands
Tenth rib
Right kidney
Ureter
Urinary bladder

Spleen
Renal artery
Renal vein
Left kidney
Inferior vena cava
Abdominal aorta
Common iliac vein
Common iliac artery

B

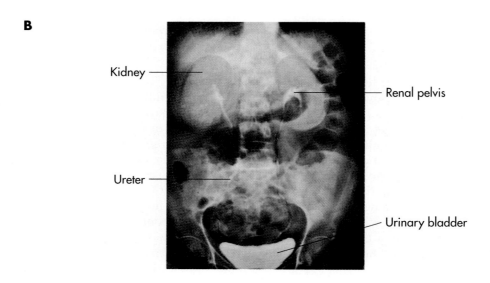

Kidney
Ureter

Renal pelvis
Urinary bladder

FIGURE 16–2 *Kidney.* Coronal section through the right kidney.

Cortex

Medulla

Arteries and veins

Renal capsule

Renal artery

Renal vein

Renal pelvis

Calyx

Ureter

Renal pyramid

1. **Renal corpuscle**
   a. **Bowman's capsule**—the cup-shaped top of a nephron (The saclike Bowman's capsule surrounds the glomerulus.)
   b. **Glomerulus** (glo-MAIR-yoo-lus) (pl. *glomeruli*)—a network of blood capillaries tucked into Bowman's capsule (Note in Figure 16-3 that the small artery that delivers blood to the glomerulus, **[afferent arteriole]** is larger in diameter than the blood vessel that drains blood from it **[efferent arteriole]** and that it is relatively short. This explains the high blood pressure that exists in the glomerular capillaries. This high pressure is required to filter wastes from the blood.)
2. **Renal tubule**
   a. **Proximal convoluted tubule**—the first segment of a renal tubule (It is called *proximal* because it lies nearest the tubule's origin from Bowman's capsule, and it is called *convoluted* because it has several bends.)
   b. **Loop of Henle** (HEN-lee)—the extension of the proximal tubule (Observe that the loop of Henle consists of a straight descending limb, a hairpin loop, and a straight ascending limb.)
   c. **Distal convoluted tubule**—the part of the tubule distal to the ascending limb of the loop of Henle (It is the extension of the ascending limb.)
   d. **Collecting tubule**—a straight (that is, not convoluted) part of a renal tubule (Distal tubules of several nephrons join to form a single collecting tubule or duct.)

FIGURE 16–3 *Location of the Nephron.* **A,** Magnified wedge cut from a renal pyramid. **B,** Scanning electron micrograph showing several glomeruli and their associated blood vessels.

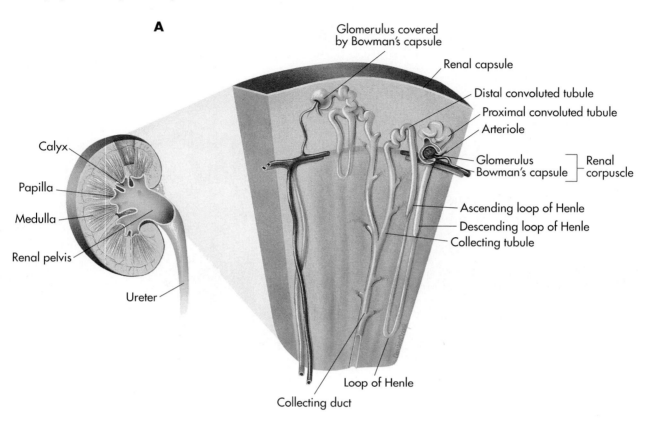

A

Glomerulus covered by Bowman's capsule

Renal capsule

Distal convoluted tubule

Proximal convoluted tubule

Arteriole

Glomerulus
Bowman's capsule — Renal corpuscle

Calyx

Papilla

Medulla

Renal pelvis

Ureter

Ascending loop of Henle

Descending loop of Henle

Collecting tubule

Loop of Henle

Collecting duct

B

Glomerular capillaries

Efferent arteriole

Afferent arteriole

FIGURE 16–4 *The Nephron Unit.* Cross sections from the four segments of the renal tubule are shown. The differences in appearance in tubular cells seen in a cross section reflect the differing functions of each nephron segment.

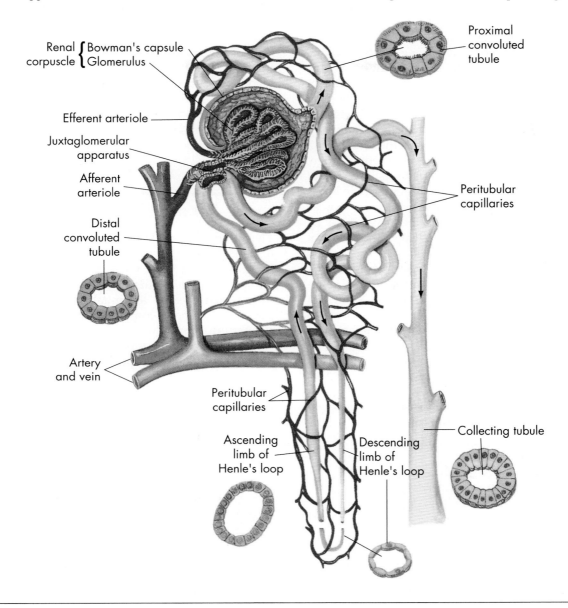

Look again at Figure 16-3. Note that the renal corpuscles (glomeruli surrounded by Bowman's capsule) and both proximal and distal convoluted tubules are located in the cortex of the kidney. The medulla contains the loop of Henle and collecting tubules. Urine from the collecting tubules exits from the pyramid through the papilla and enters the calyx and renal pelvis before flowing into the ureter.

FUNCTION

The kidneys are vital organs. The function they perform, that of forming urine, is essential for homeostasis and maintenance of life. Early in the process of urine formation, fluid, electrolytes, and wastes from metabolism are *filtered* from the blood and enter the nephron. Additional wastes may be *secreted* into the tubules of the nephron as substances useful to the body are reabsorbed into the blood. Table 16-1 lists the components of normal versus abnormal urine. Normally the kidneys balance the amount of many substances entering and leaving the blood over time so that normal concentrations can be maintained. In short, the kidneys adjust their output to equal the intake of the body. By eliminating wastes and adjusting fluid balance, the kidneys play an essential part in maintaining homeostasis. Homeostasis cannot be maintained—nor can life itself—if the kidneys fail and the condition is not soon corrected. Nitrogenous waste products accumulate as a result of protein breakdown and quickly reach toxic levels if not excreted. If kidney function ceases because of injury or disease, life can be maintained by using an artificial kidney to cleanse the blood of wastes.

Excretion of toxins and of waste products containing nitrogen such as urea and ammonia represents only one of the important responsibilities of the kidney. The kidney also plays a key role in regulating the levels of many chemical substances in the blood such as chloride, sodium, potassium, and bicarbonate. The kidneys also regulate the proper balance between body water content and salt by selectively retaining or excreting both substances as requirements demand. In addition, the cells of the *juxtaglomerular apparatus* (Figure 16-4) function in blood pressure regulation. When blood pressure is low, these cells secrete a hormone that initiates constriction of blood vessels and thus raises blood pressure. It is easy to understand why the kidneys are often considered to be the most important homeostatic organs in the body.

 Formation of Urine

The kidney's 2 million or more nephrons form urine by a series of three processes: (1) filtration, (2) reabsorption, and (3) secretion (Figure 16-5). Urine formation begins with the process of **filtration,** which goes on continually in the renal corpuscles (Bowman's capsules plus their encased glomeruli). Blood flowing through the glomeruli exerts pressure, and this glomerular blood pressure is high enough to push water and dissolved

TABLE 16–1 Characteristics of Urine	
NORMAL CHARACTERISTICS	**ABNORMAL CHARACTERISTICS**
Color	
Transparent yellow, amber, or straw colored	Abnormal colors or cloudiness, which may indicate presence of blood, bile, bacteria, drugs, food pigments, or high-solute concentration
Compounds	
Mineral ions (for example, Na, Cl, K)	Acetone
Nitrogenous wastes: ammonia, creatinine, urea, uric acid	Albumin
Suspended solids (sediment)*: bacteria, blood cells, casts (solid matter)	Bile
Urine pigments	Glucose
Odor	
Slight odor	Acetone odor, which is common in diabetes mellitus
pH	
4.6-8.0	High in alkalosis; low in acidosis
Specific Gravity	
1.001-1.035	High specific gravity can cause precipitation of solutes and formation of kidney stones

*Occasional trace amounts.

substances out of the glomeruli into the Bowman's capsule. Briefly, glomerular blood pressure causes filtration through the glomerular-capsular membrane. If the glomerular blood pressure drops below a certain level, filtration and urine formation cease. Hemorrhage, for example, may cause a precipitous drop in blood pressure followed by kidney failure.

Glomerular filtration normally occurs at the rate of 125 ml per minute. As a result, about 180 liters (almost 190 quarts) of **glomerular filtrate** is produced by the kidneys every day.

Obviously no one ever excretes anywhere near 180 L of urine per day. Why? Because most of the fluid that leaves the blood by glomerular filtration, the first process in urine formation, returns to the blood by the second process—reabsorption.

Reabsorption is the movement of substances out of the renal tubules into the blood capillaries located around the tubules (peritubular capillaries). Water, glucose and other nutrients, and sodium and other ions are substances that are reabsorbed. Reabsorption begins in the proximal convoluted tubules and continues in the loop of Henle, distal convoluted tubules, and collecting tubules.

Large amounts of water—approximately 178 L per day—are reabsorbed by osmosis from the proximal tubules. In other words, nearly 99% of the 180 L of water that leave the blood each day by glomerular filtration returns to the blood by proximal tubule reabsorption.

The nutrient glucose is only reabsorbed from the proximal tubules. It is actively transported out of them into peritubular capillary blood. None of this valuable nutrient is wasted by being lost in the urine. However, exceptions occur. For example, in *diabetes mellitus*, if blood glucose concentration increases above a certain level, the tubular filtrate then contains more glucose than kidney tubule cells can reabsorb. Some of the glucose therefore remains behind in the urine. Glucose in the urine—**glycosuria** (glye-ko-SOO-ree-ah)—is a well-known sign of the condition, diabetes mellitus (Figure 16-6).

Sodium ions and other ions are only partially reabsorbed from renal tubules. For the most part, sodium ions are actively transported back into blood from the tubular urine. The amount of sodium reabsorbed varies from time to time; it

---

FIGURE 16–5 *Formation of Urine.* Diagram shows the steps in urine formation in successive parts of a nephron: filtration, reabsorption, and secretion.

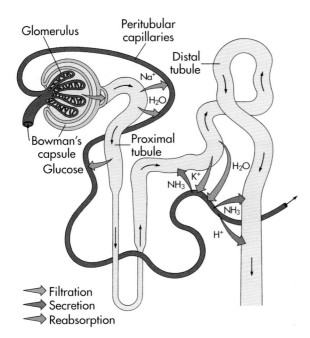

FIGURE 16–6 *Glycosuria.* Using a reagent strip to check the level of glucose in the urine of a diabetic patient.

depends largely on salt intake. In general the greater the amount of salt intake, the less the amount of salt reabsorption and therefore the greater the amount of salt excreted in the urine. Also, the less the salt intake, the greater the salt reabsorption and the less salt excreted in the urine. By varying the amount of salt reabsorbed, the body usually can maintain homeostasis of the blood's salt concentration. This is an extremely important matter because cells are damaged by either too much or too little salt in the fluid around them (see box about tonicity, p. 24).

Secretion is the process by which substances move into urine in the distal and collecting tubules from blood in the capillaries around these tubules. In this respect, secretion is reabsorption in reverse. Whereas reabsorption moves substances out of the urine into the blood, secretion moves substances out of the blood into the urine. Substances secreted are hydrogen ions, potassium ions, ammonia, and certain drugs. Hydrogen ions, potassium ions, and drugs are secreted by being actively transported out of the blood into tubular urine. Ammonia is secreted by diffusion. Kidney tubule secretion plays a crucial role in maintaining the body's acid-base balance (see Chapter 18).

In summary, the following processes occurring in successive portions of the nephron accomplish the function of urine formation (Table 16-2):

1. **Filtration**—of water and dissolved substances out of the blood in the glomeruli into Bowman's capsule

2. **Reabsorption**—of water and dissolved substances out of kidney tubules back into blood (This prevents substances needed by the body from being lost in urine. Usually, 97% to 99% of water filtered out of glomerular blood is retrieved from tubules.)

3. **Secretion**—of hydrogen ions, potassium ions, and certain drugs

CONTROL OF URINE VOLUME

The body has ways to control the amount and composition of the urine that it excretes. It does this mainly by controlling the amount of water and dissolved substances that are reabsorbed by the convoluted tubules. For example, a hormone (antidiuretic hormone or ADH) from the posterior pituitary gland decreases the amount of urine by making collecting tubules permeable to water. If no ADH is present, the tubules are practically impermeable to water, so little or no water is reabsorbed from them. When ADH is present in the blood, collecting tubules are permeable to water and water is reabsorbed from them. As a result, less water is lost from the body as urine, or more water is retained from the tubules—whichever way you wish to say it. At any rate, for this reason ADH is accurately described as the "water-retaining hormone." You might also think of it as the "urine-decreasing hormone."

The hormone aldosterone, secreted by the adrenal cortex, plays an important part in controlling the kidney tubules' reabsorption of salt. Primarily it stimulates the tubules to reabsorb

PART OF NEPHRON	PROCESS IN URINE FORMATION	SUBSTANCES MOVED
Glomerulus	Filtration	Water and solutes (for example, sodium and other ions, glucose and other nutrients filtering out of glomeruli into Bowman's capsules)
Probimal tubule	Reabsorption	Water and solutes
Loop of Henle	Reabsorption	Sodium and chloride ions
Distal and collecting tubules	Reabsorption	Water, sodium, and chloride ions
	Secretion	Ammonia, potassium ions, hydrogen ions, and some drugs

TABLE 16–2 Functions of Parts of Nephron in Urine Formation

sodium salts at a faster rate. Secondarily, aldosterone also increases tubular water reabsorption. The term *salt- and water-retaining hormone* therefore is a descriptive nickname for aldosterone. This mechanism is discussed in the next chapter.

Another hormone, atrial natriuretic hormone (ANH) secreted from the heart's atrial wall, has the opposite effect of aldosterone. ANH stimulates kidney tubules to secrete more sodium and thus lose more water. Thus ANH is a *salt- and water-losing hormone.* The body secretes ADH, aldosterone, and ANH in different amounts, depending on the homeostatic balance of body fluids at any particular moment.

Sometimes the kidneys do not excrete normal amounts of urine as a result of kidney disease, cardiovascular disease, or stress. Here are some terms associated with abnormal amounts of urine:

1. **Anuria** (ah-NOO-ree-ah)—absence of urine
2. **Oliguria** (ol-i-GOO-ree-ah)—scanty amounts of urine
3. **Polyuria** (pol-e-YOO-ree-ah)—an unusually large amount of urine

Ureters

Urine drains out of the collecting tubules of each kidney into the renal pelvis and down the ureter into the urinary bladder (Figure 16-1). The **renal pelvis** is the basinlike upper end of the ureter located inside the kidney. Ureters are narrow tubes less than 6 mm (¼ inch) wide and 25 to 30 cm (10 to 12 inches) long. Mucous membranes line both ureters and each renal pelvis. Note in Figure 16-7 that the ureter has a thick, muscular wall. Contraction of the muscular coat produces peristaltic-type movements that assist in moving urine down the ureters into the bladder. The lining membrane of the ureters is richly supplied with sensory nerve endings.

FIGURE 16–7 *Ureter Cross Section.* Note the thick layer of muscle around the tube.

Proteinuria After Exercise

Proteinuria is the presence of abnormally large amounts of plasma proteins in the urine. Proteinuria usually indicates kidney disease **(nephropathy)** because only damaged nephrons consistently allow many plasma protein molecules to leave the blood. However, intense exercise causes temporary proteinuria in many individuals. Some exercise physiologists believed that intense athletic activities cause kidney damage, but subsequent research has ruled out that explanation. One current hypothesis is that hormonal changes during strenuous exercise increase the permeability of the nephron's filtration membrane, allowing more plasma proteins to enter the filtrate. Some postexercise proteinuria is usually considered normal.

Adipose tissue   Muscle layer

Connective tissue   Transitional epithelium

Artificial Kidney

The artificial kidney is a mechanical device that uses the principle of dialysis to remove or separate waste products from the blood. In the event of kidney failure, the process, appropriately called **hemodialysis** (Greek *haima* "blood" and *lysis* "separate"), is a reprieve from death for the patient. During a hemodialysis treatment, a semipermeable membrane is used to separate large (nondiffusible) particles such as blood cells from small (diffusible) ones such as urea and other wastes. Figure *A* shows blood from the radial artery passing through a porous (semipermeable) cellophane tube that is housed in a tanklike container. The tube is surrounded by a bath or dialyzing solution containing varying concentrations of elec-

trolytes and other chemicals. The pores in the membrane are small and allow only very small molecules, such as urea, to escape into the surrounding fluid. Larger molecules and blood cells cannot escape and are returned through the tube to reenter the patient via a wrist vein. By constantly replacing the bath solution in the dialysis tank with freshly mixed solution, levels of waste materials can be kept at low levels. As a result, wastes such as urea in the blood rapidly pass into the surrounding wash solution. For a patient with complete kidney failure, two or three hemodialysis treatments a week are required.

Dialysis treatments are now being monitored and controlled by sophisticated com-

A

puter components and software integrated into modern hemodialysis equipment. New and dramatic advances in both treatment techniques and equipment are expected to continue. Although most hemodialysis treatments occur in hospital or clinical settings, equipment designed for use in the home is now available and appropriate for some individuals. Patients and their families using this equipment are initially instructed in its use and then monitored and supported on an ongoing basis by home health care professionals.

Another technique used in the treatment of renal failure is called **continuous ambulatory peritoneal dialysis (CAPD).** In this procedure, 1 to 3 L of sterile dialysis fluid is introduced directly into the peritoneal cavity through an opening in the abdominal wall (Figure *B*). Peritoneal membranes in the abdominal cavity transfer waste products from the blood into the dialysis fluid, which is then drained back into a plastic container after about 2 hours. This technique is less expensive than hemodialysis and does not require the use of complex equipment. CAPD is the more frequently used home-based dialysis treatment for patients with chronic renal failure. Successful longterm treatment is greatly enhanced by support from professionals trained in home health care services.

B

Dialysis fluid

Abdominal cavity

FIGURE 16–8 *The Male Urinary Bladder.* This view (with bladder cut to show the interior) shows how the prostate gland surrounds the urethra as it exits from the bladder. The glands associated with the male reproductive system are further discussed in Chapter 19.

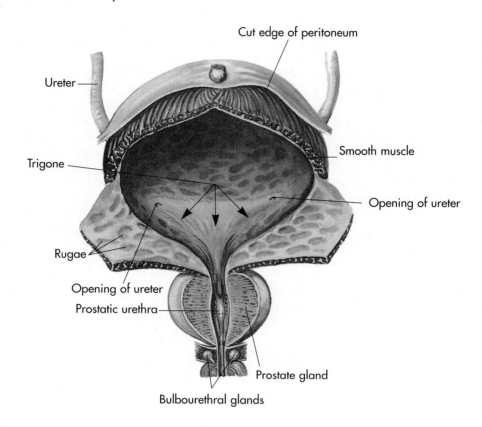

Urinary Bladder

The empty urinary bladder lies in the pelvis just behind the pubic symphysis. When full of urine, it projects upward into the lower portion of the abdominal cavity.

Elastic fibers and involuntary muscle fibers in the wall of the urinary bladder make it well suited for expanding to hold variable amounts of urine and then contracting to empty itself. Mucous membrane lines the urinary bladder. The lining is loosely attached to the deeper muscular layer so that the bladder is very wrinkled and lies in folds called *rugae* when it is empty. When the bladder is filled, its inner surface is smooth. Note in Figure 16-8 that one triangular area on the back or posterior surface of the bladder is free of rugae. This area, called the *trigone,* is always smooth. There, the lining membrane is tightly fixed to the deeper muscle coat. The trigone extends between the openings of the two ureters above and the point of exit of the urethra below.

Attacks of *renal colic*—pain caused by the passage of a kidney stone—have been described in medical writings since antiquity. Kidney stones cause intense pain if they have sharp edges or are large enough to distend the walls or cut the lining of the ureters or urethra as they pass from the kidneys to the exterior of the body.

Urinary Catheterization

Urinary catheterization is the passage or insertion of a hollow tube or catheter through the urethra into the bladder for the withdrawal of urine. It is a medical procedure commonly performed on many patients who undergo prolonged surgical or diagnostic procedures or who experience problems with urinary retention.

Correct catheterization procedures require aseptic techniques to prevent the introduction of infectious bacteria into the urinary system. Clinical studies have proved that improper catheterization techniques cause bladder infections (cystitis) in hospitalized patients. One landmark study confirmed the high percentage of catheterized patients who develop cystitis (almost 8%) and found that, of those who developed such infections, a significant number died. The results of the study are sobering and point out the need for extensive training of health professionals in this area.

 ## Urethra

To leave the body, urine passes from the bladder, down the urethra, and out of its external opening, the **urinary meatus.** In other words, the urethra is the lowest part of the urinary tract. The same sheet of mucous membrane that lines each renal pelvis, the ureters, and the bladder extends down into the urethra, too; this is a structural feature worth noting because it accounts for the fact that an infection of the urethra may spread upward through the urinary tract. The urethra is a narrow tube; it is only about 4 cm (1½ inches) long in a woman, but it is about 20 cm (8 inches) long in a man. In a man, the urethra has two functions: (1) it is the terminal portion of the urinary tract, and (2) it is the passageway for movement of the reproductive fluid (semen) from the body. In a woman, the urethra is a part of only the urinary tract.

 ## Micturition

The terms **micturition** (mik-too-RISH-un), **urination** (yoor-i-NAY-shun), and **voiding** refer to the passage of urine from the body or the emptying of the bladder. This is a reflex action in infants or very young children. Although there is considerable variation between individuals, most children between 2 and 3 years of age learn to urinate voluntarily and also to inhibit voiding if the urge comes at an inconvenient time.

Two **sphincters** (SFINGK-ters) or rings of muscle tissue guard the pathway leading from the bladder. The **internal urethral sphincter** is located at the bladder exit, and the **external urethral sphincter** circles the urethra just below the neck of the bladder. When contracted, both sphincters seal off the bladder and allow urine to accumulate without leaking to the exterior. The internal urethral sphincter is involuntary, and the external urethral sphincter is composed of striated muscle and is under voluntary control.

The muscular wall of the bladder permits this organ to accommodate a considerable volume of urine with very little increase in pressure until a volume of 300 to 400 ml is reached. As the volume of urine increases, the need to void may be noticed at volumes of 150 ml, but micturition in adults does not normally occur much below volumes of 350 ml. As the bladder wall stretches, nervous impulses are transmitted to the second, third, and fourth sacral segments of the spinal cord, and an **emptying reflex** is initiated. The reflex causes contraction of the muscle of the bladder wall and relaxation of the internal sphincter. Urine then enters the urethra. If the external sphincter, which is under voluntary control, is relaxed, micturition occurs. Voluntary contraction of the external sphincter suppresses the emptying reflex until the bladder is filled to capacity with urine and loss of control occurs. Contraction of this powerful sphincter also abruptly terminates urination voluntarily.

Higher centers in the brain also function in micturition by integrating bladder contraction and internal and external sphincter relaxation, with the cooperative contraction of pelvic and abdominal muscles. Urinary **retention** is a condition in which no urine is voided. The kidneys produce urine, but the bladder for one reason or

Removal of Kidney Stones Using Ultrasound

Statistics suggest that approximately 1 in every 1000 adults in the United States suffers from kidney stones or **renal calculi** (KAL-kyoo-lye) at some point in their life. Although symptoms of excruciating pain are common, many kidney stones are small enough to pass out of the urinary system spontaneously. If this is possible, no therapy is required other than treatment for pain and antibiotics if the calculi are associated with infection. Larger stones, however, may obstruct the flow of urine and are much more serious and difficult to treat.

Until recently, only traditional surgical procedures were effective in removing relatively large stones that formed in the calyces and renal pelvis of the kidney. In addition to the risks that always accompany major medical procedures, surgical removal of stones from the kidneys frequently requires rather extensive hospital and home recovery periods, lasting 6 weeks or more.

A technique that uses ultrasound to pulverize the stones so that they can be flushed out of the urinary tract without surgery is used in hospitals across the United States. The specially designed ultrasound generator required for the procedure is called a **lithotriptor** (li-tho-TRIP-ter). Using a lithotriptor, physicians break up the stones with ultrasound waves without making an incision. Recovery time is minimal and patient risk and costs are reduced.

another cannot empty itself. In urinary **suppression** the opposite is true. The kidneys do not produce any urine, but the bladder retains the ability to empty itself.

Incontinence (in-KON-ti-nens) is a condition in which the patient voids urine involuntarily. It frequently occurs in patients who have suffered a stroke or spinal cord injury. If the sacral segments of the spinal cord are injured, some loss of bladder function always occurs. Although the voiding reflex may be reestablished to some degree, the bladder does not empty completely. In these individuals the residual urine is often the cause of repeated bladder infections or **cystitis** (sis-TIE-tis). Complete destruction or transection of the sacral cord produces a condition called an *automatic bladder.* Totally cut off from spinal innervation, the bladder musculature acquires some automatic action and periodic but unpredictable voiding occurs.

KIDNEYS

A Location—under back muscles, behind parietal peritoneum, just above waistline; right kidney usually a little lower than left (Figure 16-1)
B Internal structure (Figure 16-2)
  1 Cortex—outer layer of kidney substance
  2 Medulla—inner portion of kidney
  3 Pyramids—triangular divisions of medulla
  4 Papilla—narrow, innermost end of pyramid
  5 Pelvis—expansion of upper end of ureter; lies inside kidney
  6 Calyces—divisions of renal pelvis
C Microscopic structure—nephrons are microscopic units of kidneys; consist of (Figure 16-3):
  1 Renal corpuscle—Bowman's capsule with its glomerulus
    a Bowman's capsule—the cup-shaped top
    b Glomerulus—network of blood capillaries surrounded by Bowman's capsule
  2 Renal tubule
    a Proximal convoluted tubule—first segment
    b Loop of Henle—extension of proximal tubule; consists of descending limb, loop, and ascending limb
    c Distal convoluted tubule—extension of ascending limb of loop of Henle
    d Collecting tubule—straight extension of distal tubule
D Functions
  1 Excretes toxins and nitrogenous wastes
  2 Regulates levels of many chemicals in blood
  3 Maintains water balance
  4 Helps regulate blood pressure via secretion of renin

FORMATION OF URINE (Figure 16-5)

A Occurs by a series of three processes that take place in successive parts of nephron
  1 Filtration—goes on continually in renal corpuscles; glomerular blood pressure causes water and dissolved substances to filter out of glomeruli into Bowman's capsule; normal glomerular filtration rate 125 ml per minute
  2 Reabsorption—movement of substances out of renal tubules into blood in peritubular capillaries; water, nutrients, and ions are reabsorbed; water is reabsorbed by osmosis from proximal tubules
  3 Secretion—movement of substances into urine in the distal and collecting tubules from blood in peritubular capillaries; hydrogen ions, potassium ions, and certain drugs are secreted by active transport; ammonia is secreted by diffusion

B Control of urine volume—mainly by posterior pituitary hormone's ADH, which decreases it

URETERS

A Structure—narrow long tubes with expanded upper end (renal pelvis) located inside kidney and lined with mucous membrane
B Function—drain urine from renal pelvis to urinary bladder

URINARY BLADDER

A Structure (Figure 16-8)
  1 Elastic muscular organ, capable of great expansion
  2 Lined with mucous membrane arranged in rugae, like stomach mucosa
B Functions
  1 Storage of urine before voiding
  2 Voiding

URETHRA

A Structure
  1 Narrow tube from urinary bladder to exterior
  2 Lined with mucous membrane
  3 Opening of urethra to the exterior called *urinary meatus*
B Functions
  1 Passage of urine from bladder to exterior of the body
  2 Passage of male reproductive fluid (semen) from the body

MICTURITION

A Passage of urine from body (also called *urination* or *voiding*)
B Regulatory sphincters
  1 Internal urethral sphincter (involuntary)
  2 External urethral sphincter (voluntary)
C Bladder wall permits storage of urine with little increase in pressure
D Emptying reflex
  1 Initiated by stretch reflex in bladder wall
  2 Bladder wall contracts
  3 Internal sphincter relaxes
  4 External sphincter relaxes, and urination occurs
E Urinary retention—urine produced but not voided
F Urinary suppression—no urine produced but bladder is normal
G Incontinence—urine is voided involuntarily
  1 May be caused by spinal injury or stroke
  2 Retention of urine may cause cystitis

N E W   W O R D S

anuria	cystitis	micturition	renal colic
atrial natriuretic	glomerulus	oliguria	trigone
hormone (ANH)	glycosuria	papilla	uremia
Bowman's capsule	incontinence	polyuria	urination
calyx	lithotriptor	pyramid	voiding
catheterization			

C H A P T E R   T E S T

1. Failure of the urinary system results in rapid accumulation of toxic wastes, a condition called
   _____.

2. Normally a little over _____ of the total blood pumped by the heart enters the kidneys.

3. The outer portion of the kidney is called the _____, and the inner portion is called the _____.

4. The functional unit of the kidney is called the _____.

5. The renal corpuscle is composed of the cup-shaped _____, and a network of blood capillaries called the _____.

6. The first segment of the renal tubule is called the _____ convoluted tubule.

7. Urine formation begins with the process of _____ of wastes from the blood.

8. Water is reabsorbed from the proximal convoluted tubules by _____.

9. Glucose is entirely reabsorbed from the _____ tubules.

10. Production of unusually large amounts of urine is called _____.

11. The basinlike upper end of the ureter located inside the kidney is called the renal _____.

12. Attacks of pain caused by the passage of a kidney stone are called _____ _____.

13. Urine passes from the bladder down the _____ and out its external opening, called the urinary _____.

14. Voiding urine involuntarily is called _____.

15. The medical term for bladder infection is _____.

16. Glomerular filtration normally occurs at the rate of about _____ ml per minute.

Select the most correct answer from Column B for each statement in Column A. (Only one answer is correct.)

COLUMN A		COLUMN B
17. _____	High waste levels	a. Glycosuria
18. _____	Contain renal corpuscles	b. Glomerulus
		c. ADH
19. _____	Functional unit of kidney	d. Nephron
		e. Uremia
20. _____	Tuft of capillaries	f. Renal calculi
21. _____	Glucose in urine	g. Anuria
22. _____	Absence of urine	h. Micturition
23. _____	Water-retaining hormone	i. Cortex
24. _____	Kidney stones	
25. _____	Urination	

REVIEW QUESTIONS

1. What organs form the urinary system?
2. Name the parts of a nephron.
3. What and where are the glomeruli and Bowman's capsules?
4. Explain briefly the functions of the glomeruli and Bowman's capsules.
5. Explain briefly the function of the renal tubules.
6. What kind of membrane lines the urinary tract?
7. Explain briefly the function of ADH. What is the full name of this hormone? What gland secretes it?
8. What hormone might appropriately be nicknamed the "water-retaining hormone"?
9. What hormone might appropriately be nicknamed the "salt- and water-retaining hormone"?
10. What hormone might be called the "salt- and water-losing hormone"?
11. What is the urinary meatus?
12. What and where are the ureters and the urethra?

CRITICAL THINKING

13. To operate on a kidney, does a surgeon have to cut through the peritoneum? Explain your answer.
14. Suppose that ADH secretion increases noticeably. Would this increase or decrease urine volume? Why?
15. Explain the process of micturition.

Fluid and
Electrolyte Balance

OBJECTIVES

*After you have completed this chapter,
you should be able to:*

1. List, describe, and compare the body fluid compartments and their subdivisions.
2. Discuss avenues by which water enters and leaves the body and the mechanisms that maintain fluid balance.
3. Discuss the nature and importance of electrolytes in body fluids and explain the aldosterone mechanism of extracellular fluid volume control.
4. Explain the interaction between capillary blood pressure and blood proteins.
5. Give examples of common fluid imbalances.

Have you ever wondered why you sometimes excrete great volumes of urine and sometimes excrete almost none at all? Why sometimes you feel so thirsty that you can hardly get enough to drink and other times you want no liquids at all? These conditions and many more relate to one of the body's most important functions—that of maintaining its **fluid** and **electrolyte balance.**

The term *fluid balance* means several things. Of course, it means the same thing as homeostasis of fluids. To say that the body is in a state of fluid balance is to say that the total amount of water in the body is normal and that it remains relatively constant. Electrolytes are substances such as salts that dissolve or break apart in water solution. Health and sometimes even survival itself depend on maintaining proper balance of water and the electrolytes within it.

In this chapter you will find a discussion of body fluids and electrolytes, their normal values, the mechanisms that operate to keep them normal, and some of the more common types of fluid and electrolyte imbalances.

Body Fluids

If you are a healthy young person and you weigh 120 pounds, there is a good chance that, of the hundreds of compounds present in your body, one substance alone accounts for about 50% to 60% of your total weight. This, the body's most abundant compound, is water. It occupies three main locations known as **fluid compartments.** Look now at Figure 17-1. Note that the largest volume of water by far lies inside cells and that it is called, appropriately, **intracellular fluid (ICF).** Note, too, that the water outside of cells—**extracellular fluid (ECF)**—is located in two compartments: in the microscopic spaces between cells, where it is called **interstitial fluid (IF),** and in the blood vessels, where it is called **plasma.** Plasma is the liquid part of the blood, constituting a little more than half of the total blood volume (about 55%); blood cells make up the rest of the volume.

A normal body maintains *fluid balance.* The term fluid balance means that the volumes of ICF, IF, and plasma and the total volume of water in the body remain relatively constant. Of course,

not all bodies contain the same amount of water. The more a person weighs, the more water the body contains. This is true because, excluding fat or adipose tissue, about 55% of the body weight is water. Because fat is almost water free, the more fat present in the body, the less the total water content is per unit of weight. In other words, fat people have a lower water content per pound of body weight than slender people. The body of a slender adult man, for instance, typically consists of about 60% water. An obese male body, in contrast, may consist of only 50% water or even less.

Sex and age also influence the amount of water in a body. Infants have more water compared to body weight than adults of either sex. In a newborn, water may account for up to 80% of total body weight. There is a rapid decline in the proportion of body water to body weight during the first year of life. Figure 17-2 illustrates the proportion of body weight represented by water in newborn infants, men, and women. The female body contains slightly less water per pound of weight because it contains slightly more fat than the male body. Age and the body's water content are inversely related. In general, as age increases, the amount of water per pound of body weight decreases.

Mechanisms That Maintain Fluid Balance

Under normal conditions, homeostasis of the total volume of water in the body is maintained or restored primarily by devices that adjust output (urine volume) to intake and secondarily by mechanisms that adjust fluid intake. There is no question about which of the two mechanisms is more important; the body's chief mechanism, by far, for maintaining fluid balance is to adjust its fluid output so that it equals its fluid intake.

Obviously, as long as output and intake are equal, the total amount of water in the body does not change. Figure 17-3 shows the three sources of fluid intake: the liquids we drink, the water in the foods we eat, and the water formed by catabolism of foods. Table 17-1 gives their normal volumes. However, these can vary a great deal and still be considered normal. Table 17-1 also indicates that fluid output from the body occurs through four

FIGURE 17–1 *Fluid Compartments of the Body.* Percentages and volumes are given for young adults weighing 55 kg (120 lbs).

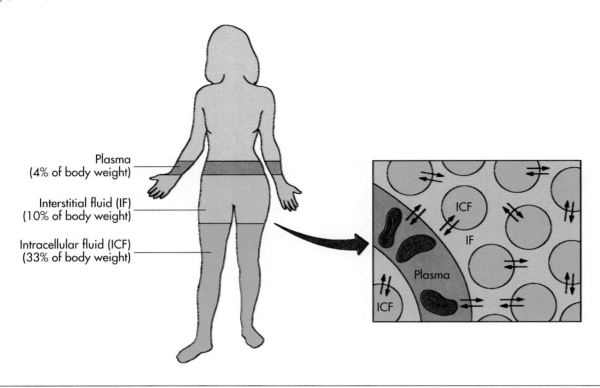

Plasma
(4% of body weight)

Interstitial fluid (IF)
(10% of body weight)

Intracellular fluid (ICF)
(33% of body weight)

FIGURE 17–2 *Proportion of Body Weight Represented by Water.*

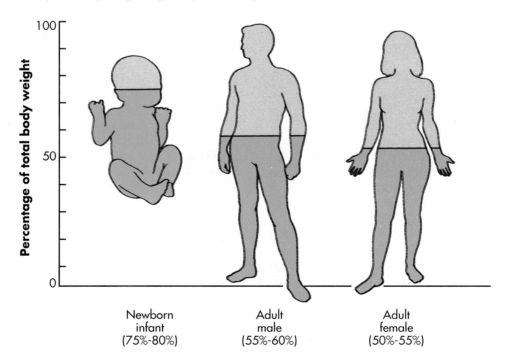

Newborn
infant
(75%-80%)

Adult
male
(55%-60%)

Adult
female
(50%-55%)

FIGURE 17–3 *Sources of Fluid Intake and Output.*

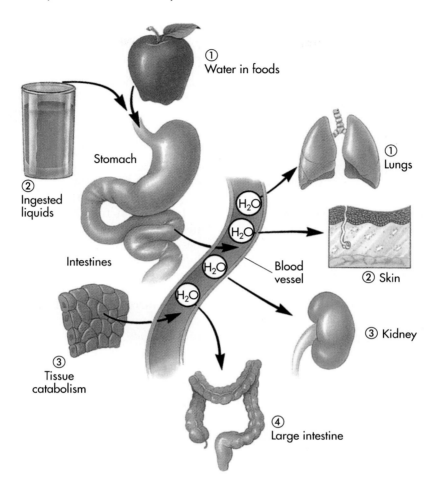

TABLE 17–1 Typical Normal Values for Each Portal of Water Entry and Exit (24 Hours)

INTAKE	AMOUNT*	OUTPUT	AMOUNT*
Water in foods	700 ml	Lungs (water in expired air)	350 ml
Ingested liquids	1500 ml	Skin	
Water formed by catabolism	200 ml	By diffusion	350 ml
		By sweat	100 ml
		Kidneys (urine)	1400 ml
		Intestines (in feces)	200 ml
TOTALS	2400 ml		2400 ml

*Amounts vary widely.

organs: the kidneys, lungs, skin, and intestines. The fluid output that changes the most is that from the kidneys. The body maintains fluid balance mainly by changing the volume of urine excreted to match changes in the volume of fluid intake. Everyone knows this from experience. The more liquid one drinks, the more urine one excretes. Conversely, the less the fluid intake, the less the urine volume. How changes in urine volume come about was discussed on pp. 348–349. This would be a good time to review these paragraphs.

It is important to remember from your study of the urinary system that the rate of water and salt resorption by the renal tubules is the most important factor in determining urine volume. Urine volume is regulated chiefly by hormones secreted by the posterior lobe of the pituitary gland (antidiuretic hormone or ADH) and the adrenal cortex (aldosterone). Atrial natriuretic hormone (ANH) from the atrial wall of the heart also affects urine volume. See p. 348 for a review of the hormonal control of urine volume.

Several factors act as mechanisms for controlling plasma, IF, and ICF volumes. We shall limit our discussion to naming only three of these factors, stating their effects on fluid volumes, and giving some specific examples of them. Three of the main factors are:

1. The concentration of electrolytes in ECF
2. The capillary blood pressure
3. The concentration of proteins in blood

REGULATION OF FLUID INTAKE

Physiologists disagree about the details of the mechanism for controlling and regulating fluid intake to compensate for factors that would lead to dehydration. In general it appears to operate in this way: when dehydration starts to develop—that is, when fluid loss from the body exceeds fluid intake—salivary secretion decreases, producing a "dry-mouth feeling" and the sensation of thirst. The individual then drinks water, thereby increasing fluid intake and compensating for previous fluid losses. This tends to restore fluid balance (Figure 17-4). If an individual takes nothing by mouth for days, can his fluid output decrease to zero? The answer—no—becomes obvious after reviewing the information in Table 17-1. Despite every effort of homeostatic mechanisms to compensate for zero intake, some output (loss) of fluid occurs as long as life continues. Water is continually lost from the body through expired air and diffusion through skin.

Although the body adjusts fluid intake, factors that adjust fluid output, such as electrolytes and blood proteins, are far more important.

IMPORTANCE OF ELECTROLYTES IN BODY FLUIDS

The bonds that hold together the molecules of certain organic substances such as glucose are such that they do not permit the compound to break up or **dissociate** in water solution. Such compounds are called **nonelectrolytes.** Compounds such as ordinary table salt or sodium chloride (NaCl) that have molecular bonds that permit them to break up or dissociate in water solution into separate particles (Na$^+$ and Cl$^-$) are **electrolytes.** The dissociated particles of an electrolyte are **ions** and carry an electrical charge.

 ## Making Weight

Some sports require athletes to "make weight" or weigh in within certain limits just before a competition. Some athletes add muscle mass until they are over their weight limit and then fast or take diuretics just before weigh-in to lose weight. Their theory is that the increased muscle mass improves their competitive edge. Unfortunately, dehydration and loss of glycogen always result from these techniques. Because water and glycogen cannot possibly be replaced in time for the competition, such an athlete is at a physiological disadvantage. Performance and thus the competitive edge is much greater if an athlete can make weight without resorting to "quick loss" methods.

FIGURE 17–4 *Homeostasis of the Total Volume of Body Water.* A basic mechanism for adjusting intake to compensate for excess output of body fluid is diagrammed.

 Diuretics

The word **diuretic** is from the Greek word **diouretikos** meaning "causing urine." By definition a diuretic drug is a substance that promotes or stimulates the production of urine.

As a group, diuretics are among the most commonly used drugs in medicine. They are used because of their role in influencing water and electrolyte balance, especially sodium, in the body. Diuretics have their effect on tubular function in the nephron, and the differing types of diuretics are often classified according to their major site of action. Examples would include: (1) *proximal tubule diuretics* such as acetazolamide (Diamox), (2) *loop of Henle diuretics* such as ethacrynic acid (Edecrin) or furosemide (Lasix), and (3) *distal tubule diuretics* such as chlorothiazide (Diuril).

Classification can also be made according to the effect the drug has on the level or concentration of sodium (Na^+), chloride (Cl^-), potassium (K^+), and bicarbonate (HCO_3^-) ions in the tubular fluid.

Nursing implications for caregivers monitoring patients receiving diuretics both in hospitals and home health care environments include keeping a careful record of fluid intake and output and assessing the patient for signs and symptoms of electrolyte and water imbalance. For example, diuretic induced dehydration resulting in a loss of only 6% of initial body weight will cause tingling in the extremities, stumbling gait, headache, fever, and an increase in both pulse and respiratory rates.

Edema

Edema may be defined as the presence of abnormally large amounts of fluid in the intercellular tissue spaces of the body. The condition is a classic example of fluid imbalance and may be caused by disturbances in any factor that governs the interchange between blood plasma and IF compartments. Examples include the following: (1) **Retention of electrolytes (especially Na$^+$)** in the extracellular fluid as a result of increased aldosterone secretion or after serious kidney disease. (2) **An increase in capillary blood pressure.** Normally, fluid is drawn from the tissue spaces into the venous end of a tissue capillary because of the low venous pressure and the relatively high water-pulling force of the plasma proteins. This balance is upset by anything that increases the capillary hydrostatic pressure. The generalized venous congestion of heart failure is the most common cause of widespread edema. In patients with this condition, blood cannot flow freely through the capillary beds, and therefore the pressure will increase until venous return of blood improves. (3) **A decrease in the concentration of plasma proteins** caused by "leakage" into the interstitial spaces of proteins normally retained in the blood. This may occur as a result of increased capillary permeability caused by infection, burns, or shock.

TABLE 17–2   Common Electrolytes Found in Blood Plasma

POSITIVELY CHARGED IONS	NEGATIVELY CHARGED IONS
142 mEq Na$^+$	102 mEq Cl$^-$
4 mEq K$^+$	26 mEq HCO$_3^-$
5 mEq Ca^{++}	17 mEq protein$^-$
2 mEq Mg^{++}	6 mEq other
	2 mEq HPO$_4^=$
153 mEq/L plasma	153 mEq/L plasma

Important positively charged ions include sodium (Na$^+$), calcium (Ca^{++}), potassium (K$^+$), and magnesium (Mg^{++}). Important negatively charged ions include chloride (Cl$^-$), bicarbonate (HCO$_3^-$), phosphate (HPO$_4^=$), and many proteins. Table 17-2 shows that although blood plasma contains a number of important electrolytes, by far the most abundant one is sodium chloride (ordinary table salt, Na$^+$ Cl$^-$).

A variety of electrolytes have important nutrient or regulatory roles in the body. Many ions are major or important "trace" elements in the body (see Appendix A). Iron, for example, is required for hemoglobin production, and iodine must be available for synthesis of thyroid hormones. Electrolytes are also required for many cellular activities such as nerve conduction and muscle contraction.

Additionally, electrolytes influence the movement of water among the three fluid compartments of the body. To remember how ECF electrolyte concentration affects fluid volumes, remember this one short sentence: where sodium goes, water soon follows. If, for example, the concentration of sodium in blood increases, the volume of blood soon increases. Conversely, if blood sodium concentration decreases, blood volume soon decreases.

Figure 17-5 traces one mechanism that tends to maintain fluid homeostasis. Aldosterone, secreted by the adrenal cortex, increases Na$^+$ reabsorption by the kidney tubules. Water reabsorption also increases, causing an increase in ECF volume. Begin in the upper right of the diagram and follow, in sequence, each of the informational steps. In summary:

FIGURE 17–5 *Aldosterone Mechanism.* Aldosterone restores normal ECF volume when such levels decrease below normal. Excess aldosterone, however, leads to excess ECF volume—that is, excess blood volume (hypervolemia) and excess interstitial fluid volume (edema)—and also to an excess of the total Na^+ content of the body.

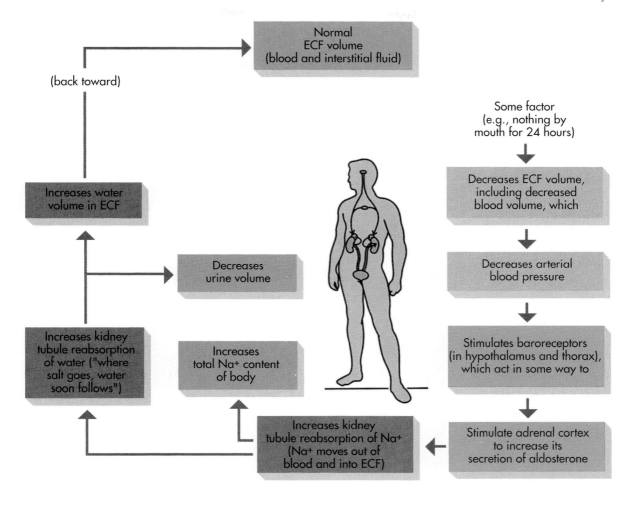

1. Overall fluid balance requires that fluid output equal fluid intake.
2. The type of fluid output that changes most is urine volume.
3. Renal tubule regulation of salt and water is the most important factor in determining urine volume.
4. Aldosterone controls sodium reabsorption in the kidney.
5. The presence of sodium forces water to move (where sodium goes, water soon follows).

The flow chart diagram in Figure 17-5 explains, in a very concise and brief way, the aldosterone mechanism that helps to restore normal ECF volume when it decreases below normal. Can you construct a similar diagram to show the effect of ADH secretion on ECF volume?

Although wide variations are possible, the average daily diet contains about 100 milliequivalents of sodium. The milliequivalent (mEq) (Table 17-2) is a unit of measurement related to reactivity. In a healthy individual, sodium excretion from the body by the kidney is about the same as

FIGURE 17–6   *Sodium-Containing Internal Secretions.* The total volume of these secretions may reach 8000 ml or more in 24 hours.

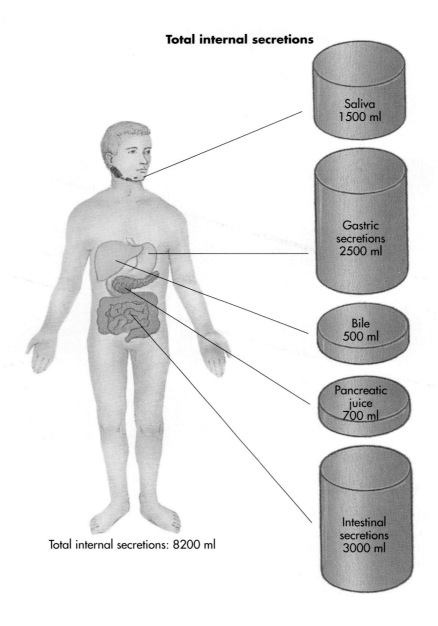

Total internal secretions

Saliva
1500 ml

Gastric
secretions
2500 ml

Bile
500 ml

Pancreatic
juice
700 ml

Intestinal
secretions
3000 ml

Total internal secretions: 8200 ml

intake. The kidney acts as the chief regulator of sodium levels in body fluids. It is important to know that many electrolytes such as sodium not only pass into and out of the body but also move back and forth between a number of body fluids during each 24-hour period. Figure 17-6 shows the large volumes of sodium-containing internal secretions produced each day. During a 24-hour period, over 8 liters of fluid containing 1000 to 1300 mEq of sodium are poured into the digestive system as part of saliva, gastric secretions, bile, pancreatic juice, and IF secretions. This sodium, along with most of that contained in the diet, is almost completely reabsorbed in the large intestine. Very little sodium is lost in the feces. Precise regulation and control of sodium levels are required for survival.

CAPILLARY BLOOD PRESSURE AND BLOOD PROTEINS

Capillary blood pressure is a "water-pushing" force. It pushes fluid out of the blood in capillaries into the IF. Therefore if capillary blood pressure increases, more fluid is pushed—filtered—out of blood into the IF. The effect of an increase in capillary blood pressure, then, is to transfer fluid from blood to IF. In turn this fluid shift, as it is called, changes blood and IF volumes. It decreases blood volume by increasing IF volume. If, on the other hand, capillary blood pressure decreases, less fluid filters out of blood into IF.

Water continually moves in both directions through the membranous walls of capillaries (Figure 17-1). The amount that moves out of capillary blood into IF depends largely on capillary blood pressure, a water-pushing force. The amount that moves in the opposite direction (that is, into blood from IF) depends largely on the concentration of proteins in blood plasma. Plasma proteins act as a water-pulling or water-holding force. They hold water in the blood and pull it into the blood from IF. If, for example, the concentration of proteins in blood decreases appreciably—as it does in some abnormal conditions such as dietary deficiency—less water moves into blood from IF. As a result, blood volume decreases and IF volume increases. Of the three main body fluids, IF volume varies the most. Plasma volume usually fluctuates only slightly and briefly. If a pronounced change in its volume occurs, adequate circulation cannot be maintained.

Fluid Imbalances

Fluid imbalances are common ailments. They take several forms and stem from a variety of causes, but they all share a common characteristic—that of abnormally low or abnormally high volumes of one or more body fluids.

Dehydration is the fluid imbalance seen most often. In this potentially dangerous condition, IF volume decreases first, but eventually, if treatment has not been given, ICF and plasma volumes also decrease below normal levels. Either too small a fluid intake or too large a fluid output causes dehydration. Prolonged diarrhea or vomiting may result in dehydration due to the loss of body fluids. This is particularly true in infants where the total fluid volume is much smaller than it is in adults. Loss of skin elasticity is a clinical sign of dehydration (Figure 17-7).

Overhydration can also occur but is much less common than dehydration. The grave danger of giving intravenous fluids too rapidly or in too large amounts is overhydration, which can put too heavy a burden on the heart.

FIGURE 17–7 *Testing for Dehydration.* Loss of skin elasticity is a sign of dehydration. Skin that does not return quickly to its normal shape after being pinched indicates interstitial water loss.

O U T L I N E   S U M M A R Y

BODY FLUIDS

A Major locations—inside cells (ICF) and outside of cells in extracellular fluid (ECF) located between the cells as interstitial fluid (IF) and in blood vessels (plasma) (Figure 17-1)

B Total body fluid is approximately 50% to 60% of body weight in a healthy adult.

C Variation in total body water related to:
  1 Total body weight—the more a person weighs, the more water the body contains
  2 Fat content of body—the more fat present, the less total water content per unit of weight (Fat is almost water free.)
  3 Sex—proportion of body weight represented by water is about 10% less in women than in men (Figure 17-2)
  4 Age—in newborn infant water may account for 80% of total body weight

MECHANISMS THAT MAINTAIN FLUID BALANCE

A Fluid output, mainly urine volume, adjusts to fluid intake; ADH from posterior pituitary gland acts to increase kidney tubule reabsorption of sodium and water from tubular urine into blood, thereby tending to increase ECF (and total body fluid) by decreasing urine volume (Figure 17-5)

B ECF electrolyte concentration (mainly Na^+ concentration) influences ECF volume; an increase in ECF Na^+ tends to increase ECF volume by increasing movement of water out of ICF and by increasing ADH secretion, which decreases urine volume, and this, in turn, increases ECF volume

C Capillary blood pressure pushes water out of blood, into IF; blood protein concentration pulls water into blood from IF; hence these two forces regulate plasma and IF volume under usual conditions

D Importance of electrolytes in body fluids
  1 Nonelectrolytes—organic substances that do not break up or dissociate when placed in water solution (for example, glucose)
  2 Electrolytes—compounds that break up or dissociate in water solution into separate particles called ions (for example, ordinary table salt or sodium chloride)
  3 Ions—the dissociated particles of an electrolyte that carry an electrical charge (for example, sodium ion [Na^+])
  4 Positively charged ions (for example, potassium [K^+] and sodium [Na^+]
  5 Negatively charged particles (ions) (for example, chloride [Cl^-] and bicarbonate [HCO_3^-])
  6 Electrolyte composition of blood plasma—Table 17-2
  7 Sodium—most abundant and important positively charged ion of plasma
    a Normal plasma level—142 mEq/L
    b Average daily intake (diet)—100 mEq
    c Chief method of regulation—kidney
    d Aldosterone increases Na^+ reabsorption in kidney tubules (Figure 17-5)
    e Sodium containing internal secretions—Figure 17-6

E Capillary blood pressure and blood proteins

FLUID IMBALANCES

A Dehydration—total volume of body fluids smaller than normal; IF volume shrinks first, and then if treatment is not given, ICF volume and plasma volume decrease; dehydration occurs when fluid output exceeds intake for an extended period

B Overhydration—total volume of body fluids larger than normal; overhydration occurs when fluid intake exceeds output; various factors may cause this (for example, giving excessive amounts of intravenous fluids or giving them too rapidly may increase intake above output)

N E W   W O R D S

dehydration
dissociate
diuretic

edema
electrolyte
extracellular fluid (ECF)

interstitial fluid (IF)
intracellular fluid (ICF)

ions
overhydration

CHAPTER TEST

1. The largest volume of water in the body is classified as _____ fluid.
2. ECF can be subdivided into the fluid between cells, called the _____ fluid, and the fluid in blood, called _____.
3. Substances that do not break up or dissociate in water solution are called _____.
4. The dissociated particles of an electrolyte in water solution are called _____.
5. The largest quantity of water leaving the body exits as _____ produced by the kidneys.
6. The most abundant positively charged ion of blood plasma is _____, and the most abundant negatively charged ion is _____.
7. A drug that promotes or stimulates the production of urine is called a _____.
8. The presence of abnormally large amounts of fluid in the intercellular tissue spaces of the body is called _____.

Circle the "T" before each true statement and the "F" before each false statement.

T F **9.** About 60% of total body weight is water.
T F **10.** Blood plasma is one type of ECF.
T F **11.** There is more IF in the body than ICF.
T F **12.** The more fat in the body the more water per pound of weight.
T F **13.** The chief mechanism for maintaining fluid balance is to adjust fluid output to equal fluid intake.
T F **14.** Capillary blood pressure does not influence ECF or ICF volume.
T F **15.** Water is continually lost from the body by way of expired air and diffusion through the skin.
T F **16.** Electrolytes dissociate in solution to yield charged particles called ions.
T F **17.** Edema is a serious side effect of diuretic drugs.

REVIEW QUESTIONS

1. Suppose a person who had never heard the term *fluid balance* were to ask you what it meant. How would you explain it briefly and simply?
2. Approximately what percentage of a slender man's body weight consists of water? How does that compare to the percentage of body weight consisting of water in a slender woman?
3. The volume of blood plasma in a normal-sized adult weighs approximately what percentage of body weight?
4. The proportion of body weight represented by water is about 10% higher in men than in women. Why?
5. ICF makes up approximately what percentage of adult body weight?
6. IF makes up approximately what percentage of adult body weight?
7. To maintain fluid balance, does output usually change to match intake or does intake usually adjust to output?
8. Explain in words and by a diagram how ADH functions to maintain fluid balance.
9. Define electrolyte, ion, and nonelectrolyte.
10. List the important negatively and positively charged ions that are in blood plasma.
11. List the sodium-containing internal secretions.
12. Explain by words or diagram how capillary blood pressure and blood protein concentration function to maintain fluid balance.
13. If an individual becomes dehydrated, which fluid volume decreases first?

CRITICAL THINKING

14. Use the phrase, *where sodium goes, water soon follows,* to explain how ECF electrolyte concentration affects fluid volumes.
15. Suppose that an individual has suffered a hemorrhage and that, as a result, her capillary blood pressure has decreased below normal. What change would occur in blood and IF volumes as a result of this decrease in capillary blood pressure?
16. Suppose that an individual has a type of kidney disease that allows plasma proteins to be lost in the urine and that, as a result, his plasma protein concentration decreases. How would this tend to change blood and IF volumes?

Acid-Base Balance

OBJECTIVES

*After you have completed this chapter,
you should be able to:*

1. Discuss the concept of pH and define the phrase *acid-base balance.*

2. Define the terms *buffer* and *buffer pair* and contrast strong and weak acids and bases.

3. Contrast the respiratory and urinary mechanisms of pH control.

4. Discuss compensatory mechanisms that may help return blood pH to near-normal levels in cases of pH imbalances.

5. Compare and contrast metabolic and respiratory types of pH imbalances.

One of the requirements for homeostasis and healthy survival is that the body maintain, or quickly restore, the **acid-base balance** of its fluids. Maintaining acid-base balance means keeping the concentration of hydrogen ions in body fluids relatively constant. This is of vital importance. If the hydrogen ion concentration veers away from normal even slightly, cellular chemical reactions cannot take place normally, and survival is thereby threatened.

● pH of Body Fluids

Water and all water solutions contain **hydrogen ions (H^+)** and **hydroxide ions (OH^-)** The term *pH* followed by a number indicates a solution's hydrogen ion concentration. More specifically, pH 7.0 means that a solution contains an equal concentration of hydrogen and hydroxide ions. Therefore pH 7.0 also means that a fluid is neutral in reaction (that is, neither acid nor alkaline) (Figure 18-1). The pH of water, for example, is 7.0. A pH higher than 7.0 indicates an alkaline or basic solution (that is, one with a lower concentration of hydrogen than hydroxide ions). The more alkaline a solution, the higher is its pH. A pH lower than 7.0 indicates an acid solution (that is, one with a higher hydrogen ion concentration than hydroxide ion concentration). The higher the hydrogen ion concentration, the lower the pH and the more acid a solution is. With a pH of about 1.6, gastric juice is the most acid substance in the body. Saliva has a pH of 7.7, on the alkaline side. Normally, the pH of arterial blood is about 7.45, and the pH of venous blood is about 7.35. By applying the information given in the last few sentences, you can deduce the answers to the following questions. Is arterial blood slightly acid or slightly alkaline? Is venous blood slightly acid or slightly alkaline? Which is a more accurate statement—venous blood is more acid than arterial blood or venous blood is less alkaline than arterial blood?

Arterial and venous blood are both slightly alkaline because both have a pH slightly higher than 7.0. Venous blood, however, is less alkaline than arterial blood because venous blood's pH of about 7.35 is slightly lower than arterial blood's pH of 7.45.

● Mechanisms That Control pH of Body Fluids

The body has three mechanisms for regulating the pH of its fluids. They are (1) the buffer mechanism, (2) the respiratory mechanism, and (3) the urinary mechanism. Together, they constitute the complex pH homeostatic mechanism—the machinery that normally keeps blood slightly alkaline with a pH that stays remarkably constant. Its usual limits are very narrow, about 7.35 to 7.45.

The slightly lower pH of venous blood compared with arterial blood results primarily from carbon dioxide (CO_2) entering venous blood as a waste product of cellular metabolism. As carbon dioxide enters the blood, some of it combines with water (H_2O) and is converted into carbonic acid by **carbonic anhydrase,** an enzyme found in red blood cells. The following chemical equation represents this reaction. If you need to review chemical formulas and equations, please refer to Appendix A.

$$CO_2 + H_2O \xrightarrow{\text{carbonic anhydrase}} H_2CO_3$$

The lungs remove the equivalent of over 30 L of carbonic acid each day from the venous blood by elimination of CO_2. This almost unbelievable quantity of acid is so well buffered that a liter of venous blood contains only about $^1/_{100,000,000}$ g more H^+ than does 1 L of arterial blood. What incredible constancy! The pH homeostatic mechanism does indeed control effectively—astonishingly so.

BUFFERS

Buffers are chemical substances that prevent a sharp change in the pH of a fluid when an acid or base is added to it. Strong acids and bases, if added to blood, would "dissociate" almost completely and release large quantities of H^+ or OH^- ions. The result would be drastic changes in blood pH. Survival itself depends on protecting the body from such drastic pH changes.

FIGURE 18–1 *The pH Range.* The overall pH range is expressed numerically on what is called a *logarithmic scale* of 1 to 14. This means that a change of 1 pH represents a tenfold difference in actual concentration of hydrogen ions. Note that, as the concentration of H^+ ions increases, the solution becomes increasingly acidic and the pH value decreases. As OH^- concentration increases, the pH value also increases, and the solution becomes more and more basic or alkaline. A pH of 7 is neutral; a pH of 1 is very acidic, and a pH of 13 is very basic.

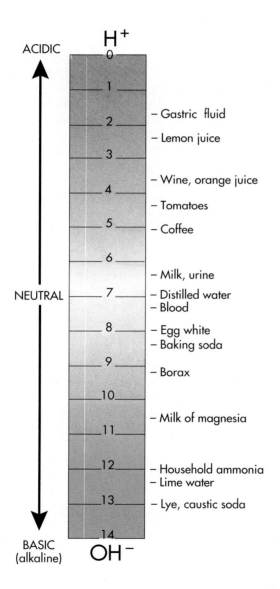

More acids than bases are usually added to body fluids. This is because catabolism, a process that goes on continually in every cell of the body, produces acids that enter blood as it flows through tissue capillaries. Almost immediately, one of the salts present in blood—a buffer, that is—reacts with these relatively strong acids to change them to weaker acids. The weaker acids decrease blood pH only slightly, whereas the stronger acids formed by catabolism would have decreased it greatly if they were not buffered.

Buffers consist of two kinds of substances and are therefore often called **buffer pairs.** One of the main blood buffer pairs is ordinary baking soda (sodium bicarbonate or $NaHCO_3$) and carbonic acid (H_2CO_3).

Let us consider, as a specific example of buffer action, how the $NaHCO_3 — H_2CO_3$ system works with a strong acid or base.

Addition of a strong acid, such as hydrochloric acid (HCl), to the $NaHCO_3 — H_2CO_3$ buffer system would initiate the reaction shown in Figure 18-2. Note how this reaction between HCl and $NaHCO_3$ applies the principle of buffering. As a result of the buffering action of $NaHCO_3$, the weak acid, H • HCO_3, replaces the very strong acid, HCl, and therefore the H^+ concentration of the blood increases much less than it would have if HCl were not buffered.

If, on the other hand, a strong base, such as sodium hydroxide (NaOH), were added to the same buffer system, the reaction shown in Figure 18-3 would take place. The H^+ of H_2CO_3 (H • HCO_3), the weak acid of the buffer pair, combines with the OH^- of the strong base NaOH to form H_2O. Note what this accomplishes. It decreases the number of OH^- added to the solution, and this in turn prevents the drastic rise in pH that would occur without buffering.

Figure 18-2 shows how a buffer system works with a strong acid. Although useful in demonstrating the principles of buffer action, HCl or similar strong acids are never introduced directly into body fluids under normal circumstances. Instead, the $NaCO_3$ buffer system is most often called on to buffer a number of weaker acids produced during catabolism. Lactic acid is a good example. As a weak acid, it does not "dissociate" as completely as HCl. Incomplete dissociation of lactic acid results in fewer hydrogen ions being

FIGURE 18–2 *Buffering Action of Sodium Bicarbonate.* Buffering of acid HCl by $NaHCO_3$. As a result of the buffer action, the strong acid (HCl) is replaced by a weaker acid (H • HCO_3). Note that HCl as a strong acid "dissociates" almost completely and releases more H^+ than H_2CO_3. Buffering decreases the number of H^+ in the system.

FIGURE 18–3 *Buffering Action of Carbonic Acid.* Buffering of base NaOH by H_2CO_3. As a result of buffer action, the strong base (NaOH) is replaced by $NaHCO_3$ and H_2O. As a strong base, NaOH "dissociates" almost completely and releases large quantities of OH^-. Dissociation of H_2O is minimal. Buffering decreases the number of OH^- in the system.

Bicarbonate Loading

The buildup of lactic acid in the blood, released as a waste product from working muscles, has been blamed for the soreness and fatigue that sometimes accompanies strenuous exercise. Some athletes have adopted a technique called **bicarbonate loading,** ingesting large amounts of sodium bicarbonate ($NaHCO_3$) to counteract the effects of lactic acid buildup. Their theory is that fatigue is avoided because the $NaHCO_3$, a base, buffers the lactic acid. Unfortunately, the diarrhea that often results can trigger fluid and electrolyte imbalances. Long-term $NaHCO_3$ abuse can lead to disruption of acid-base balance and its disastrous effects.

added to the blood and a less drastic lowering of blood pH than would occur if HCl were added in an equal amount. Without buffering, however, lactic acid buildup results in significant H^+ accumulation over time. The resulting decrease of pH can produce serious acidosis. Ordinary baking soda (sodium bicarbonate or $NaHCO_3$) is one of the main buffers of the normally occurring "fixed" acids in blood. Lactic acid is one of the most abundant of the "fixed" acids (acids that do not break down to form a gas). Figure 18-4 shows the compounds formed by buffering of lactic acid (a "fixed" acid), produced by normal catabolism. The following changes in blood result from buffering of fixed acids in tissue capillaries:

1. The amount of H_2CO_3 in blood increases slightly because an acid (such as lactic acid) is converted to H_2CO_3.

FIGURE 18–4 *Lactic Acid Buffered by Sodium Bicarbonate.* Lactic acid (H • lactate) and other "fixed" acids are buffered by $NaHCO_3$ in the blood. Carbonic acid (H • HCO_3 or H_2CO_3, a weaker acid than lactic acid) replaces lactic acid. As a result, fewer H^+ are added to blood than would be if lactic acid were not buffered.

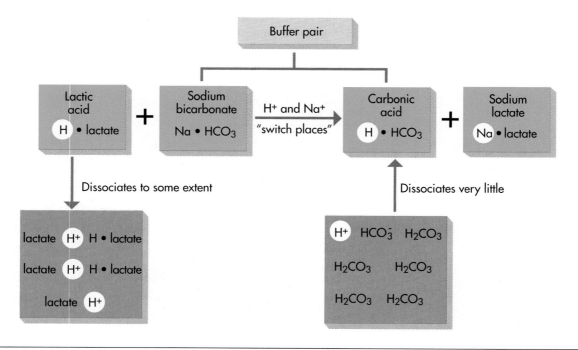

2. The amount of bicarbonate in blood (mainly $NaHCO_3$) decreases because bicarbonate ions become part of the newly formed H_2CO_3. Normal arterial blood with a pH of 7.45 contains 20 times more $NaHCO_3$ than H_2CO_3. If this ratio decreases, blood pH decreases below 7.45.

3. The H^+ concentration of blood increases slightly. H_2CO_3 adds hydrogen ions to blood, but it adds fewer of them than lactic acid would have because it is a weaker acid than lactic acid. In other words the buffering mechanisms do not totally prevent blood hydrogen ion concentration from increasing. It simply minimizes the increase.

4. Blood pH decreases slightly because of the small increase in blood concentration.

H_2CO_3 is the most abundant acid in body fluids because it is formed by the buffering of fixed acids and also because CO_2 forms it by combining with H_2O. Large amounts of CO_2, an end product of catabolism, continually pour into tissue capillary blood from cells. Much of the H_2CO_3 formed in blood diffuses into red blood cells where it is buffered by the potassium salt of hemoglobin. H_2CO_3 breaks down to form the gas, CO_2, and H_2O. This takes place in the blood as it moves through the lung capillaries. Read the next several paragraphs to find out how this affects blood pH.

RESPIRATORY MECHANISM OF pH CONTROL

Respirations play a vital part in controlling pH. With every expiration, CO_2 and H_2O leave the body in the expired air. The CO_2 has diffused out of the venous blood as it moves through the lung capillaries. Less CO_2 therefore remains in the arterial blood leaving the lung capillaries, so less of it is available for combining with water to form H_2CO_3. Hence the arterial blood contains less H_2CO_3 and fewer hydrogen ions and has a higher pH (7.45) than does the venous blood (pH 7.35).

Let us consider now how a change in respirations can change blood pH. Suppose you were to pinch your nose shut and hold your breath for a full minute or a little longer. Obviously, no CO_2 would leave your body by way of the expired air during that time, and the blood's CO_2 content would necessarily increase. This would increase

the amount of H_2CO_3 and the hydrogen-ion concentration of blood, which in turn would decrease blood pH. Here then are two useful facts to remember. Anything that causes an appreciable decrease in respirations will in time produce **acidosis**. Conversely, anything that causes an excessive increase in respirations will in time produce **alkalosis.**

URINARY MECHANISM OF pH CONTROL

Most people know that the kidneys are vital organs and that life soon ebbs away if they stop functioning. One reason is that the kidneys are the body's most effective regulators of blood pH. They can eliminate much larger amounts of acid than can the lungs and, if it becomes necessary, they can also excrete excess base. The lungs cannot. In short, the kidneys are the body's last and best defense against wide variations in blood pH. If they fail, homeostasis of pH—acid-base balance—fails.

Because more acids than bases usually enter blood, more acids than bases are usually excreted by the kidneys. In other words, most of the time the kidneys acidify urine; that is, they excrete

 Ketosis

An important part of home care for diabetics involves monitoring the level of glucose in the blood and, especially for patients taking insulin, carefully watching for the appearance of **ketone bodies** in the urine. Accumulation of these acidic substances in the blood results from the excessive metabolism of fats in uncontrolled diabetics. These individuals have trouble metabolizing carbohydrates and instead burn fat as a primary energy source. The accumulation of ketone bodies results in a condition called **ketosis** that causes the blood to become dangerously acidic. As blood levels of ketones increase they "spill over" into the urine and can be detected by use of appropriate reagent strips.

FIGURE 18–5   *Acidification of Urine and Conservation of Base by Distal Renal Tubule Secretion of H⁺ Ions.*

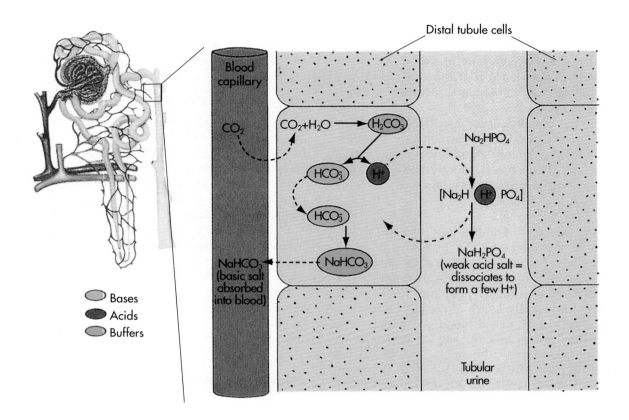

enough acid to give urine an acid pH frequently as low as 4.8. (How does this compare with normal blood pH?) The distal tubules of the kidneys rid the blood of excess acid and at the same time conserve the base present in it by the two mechanisms illustrated by Figures 18-5 and 18-6. To understand these figures fully, you need to have some grasp of basic chemistry. If necessary, refer to Appendix A before proceeding. Then look at Figure 18-5 and find the CO_2 leaving the blood (as it flows through a kidney capillary) and entering one of the cells that helps form the wall of a distal kidney tubule. Note that in this cell the CO_2 combines with water to form H_2CO_3. This occurs rapidly because the cell contains carbonic anhydrase, an enzyme that accelerates this reaction. As soon as H_2CO_3 forms, some of it dissociates to

yield hydrogen ions and bicarbonate ions. Note what happens to these ions. Hydrogen ions diffuse out of the tubule cell into the urine trickling down the tubule. There, it replaces one of the sodium ions (Na^+) in a salt (Na_2HPO_4) to form another salt (NaH_2PO_4), which leaves the body in the urine. Notice next that the Na^+ displaced from Na_2HPO_4 by the H^+ moves out of the tubular urine into a tubular cell. Here it combines with a bicarbonate (HCO_3^-) ion to form sodium bicarbonate, which then is resorbed into the blood. What this complex of reactions has accomplished is to add hydrogen ions to the urine—that is, acidify it—and to conserve $NaHCO_3$ by reabsorbing it into the blood.

Figure 18-6 illustrates another method of acidifying urine, as explained in the legend.

FIGURE 18–6 *Acidification of Urine by Tubule Secretion of Ammonia (NH₃).* An amino acid (glutamine) moves into the tubule cell and loses an amino group (NH₂) to form ammonia, which is secreted into urine. In exchange, the tubule cell reabsorbs a basic salt (mainly NaHCO₃) into blood from urine.

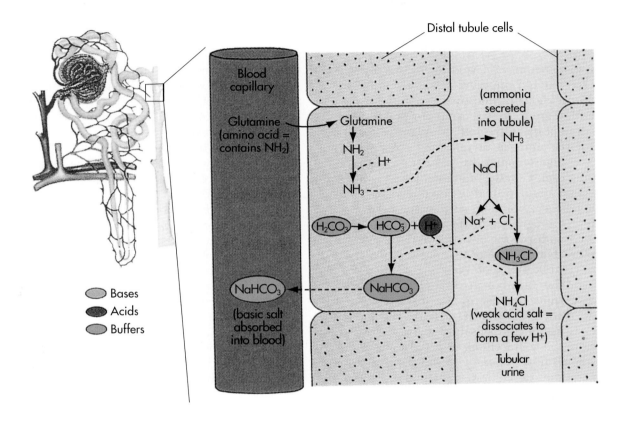

pH Imbalances

Acidosis and **alkalosis** are the two kinds of pH or acid-base imbalance. In acidosis the blood pH falls as H⁺ ion concentration increases. Only rarely does it fall as low as 7.0 (neutrality) and almost never does it become even slightly acid, because death usually intervenes before the pH drops this much. In alkalosis, which develops less often than acidosis, the blood pH is higher than normal.

From a clinical standpoint, disturbances in acid-base balance can be considered dependent on the relative quantities (ratio) of H_2CO_3 and $NaHCO_3$ in the blood. Components of this important buffer pair must be maintained at the proper ratio (20 times more $NaHCO_3$ than H_2CO_3) if acid-base balance is to remain normal. It is fortunate that the body can regulate both chemicals in the $NaHCO_3$ — H_2CO_3 buffer system. Blood levels of $NaHCO_3$ can be regulated by the kidneys and H_2CO_3 levels by the respiratory system (lungs).

METABOLIC AND RESPIRATORY DISTURBANCES

Two types of disturbances, metabolic and respiratory, can alter the proper ratio of these components. Metabolic disturbances affect the bicarbonate ($NaHCO_3$) element of the buffer pair, and respiratory disturbances affect the H_2CO_3 element, as follows:

1. **Metabolic disturbances**
   a. *Metabolic acidosis* (bicarbonate deficit)
   b. *Metabolic alkalosis* (bicarbonate excess)
2. **Respiratory disturbances**
   a. *Respiratory acidosis* (H_2CO_3 excess)
   b. *Respiratory alkalosis* (H_2CO_3 deficit)

VOMITING

Vomiting, sometimes referred to as *emesis* (EM-e-sis), is the forcible emptying or expulsion of gastric and occasionally intestinal contents through the mouth (Figure 18-7). It can occur as a result of many stimuli, including foul odors or tastes, irritation of the stomach or intestinal mucosa, and some vomitive (emetic) drugs such as ipecac. A "vomiting center" in the brain regulates the many coordinated (but primarily involuntary) steps involved. Severe vomiting such as the pernicious vomiting of pregnancy or the repeated vomiting associated with pyloric obstruction in infants can be life threatening. One of the most frequent and serious complications of vomiting is metabolic alkalosis. The bicarbonate excess of metabolic alkalosis results because of the massive loss of chloride from the stomach as HCl. The loss of chloride causes a compensatory increase of bicarbonate in the extracellular fluid. The result is metabolic alkalosis. Therapy includes intravenous administration of chloride-containing solutions such as **normal saline**. The chloride ions of the solution replace bicarbonate ions and thus help relieve the bicarbonate excess responsible for the imbalance.

The *ratio* of $NaHCO_3$ to H_2CO_3 levels in the blood is the key to acid-base balance. If the normal ratio (20:1 $NaHCO_3/H_2CO_3$) can be maintained, the acid-base balance and pH remain normal despite changes in the absolute amounts of either component of the buffer pair in the blood.

As a clinical example, in a person suffering from untreated diabetes, abnormally large amounts of acids enter the blood. The normal 20:1 ratio is altered as the $NaHCO_3$ component of the buffer pair reacts with the acids. Blood levels of $NaHCO_3$ decrease rapidly in these patients. The result is a lower ratio of $NaHCO_3$ to H_2CO_3 (perhaps 10:1) and lower blood pH. The condition is called **uncompensated metabolic acidosis.** The body attempts to correct or *compensate* for the acidosis by altering the *ratio* of $NaHCO_3$ to H_2CO_3. Acidosis in a diabetic patient is often accompanied by rapid breathing or hyperventilation. This compensatory action of the respiratory system results in a "blow-off" of CO_2. Decreased blood levels of CO_2 result in lower H_2CO_3 levels. A new compensated ratio of $NaHCO_3$ to H_2CO_3 (perhaps 10:0.5) may result. In such individuals the blood pH returns to normal or near-normal levels. The condition is called **compensated metabolic acidosis.**

FIGURE 18–7   *The Vomit Reflex.* Severe vomiting results in significant loss of HCl and often leads to metabolic alkalosis.

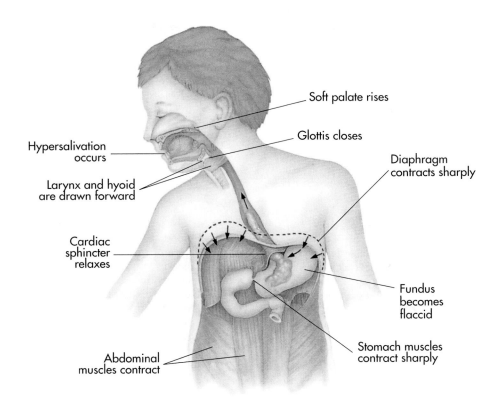

pH OF BODY FLUIDS

A Definition of pH—a number that indicates the hydrogen ion (H^+) concentration of a fluid; pH 7.0 indicates neutrality, pH higher than 7.0 indicates alkalinity, and pH less than 7.0 indicates acidity—see Figure 18-1

B Normal arterial blood pH—about 7.45

C Normal venous blood pH—about 7.35

MECHANISMS THAT CONTROL pH OF BODY FLUIDS

A Buffers

  1 Definition—substances that prevent a sharp change in the pH of a fluid when an acid or base is added to it—see Figures 18-2 and 18-3

  2 "Fixed" acids are buffered mainly by sodium bicarbonate ($NaHCO_3$)

  3 Changes in blood produced by buffering of "fixed" acids in the tissue capillaries

  a Amount of carbonic acid (H_2CO_3) in blood increases slightly

  b Amount of $NaHCO_3$ in blood decreases; ratio of amount of $NaHCO_3$ to the amount of H_2CO_3 does not normally change; normal ratio is 20:1

  c H^+ concentration of blood increases slightly

  d Blood pH decreases slightly below arterial level

B Respiratory mechanism of pH control—respirations remove some CO_2 from blood as blood flows through lung capillaries, the amount of H_2CO_3 in blood is decreased and thereby its H^+ concentration is decreased, and this in turn increases blood pH from its venous to its arterial level

C Urinary mechanism of pH control—the body's most effective regulator of blood pH, kidneys usually acidify urine by the distal tubules secreting hydro-gen ions and ammonia (NH_3) into the urine from blood in exchange for $NaHCO_3$ being reabsorbed into the blood

pH IMBALANCES

A Acidosis and alkalosis are the two kinds of pH or acid-base imbalances

B Disturbances in acid-base balance depend on relative quantities of $NaHCO_3$ and H_2CO_3 in the blood

C Body can regulate both of the components of the $NaHCO_3$ — H_2CO_3 buffer system

  1 Blood levels of $NaHCO_3$ regulated by kidneys

  2 H_2CO_3 levels regulated by lungs

D Two basic types of pH disturbances, metabolic and respiratory, can alter the normal 20:1 ratio of $NaHCO_3$ to H_2CO_3 in blood

  1 Metabolic disturbances affect the $NaHCO_3$ levels in blood

  2 Respiratory disturbances affect the H_2CO_3 levels in blood

E Types of pH or acid-base imbalances

  1 Metabolic disturbances

  a Metabolic acidosis—bicarbonate ($NaHCO_3$) deficit

  b Metabolic alkalosis—bicarbonate ($NaHCO_3$) excess; complication of severe vomiting

  2 Respiratory disturbances

  a Respiratory acidosis (H_2CO_3 excess)

  b Respiratory alkalosis (H_2CO_3 deficit)

F In uncompensated metabolic acidosis, the normal ratio of $NaHCO_3$ to H_2CO_3 is changed; in compensated metabolic acidosis, the ratio remains at 20:1, but the total amount of $NaHCO_3$ and H_2CO_3 changes

---

N E W   W O R D S

acid solution
acidosis (metabolic and respiratory)

alkaline solution
alkalosis (metabolic and respiratory)

base
buffer
buffer pairs

carbonic anhydrase
emesis
pH

CHAPTER TEST

1. A fluid having a pH of 7.0 would be described as _____ in reaction.
2. A pH higher than 7.0 indicates an _____ solution, and one with a pH lower than 7.0 indicates an _____ solution.
3. As CO_2 enters blood, some of it is converted into H_2CO_3 by an enzyme called _____ found in red blood cells.
4. Substances that prevent a sharp change in pH of a fluid when an acid or base is added are called

   _____.
5. The most abundant acid in the body is _____ acid.
6. Anything that causes an appreciable decrease in respirations will in time produce the condition called _____.
7. Acidification of urine can occur by renal tubule excretion of _____.
8. One of the most frequent complications of vomiting is metabolic _____.
9. A body deficit of $NaHCO_3$ results in metabolic

   _____.
10. The ability of the body to maintain a constant blood pH during changing conditions is an example of _____.

Select the most correct answer from Column B for each statement in Column A. (Only one answer is correct.)

COLUMN A	COLUMN B
11. _____ pH of arterial blood	a. Lactic acid
	b. pH 7.0
12. _____ Neutral in reaction	c. Excretion of ammonia
	d. Buffers
13. _____ Lost as CO_2	e. Metabolic alkalosis
14. _____ Prevent pH changes	f. pH 7.45
	g. Respiratory acidosis
15. _____ "Fixed" acid	h. pH
16. _____ Hydrogen ion concentration	i. Hyperventilation
	j. H_2CO_3
17. _____ Acidification of urine	
18. _____ Vomiting	
19. _____ H_2CO_3 excess	
20. _____ Compensates for acidosis	

REVIEW QUESTIONS

1. Explain briefly what the term *pH* means.
2. What is a typical normal pH for venous blood? For arterial blood?
3. What are the functions of buffers?
4. Explain how respirations affect blood pH.
5. Explain how the kidneys maintain normal blood pH.
6. Which is the most important for maintaining acid-base balance buffers, respirations or kidney functioning?
7. Briefly, how does compensated acidosis differ from uncompensated acidosis?
8. List the types of pH or acid-base imbalances.

CRITICAL THINKING

9. When the body is in acid-base balance, arterial blood contains how many times more sodium bicarbonate than carbonic acid? In other words, what is the normal $NaHCO_3/H_2CO_3$ ratio in blood?
10. How does prolonged hyperventilation (abnormally increased respirations) affect blood pH?

The Reproductive Systems

19

OBJECTIVES

After you have completed this chapter, you should be able to:

1. List the essential and accessory organs of the male and female reproductive systems and give the generalized function of each.

2. Describe the gross and microscopic structure of the gonads in both sexes and explain the developmental steps in spermatogenesis and oogenesis.

3. Discuss the primary functions of the sex hormones and identify the cell type or structure responsible for their secretion.

4. Identify and describe the structures that constitute the external genitals in both sexes.

5. Identify and discuss the phases of the endometrial or menstrual cycle and correlate each phase with its occurrence in a typical 28-day cycle.

The offspring of many one-celled plants and bacteria come from a single parent. These organisms are said to be *asexual* because they do not produce specialized reproductive or sex cells called **gametes** (GAM-eets). In humans, gametes, called **ova** and **sperm,** fuse during the process of fertilization to produce a cell called the **zygote** (ZYE-gote), which ultimately develops into the new individual. The zygote, which contains an intermingling of genetic messages from the sex cells of both parents, ultimately permits development of new human life. Reproduction in humans therefore is said to be *sexual.* As with all sexually produced offspring, new human life results from the equal contribution of not one but two parent cells—the female ovum and male sperm.

This chapter deals with the structure and function of the reproductive system in men and women. We are truly "fearfully and wonderfully made." Almost any one of the body's organ systems might have inspired this statement, but of them all, perhaps the reproductive systems best deserve such praise. Their awesome achievement is the creation of nature's most complex and beautiful structure—the human body; their ultimate goal is the survival of the human species. After study of the reproductive system in both sexes, Chapter 20 will cover the topic of human development—a process extending from fertilization until death.

Common Structural and Functional Characteristics Between the Sexes

Although the organs and specific functions of the male and female reproductive systems will be discussed separately, it is important to understand that a common general structure and function can be identified between the systems in both sexes and that both sexes contribute in uniquely important ways to overall reproductive success.

In both men and women, the organs of the reproductive system are adapted for the specific sequence of functions that permit development of sperm or ova followed by successful fertilization and then the normal development and birth of a baby. In addition, production of hormones that permits development of secondary sex characteristics, such as breast development in women and beard growth in men, occurs as a result of normal reproductive system activity.

As you study the specifics of each system keep in mind that the male organs function to produce, store, and ultimately introduce mature sperm into the female reproductive tract and that the female system is designed to produce ova, receive the sperm, and permit fertilization. In addition, the highly developed and specialized reproductive system in women permits the fertilized ovum to develop and mature until birth. The complex and cyclic control of reproductive functions in women are particularly crucial to overall reproductive success in humans. The production of sex hormones is required not only for development of the secondary sexual characteristics but also for normal reproductive functions in both sexes. This chapter will end with a table comparing reproductive structures and functions in women and men.

Male Reproductive System

STRUCTURAL PLAN

So many organs make up the male reproductive system that we need to look first at the structural plan of the system as a whole. Reproductive organs can be classified as **essential** or **accessory.**

Essential Organs

The essential organs of reproduction in men and women are called the **gonads.** The gonads of men consist of a pair of main sex glands called the **testes** (TES-teez). The testes produce the male sex cells or **spermatozoa** (sper-ma-toe-ZO-ah).

Accessory Organs

The accessory organs of reproduction in men consist of the following structures:

1. A series of passageways or ducts that carry the sperm from the testes to the exterior
2. Additional sex glands that provide secretions that protect and nurture sperm
3. The external reproductive organs called the external genitals

Table 19-1 lists the names of the essential and accessory organs of reproduction in men, and Figure 19-1 shows the location of most of them. The table and the illustration are included very early in the chapter to provide a preliminary but important overview. Refer back to this table and illustration frequently as you learn about each organ in the pages that follow.

TESTES

Structure and Location

The paired **testes** are the gonads of men. They are located in the pouchlike **scrotum** (SKRO-tum), which is suspended outside of the body cavity behind the penis (Figure 19-1). This exposed location provides an environment about 1°C (3°F) cooler than normal body temperature, an important requirement for the normal production and survival of sperm. Each testis is a small, oval gland about 3.8 cm (1½ inches) long and 2.5 cm (1

TABLE 19–1 Male Reproductive Organs	
ESSENTIAL ORGANS	**ACCESSORY ORGANS**
Gonads: testes (right testis and left testis)	Ducts: epididymis (two), vas deferens (two), ejaculatory duct (two), and urethra
	Supportive sex glands: seminal vesicle (two), bulbourethral or Cowper's gland (two), and prostate gland
	External genitals: scrotum and penis

FIGURE 19–1 *Organization of the Male Reproductive Organs.*

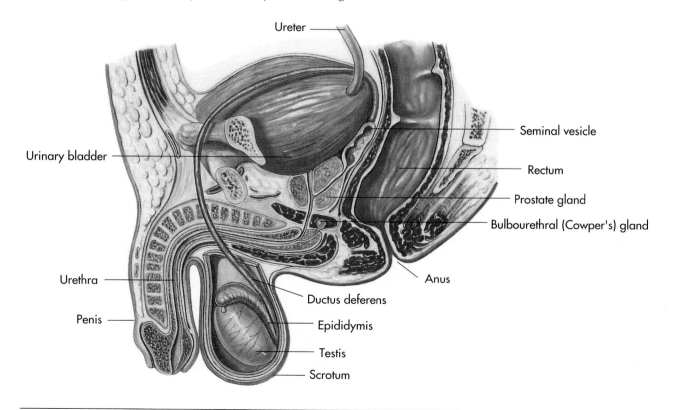

inch) wide. The testis is shaped like an egg that has been flattened slightly from side to side. Note in Figure 19-2 that each testis is surrounded by a tough, whitish membrane called the **tunica** (TOO-ni-kah) **albuginea** (al-byoo-JIN-ee-ah). This membrane covers the testicle and then enters the gland to form the many septa that divide it into sections or lobules. As you can see in Figure 19-2, each lobule consists of a narrow but long and coiled **seminiferous** (se-mi-NIF-er-us) **tubule**. These coiled structures form the bulk of the testicular tissue mass. Small, specialized cells lying near the septa that separate the lobules can be seen in Figure 19-3. These are the **interstitial cells** of the testes that secrete the male sex hormone **testosterone** (tes-TOS-te-rone).

Each seminiferous tubule is a long duct with a central lumen or passageway (Figure 19-3). Sperm develop in the walls of the tubule and are then released into the lumen and begin their journey to the exterior of the body.

FIGURE 19–3 *Testis Tissue.* Several seminiferous tubules surrounded by septa containing interstitial cells are shown.

Tunica albuginea

Interstitial cells

Seminiferous tubule

Developing sperm cells

FIGURE 19–2 *Tubules of the Testis and Epididymis.* The ducts and tubules are exaggerated in size. In the photograph, the testicle is the darker sphere in the center.

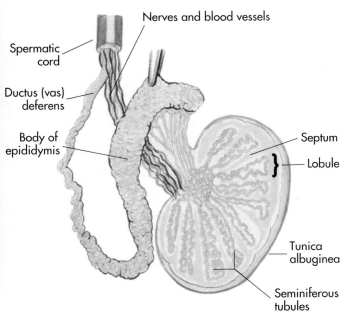

Nerves and blood vessels

Spermatic cord

Ductus (vas) deferens

Body of epididymis

Septum

Lobule

Tunica albuginea

Seminiferous tubules

Testis Functions

Spermatogenesis. Sperm production is called **spermatogenesis** (sper-ma-toe-JEN-e-sis). From puberty on, the seminiferous tubules continuously form spermatozoa or sperm. Although the number of sperm produced each day diminish with increasing age, most men continue to produce significant numbers throughout life.

The testes prepare for sperm production before puberty by increasing the numbers of sperm precursor (stem) cells called **spermatogonia** (sper-ma-toe-GO-nee-ah). These cells are located near the outer edge of each seminiferous tubule (Figure 19-4, *A*). Before puberty, spermatogonia increase in number by the process of mitotic cell division, which was described in Chapter 2. Recall that mitosis results in the division of a "parent" cell into two "daughter" cells, each identical to the parent and each containing a complete copy of the genetic material represented in the normal number of 46 chromosomes.

When a boy enters puberty, circulating levels of follicle-stimulating hormone (FSH) cause a spermatogonium to undergo a unique type of cell division. When the spermatogonium undergoes cell division and mitosis under the influence of FSH, it produces two daughter cells. One of these cells remains as a spermatogonium and the other forms another, more specialized cell called a **primary spermatocyte** (SPER-ma-toe-cyte). These primary spermatocytes then undergo a specialized type of division called **meiosis** (my-O-sis), which ultimately results in sperm formation. Note in Figure 19-4, *B,* that during meiosis two cell divisions occur (not one as in mitosis) and that four daughter cells (not two as in mitosis) are formed. The daughter cells are called **spermatids** (SPER-ma-tids). Unlike the two daughter cells that result from mitosis, the four spermatids, which will develop into spermatozoa, have only half the genetic material and half of the chromosomes (23) of other body cells.

In women, meiosis results in a single ovum, which also has 23 chromosomes. This will be discussed in more detail later in the chapter.

Look again at the diagram of meiosis in Figure 19-4, *B.* It shows that each primary spermatocyte ultimately produces four sperm cells. Note that, in the portion of a seminiferous tubule shown in Figure 19-4, *B,* spermatogonia are found at the outer surface of the tubule, primary and secondary spermatocytes lie deeper in the tubule wall, and mature but immotile sperm are seen about to enter the lumen of the tube and begin their journey through the reproductive ducts to the exterior of the body.

Cryptorchidism

Early in fetal life the testes are located in the abdominal cavity but normally descend into the scrotum about 2 months before birth. Occasionally a baby is born with undescended testes, a condition called **cryptorchidism** (krip-TOR-ki-dizm), which is readily observed by palpation of the scrotum at delivery. The word *cryptorchidism* is from the Greek words *kryptikos* (hidden) and *orchis* (testis). Failure of the testes to descend may be caused by hormonal imbalances in the developing fetus or by a physical deficiency or obstruction. Regardless of cause, in the cryptorchid infant the testes remain "hidden" in the abdominal cavity. Because the higher temperature inside the body cavity inhibits spermatogenesis, measures must be taken to bring the testes down into the scrotum to prevent permanent sterility. Early treatment of this condition by surgery or by injection of testosterone, which stimulates the testes to descend, may result in normal testicular and sexual development.

FIGURE 19–4 *Spermatogenesis.* **A,** Cross section of tubule shows progressive meiotic cell types in the wall of a seminiferous tubule. **B,** Diagram of meiotic events and cell types leading to sperm formation.

Spermatozoa. Spermatozoa are among the smallest and most highly specialized cells in the body (Figure 19-5, *A*). All of the characteristics that a baby will inherit from its father at fertilization are contained in the condensed nuclear (genetic) material found in each sperm head. However, this genetic information from the father can fuse with genetic material contained in the mother's ovum only if successful fertilization occurs. Ejaculation of sperm into the female vagina during sexual intercourse is only one step in the long journey that these sex cells must make before they can meet and fertilize an ovum. To accomplish their task, these specialized packages of genetic information are equipped with tails for motility and are designed to penetrate the outer membrane of the ovum when contact occurs with it.

The structure of a mature sperm is diagrammed in Figure 19-5, *B.* Note the sperm head containing the nucleus with its genetic material from the father. The nucleus is covered by the **acrosome** (AK-ro-sohm)—a specialized structure containing enzymes that enable the sperm to break down the covering of the ovum and permit entry if contact occurs. In addition to the head with its covering acrosome, each sperm has a midpiece and an elongated tail. Mitochondria in the midpiece break down adenosine triphosphate (ATP) to provide energy for the tail movements required to propel the sperm and allow them to "swim" for relatively long distances through the female reproductive ducts.

Production of testosterone. In addition to spermatogenesis, the other function of the testes is to secrete the male hormone, testosterone. This function is carried on by the interstitial cells of the testes, not by their seminiferous tubules. Testosterone serves the following general functions:

1. It masculinizes. The various characteristics that we think of as "male" develop because of testosterone's influence. For instance, when a young boy's voice changes, it is testosterone that brings this about.
2. It promotes and maintains the development of the male accessory organs (prostate gland, seminal vesicles, and so on).

FIGURE 19–5 *Human Sperm.* **A,** Micrograph shows the heads and long, slender tails of several spermatozoa. **B,** Illustration shows the components of a mature sperm cell and an enlargement of a sperm head.

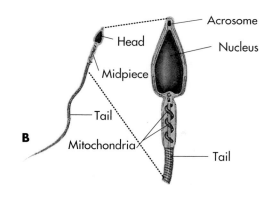

3. It has a stimulating effect on protein anabolism. Testosterone thus is responsible for the greater muscular development and strength of the male.

A good way to remember testosterone's functions is to think of it as "the masculinizing hormone" and the "anabolic hormone."

REPRODUCTIVE DUCTS

The ducts through which sperm must pass after exiting from the testes until they reach the exterior of the body are important components of the accessory reproductive structures. The other two components included in the listing of accessory organs of reproduction in the male—the supportive sex glands and external genitals—will be discussed separately.

Sperm are formed within the walls of the seminiferous tubules of the testes. When they exit from these tubules within the testis, they enter and then pass, in sequence, through the epididymis, ductus (vas) deferens, ejaculatory duct, and the urethra on their journey out of the body.

Epididymis

Each **epididymis** (ep-i-DID-i-mis) consists of a single and very tightly coiled tube about 6 m (20 feet) in length. It is a comma-shaped structure (see Figure 19-2) that lies along the top and behind the testes inside the scrotum. Sperm mature and develop their ability to move or swim as they pass through the epididymis.

Ductus (Vas) Deferens

The **ductus** (DUK-tus) **deferens** (DEF-er-enz) or **vas deferens** is the tube that permits sperm to exit from the epididymis and pass from the scrotal sac upward into the abdominal cavity. Each ductus deferens is a thick, smooth, very muscular, and movable tube that can easily be felt or "palpated" through the thin skin of the scrotal wall. It passes through the inguinal canal into the abdominal cavity as part of the *spermatic cord,* a connective tissue sheath that also encloses blood vessels and nerves.

Ejaculatory Duct and Urethra

Once in the abdominal cavity, the ductus deferens extends over the top and down the posterior surface of the bladder, where it joins the duct from the seminal vesicle to form the **ejaculatory** (ee-JAK-yoo-lah-toe-ree) **duct.** Note in Figure 19-1 that the ejaculatory duct passes through the substance of the prostate gland and permits sperm to empty into the **urethra,** which eventually passes through the penis and opens to the exterior at the external urethral orifice.

ACCESSORY OR SUPPORTIVE SEX GLANDS

The term **semen** (SEE-men) or **seminal fluid** is used to describe the mixture of sex cells or sperm produced by the testes and the secretions of the accessory or supportive sex glands. The accessory glands, which contribute over 95% of the secretions to the gelatinous fluid part of the semen, include the two seminal vesicles, one prostate gland, and two bulbourethral (Cowper's) glands. In addition to the production of sperm, the seminiferous tubules of the testes contribute somewhat less than 5% of the seminal fluid volume. Usually 3 to 5 ml (about 1 teaspoon) of semen is ejaculated at one time, and each milliliter normally contains about 100 million sperm. Semen is alkaline and protects sperm from the acidic environment of the female reproductive tract.

Seminal Vesicles

The paired **seminal vesicles** are pouchlike glands that contribute about 60% of the seminal fluid volume. Their secretions are yellowish, thick, and rich in the sugar fructose. This fraction of the seminal fluid helps provide a source of energy for the highly motile sperm.

Prostate Gland

The **prostate gland** lies just below the bladder and is shaped like a doughnut. The urethra passes through the center of the prostate before traversing the penis to end at the external urinary orifice. The prostate secretes a thin, milk-colored fluid that constitutes about 30% of the total seminal fluid volume. This fraction of the ejaculate helps to activate the sperm and maintain their motility.

Bulbourethral Glands

Each of the two **bulbourethral** (BUL-bo-yoo-REE-thral) glands (also called *Cowper's glands*) resemble peas in size and shape. They are located just

FIGURE 19–6 *The Penis.* **A,** In this sagittal section of the penis viewed from above, the urethra is exposed throughout its length and can be seen exiting from the bladder and passing through the prostate gland before entering the penis to end at the external urethral orifice. **B,** Photograph of a cross section of the shaft of the penis showing the three columns of erectile or cavernous tissue. Note the urethra within the substance of the corpus spongiosum.

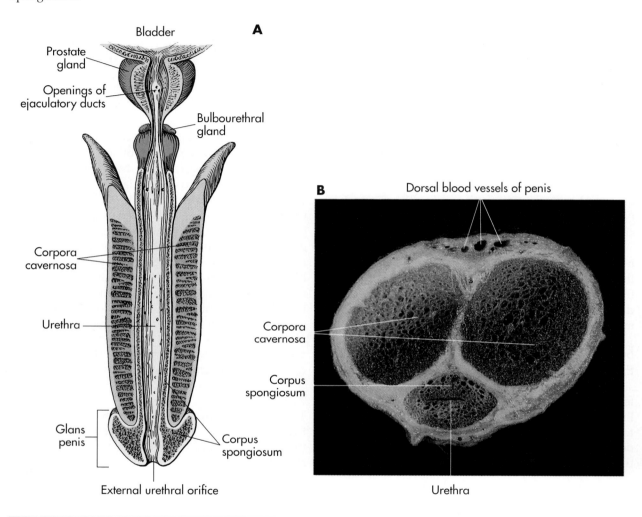

below the prostate gland and empty their secretions into the penile portion of the urethra. The mucus-like secretions of these glands lubricate the terminal portion of the urethra and contribute less than 5% of the seminal fluid volume.

EXTERNAL GENITALS

The **penis** (PEE-nis) and **scrotum** constitute the external reproductive organs or **genitalia** (jen-i-TAL-ee-ah) of men. The penis (Figure 19-6) is the organ that, when made stiff and erect by the filling of its spongy or erectile tissue components

with blood during sexual arousal, can enter and deposit sperm in the vagina during intercourse. The penis has three separate columns of erectile tissue in its shaft: one **corpus** (KOR-pus) **spongiosum** (spun-jee-O-sum), which surrounds the urethra, and two **corpora** (KOR-por-ah) **cavernosa** (kav-er-NO-sa), which lie above. The spongy nature of erectile tissue is apparent in Figure 19-6. At the distal end of the shaft of the penis is the enlarged **glans,** over which the skin is folded doubly to form a loose-fitting retractable casing called the **foreskin** or **prepuce** (PRE-pus). If the foreskin

Prostatic Hypertrophy

A noncancerous condition called **benign** (bee-nine) **prostatic hypertrophy** (hye-PER-tro-fee) is a common problem in older men. The condition is characterized by an enlargement or hypertrophy of the prostate gland (see p. 389). The fact that the urethra passes through the center of the prostate after exiting from the bladder is a matter of considerable clinical significance in this condition. As the prostate enlarges, it squeezes the urethra, frequently closing it so completely that urination becomes very difficult or even impossible. In such cases, surgical removal of a part or all of the gland, a procedure called **prostatectomy** (pros-ta-TEC-toe-me), is sometimes performed. Other options, especially for treatment of cancerous prostatic growths, include systemic chemotherapy, cryotherapy (freezing) of prostatic tissue, microwave therapy, hormonal therapy, the placing of radioactive "seeds" directly into the gland, and various types of external beam x-ray radiation treatments.

fits too tightly about the glans, a **circumcision** or surgical removal of the foreskin is usually performed to prevent irritation. The external urethral orifice is the opening of the urethra at the tip of the glans.

The scrotum is a skin-covered pouch suspended from the groin. Internally, it is divided into two sacs by a septum; each sac contains a testis, epididymis, the lower part of the ductus deferens, and the beginning of the spermatic cords.

Female Reproductive System

STRUCTURAL PLAN

The structural plan of the reproductive system in both sexes is similar in that organs are characterized as **essential** or **accessory.**

Essential Organs

The essential organs of reproduction in women, the **gonads,** are the paired **ovaries.** The female sex cells or **ova** are produced here.

Accessory Organs

The accessory organs of reproduction in women consist of the following structures:

1. A series of ducts or modified duct structures that extend from near the ovaries to the exterior
2. Additional sex glands, including the mammary glands, which have an important reproductive function only in women
3. The external reproductive organs or external genitals

Table 19-2 lists the names of the essential and accessory organs of reproduction, and Figure 19-7 shows the location of most of them. Refer back to this table and illustration as you read about each structure in the pages that follow.

TABLE 19–2 Female Reproductive Organs	
ESSENTIAL ORGANS	**ACCESSORY ORGANS**
Gonads: ovaries (right ovary and left ovary)	Ducts: uterine tubes (two), uterus, vagina
	Accessory sex glands: Bartholin's glands (two), breasts (two)
	External genitals: vulva

FIGURE 19–7   *Organization of the Female Reproductive Organs.*

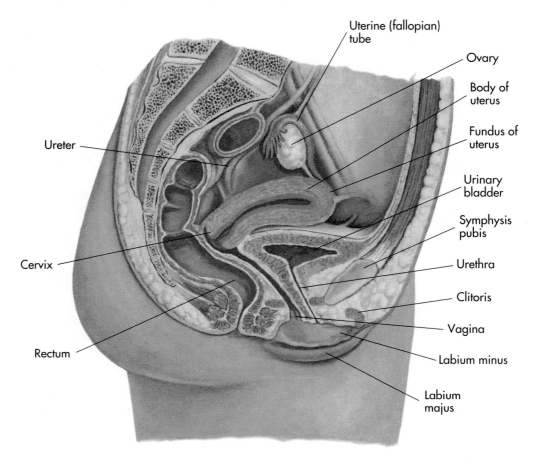

OVARIES

Structure and Location

The paired ovaries are the gonads of women. They have a puckered, uneven surface; each weighs about 3 g. The ovaries resemble large almonds in size and shape and are attached to ligaments in the pelvic cavity on each side of the uterus.

Embedded in a connective tissue matrix just below the outer layer of each ovary in a newborn baby girl are about 1 million **ovarian follicles;** each contains an **oocyte,** an immature stage of the female sex cell. By the time a girl reaches puberty, however, further development has resulted in the formation of a reduced number (about 400,000) of

what are now called **primary follicles.** Each primary follicle has a layer of **granulosa cells** around the oocyte. During the reproductive lifetime of most women, only about 350 to 500 of these primary follicles fully develop into **mature follicles,** which ovulate and release an ovum for potential fertilization. Follicles that do not mature degenerate and are reabsorbed into the ovarian tissue. A mature ovum in its sac is sometimes called a **Graafian** (GRAHF-ee-an) **follicle,** in honor of the Dutch anatomist who discovered them some 300 years ago.

The progression of development from primary follicle to ovulation is shown in Figure 19-8. As the thickness of the granulosa cell layer around

FIGURE 19–8 *Diagram of Ovary and Oogenesis.* Cross section of mammalian ovary shows successive stages of ovarian (Graafian) follicle and ovum development. Begin with the first stage (primary follicle) and follow around clockwise to the final state (degenerating corpus luteum).

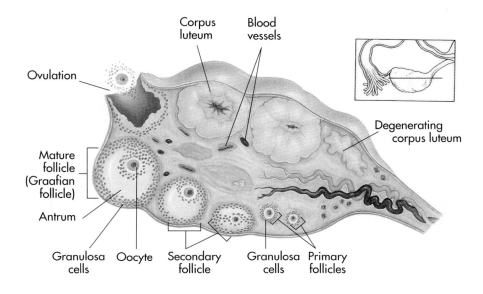

the oocyte increases, a hollow chamber called an *antrum* (AN-trum) appears, and a **secondary follicle** is formed. Development continues, and, after ovulation, the ruptured follicle is transformed into a hormone-secreting glandular structure called the **corpus** (KOR-pus) **luteum** (LOO-tee-um), which is described later. Corpus luteum is from the Latin word meaning "yellow body," an appropriate name to describe the yellow appearance of this glandular structure.

Ovary Functions

Oogenesis. The production of female gametes or sex cells is called **oogenesis** (o-o-JEN-e-sis). The specialized type of cell division that results in sperm formation, meiosis, is also responsible for development of ova. During the developmental phases experienced by the female sex cell from its earliest stage to just after fertilization, two meiotic divisions occur. As a result of meiosis in the female sex cell, the number of chromosomes is reduced equally in each daughter cell to half the number (23) found in other body cells (46). However, the amount of cytoplasm is divided unequally. The result is formation of one large ovum and small daughter cells called *polar bodies* that degenerate.

The ovum, with its large supply of cytoplasm, is one of the body's largest cells and is uniquely designed to provide nutrients for rapid development of the embryo until implantation in the uterus occurs. At fertilization, the sex cells from both parents fuse, and the normal chromosome number (46) is achieved.

Production of estrogen and progesterone. The second major function of the ovary, in addition to oogenesis, is secretion of the sex hormones, **estrogen** and **progesterone.** Hormone production in the ovary begins at puberty with the cyclic development and maturation of the ovum. The granulosa cells around the oocyte in the growing and mature follicle secrete estrogen. The corpus luteum, which develops after ovulation, chiefly secretes progesterone but also some estrogen.

Estrogen is the sex hormone that causes the development and maintenance of the female *secondary sex characteristics* and stimulates growth of the epithelial cells lining the uterus. Some of the actions of estrogen include the following:

1. Development and maturation of female reproductive organs, including the external genitals
2. Appearance of pubic hair and breast development
3. Development of female body contours by deposition of fat below the skin surface and in the breasts and hip region
4. Initiation of the first menstrual cycle

Progesterone is produced by the corpus luteum, which is a glandular structure that develops from a follicle that has just released an ovum. If stimulated by the appropriate anterior pituitary hormone, the corpus luteum produces progesterone for about 11 days after ovulation. Progesterone stimulates proliferation and vascularization of the epithelial lining of the uterus and acts with estrogen to initiate the menstrual cycle in girls entering puberty.

REPRODUCTIVE DUCTS

Uterine Tubes

The two **uterine tubes,** also called **fallopian** (fal-LO-pee-an) **tubes** or **oviducts** (O-vi-dukts), serve as ducts for the ovaries, even though they are not attached to them. The outer end of each tube terminates in an expanded, funnel-shaped structure that has fringelike projections called **fimbriae** (FIM-bree-ee) along its edge. This part of the tube curves over the top of each ovary (Figure 19-9) and opens into the abdominal cavity. The inner end of each uterine tube attaches to the uterus, and the cavity inside the tube opens into the cavity in the uterus. Each tube is about 10 cm (4 inches) in length.

After ovulation the discharged ovum first enters the abdominal cavity and then enters the uterine tube assisted by the wavelike movement of the fimbriae and the beating of the cilia on their surface. Once in the tube the ovum begins its journey to the uterus. Some ova never find their way into the oviduct and remain in the abdominal cavity where they are reabsorbed. In Chapter 20 the details of fertilization, which normally occurs in

Ectopic Pregnancy

The term **ectopic** (ek-TOP-ic) **pregnancy** is used to describe a pregnancy resulting from the implantation of a fertilized ovum in any location other than the uterus. Occasionally, because the outer ends of the uterine tubes open into the pelvic cavity and are not actually connected to the ovaries, an ovum does not enter an oviduct but becomes fertilized and remains in the abdominal cavity. Although rare, if implantation occurs on the surface of an abdominal organ or on one of the mesenteries, development may continue to term. In such cases delivery by caesarean section is required. Most ectopic pregnancies involve implantation in the uterine tube and are therefore called *tubal pregnancies.* They result in tubal rupture and fetal death.

the outer one third of the uterine tube, will be discussed.

The mucosal lining of the uterine tubes is directly continuous with the lining of the abdominal cavity on one end and with the lining of the uterus and vagina on the other. This is of great clinical significance because infections of the vagina or uterus such as gonorrhea may pass into the abdominal cavity, where they may become life threatening.

Uterus

The **uterus** (YOO-ter-us) is a small organ—only about the size of a pear—but it is extremely strong. It is almost all muscle or **myometrium** (my-o-ME-tree-um), with only a small cavity inside. During pregnancy the uterus grows many times larger so that it becomes big enough to hold a baby and a considerable amount of fluid. The uterus is composed of two parts: an upper portion, the **body,** and a lower narrow section, the **cervix.** Just above the level where the uterine tubes attach to the body of the uterus, it rounds out to form a bulging prominence called the **fun-**

FIGURE 19–9   *The Uterus.* Sectioned view shows muscle layers of the uterus and its relationship to the ovaries and vagina.

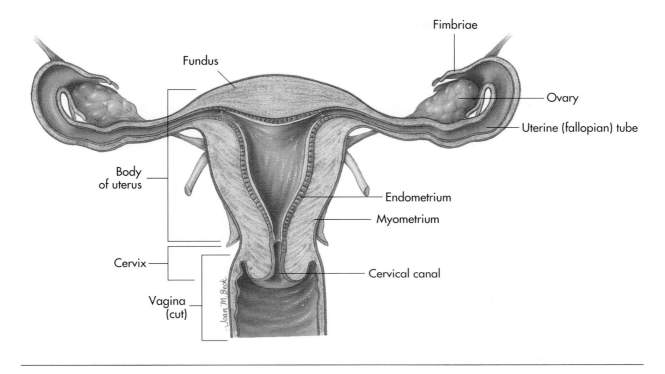

dus (see Figure 19-9). Except during pregnancy, the uterus lies in the pelvic cavity just behind the urinary bladder. By the end of pregnancy, it becomes large enough to extend up to the top of the abdominal cavity. It then pushes the liver against the underside of the diaphragm—a fact that explains such a comment as "I can't seem to take a deep breath since I've gotten so big," made by many women late in their pregnancies.

The uterus functions in three processes—menstruation, pregnancy, and labor. The corpus luteum stops secreting progesterone and decreases its secretion of estrogens about 11 days after ovulation. About 3 days later, when the progesterone and estrogen concentrations in the blood are at their lowest, menstruation starts. Small pieces of the mucous membrane lining of the uterus, or the **endometrium** (en-doe-ME-tree-um) pull loose, leaving torn blood vessels underneath. Blood and

bits of endometrium trickle out of the uterus into the vagina and out of the body. Immediately after menstruation the endometrium starts to repair itself. It again grows thick and becomes lavishly supplied with blood in preparation for pregnancy. If fertilization does not take place, the uterus once more sheds the lining made ready for a pregnancy that did not occur. Because these changes in the uterine lining continue to repeat themselves, they are spoken of as the **menstrual cycle** (see p. 397).

If fertilization occurs, pregnancy begins, and the endometrium remains intact. The events of pregnancy are discussed in Chapter 20.

Menstruation first occurs at puberty, often around the age of 12 years. Normally it repeats itself about every 28 days or 13 times a year for some 30 to 40 years before it ceases at *menopause* (MEN-o-pawz), when a woman is somewhere around the age of 50 years.

Vagina

The **vagina** (vah-JYE-nah) is a distensible tube about 10 cm (4 inches) long made mainly of smooth muscle and lined with mucous membrane. It lies in the pelvic cavity between the urinary bladder and the rectum (Figure 19-7). As the part of the female reproductive tract that opens to the exterior, the vagina is the organ that sperm enter during their journey to meet an ovum, and it is also the organ from which a baby emerges to meet its new world.

ACCESSORY OR SUPPORTIVE SEX GLANDS

Bartholin's Glands

One of the small **Bartholin's** (BAR-toe-linz) or **greater vestibular** (ves-TIB-yoo-lar) **glands** lies to the right of the vaginal outlet, and one lies to the left of it. Secretion of a mucuslike lubricating fluid is their function. Their ducts open into the space between the labia minora and the vaginal orifice called the **vestibule** (see Figure 19-11).

Breasts

The **breasts** lie over the pectoral muscles and are attached to them by connective tissue ligaments (of Cooper). Breast size is determined more by the amount of fat around the glandular (milk-secreting) tissue than by the amount of glandular tissue itself. Hence the size of the breast has little to do with its ability to secrete adequate amounts of milk after the birth of a baby.

Each breast consists of 15 to 20 divisions or lobes that are arranged radially (Figure 19-10). Each lobe consists of several lobules, and each lobule consists of milk-secreting glandular cells. The milk-secreting cells are arranged in grapelike clusters called *alveoli.* Small **lactiferous** (lak-TIF-er-us) **ducts** drain the alveoli and converge toward the nipple like the spokes of a wheel. Only one lactiferous duct leads from each lobe to an opening in the nipple. The colored area around the nipple is the **areola** (ah-REE-o-lah).

A knowledge of the lymphatic drainage of the breast is important because cancerous cells from breast tumors often spread to other areas of the body through the lymphatic system. This lymphatic drainage is discussed in Chapter 12 (see also Figure 12-5).

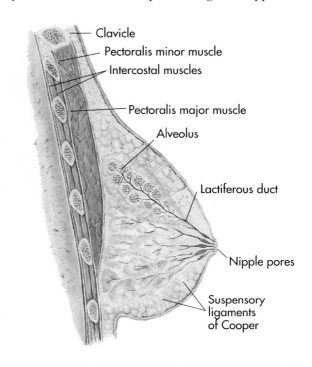

FIGURE 19–10 *Lateral View of the Breast.* Sagittal section shows the gland fixed to the overlying skin and the pectoral muscles by the suspensory ligaments of Cooper. Each lobule of secretory tissue is drained by a lactiferous duct that opens through the nipple.

- Clavicle
- Pectoralis minor muscle
- Intercostal muscles
- Pectoralis major muscle
- Alveolus
- Lactiferous duct
- Nipple pores
- Suspensory ligaments of Cooper

EXTERNAL GENITALS

The **external genitalia** or **vulva** (VUL-vah) of women consist of the following:

1. Mons pubis
2. Clitoris
3. Orifice of urethra
4. Labia minora (singular, *labium minus*) (small lips)
5. Hymen
6. Orifice, duct of Bartholin's gland
7. Orifice of vagina
8. Labia majora (singular, *labium majus*) (large lips)

The **mons pubis** is a skin-covered pad of fat over the symphysis pubis. Hair appears on this structure at puberty and persists throughout life. Extending downward from the elevated mons

FIGURE 19–11 *External Genitals of the Female.*

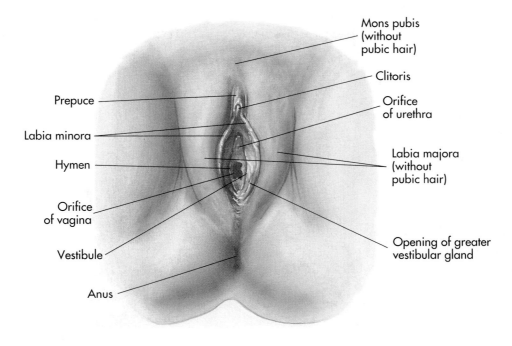

Mons pubis (without pubic hair)

Clitoris

Prepuce

Orifice of urethra

Labia minora

Hymen

Labia majora (without pubic hair)

Orifice of vagina

Vestibule

Opening of greater vestibular gland

Anus

pubis are the **labia** (LAY-bee-ah) **majora** (ma-JO-rah) or "large lips." These elongated folds, which are composed mainly of fat and numerous glands, are covered with pigmented skin and hair on the outer surface and are smooth and free from hair on the inner surface. The **labia minora** or "small lips" are located within the labia majora and are covered with modified skin. These two lips join anteriorly at the midline. The area between the labia minora is the vestibule (Figure 19-11). Several genital structures are located in the vestibule. The **clitoris** (KLIT-o-ris), which is composed of erectile tissue, is located just behind the anterior junction of the labia minora. Situated between the clitoris above and the vaginal opening below is the orifice of the urethra. The vaginal orifice is sometimes partially closed by a membranous **hymen** (HYE-men). The ducts of Bartholin's glands open on either side of the vaginal orifice inside the labia minora.

The term **perineum** (pair-i-NEE-um) is used to describe the area between the vaginal opening and anus. This area is sometimes cut in a surgical procedure called an **episiotomy** (e-piz-ee-OT-o-me) to prevent tearing of tissue during childbirth.

MENSTRUAL CYCLE
Phases and Events
The menstrual cycle consists of many changes in the uterus, ovaries, vagina, and breasts and in the anterior pituitary gland's secretion of hormones (Figure 19-12). In the majority of women, these changes occur with almost precise regularity throughout their reproductive years. The first indication of changes comes with the first menstrual period. The first **menses** (MEN-seez) or menstrual flow is referred to as the **menarche** (me-NAR-kee).

A typical menstrual cycle covers a period of about 28 days. The length of the cycle varies

FIGURE 19–12 *The 28-day Menstrual Cycle.*

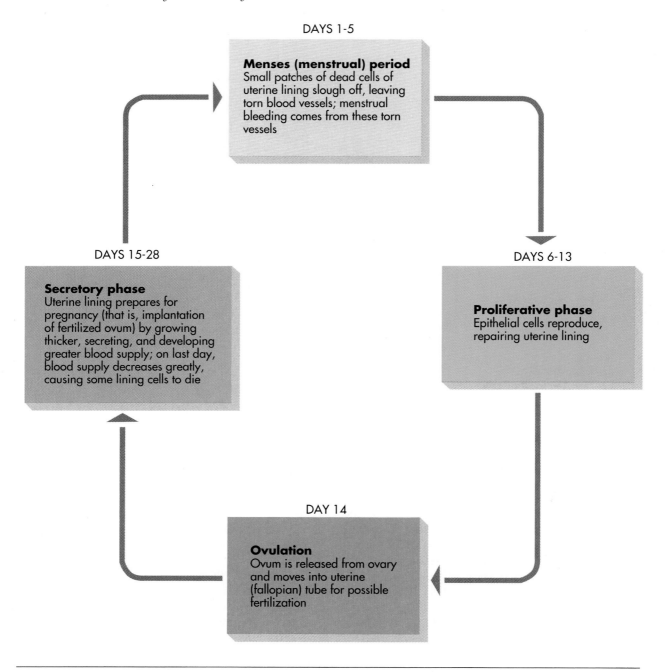

DAYS 1-5

Menses (menstrual) period
Small patches of dead cells of
uterine lining slough off, leaving
torn blood vessels; menstrual
bleeding comes from these torn
vessels

DAYS 15-28

Secretory phase
Uterine lining prepares for
pregnancy (that is, implantation
of fertilized ovum) by growing
thicker, secreting, and developing
greater blood supply; on last day,
blood supply decreases greatly,
causing some lining cells to die

DAYS 6-13

Proliferative phase
Epithelial cells reproduce,
repairing uterine lining

DAY 14

Ovulation
Ovum is released from ovary
and moves into uterine
(fallopian) tube for possible
fertilization

among women. Some women, for example, may
have a regular cycle that covers about 24 days.
The length of the cycle also varies within one
woman. Some women, for example, may have
irregular cycles that range from 21 to 28 days,
whereas others may be 2 to 3 months long. Each
cycle consists of three phases. The three periods of
time in each cycle are called the **menses,** the **proliferative phase,** and the **secretory phase.** Refer
often to Figure 19-13 as you read about the events
occurring during each phase of the cycle in the
pituitary gland, the ovary, and in the uterus. Be
sure that you do not overlook the event that
occurs around day 14 of a 28-day cycle.

FIGURE 19–13 *The Human Menstrual Cycle.* Diagram illustrates the interrelationship of pituitary, ovarian, and uterine functions throughout a usual 28-day cycle. A sharp increase in LH levels causes ovulation, whereas menstruation (sloughing off of the endometrial lining) is initiated by lower levels of progesterone.

Hysterectomy

The word **hysterectomy** (his-te-REK-toe-me) comes from the combination of two Greek words: *hystera*, meaning "uterus," and *ektome*, meaning "to cut out." By definition it is the surgical removal of the uterus. *Hysterectomy* is a term that is often misused, however, by incorrectly expanding its definition to include the removal of the ovaries or other reproductive structures. Only the uterus is removed in a hysterectomy. If the total uterus, including the cervix, is removed, the terms *total hysterectomy* or *panhysterectomy* may be used. If the cervical portion of the uterus is left in place and only the body of the organ is removed, the term *subtotal hysterectomy* is appropriate. The actual removal of the uterus may be performed through an incision made in the abdominal wall—*abdominal hysterectomy*—or through the vagina—*vaginal hysterectomy*. The term **oophorectomy** (o-off-o-REK-toe-me) is used to describe removal of the ovaries. Although the two surgical procedures may take place during the same operation—for a woman with uterine or ovarian cancer, for example—the terms used to describe them should not be used interchangeably.

The menses is a period of 4 or 5 days characterized by menstrual bleeding. The first day of menstrual flow is considered day 1 of the menstrual cycle. The proliferative phase begins after the menstrual flow ends and lasts until ovulation. During this period the follicles mature, the uterine lining thickens (proliferates), and estrogen secretion increases to its highest level. The secretory phase of the menstrual cycle begins at ovulation and lasts until the next menses begins. It is during this phase of the menstrual cycle that the uterine lining reaches its greatest thickness and the ovary secretes its highest levels of progesterone.

As a general rule, during the 30 or 40 years that a woman has periods, only one ovum matures each month. However, there are exceptions to this rule. Some months, more than one matures, and some months no ovum matures. Ovulation occurs 14 days before the next menses begins. In a 28-day cycle, this means that ovulation occurs around day 14 of the cycle, as shown in Figure 19-12. (Recall that the first day of the menses is considered the first day of the cycle.) In a 30-day cycle, however, ovulation would not occur on the fourteenth cycle day, but the sixteenth. And in a 25-day cycle, ovulation would occur the eleventh cycle day.

This matter of the time of ovulation has great practical importance. An ovum lives only a short time after it is ejected from its follicle, and sperm live only a short time after they enter the female body. Fertilization of an ovum by a sperm therefore can occur only around the time of ovulation. In other words, a woman's fertile period lasts only a few days each month.

Control of Menstrual Cycle Changes

The anterior pituitary gland plays a critical role in regulating the cyclic changes that characterize the functions of the female reproductive system (see Chapter 9). From day 1 to about day 7 of the menstrual cycle, the anterior pituitary gland secretes increasing amounts of FSH. A high blood concentration of FSH stimulates several immature ovarian follicles to start growing and secreting estrogens (Figure 19-13). As the estrogen content of blood increases, it stimulates the anterior pituitary gland to secrete another hormone, luteinizing hormone (LH). LH causes maturing of a follicle and its ovum, ovulation (rupturing of mature follicle with ejection of ovum), and luteinization (formation of a yellow body, the corpus luteum, from the ruptured follicle).

Which hormone—FSH or LH—would you call the "ovulating hormone"? Do you think ovulation could occur if the blood concentration of FSH remained low throughout the menstrual cycle? If you answered LH to the first question and no to the second, you answered both questions correctly. Ovulation cannot occur if the blood level of FSH stays low because a high concentration of this hormone is essential to stimulation of ovarian follicle growth and maturation. With a low level

TABLE 19–3   Analogous Features of the Reproductive Systems

FEATURE	FEMALE	MALE
Essential organs	Ovaries	Testes
Sex cells	Ova (eggs)	Sperm
Hormones	Estrogen and progesterone	Testosterone
Hormone-producing cells	Granulosa cells and corpus luteum	Interstitial cells
Duct systems	Uterine (fallopian) tubes, uterus, and vagina	Epididymis, urethra, and ductus (vas) deferens
External genitals	Clitoris and vulva	Penis and scrotum

of FSH, no follicles start to grow, and therefore none become ripe enough to ovulate. Ovulation is caused by the combined actions of FSH and LH. Birth control pills that contain estrogen substances suppress FSH secretion. This indirectly prevents ovulation.

Ovulation occurs, as we have said, because of the combined actions of the two anterior pituitary hormones, FSH and LH. The next question is: what causes menstruation? A brief answer is this: a sudden, sharp decrease in estrogen and progesterone secretion toward the end of the premenstrual period causes the uterine lining to break down and another menstrual period to begin.

○ *Summary of Male and Female Reproductive Systems*

The reproductive systems in both sexes are centered around the production of highly specialized reproductive cells or gametes (sperm and ova), as well as mechanisms to ensure union of these two cells; the fusion of these cells enables transfer of parental genetic information to the next generation. Table 19-3 compares several analogous components of the reproductive systems in both sexes. You can see that men and women have similar

 Amenorrhea in Female Athletes

Failure to have a menstrual period is called **amenorrhea**. Amenorrhea occurs in some female athletes, probably resulting from a body fat composition that is too low to sustain normal reproductive function. Although it keeps the hematocrit (RBC level) higher than during menstruation, it is not considered a desirable condition. Besides infertility, amenorrhea may cause other problems. For example, the low blood levels of estrogen associated with long-term amenorrhea may cause osteoporosis (loss of bone mass).

structures to accomplish complementary functions. In addition, the female reproductive system permits development and birth of the offspring—and this will be the first subject of our last chapter.

COMMON STRUCTURAL AND FUNCTIONAL CHARACTERISTICS BETWEEN THE SEXES

A Common general structure and function can be identified between the systems in both sexes

B Systems adapted for development of sperm or ova followed by successful fertilization, development, and birth of offspring

C Sex hormones in both sexes important in development of secondary sexual characteristics and normal reproductive system activity

MALE REPRODUCTIVE SYSTEM

A Structural plan—organs classified as essential or accessory

  1 Essential organs of reproduction are the gonads (testes), which produce sex cells (sperm)

  2 Accessory organs of reproduction

    a Ducts—passageways that carry sperm from testes to exterior

    b Sex glands—produce protective and nutrient solution for sperm

    c External genitals

B Testes—the gonads of men

  1 Structure and location (Figure 19-2)

    a Testes in scrotum—lower temperature

    b Covered by tunica albuginea, which divides testis into lobules containing seminiferous tubules

    c Interstitial cells produce testosterone

  2 Functions

    a Spermatogenesis is process of sperm production (Figure 19-4)

      (1) Sperm precursor cells called *spermatogonia*

      (2) Meiosis produces primary spermatocyte, which forms four spermatids with 23 chromosomes

      (3) Spermatozoa—highly specialized cell

        (a) Head contains genetic material

        (b) Acrosome contains enzymes to assist sperm in penetration of ovum

        (c) Mitochondria provide energy for movement

    b Production of testosterone by interstitial cells

      (1) Testosterone "masculinizes" and promotes development of male accessory organs

      (2) Stimulates protein anabolism and development of muscle strength

C Reproductive ducts—ducts through which sperm pass after exiting testes until they exit from the body

  1 Epididymis—single, coiled tube about 6 m in length; lies along the top and behind the testis in the scrotum

    a Sperm mature and develop the capacity for motility as they pass through epididymis

  2 Ductus (vas) deferens—receives sperm from the epididymis and transports them from scrotal sac through the abdominal cavity

    a Passes through inguinal canal

    b Joins duct of seminal vesicle to form the ejaculatory duct

D Accessory or supportive sex glands—semen: mixture of sperm and secretions of accessory sex glands. Averages 3 to 5 ml per ejaculation, with each milliliter containing about 100 million sperm

  1 Seminal vesicles

    a Pouchlike glands that produce about 60% of seminal fluid volume

    b Secretion is yellowish, thick, and rich in fructose to provide energy needed by sperm for motility

  2 Prostate gland

    a Shaped like a doughnut and located below bladder

    b Urethra passes through the gland

    c Secretion represents 30% of seminal fluid volume—is thin and milk-colored

    d Activates sperm and is needed for ongoing sperm motility

  3 Bulbourethral (Cowper's) glands

    a Resemble peas in size and shape

    b Secrete mucuslike fluid constituting less than 5% of seminal fluid volume

E External genitals

  1 Penis and scrotum called *genitalia*

  2 Penis has three columns of erectile tissue—two dorsal columns called *corpora cavernosa* and one ventral column surrounding urethra called *corpus spongiosum*

  3 Glans penis covered by foreskin

  4 Surgical removal of foreskin called *circumcision*

FEMALE REPRODUCTIVE SYSTEM

A Structural plan—organs classified as essential or accessory

  1 Essential organs are gonads (ovaries), which produce sex cells (ova)

  2 Accessory organs of reproduction

    a Ducts or modified ducts—including oviducts, uterus, and vagina

    b Sex glands—including those in the breasts

    c External genitals

B Ovaries
  1 Structure and location
    a Paired glands weighing about 3 g each
    b Resemble large almonds
    c Attached to ligaments in pelvic cavity on each side of uterus
    d Microscopic structure (Figure 19-8)
      (1) Ovarian follicles—contain oocyte, which is immature sex cell (about 1 million at birth)
      (2) Primary follicles—about 400,000 at puberty are covered with granulosa cells
      (3) About 350 to 500 mature follicles ovulate during the reproductive lifetime of most women—sometimes called *Graafian follicles*
      (4) Secondary follicles have hollow chamber called *antrum*
      (5) Corpus luteum forms after ovulation
  2 Functions
    a Oogenesis—this meiotic cell division produces daughter cells with equal chromosome numbers (23) but unequal cytoplasm. Ovum is large; polar bodies are small and degenerate
    b Production of estrogen and progesterone
      (1) Granulosa cells surrounding the oocyte in the mature and growing follicles produce estrogen
      (2) Corpus luteum produces progesterone
      (3) Estrogen causes development and maintenance of secondary sex characteristics
      (4) Progesterone stimulates secretory activity of uterine epithelium and assists estrogen in initiating menses
C Reproductive ducts
  1 Uterine (fallopian) tubes
    a Extend about 10 cm from uterus into abdominal cavity
    b Expanded distal end surrounded by fimbriae
    c Mucosal lining of tube is directly continuous with lining of abdominal cavity
  2 Uterus—composed of body, fundus, and cervix (Figure 19-9)
    a Lies in pelvic cavity just behind urinary bladder
    b Myometrium is muscle layer
    c Endometrium lost in menstruation
    d Menopause—end of repetitive menstrual cycles (about 45 years of age)
  3 Vagina
    a Distensible tube about 10 cm long
    b Located between urinary bladder and rectum in the pelvis
    c Receives penis during sexual intercourse and is birth canal for normal delivery of baby at termination of pregnancy
D Accessory or supportive sex glands
  1 Bartholin's (greater vestibular) glands
    a Secrete mucuslike lubricating fluid
    b Ducts open between labia minora
  2 Breasts (Figure 19-10)
    a Located over pectoral muscles of thorax
    b Size determined by fat quantity more than amount of glandular (milk-secreting) tissue
    c Lactiferous ducts drain at nipple, which is surrounded by pigmented areola
    d Lymphatic drainage important in spread of cancer cells to other body areas
E External genitals (Figure 19-11)
  1 Include mons pubis, clitoris, orifice of urethra, Bartholin's gland, vagina, labia minora and majora, and hymen
  2 Perineum—area between vaginal opening and anus
    a Surgical cut during birth called *episiotomy*
F Menstrual cycle—involves many changes in the uterus, ovaries, vagina, and breasts (Figures 19-12 and 19-13)
  1 Length—about 28 days, varies from month to month in individuals and in the same individual
  2 Phases
    a Menses—about the first 4 or 5 days of the cycle, varies somewhat; characterized by sloughing of bits of endometrium (uterine lining) with bleeding
    b Proliferative phase—days between the end of menses and secretory phase; varies in length; the shorter the cycle, the shorter the proliferative phase; the longer the cycle, the longer the proliferative phase; examples: in 28-day cycle, proliferative phase ends on day 13, but in 26-day cycle, it ends on the 11th day and in 32-day cycle, it ends on day 17; characterized by repair of endometrium
    c Secretory phase—days between ovulation and beginning of next menses; secretory about 14 days before next menses; characterized by further thickening of endometrium and secretion by its glands in preparation for implantation of fertilized ovum; combined actions of the anterior pituitary hormones FSH and LH cause ovulation; sudden sharp decrease in estrogens and progesterone bring on menstruation if pregnancy does not occur.

SUMMARY OF MALE AND FEMALE REPRODUCTIVE SYSTEMS

A In men and women the organs of the reproductive system are adapted for the specific sequence of functions that permit development of sperm or ova followed by the successful fertilization and then the normal development and birth of offspring

B The male organs produce, store, and ultimately introduce mature sperm into the female reproductive tract

C The female system produces ova, receives the sperm, and permits fertilization followed by fetal development and birth, with lactation afterward

D Production of sex hormones is required for development of secondary sex characteristics and for normal reproductive functions in both sexes

---

N E W   W O R D S

areola	estrogen	menses	semen
circumcision	fimbriae	oocyte	seminiferous tubule
clitoris	gametes	ovulation	spermatogenesis
corpus luteum	genitals	perineum	spermatozoa
ejaculation	gonads	polar body	testes
endometrium	Graafian follicle	prepuce	testosterone
epididymis	meiosis	progesterone	vulva
episiotomy	menopause	scrotum	zygote

C H A P T E R   T E S T

1. The gonads of the male are the _____.
2. In men the sex cells are called _____.
3. The essential organs of reproduction in both sexes are called the _____.
4. Each lobule of the testis consists of a narrow but long and coiled tube called the _____ tubule.
5. Testosterone formation is the function of the _____ cells of the testis.
6. In men the urethra passes through the center of the doughnut-shaped _____ gland.
7. The narrow section of the uterus that opens into the vagina is called the _____.
8. The corpus luteum secretes _____.
9. The mucous membrane lining the uterus is called the _____.
10. The colored area around the nipple is called the _____.
11. The scientific name for the beginning of the menses is called _____.
12. The average length of a typical menstrual cycle is _____ days.

Select the most correct answer from Column B for each statement in Column A. (Only one answer is correct.)

COLUMN A

13. _____ Female gonads
14. _____ External genitals
15. _____ Secrete estrogen
16. _____ Corpus luteum
17. _____ Sperm formation
18. _____ "Ovulating hormone"
19. _____ Surgical removal of foreskin
20. _____ Feminizing hormone
21. _____ Female sex cell
22. _____ Male sex hormone

COLUMN B

a. Ovarian follicles
b. Testosterone
c. LH
d. Secrete progesterone
e. Circumcision
f. Ovaries
g. Estrogen
h. Spermatogenesis
i. Vulva
j. Ovum

REVIEW QUESTIONS

1. Identify the essential and accessory organs of reproduction in men.
2. What organs are included in the external genitals of the male?
3. Discuss the structure of the testes. Explain the function of the seminiferous tubules and the interstitial cells of the testes.
4. What is the name of the masculinizing hormone? What secretes it? What are its general functions?
5. Discuss the anatomy of a sperm cell. Why is it motile? What parts of the sperm cell are designed to provide motility?
6. How many sperm are normally present in one ejaculation of semen? What is the usual volume of semen ejaculated at one time?
7. Discuss the functions of the accessory glands in men.
8. Name the reproductive ducts in men and women.
9. Identify the feminizing hormone by its scientific name. What gland secretes it?
10. Identify the ovulating hormone by its scientific name. What glands secrete it?
11. Identify and locate the alveoli of breast, areola, cervix, clitoris, fundus of uterus, Graafian follicle, labia majora, mons pubis, ovum, and uterine tube.
12. What causes ovulation?
13. What is menstruation? What causes it?

CRITICAL THINKING

14. Trace a sperm cell from its point of formation in the testes through the male reproductive ducts to ejaculation.
15. Castration is an operation that removes the testes. Would this sterilize the male? Why? What other effects would you expect to see as a result of castration? Why?
16. How many female sex cells are usually formed each month? How does this compare with the number of male sex cells formed each month?

Growth and Development

OBJECTIVES

*After you have completed this chapter,
you should be able to:*

1. Discuss the concept of development as a biological process characterized by continuous modification and change.

2. Discuss the major developmental changes characteristic of the prenatal stage of life from fertilization to birth.

3. Discuss the three stages of labor that characterize a normal, vaginal birth.

4. Identify the three primary germ layers and several derivatives in the adult body that develop from each layer.

5. List and discuss the major developmental changes characteristic of the four postnatal periods of life.

6. Discuss the effects of aging on the major body organ systems.

Many of your fondest and most vivid memories are probably associated with your birthdays. The day of birth is an important milestone of life. Most people continue to remember their birthday in some special way each year; birthdays serve as pleasant and convenient reference points to mark periods of transition or change in our lives. The actual day of birth marks the end of one phase of life called the **prenatal period** and the beginning of a second called the **postnatal period.** The prenatal period begins at conception and ends at birth; the postnatal period begins at birth and continues until death. Although important periods in our lives such as childhood and adolescence are often remembered as a series of individual and isolated events, they are in reality part of an ongoing and continuous process. In reviewing the many changes that occur during the cycle of life from conception to death, it is often convenient to isolate certain periods such as infancy or adulthood for study. It is important to remember, however, that life is not a series of stop-and-start events or individual and isolated periods of time. Instead, it is a biological process that is characterized by continuous modification and change.

This chapter discusses some of the events and changes that occur in the development of the individual from conception to death. Study of development during the prenatal period is followed by a discussion of the birth process and a review of changes occurring during infancy and adulthood. Finally some important changes that occur in the individual organ systems of the body as a result of aging are discussed.

⬤ *Prenatal Period*

The **prenatal stage of development** begins at the time of conception or fertilization (that is, at the moment the female ovum and the male sperm cells unite) (Figure 20-1). The period of prenatal development continues until the birth of the child about 39 weeks later. The science of the development of the individual before birth is called **embryology** (em-bree-OL-o-jee). It is a story of miracles, describing the means by which a new human life is started and the steps by which a single microscopic cell is transformed into a complex human being.

FERTILIZATION TO IMPLANTATION

After ovulation the discharged ovum first enters the abdominal cavity and then finds its way into the uterine (fallopian) tubes. Sperm cells "swim" up the uterine tubes toward the ovum. Look at the relationship of the ovary, the two uterine tubes, and the uterus in Figure 20-2. Recall from Chapter 19 that each uterine tube extends outward from the uterus for about 10 cm. It then ends in the abdominal cavity near the ovary, as you can see in Figure 20-2, in an opening surrounded by fringelike processes, the *fimbriae.* Using the uterus as a point of reference, anatomists divide each uterine tube into three parts. The innermost part of the tube actually extends through the uterine wall, the middle third extends out into the abdominal cavity, and the outermost third of the tube ends near the ovary in the dilated, funnel-shaped opening described above.

Sperm cells that are deposited in the vagina must enter and "swim" through the uterus and then move out of the uterine cavity and through the uterine tube to meet the ovum. Fertilization most often occurs in the outer one third of the oviduct as shown in Figure 20-2. The fertilized ovum or **zygote** (ZYE-gote) is genetically complete; it represents a new single-celled individual. Time and nourishment are all that is needed for expression of characteristics such as sex, body build, and skin color that were determined at the time of fertilization. As you can see in the figure, the zygote immediately begins mitotic division, and in about 3 days a solid mass of cells called a **morula** (MOR-yoo-lah) is formed (Figure 20-2). The cells of the morula continue to divide, and by the time the developing embryo reaches the uterus, it is a hollow ball of cells called a **blastocyst** (BLAS-toe-sist).

During the 10 days from the time of fertilization to the time when the blastocyst is completely implanted in the uterine lining, no nutrients from the mother are available. The rapid cell division taking place up to the blastocyst stage occurs with no significant increase in total mass compared to

FIGURE 20–1   *Fertilization.* Fertilization is a specific biological event. It occurs when the male and female sex cells fuse. After union between a sperm cell and the ovum has occurred, the cycle of life begins. The scanning electron micrograph shows spermatozoa attaching themselves to the surface of an ovum. Only one will penetrate and fertilize the ovum.

FIGURE 20–2   *Fertilization and Implantation.* At ovulation, an ovum is released from the ovary and begins its journey through the uterine tube. While in the tube, the ovum is fertilized by a sperm to form the single-celled zygote. After a few days of rapid mitotic division, a ball of cells called a *morula* is formed. After the morula develops into a hollow ball called a *blastocyst,* implantation occurs.

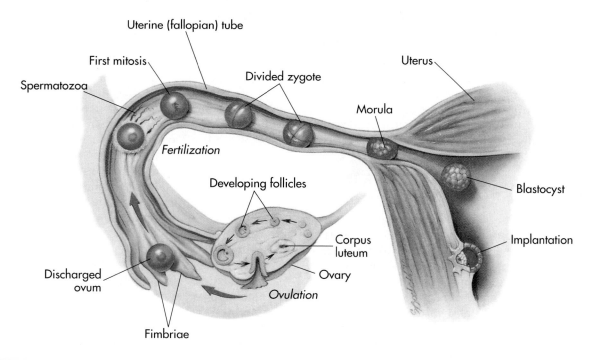

FIGURE 20–3 *Early Stages of Human Development.* **A,** Fertilized ovum or zygote. **B** to **D,** Early cell divisions produce more and more cells. The solid mass of cells shown in **D** forms the morula—an early stage in embryonic development.

the zygote (Figure 20-3). One of the specializations of the ovum is its incredible store of nutrients that help support this embryonic development until implantation has occurred.

Note in Figure 20-4 that the blastocyst consists of an outer layer of cells and an inner cell mass. As the blastocyst develops, it forms a structure with two cavities, the **yolk sac** and **amniotic** (am-nee-OT-ik) **cavity.** The yolk sac is most important in animals, such as birds, that depend heavily on yolk as the sole source of nutrients for the developing embryo. In these animals the yolk sac digests the yolk and provides the resulting nutrients to the embryo. Because uterine fluids provide nutrients to the developing embryo in humans until the placenta develops, the function of the yolk sac is not a nutritive one. Instead, it has other functions, including production of blood cells.

FIGURE 20–4 *Implantation and Early Development.* The hollow blastocyst implants itself in the uterine lining about 10 days after ovulation. Until the placenta is functional, nutrients are obtained by diffusion from uterine fluids. Notice the developing chorion and how the blastocyst eventually forms a yolk sac and amniotic cavity.

In Vitro Fertilization

The Latin term *in vitro* means, literally, "within a glass." In the case of in vitro fertilization, it refers to the glass laboratory dish where an ovum and sperm are mixed and where fertilization occurs.

In the classic technique, the ovum is obtained from the mother by first inserting a fiberoptic viewing instrument called a **laparoscope** through a very small incision in the woman's abdomen. Once in the abdominal cavity the device allows the physician to view the ovary and then puncture and "suck up" an ovum from a mature follicle. Over the years refinements to this technique have been made and less invasive procedures are currently being used. After about 2½ days growth in a tempera-ture-controlled environment the developing zygote, which by then has reached the 8- or 16-cell stage, is returned by the physician to the mother's uterus. If implantation is successful, growth will continue and the subsequent pregnancy will progress. In the most successful fertility clinics in the U.S., a normal term birth will occur in about 30% of in vitro fertilization attempts.

The first "test tube" baby was born in England nearly 20 years ago. Since that time, improved techniques of in vitro fertilization and advances in our knowledge of reproductive biology have made it possible for growing numbers of infertile couples to conceive and deliver healthy infants.

FIGURE 20–5   *The Placenta: Interface Between Maternal and Fetal Circulation.* **A,** Relationship of uterus, developing infant, and placenta. **B,** The close placement of the fetal blood supply and the maternal blood in the lacunae of the placenta permits diffusion of nutrients and other substances. It also forms a thin barrier to prevent diffusion of most harmful substances. No mixing of fetal and maternal blood occurs.

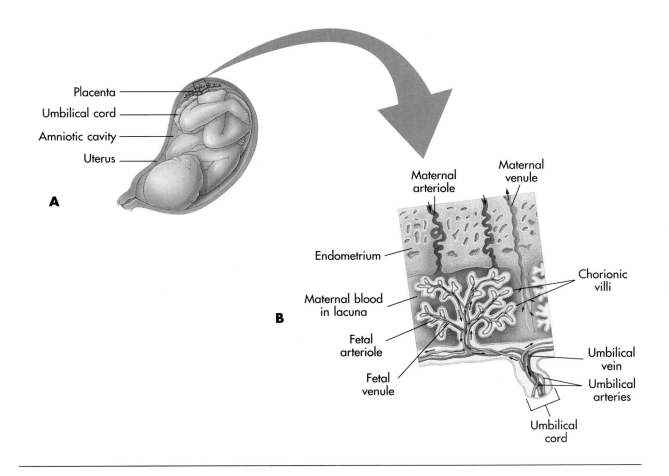

The amniotic cavity becomes a fluid-filled, shock-absorbing sac, sometimes called the *bag of waters,* in which the embryo floats during development. The **chorion** (KO-ree-on), shown in Figures 20-4 and 20-5, develops into an important fetal membrane in the **placenta** (plah-SEN-tah). The *chorionic villi* shown in Figure 20-5 connect the blood vessels of the chorion to the placenta. The placenta (Figure 20-5) anchors the developing fetus to the uterus and provides a "bridge" for the exchange of nutrients and waste products between mother and baby.

The placenta is a unique and highly specialized structure that has a temporary but very important series of functions during pregnancy. It is composed of tissues from mother and child and functions not only as a structural "anchor" and nutritive bridge, but also as an excretory, respiratory, and endocrine organ (Figure 20-5).

Placental tissue normally separates the maternal blood, which fills the lacunae of the placenta, from the fetal blood so that no intermixing occurs. The very thin layer of placental tissue that separates maternal and fetal blood also serves as an effective "barrier" that can protect the developing baby from many harmful substances that may enter the mother's bloodstream. Unfortunately, toxic substances, such as alcohol and some infectious organisms, may penetrate this protective placental barrier and injure the developing baby. The virus responsible for German measles (rubella), for example, can easily pass through the placenta and cause tragic developmental defects in the fetus.

PERIODS OF DEVELOPMENT

The length of pregnancy (about 39 weeks)—called the *gestation period*—is divided into three 3-month segments called *trimesters*. A number of terms are used to describe development during these periods known as the first, second, and third trimesters of pregnancy.

During the first trimester or 3 months of pregnancy, many terms are used. *Zygote* describes the ovum just after fertilization by a sperm cell. After about 3 days of constant cell division, the solid mass of cells, identified earlier as the *morula*, enters the uterus. Continued development transforms the morula into the hollow blastocyst, which then implants into the uterine wall.

The embryonic phase of development extends from fertilization until the end of week 8 of gestation. During this period in the first trimester, the term *embryo* is used to describe the developing individual. The fetal phase is used to indicate the period of development extending from week 9 to week 39. During this period, the term *embryo* is replaced by *fetus.*

By day 35 of gestation (Figure 20-6, *A*), the heart is beating and, although the embryo is only 8 mm (about ⅜ inch) long, the eyes and so-called limb buds, which ultimately form the arms and legs, are clearly visible. Figure 20-6, *C* shows the stage of development of the fetus at the end of the first trimester of gestation. Body size is about 7 to 8 cm (3.2 inches) long. The facial features of the fetus are apparent, the limbs are complete, and gender can be identified. By month 4 (Figure 20-6, *D,*) all organ systems are complete and in place.

FORMATION OF THE PRIMARY GERM LAYERS

Early in the first trimester of pregnancy, three layers of specialized cells develop that embryologists call the **primary germ layers** (Table 20-1). Each layer gives rise to definite structures such as the skin, nervous tissue, muscles, or digestive organs. Table 20-1 lists a number of structures derived from each primary germ layer called, respectively, **endoderm** (EN-doe-derm) or inside layer, **ectoderm** (EK-toe-derm) or outside layer, and **mesoderm** (MEZ-o-derm) or middle layer.

TABLE 20–1   Primary Germ Layer Derivatives

ENDODERM	ECTODERM	MESODERM
Lining of gastrointestinal tract	Epidermis of skin	Dermis of skin
Lining of lungs	Tooth enamel	Circulatory system
Lining of hepatic and pancreatic ducts	Lens and cornea of eye	Many glands
Kidney ducts and bladder	Outer ear	Kidneys
Anterior pituitary gland (adenohypophysis)	Nasal cavity	Gonads
Thymus gland	Facial bones	Muscle
Thyroid gland	Skeletal muscles in head	Bones (except facial)
Parathyroid gland	Brain and spinal cord	
Tonsils	Sensory neurons	
	Adrenal medulla	

FIGURE 20–6 *Human Embryos and Fetuses.* **A,** At 35 days. **B,** At 49 days. **C,** At the end of the first trimester. **D,** At 4 months.

Quickening

Pregnant women usually notice fetal movement for the first time between weeks 16 and 18 of pregnancy. The term **quickening** has been used for generations to describe these first recognizable movements of the fetus. From an occasional "kick" during months 4 and 5 of pregnancy, the frequency of fetal movements steadily increases as gestation progresses. The frequency of fetal movements is an excellent indicator of the unborn baby's health.

Recent studies have shown that simply by recording the number of fetal movements each day after week 28 of pregnancy, a woman can provide her physician with extremely useful information about the health of her unborn child. Ten or more movements during a daily measurement period are considered normal.

Educating pregnant women about fetal movements and how to monitor their frequency is but one example of expanded interest in prenatal home care. Assisting pregnant women with making informed judgments about nutrition, exercise, lifestyle adjustments, and birthing options before they enter the hospital for delivery of their baby is an important and growing part of home health care services.

HISTOGENESIS AND ORGANOGENESIS

The study of how the primary germ layers develop into many different kinds of tissues is called **histogenesis** (his-toe-JEN-e-sis). The way that those tissues arrange themselves into organs is called **organogenesis** (or-ga-no-JEN-e-sis). The fascinating story of histogenesis and organogenesis in human development is long and complicated; its telling belongs to the science of embryology. But for the beginning student of anatomy, it seems sufficient to appreciate that life begins when two sex cells unite to form a single-celled zygote and that the new human body evolves by a series of processes consisting of cell differentiation, multiplication, growth, and rearrangement, all of which take place in definite, orderly sequence. Development of structure and function go hand in hand, and from 4 months of gestation, when every organ system is complete and in place, until term (about 280 days), fetal development is mainly a matter of growth. Figure 20-7, *A*, shows the normal intrauterine placement of a fetus just before birth in a full-term pregnancy.

Birth or Parturition

The process of birth or **parturition** (par-too-RISH-un) is the point of transition between the prenatal and postnatal periods of life. As pregnancy draws to a close, the uterus becomes "irritable" and, ultimately, muscular contractions begin and cause the cervix to dilate or open, thus permitting the fetus to move from the uterus through the vagina or "birth canal" to the exterior. The process normally begins with the fetus taking a head-down position against the cervix (Figure 20-7, *A*). When contractions occur, the amniotic sac or "bag of waters" ruptures, and labor begins.

STAGES OF LABOR

Labor is the process that results in the birth of a baby. It has three stages (Figure 20-7, *B* to *E*):

1. Stage one—period from onset of uterine contractions until dilation of the cervix is complete
2. Stage two—period from the time of maximal cervical dilation until the baby exits through the vagina
3. Stage three—process of expulsion of the placenta through the vagina

The time required for normal vaginal birth varies widely and may be influenced by many variables, including whether the woman has previously had a child. In most cases, stage one of labor lasts from 6 to 24 hours, and stage two lasts from a few minutes to an hour. Delivery of the placenta (stage three) normally occurs within 15 minutes after the birth of the baby.

FIGURE 20–7 *Parturition.* **A,** The relation of the fetus to the mother. **B,** The fetus moves into the opening of the birth canal, and the cervix begins to dilate. **C,** Dilation of the cervix is complete. **D,** The fetus is expelled from the uterus. **E,** The placenta is expelled.

Antenatal Diagnosis and Treatment

Advances in **antenatal** (from the Latin *ante*, "before," *natus*, "birth") **medicine** now permit extensive diagnosis and treatment of disease in the fetus much like any other patient. This new dimension in medicine began with techniques by which Rh$^+$ babies could be given transfusions before birth.

Current procedures using images provided by ultrasound equipment (Figures *A* and *B*) allow physicians to prepare for and

perform, before the birth of a baby, corrective surgical procedures such as bladder repair. These procedures also allow physicians to monitor the progress of other types of treatment on a developing fetus. Figure *A* shows placement of the ultrasound transducer on the abdominal wall. The resulting image (Figure *B*), called an *ultrasonogram*, shows a 29-week embryo. The image plane is showing the head.

Postnatal Period

The **postnatal period** begins at birth and lasts until death. Although it is often divided into major periods for study, we need to understand and appreciate the fact that growth and development are continuous processes that occur throughout the life cycle. Gradual changes in the physical appearance of the body as a whole and in the relative proportions of the head, trunk, and limbs are quite noticeable between birth and ado-

lescence. Note in Figure 20-8 the obvious changes in the size of bones and in the proportionate sizes between different bones and body areas. The head, for example, becomes proportionately smaller. Whereas the infant head is approximately one fourth the total height of the body, the adult head is only about one eighth the total height. The facial bones also show several changes between infancy and adulthood. In an infant the face is one eighth of the skull surface, but in an adult the face is half of the skull surface. Another change in pro-

FIGURE 20–8 *Changes in the Proportions of Body Parts from Birth to Maturity.* Note the dramatic differences in head size.

| Newborn | 2-year-old | 5-year-old | 13-year-old | Adult |

portion involves the trunk and lower extremities. The legs become proportionately longer and the trunk proportionately shorter. In addition, the thoracic and abdominal contours change, roughly speaking, from round to elliptical.

Such changes are good examples of the ever-changing and ongoing nature of growth and development. It is unfortunate that many of the changes that occur in the later years of life do not result in increased function. These degenerative changes are certainly important, however, and will be discussed later in this chapter. The following are the most common postnatal periods: (1) **infancy,** (2) **childhood,** (3) **adolescence** and **adulthood,** and (4) **older adulthood.**

INFANCY

The period of infancy begins abruptly at birth and lasts about 18 months. The first 4 weeks of infancy are often referred to as the **neonatal** (nee-o-NAY-tal) **period** (Figure 20-9). Dramatic changes occur at a rapid rate during this short but critical period. **Neonatology** (nee-o-nay-TOL-o-jee) is the medical and nursing specialty concerned with the diagnosis and treatment of disorders of the new-

FIGURE 20–9 *The Neonate Infant.* The umbilical cord has been cut.

born. Advances in this area have resulted in dramatically reduced infant mortality.

Many of the changes that occur in the cardiovascular and respiratory systems at birth are necessary for survival. Whereas the fetus totally depended on the mother for life support, the newborn infant must become totally self-supporting in terms of blood circulation and respiration immediately after birth. A baby's first breath is deep and forceful. The stimulus to breathe results primarily from the increasing amounts of carbon dioxide (CO_2) that accumulate in the blood after the umbilical cord is cut following delivery.

Many developmental changes occur between the end of the neonatal period and 18 months of age. Birth weight doubles during the first 4 months and then triples by 1 year. The baby also increases in length by 50% by the twelfth month. The "baby fat" that accumulated under the skin during the first year begins to decrease, and the plump infant becomes leaner.

Early in infancy the baby has only one spinal curvature (Figure 20-10). The lumbar curvature appears between 12 and 18 months, and the once-helpless infant becomes a toddler who can stand (Figure 20-11). One of the most striking changes to occur during infancy is the rapid development of the nervous and muscular systems. This permits the infant to follow a moving object with the eyes (2 months); lift the head and raise the chest (3 months); sit when well supported (4 months); crawl (10 months); stand alone (12 months); and run, although a bit stiffly (18 months).

FIGURE 20–10 *Normal Curvature of the Infant's Spine.*

FIGURE 20–11 *Normal Lumbar Curvature of a Toddler's Spine.*

CHILDHOOD

Childhood extends from the end of infancy to sexual maturity or puberty—12 to 14 years in girls and 14 to 16 years in boys. Overall, growth during early childhood continues at a rather rapid pace, but month-to-month gains become less consistent. By the age of 6 the child appears more like a preadolescent than an infant or toddler. The child becomes less chubby, the potbelly becomes flatter, and the face loses its babyish look. The nervous and muscular systems continue to develop rapidly during the middle years of childhood; by 10 years of age the child has developed numerous motor and coordination skills.

Fetal Alcohol Syndrome

Consumption of alcohol during pregnancy can have tragic effects on a developing fetus. Educational efforts to inform pregnant women about the dangers of alcohol are now receiving national attention. Even very limited consumption of alcohol during pregnancy poses significant hazards to the developing baby because alcohol can easily cross the placental barrier and enter the fetal bloodstream.

When alcohol enters the fetal blood, the potential result, called **fetal alcohol syndrome (FAS),** can cause tragic congenital abnormalities such as "small head" or **microcephaly** (my-kro-SEF-ah-lee), low birth weight, developmental disabilities such as mental retardation, and even fetal death.

The *deciduous teeth*, which began to appear at about 6 months of age, are lost during childhood, beginning at about 6 years of age. The *permanent teeth*, with the possible exception of the third molars or wisdom teeth, have all erupted by age 14.

ADOLESCENCE AND ADULTHOOD

The average age range of **adolescence** varies but generally the teenage years (13 to 19) are used. The period is marked by rapid and intense physical growth, which ultimately results in sexual maturity. Many of the developmental changes that occur during this period are controlled by the secretion of sex hormones and are classified as **secondary sex characteristics.** Breast development is often the first sign of approaching puberty in girls, beginning about age 10. Most girls begin to menstruate at 12 to 13 years of age. In boys the first sign of puberty is often enlargement of the testicles, which begins between 10 and 13 years of age. Both sexes show a spurt in height during adolescence. In girls the spurt in height begins between the ages of 10 and 12 and is nearly complete by 14 or 15. In boys the period of rapid growth begins between 12 and 13 and is generally complete by 16.

Many developmental changes that began early in childhood are not completed until the early or middle years of **adulthood.** Examples include the maturation of bone, resulting in the full closure of the growth plates, and changes in the size and placement of other body components such as the sinuses. Many body traits do not become apparent for years after birth. Normal balding patterns, for example, are determined at the time of fertilization by heredity but do not appear until maturity As a general rule, adulthood is characterized by maintenance of existing body tissues. With the passage of years the ongoing effort of maintenance and repair of body tissues become more and more difficult. As a result, degeneration begins. It is the process of aging, and it culminates in death.

OLDER ADULTHOOD

Most body systems are in peak condition and function at a high level of efficiency during the early years of adulthood. As a person grows older, a gradual but certain decline takes place in the functioning of every major organ system in the body. The study of aging is called *gerontology.* The remainder of this chapter deals with a number of the more common degenerative changes that frequently characterize **senescence** (se-NES-ens) or older adulthood. Many of the biological changes associated with advancing age are shown in Figure 20-12. The illustration highlights the proportion of remaining function in a number of organs in older adulthood when compared with a 20-year-old person.

 Effects of Aging

SKELETAL SYSTEM

In older adulthood, bones undergo changes in texture, degree of calcification, and shape. Instead of clean-cut margins, older bones develop indistinct and shaggy-appearing margins with spurs—a process called *lipping.* This type of degenerative change restricts movement because of the piling up of bone tissue around the joints. With advancing age, changes in calcification may result in

reduction of bone size and in bones that are porous and subject to fracture. The lower cervical and thoracic vertebrae are the site of frequent fractures. The result is curvature of the spine and the shortened stature so typical of late adulthood. Degenerative joint diseases such as **osteoarthritis** (OS-tee-o-ar-THRYE-tis) are also common in elderly adults.

INTEGUMENTARY SYSTEM (SKIN)

With advancing age the skin becomes dry, thin, and inelastic. It "sags" on the body because of increased wrinkling and skinfolds. Pigmentation changes and the thinning or loss of hair are also common problems associated with the aging process.

URINARY SYSTEM

The number of nephron units in the kidney decreases by almost 50% between the ages of 30 and 75. Also, because less blood flows through the kidneys as an individual ages, there is a reduction in overall function and excretory capacity or the ability to produce urine. In the bladder, significant age-related problems often occur because of diminished muscle tone. Muscle atrophy (wasting) in the bladder wall results in decreased capacity and inability to empty or void completely.

RESPIRATORY SYSTEM

In older adulthood the costal cartilages that connect the ribs to the sternum become hardened or calcified. This makes it difficult for the rib cage to expand and contract as it normally does during inspiration and expiration. In time the ribs gradually become "fixed" to the sternum, and chest movements become difficult. When this occurs the rib cage remains in a more expanded position, respiratory efficiency decreases, and a condition called "barrel chest" results. With advancing years a generalized atrophy or wasting of muscle tissue takes place as the contractile muscle cells are replaced by connective tissue. This loss of muscle cells decreases the strength of the muscles associated with inspiration and expiration.

CARDIOVASCULAR SYSTEM

Degenerative heart and blood vessel disease is one of the most common and serious effects of

FIGURE 20–12 *Some Biological Changes Associated with Maturity and Aging.* Insets show proportion of remaining function in the organs of a person in late adulthood compared to a 20-year-old person.

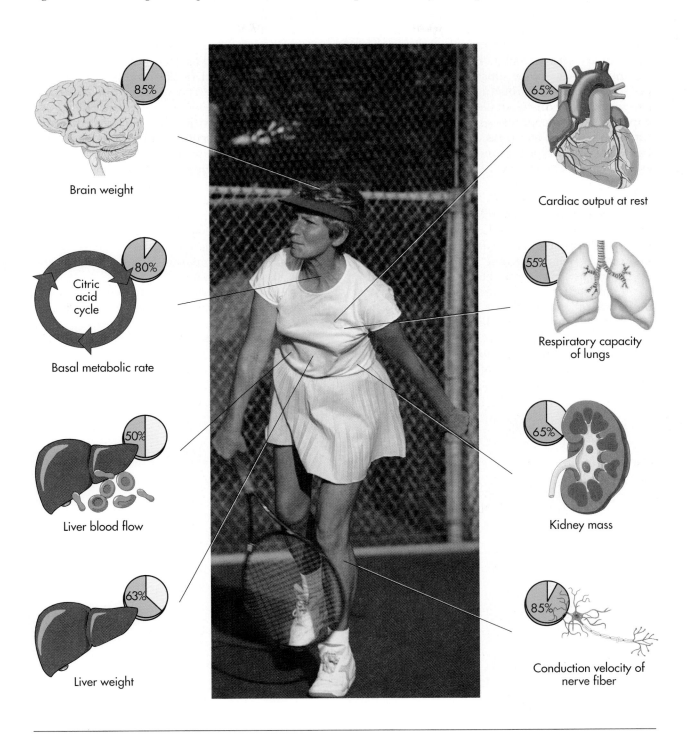

Brain weight — 85%

Basal metabolic rate (Citric acid cycle) — 80%

Liver blood flow — 50%

Liver weight — 63%

Cardiac output at rest — 65%

Respiratory capacity of lungs — 55%

Kidney mass — 65%

Conduction velocity of nerve fiber — 85%

aging. Fatty deposits build up in blood vessel walls and narrow the passageway for the movement of blood, much as the buildup of scale in a water pipe decreases flow and pressure. The resulting condition, called **atherosclerosis** (ath-er-o-skle-RO-sis), often leads to eventual blockage of the coronary arteries and a "heart attack." If fatty accumulations or other substances in blood vessels calcify, actual hardening of the arteries or **arteriosclerosis** (ar-te-ree-o-skle-RO-sis) occurs. Rupture of a hardened vessel in the brain (stroke) is a frequent cause of serious disability or death in the older adult. **Hypertension** or high blood pressure is also more common.

SPECIAL SENSES

The sense organs, as a group, all show a gradual decline in performance and capacity as a person ages. Most people are farsighted by age 65 because eye lenses become hardened and lose elasticity; the lenses cannot become curved to accommodate for near vision. This hardening of the lens is called **presbyopia** (pres-bee-O-pee-ah), which means "old eye." Many individuals first notice the change at about 40 or 45 years of age, when it becomes difficult to do close-up work or read without holding printed material at arm's length. This explains the increased need, with advancing age, for bifocals or glasses that incorporate two lenses to automatically accommodate for near and distant vision. Loss of transparency of the lens or its covering capsule is another common age-related eye change. If the lens actually becomes cloudy and significantly impairs vision, it is called a **cataract** (KAT-ah-rakt) and must be removed surgically. The incidence of **glaucoma** (glaw-KO-mah), the most serious age-related eye disorder, increases with age. Glaucoma causes an increase in the pressure within the eyeball and, unless treated, often results in blindness.

In many elderly people a very significant loss of hair cells in the organ of Corti (inner ear) causes a serious decline in the ability to hear certain frequencies. In addition, the eardrum and attached ossicles become more fixed and less able to transmit mechanical sound waves. Some degree of hearing impairment is universally present in the older adult.

The sense of taste is also decreased. This loss of appetite may be caused, at least in part, by the replacement of taste buds with connective tissue cells. Only about 40% of the taste buds present at age 30 remain in an individual at age 75.

O U T L I N E S U M M A R Y

PRENATAL PERIOD

A Prenatal period begins at conception and continues until birth (about 39 weeks)

B Science of fetal growth and development called *embryology*

C Fertilization to implantation requires about 10 days
  1 Fertilization normally occurs in outer third of oviduct (Figure 20-2)
  2 Fertilized ovum called a *zygote; zygote* is genetically complete—all that is needed for expression of hereditary traits is time and nourishment
  3 After 3 days of cell division, the zygote has developed into a solid cell mass called a *morula*
  4 Continued cell divisions of the morula produce a hollow ball of cells called a *blastocyst*
  5 Blastocyst implants in the uterine wall about 10 days after fertilization
  6 Blastocyst forms the amniotic cavity and chorion of the placenta (Figure 20-4)
  7 Placenta provides for exchange of nutrients between the mother and fetus

D Periods of development
  1 Length of pregnancy or gestation period is about 39 weeks
  2 Embryonic phase extends from fertilization to the end of week 8 of gestation
  3 Fetal phase extends from week 8 to week 39 of gestation

E Three primary germ layers appear in the developing embryo after implantation of the blastocyst (Table 20-1):
  1 Endoderm—inside layer
  2 Ectoderm—outside layer
  3 Mesoderm—middle layer
  4 All organ systems are formed and functioning by month 4 of gestation (Figure 20-6)

F Histogenesis and organogenesis
  1 Formation of new organs and tissues occurs from specific development of the primary germ layers
  2 Each primary germ layer gives rise to definite structures such as the skin and muscles
  3 Growth processes include cell differentiation, multiplication, growth, and rearrangement
  4 From 4 months of gestation until delivery, the development of the baby is mainly a matter of growth

BIRTH OR PARTURITION

A Process of birth called *parturition* (Figure 20-7)
  1 At the end of week 39 of gestation, the uterus becomes "irritable"
  2 Fetus takes head-down position against the cervix
  3 Muscular contractions begin, and labor is initiated
  4 Amniotic sac ("bag of waters") ruptures
  5 Cervix dilates
  6 Fetus moves through vagina to exterior

B Stages of labor
  1 Stage one—period from onset of uterine contractions until dilation of the cervix is complete
  2 Stage two—period from the time of maximal cervical dilation until the baby exits through the vagina
  3 Stage three—process of expulsion of the placenta through the vagina

POSTNATAL PERIOD

A Postnatal period begins at birth and lasts until death

B Divisions of postnatal period into isolated time frames can be misleading; life is a continuous process; growth and development are continuous

C Obvious changes in the physical appearance of the body in—whole and in proportion—occur between birth and maturity (Figure 20-8)

D Divisions of postnatal period
  1 Infancy
  2 Childhood
  3 Adolescence and adulthood
  4 Older adulthood

E Infancy
  1 First 4 weeks called *neonatal period* (Figure 20-9)
  2 Neonatology—medical and nursing specialty concerned with the diagnosis and treatment of disorders of the newborn
  3 Many cardiovascular changes occur at the time of birth; fetus is totally dependent on mother, whereas the newborn must immediately become totally self-supporting (respiration and circulation)
  4 Respiratory changes at birth include a deep and forceful first breath
  5 Developmental changes between the neonatal period and 18 months include:
    a Doubling of birth weight by 4 months and tripling by 1 year
    b 50% increase in body length by 12 months
    c Development of normal spinal curvature by 15 months (Figure 20-11)
    d Ability to raise head by 3 months
    e Ability to crawl by 10 months
    f Ability to stand alone by 12 months
    g Ability to run by 18 months

F Childhood
  1 Extends from end of infancy to puberty—13 years in girls and 15 in boys

2 Overall rate of growth remains rapid but decelerates

3 Continuing development of motor and coordination skills

4 Loss of deciduous or baby teeth and eruption of permanent teeth

G Adolescence and adulthood

1 Average age range of adolescence varies from 13 to 19 years

2 Period of rapid growth resulting in sexual maturity (adolescence)

3 Appearance of secondary sex characteristics regulated by secretion of sex hormones

4 Growth spurt typical of adolescence; begins in girls at about 10 and in boys at about 12

5 Growth plates fully close in adult; other structures such as the sinuses acquire adult placement

6 Adulthood characterized by maintenance of existing body tissues

7 Degeneration of body tissue begins in adulthood

H Older adulthood

1 Degenerative changes characterize older adulthood or senescence

2 Every organ system of the body undergoes degenerative changes

3 Senescence culminates in death

EFFECTS OF AGING

A Skeletal system

1 Aging causes changes in the texture, calcification, and shape of bones

2 Bone spurs develop around joints

3 Bones become porous and fracture easily

4 Degenerative joint diseases such as osteoarthritis are common

B Integumentary system (skin)

1 With age, skin "sags" and becomes:

a Thin

b Dry

c Wrinkled

2 Pigmentation problems are common

3 Frequent thinning or loss of hair occurs

C Urinary system

1 Nephron units decrease in number by 50% between ages 30 and 75

2 Blood flow to kidney and therefore ability to form urine decreases

3 Bladder problems such as inability to void completely are caused by muscle wasting in the bladder wall

D Respiratory system

1 Calcification of costal cartilages causes rib cage to remain in expanded position—barrel chest

2 Wasting of respiratory muscles decreases respiratory efficiency

3 Respiratory membrane thickens; movement of oxygen from alveoli to blood is slowed

E Cardiovascular system

1 Degenerative heart and blood vessel disease is among the most common and serious effects of aging

2 Fat deposits in blood vessels (atherosclerosis) decrease blood flow to the heart and may cause complete blockage of the coronary arteries

3 Hardening of arteries (arteriosclerosis) may result in rupture of blood vessels, especially in the brain (stroke)

4 Hypertension or high blood pressure is common in older adulthood

F Special senses

1 All sense organs show a gradual decline in performance with age

2 Eye lenses become hard and cannot accommodate for near vision; result is farsightedness in many people by age 45 (presbyopia or "old eye")

3 Loss of transparency of lens or cornea is common (cataract)

4 Glaucoma (increase in pressure in eyeball) is often the cause of blindness in older adulthood

5 Loss of hair cells in inner ear produces frequency deafness in many older people

6 Decreased transmission of sound waves caused by loss of elasticity of eardrum and fixing of the bony ear ossicles is common in older adulthood

7 Some degree of hearing impairment is universally present in the aged

8 Only about 40% of the taste buds present at age 30 remain at age 75

N E W   W O R D S

arteriosclerosis	fertilization	neonate	ectoderm
atherosclerosis	gestation	organogenesis	mesoderm
blastocyst	glaucoma	parturition	rheumatoid arthritis
cataract	histogenesis	presbyopia	senescence
chorion	implantation	primary germ layers	zygote
embryology	morula	endoderm	

C H A P T E R   T E S T

1. The prenatal period begins at _____ and ends at _____.
2. The postnatal period begins at _____ and continues until _____.
3. The science of the development of the individual before birth is called _____.
4. The fertilized ovum is also called a _____.
5. The fluid-filled, shock-absorbing sac in which the embryo floats during development is called the _____ cavity.
6. The "bridge" that permits exchange of nutrients between mother and baby before birth is called the _____.
7. The way tissues arrange themselves into organs during development is called _____.
8. The first 4 weeks of infancy are often referred to as the _____ period.
9. The developmental period that extends from the end of infancy to sexual maturity is called _____.
10. Hardening of the arteries is called _____.
11. If the lens of the eye becomes cloudy and impairs vision, the condition is called a _____.

Circle the T before each true statement and the F before each false statement.

T F 12. The fertilized ovum is called the *morula.*
T F 13. There are 5 primary germ layers in the developing embryo.
T F 14. In humans, all organ systems are formed and functioning by month 4 of pregnancy.
T F 15. The placenta develops in part from the fetal membrane called the *chorion.*
T F 16. Neonatology is the medical and nursing specialty concerned with the diagnosis and treatment of disorders of the newborn.
T F 17. Childhood extends from the end of infancy to sexual maturity.
T F 18. The term *senescence* refers to older adulthood.
T F 19. The kidney is unique in that it does not show degenerative changes with advancing age.
T F 20. The term *arteriosclerosis* refers to "hardening of the arteries."

R E V I E W   Q U E S T I O N S

1. What biological event separates the prenatal from the postnatal period of development?
2. Define embryology, gestation, fertilization, implantation, morula, and blastocyst.
3. What is the difference between an ovum and a zygote?
4. What is the difference between an embryo and a fetus?
5. Discuss the role of the placenta in fetal development.
6. Briefly define histogenesis and organogenesis.
7. At what point in development of the fetus are all the organ systems formed and functioning?
8. When does the postnatal period begin and how long does it last?
9. List the four subdivisions of the postnatal period.
10. What is the relationship of the neonatal period to infancy? What is neonatology?
11. List five developmental changes that occur during infancy.
12. Define *senescence.*
13. Describe the degenerative changes that occur in the skeletal system as a result of aging.
14. How is the skin or integumentary system affected by aging?
15. What is the difference between atherosclerosis and arteriosclerosis?
16. Define presbyopia, glaucoma, and cataract.

C R I T I C A L   T H I N K I N G

17. What are the primary germ layers and how are they related to development of the fetus?
18. Why is it necessary for a baby's first breath to be deep and forceful? What is the primary stimulus that causes a newborn to take its first breath?
19. Is the period of childhood the same length for both boys and girls? Explain the reason for your answer.
20. What is the average age range of adolescence? Functionally, how would the reproductive system in an individual at the end of adolescence be described?

Chemistry of Life

Appendix

Life is chemistry. It's not quite that simple, but the more we learn about human structure and function, the more we realize that it all boils down to interactions among chemicals. The digestion of food, the formation of bone tissue, and the contraction of a muscle are all chemical processes. Thus the basic principles of anatomy and physiology are ultimately based on principles of chemistry. A whole field of science, **biochemistry,** is devoted to studying the chemical aspects of life. To truly understand the human body, it is important to understand a few basic facts about biochemistry, the chemistry of life.

◗ *Levels of Chemical Organization*

Matter is anything that occupies space and has mass. Biochemists classify matter into several levels of organization for easier study. In the body, most chemicals are in the form of **molecules.** Molecules are particles of matter that are composed of one or more smaller units called **atoms.** Atoms in turn are composed of several kinds of *subatomic particles:* **protons, electrons,** and **neutrons.** (Refer to the Mini-Glossary on p. 440.)

ATOMS

Atoms are units that until recently could not be seen by scientists. New instruments, including *tunneling microscopes* and *atomic force microscopes,* produce pictures of atoms that confirm current models of how atoms are put together. At the core of each atom is a **nucleus** composed of positively charged protons and uncharged neutrons. The number of protons in the nucleus is an atom's atomic number. The number of protons and neutrons combined is the atom's **atomic mass.**

Negatively charged electrons surround the nucleus at a distance. In an electrically neutral atom, there is one electron for every proton. Electrons move about within certain limits called **orbitals.** Each orbital can hold two electrons. Orbitals are arranged into **energy levels** (shells), depending on their distance from the nucleus. The farther an orbital extends from the nucleus, the higher its energy level. The energy level closest to the nucleus has one orbital, so it can hold two electrons. The next energy level has up to four orbitals, so it can hold eight electrons. Figure 1 shows a carbon (C) atom. Notice that the first energy level (the innermost shell) contains two electrons and the outer energy level contains four electrons. The outer energy level of a carbon atom

FIGURE 1 *A Model of the Atom.* The nucleus—protons (+) and neutrons—is at the core. Electrons inhabit outer regions called *energy levels.* This is a carbon atom, a fact that is determined by the number of its protons. All carbon atoms (and only carbon atoms) have six protons. (One proton in the nucleus is not visible in this illustration.)

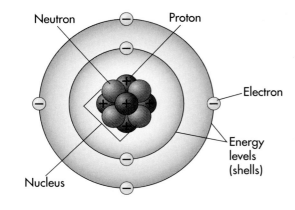

could hold up to four more electrons (for a total of eight). The number of electrons in the outer energy level of an atom determines how it behaves chemically (that is, how it may unite with other atoms). This behavior, called *chemical bonding,* will be discussed later.

ELEMENTS, MOLECULES, AND COMPOUNDS

Substances can be classified as **elements** or **compounds.** Elements are pure substances, composed of only one of over a hundred types of atoms that exist in nature. Only four kinds of atoms (**oxygen, carbon, hydrogen,** and **nitrogen**) make up about 96% of the human body. There are traces of about 20 other elements in the body. Table 1 lists some of the elements in the body. Table 1 also gives for each element its universal chemical *symbol*—the abbreviation used by chemists worldwide.

Atoms usually unite with each other to form larger chemical units called **molecules.** Some molecules are made of several atoms of the same element. *Compounds* are substances whose molecules have more than one element in them. The *formula* for a compound contains symbols for the elements in each molecule. The number of atoms of

TABLE 1   Important Elements in the Human Body

	NAME	SYMBOL	NUMBER OF ELECTRONS IN OUTER SHELL*
Major elements (over 96% of body weight)	Oxygen	O	6
	Carbon	C	4
	Hydrogen	H	1
	Nitrogen	N	5
Trace elements (examples of more than 20 trace elements found in the body)	Calcium	Ca	2
	Phosphorus	P	5
	Sodium (Latin *natrium*)	Na	1
	Potassium (Latin *kalium*)	K	1
	Chlorine	Cl	7
	Iodine	I	7

*Maximum is eight, except for hydrogen. The maximum for that element is two.

Radioactive Isotopes

Each element is unique because of the number of protons it has. In short, each element has its own *atomic number*. However, atoms of the same element can have different numbers of neutrons. Two atoms that have the same atomic number but different atomic masses are **isotopes** of the same element. An example is hydrogen. Hydrogen has three isotopes: ^1H (the most common isotope), ^2H, and ^3H. The figure shows that each different isotope has only one proton but different numbers of neutrons.

Some isotopes have unstable nuclei that radiate (give off) particles. Radiation particles include protons, neutrons, electrons, and altered versions of these normal subatomic particles. An isotope that emits radiation is called a **radioactive isotope.**

Radioactive isotopes of common elements are sometimes used to evaluate the function of body parts. Radioactive iodine (^{125}I) put into the body and taken up by the thyroid gland gives off radiation that can be easily measured. Thus the rate of thyroid activity can be determined. Images of internal organs can be formed by radiation scanners that plot out the location of injected or ingested radioactive isotopes. For example, radioactive technetium (^{99}Tc) is commonly used to image the liver and spleen. The radioactive isotopes ^{13}N, ^{15}O, and ^{11}C are often used to study the brain in a technique called the *PET scan.*

Radiation can damage cells. Exposure to high levels of radiation may cause cells to develop into cancer cells. Higher levels of radiation completely destroy tissues, causing *radiation sickness.* Low doses of radioactive substances are sometimes given to cancer patients to destroy cancer cells. The side effects of these treatments result from the unavoidable destruction of normal cells with the cancer cells.

^1H

^2H

^3H

each element in the molecule is expressed as a subscript after the elemental symbol. For example, each molecule of the compound **carbon dioxide** has one carbon (C) atom and two oxygen (O) atoms; thus its molecular formula is CO_2.

> ⬤ **Matter is composed of molecules, which are composed of atoms. Atoms are composed of protons, electrons, and neutrons. Elements are pure substances, composed of only one kind of atom. Compounds are composed of molecules having more than one kind of atom.** ⬤

⬤ *Chemical Bonding*

Chemical bonds form to make atoms more stable. An atom is said to be chemically stable when its outer energy level is "full" (that is, when its energy shells have the maximum number of electrons they can hold). All but a handful of atoms have room for more electrons in their outermost energy level. A basic chemical principle states that atoms react with one another in ways to make their outermost energy level full. To do this, atoms can share, donate, or borrow electrons.

For example, a hydrogen atom has one electron and one proton. Its single energy shell has one electron but *can* hold two—so it's not full. If two hydrogen atoms "share" their single electrons with each other, then both will have full energy shells, making them more stable *as a molecule* than either would be as an atom. This is one example of how atoms **bond** to form molecules. Other atoms may donate or borrow electrons until the outermost energy level is full.

IONIC BONDS

One common way in which atoms make their outermost energy level full is to form **ionic bonds** with other atoms. Such a bond forms between an atom that has only one or two electrons in the outermost level (that would normally hold eight) and an atom that needs only one or two electrons to fill its outer level. The atom with one or two electrons simply "donates" its outer shell electrons to the one that needs one or two.

For example, as you can see in Table 1, the sodium (Na) atom has one electron in its outer level and the chlorine (Cl) atom has seven. Both need to have eight electrons in their outer shell. Figure 2 shows how sodium and chlorine form an ionic bond when sodium "donates" the electron in its outer shell to chlorine. Now both atoms have full outer shells (although sodium's outer shell is now one energy level lower). Because the sodium atom lost an electron, it now has one more proton that it has electrons. This makes it a positive **ion,** an electrically charged atom. Chlorine has "borrowed" an electron to become a negative ion called the *chloride* ion. Because oppositely charged particles attract one another, the sodium and chloride ions are drawn together to form a sodium chloride (NaCl) molecule—common table salt. The molecule is held together by an *ionic bond.*

Ionic molecules usually dissolve easily in water because water molecules wedge between the ions and force them apart. When this happens, we say the molecules **dissociate** (dis-SO-see-ayt) to form free ions. Molecules that form ions when dissolved in water are called **electrolytes** (el-EK-tro-lites). Chapter 17 describes mechanisms that maintain the homeostasis of electrolytes in the body. Table 2 lists some of the more important ions present in body fluids.

TABLE 2 Important Ions in Human Body Fluids	
NAME	**SYMBOL**
Sodium	Na^+
Chloride	Cl^-
Potassium (Latin *kalium*)	K^+
Calcium	Ca^{++}
Hydrogen	H^+
Magnesium	Mg^{++}
Hydroxide	OH^-
Phosphate	$PO_4^=$
Bicarbonate	HCO_3^-

FIGURE 2 *Ionic Bonding.* The sodium atom donates the single electron in its outer energy level to a chlorine atom having seven electrons in its outer level. Now both have eight electrons in their outer shells. Because the electron/proton ratio changes, the sodium atom becomes a positive sodium ion. The chlorine atom becomes a negative chloride ion. The positive-negative attraction between these oppositely charged ions is called an *ionic bond.*

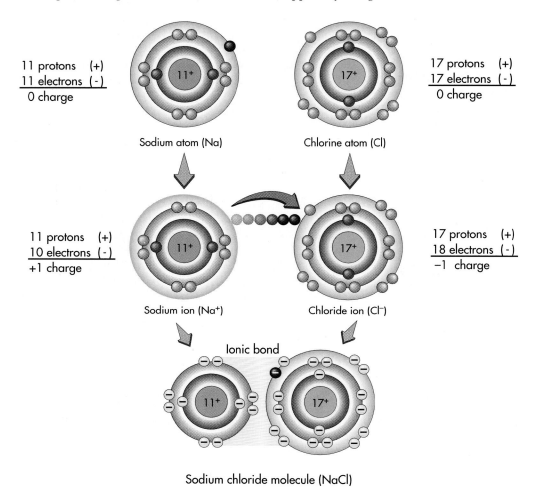

11 protons   (+)
11 electrons (-)
0 charge

Sodium atom (Na)

17 protons   (+)
17 electrons (-)
0 charge

Chlorine atom (Cl)

11 protons   (+)
10 electrons (-)
+1 charge

Sodium ion (Na+)

17 protons   (+)
18 electrons (-)
−1 charge

Chloride ion (Cl−)

Ionic bond

Sodium chloride molecule (NaCl)

The formula of an ion always shows its charge by a superscript after the chemical symbol. Thus the sodium ion is Na^+, and the chloride ion is Cl^-. Calcium (Ca) atoms lose two electrons when they form ions, so the calcium ion formula is Ca^{++}.

COVALENT BONDS

Atoms may also fill their energy levels by sharing electrons rather than donating or receiving them. When atoms share electrons, a **covalent** (ko-VAY-lent) **bond** forms. For example, Figure 3 shows how two hydrogen atoms may move together closely so that their energy levels overlap. Each energy level contributes its one electron to the sharing relationship. This way, both outer levels have access to both electrons. Because atoms involved in a covalent bond must stay close to each other, it is not surprising that covalent bonds are not easily broken. Covalent bonds normally do not break apart in water.

FIGURE 3 *Covalent Bonding.* Two hydrogen atoms move together, overlapping their energy levels. Although neither gains nor loses an electron, the atoms share the electrons, forming a covalent bond.

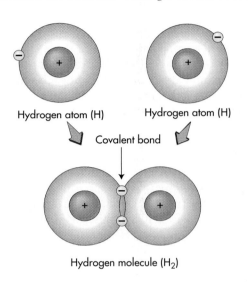

Hydrogen atom (H)   Hydrogen atom (H)

Covalent bond

Hydrogen molecule (H_2)

 Ionic bonding occurs when one atom donates an electron to another atom and the resulting ions attract each other. In covalent bonding, atoms share electrons.

Inorganic Chemistry

In living organisms, there are two kinds of compounds: **organic** and **inorganic.** Organic compounds are composed of molecules that contain carbon-carbon (C-C) covalent bonds or carbon-hydrogen (C-H) covalent bonds—or both kinds of bonds. Few inorganic compounds have carbon atoms in them and none have C-C or C-H bonds. Organic molecules are generally larger and more complex than inorganic molecules. The human body has both kinds of compounds because both are equally important to the chemistry of life. We will discuss the chemistry of inorganic compounds first, then move on to some of the important types of organic compounds.

Metric System

Scientists, many government agencies, and increasing numbers of American industries are using or moving toward the conversion of our system of English measurements to the metric system. The metric system is a decimal system in which measurement of length is based on the *meter* (39.37 inches) and weight or mass is based on the *gram* (about 454 grams equal a pound).

A micrometer is one millionth of a meter. (*Micron* is another name for micrometer.) In the metric systems the units of length are as follows:

1 meter (m) = 39.37 inches
1 centimeter (cm) = 1/100 m
1 millimeter (mm) = 1/1,000 m
1 micrometer (μ m) or micron (μ) =
    1/1,000,000 m
1 nanometer (nm) = 1/1,000,000,000 m
1 Angstrom (Å) = 1/10,000,000,000 m

Approximately equal to 1 inch:
2.5 cm
25 mm
25,000 μ m
25,000,000 nm
250,000,000 Å

WATER

One of the compounds that is most essential to life—water—is an inorganic compound. Water is the most abundant compound in the body, found in and around each cell. It is the **solvent** in which most other compounds or **solutes** are dissolved. When water is the solvent for a *mixture* (a blend of two or more kinds of molecules), the mixture is called an **aqueous solution.** An aqueous solution containing common salt (NaCl) and other molecules forms the "internal sea" of the body. Water molecules not only compose the basic internal environment of the body, but they also participate in many important *chemical reactions.* Chemical reactions are interactions among molecules in which atoms regroup into new combinations.

A common type of chemical reaction in the body is **dehydration synthesis.** In any kind of synthesis reaction, the **reactants** combine to form a larger **product.** In dehydration synthesis, reactants combine only after hydrogen (H) and oxygen (O) atoms are removed. These leftover H and O atoms come together, forming H_2O or water. As Figure 4 shows, the result is both the large product molecule and a water molecule. Just as dehydration of a cell is a loss of water from the cell and dehydration of the body is loss of fluid from the entire internal environment, dehydration synthesis is a reaction in which water is lost from the reactants.

Another common reaction in the body, **hydrolysis** (hye-DROL-i-sis), also involves water. In this reaction, water *(hydro-)* disrupts the bonds in large molecules, causing them to be broken down into smaller molecules *(lysis).* Hydrolysis is virtually the reverse of dehydration synthesis, as Figure 4 shows.

Chemical reactions always involve energy transfers. Energy is required to build the molecules. Some of that energy is stored as potential energy in the chemical bonds. The stored energy can then be released when the chemical bonds

in the molecule are later broken apart. For example, a molecule called **adenosine triphosphate (ATP)** breaks apart in the muscle cells to yield the energy needed for muscle contraction (see Figure 15-2).

Chemists often use a *chemical equation* to represent a chemical reaction. In a chemical equation, the reactants are separated from the products by an arrow ($\rightarrow$) showing the "direction" of the reaction. Reactants are separated from each other and products are separated from each other by addition signs (+). Thus the reaction *potassium and chloride combine to form potassium chloride* can be expressed as the equation:

$$K^+ + Cl^- \rightarrow KCl$$

The single arrow ($\rightarrow$) is used for equations that occur in only one direction. For example, when hydrochloric acid (HCl) is dissolved in water, *all* of it dissociates to form H^+ and Cl^-.

$$HCl \rightarrow H^+ + Cl^-$$

The double arrow ($\leftrightarrows$) is used for reactions that happen in "both directions" at the same time. When carbonic acid (H_2CO_3) dissolves in water, *some* of it dissociates into H^+ (hydrogen ion) and

---

FIGURE 4   *Water-Based Chemistry.* Dehydration synthesis is a reaction in which small molecules are assembled into large molecules by removing water (H and O atoms). Hydrolysis operates in the reverse direction; H and O from water is added as large molecules are broken down into small molecules.

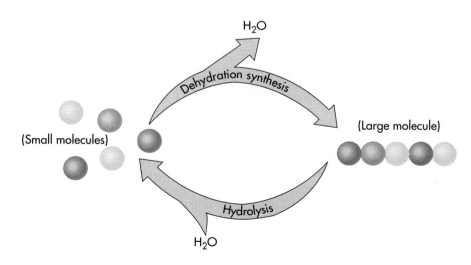

HCO_3^- (bicarbonate) but not all of it. As additional ions dissociate, previously dissociated ions bond together again forming H_2CO_3.

$$H_2CO_3 \leftrightarrows H^+ HCO_3^-$$

In short, the double arrow indicates that at any instant in time both reactants and products are present in the solution at the same time.

ACIDS, BASES, AND SALTS

Besides water, many other inorganic compounds are important in the chemistry of life. For example, **acids** and **bases** are compounds that profoundly affect chemical reactions in the body. As explained in more detail at the beginning of Chapter 18, a few water molecules dissociate to form the H^+ ion and the OH^- (hydroxide) ion:

$$\mathbf{H_2O \leftrightarrows H^+ + OH^-}$$

In pure water, the balance between these two ions is equal. However, when an acid such as hydrochloric acid (HCl) dissociates into H^+ and Cl^-, it shifts this balance in favor of excess H^+ ions. In the blood, carbon dioxide (CO_2) forms carbonic acid (H_2CO_3) when it dissolves in water. Some of the carbonic acid then dissociates to form H^+ ions and HCO_3^- (bicarbonate) ions, producing an excess of H^+ ions in the blood. Thus high CO_2 levels in the blood make the blood acidic.

Bases or **alkaline** compounds, on the other hand, shift the balance in the opposite direction. For example, sodium hydroxide (NaOH) is a base that forms OH^- ions but no H^+ ions. In short, acids are compounds that produce an excess of H^+ ions, and bases are compounds that produce an excess of OH^- ions (or a decrease in H^+).

The relative H^+ concentration is a measure of how acidic or basic a solution is. The H^+ concentration is usually expressed in units of **pH.** The formula used to calculate pH units gives a value of 7 to pure water. A higher pH value indicates a low relative concentration of H^+—a base. A lower pH value indicates a higher H^+ concentration—an acid. Figure 5 shows a scale of pH from 0 to 14. Notice that when the pH of a solution is less than 7, the scale "tips" toward the side marked "high H^+." When the pH is more than 7, the scale "tips" toward the side marked "low H^+." pH units

FIGURE 5 *The pH Scale.* The H^+ concentration is balanced with the OH^- concentration at pH 7. At values above 7 (low H^+), the scale tips in the basic direction. At values below 7 (high H^+), the scale tips toward the acid side.

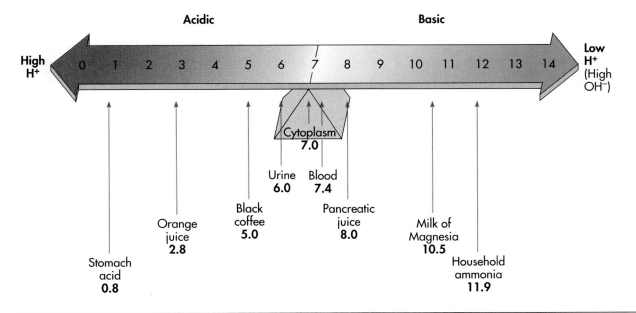

increase or decrease by factors of 10. Thus a pH 5 solution has ten times the H^+ concentration of a pH 6 solution. A pH 4 solution has a hundred times the H^+ concentration of a pH 6 solution.

A *strong acid* is an acid that completely, or almost completely, dissociates to form H^+ ions. A *weak acid*, on the other hand, dissociates very little and therefore produces few excess H^+ ions in solution.

When a strong acid and a strong base mix, excess H^+ ions may combine with the excess OH^- ions to form water. That is, they may *neutralize* each other. The remaining ions usually form neutral ionic compounds called **salts.** For example:

$$HCl + NaOH \rightarrow H^+ + Cl^- + Na^+ + OH^- \rightarrow H_2O + NaCl$$
acid   base   water   salt

The pH of body fluids affects body chemistry so greatly that normal body function can be maintained only within a narrow range of pH. The body can remove excess H^+ ions by excreting them in the urine (see Chapter 16). Another way to remove acid is by increasing the loss of CO_2 (an acid) by the respiratory system (see Chapter 13). A third way to adjust the body's pH is the use of **buffers**—chemicals in the blood that maintain pH. Buffers maintain pH balance by preventing sudden changes in the H^+ ion concentration. Buffers do this by forming a chemical system that neutralizes acids and bases as they are added to a solution. The mechanisms by which the body maintains pH homeostasis, or acid-base balance, are discussed further in Chapter 18.

Organic compounds always contain carbon (forming carbon-carbon or carbon-hydrogen bonds); few inorganic compounds even contain carbon. Biological reactions take place in aqueous solutions. In hydrolysis, chemical bonds are broken. In dehydration synthesis, bonds are formed. The pH (relative H+ concentration) of body fluids affects chemical reactions.

Organic Chemistry

Organic compounds are much more complex than inorganic compounds. In this section, we will describe the basic structure and function of each major type of organic compound found in the body: **carbohydrates, lipids** (fats), **proteins,** and **nucleic acids.** Table 3 summarizes the structure and the function of each type. Refer to this table as you read through the descriptions that follow.

CARBOHYDRATES

The name *carbohydrate* literally means "carbon (C) and water (H_2O)," signifying the types of atoms that form carbohydrate molecules. The basic unit of carbohydrate molecules is called a *monosaccharide* (mah-no-SAK-ah-ride) (Figure 6). Glucose (dextrose) is an important monosaccharide in the body; cells use it as their primary source of energy (see Chapter 15). A molecule made of two saccharide units is a double sugar or *disaccharide*. The disaccharides sucrose (table sugar) and lactose (milk sugar) are important dietary carbohydrates. After they are eaten, the body digests them to form monosaccharides that can be used as cellular fuel. Many saccharide units joined together form *polysaccharides*. Examples of polysaccharides are **glycogen** (GLY-ko-jen) and *starch*. Each glycogen molecule is a chain of glucose molecules joined together. Liver cells and muscle cells form glycogen when there is an excess of glucose in the blood, thus putting them into "storage" for later use.

FIGURE 6   *Carbohydrates.* Monosaccharides are single carbohydrate units joined by dehydration synthesis to form disaccharides and polysaccharides.

Carbohydrates

Monosaccharide

Disaccharide

Polysaccharide

TABLE 3   Major Types of Organic Compounds

EXAMPLE	COMPONENTS	FUNCTIONS
Carbohydrate		
Monosaccharide (glucose, galactose, fructose)	Single monosaccharide unit	Unit as source of energy; used to build other carbohydrates
Disaccharide (sucrose, lactose, maltose)	Two monosaccharide units	Can be broken into monosaccharides
Polysaccharide (glycogen, starch)	Many monosaccharide units	Used to store monosaccharides (thus to store energy)
Lipid		
Triglyceride	One glycerol, three fatty acids	Stores energy
Phospholipid	Phosphorus-containing unit, two fatty acids	Forms cell membranes
Cholesterol	Four carbon rings at core	Transports lipids; is basis of steroid hormones
Protein		
Structural proteins (fibers)	Amino acids	Form structures of the body
Functional proteins (enzymes, hormones)	Amino acids	Facilitate chemical reactions; send signals; regulate functions
Nucleic Acid		
Deoxyribonucleic acid (DNA)	Nucleotides (contain deoxyribose)	Contains information (genetic code) for making proteins
Ribonucleic acid (RNA)	Nucleotides (contain ribose)	Serves as a copy of a portion of the genetic code

Carbohydrates have potential energy stored in their bonds. When the bonds are broken in cells, the energy is released and then trapped by the cell's chemistry to do work. Chapter 15 explains more about the process by which the body extracts energy from carbohydrates and other food molecules.

LIPIDS

Lipids are fats and oils. Fats are lipids that are solid at room temperature, such as the fat in butter and lard. Oils, such as corn oil and olive oil, are liquid at room temperature. There are several important types of lipids in the body:

1. **Triglycerides** (try-GLIS-er-ides) are lipid molecules formed by a *glycerol* unit joined to three *fatty acids* (Figure 7). Like carbohydrates, their bonds can be broken apart to yield energy (see Chapter 15). Thus triglycerides are useful in storing energy in cells for later use.

FIGURE 7   *Triglyceride.* Each triglyceride is composed of three fatty acid units attached to a glycerol unit.

Triglyceride

Glycerol

Fatty acids

Blood Lipoproteins

A lipid such as cholesterol can travel in the blood only after it has attached to a protein molecule—forming a lipoprotein. Some of these molecules are called *high-density lipoproteins (HDLs)* because they have a high density of protein (more protein than lipid). Another type of molecule contains less protein (and more lipid), so it is called *low-density lipoprotein (LDL)*.

The cholesterol in LDLs is often called "bad" cholesterol because high blood levels of LDL are associated with **atherosclerosis,** a life-threatening blockage of arter-

ies. LDLs carry cholesterol *to cells,* including the cells that line blood vessels. HDLs, on the other hand, carry so-called "good" cholesterol *away from cells* and toward the liver for elimination from the body. A high proportion of HDL in the blood is associated with a low risk of developing atherosclerosis. Factors such as cigarette smoking decrease HDL levels and thus contribute to risk of atherosclerosis. Factors such as exercise increase HDL levels and thus decrease the risk of atherosclerosis.

---

2. **Phospholipids** are similar to triglycerides but have phosphorus-containing units in them, as their name implies. The phosphorus-containing unit in each molecule forms a "head" that attracts water. Two fatty acid "tails" repel water. Figure 8, *A,* shows the head and tail of the phospholipid molecule. This structure allows them to form a stable *bilayer* in water that forms the foundation for the cell membrane. In Figure 8, *B,* the water-attracting heads face the water and the water-repelling tails face away from the water (and toward each other).

3. **Cholesterol** is a *steroid* lipid that performs several important functions in the body. It combines with phospholipids in the cell membrane to help stabilize its bilayer structure. The body also uses cholesterol as a starting point in making steroid hormones such as estrogen, testosterone, and cortisone (see Chapter 9).

PROTEINS

Proteins are very large molecules composed of basic units called **amino acids.** In addition to carbon, hydrogen, and oxygen, amino acids contain nitrogen (N). By means of a process described fully in Chapter 2, a particular sequence of amino acids is strung together and held by **peptide bonds.** Positive-negative attractions between dif-

ferent atoms in the long amino acid strand cause it to coil on itself and maintain its shape. The complex, three-dimensional molecule that results is a protein molecule (Figure 9).

The shape of a protein molecule determines its role in body chemistry. **Structural proteins** are shaped in ways that allow them to form essential structures of the body. Collagen, a protein with a fiber shape, holds most of the body tissues together. Keratin, another structural protein, forms a network of waterproof fibers in the outer layer of the skin. **Functional proteins** participate in chemical processes of the body. Functional proteins include some of the hormones, growth factors, cell membrane channels and receptors, and enzymes.

Enzymes are chemical catalysts. This means that they help a chemical reaction occur but are not reactants or products themselves. They participate in chemical reactions but are not changed by the reactions. Enzymes are vital to body chemistry. No reaction in the body occurs fast enough unless the specific enzymes needed for that reaction are present.

Figure 10 illustrates how shape is important to the function of enzyme molecules. Each enzyme has a shape that "fits" the specific molecules it works on much as a key fits specific locks. This explanation of enzyme action is sometimes called the **lock-and-key model.**

FIGURE 8   *Phospholipids.* **A,** Each phospholipid molecule has a phosphorus-containing "head" that attracts water and a lipid "tail" that repels water. **B,** Because the tails repel water, phospholipid molecules often arrange themselves so that their tails face away from water. The stable structure that results is a bilayer sheet forming a small bubble.

A

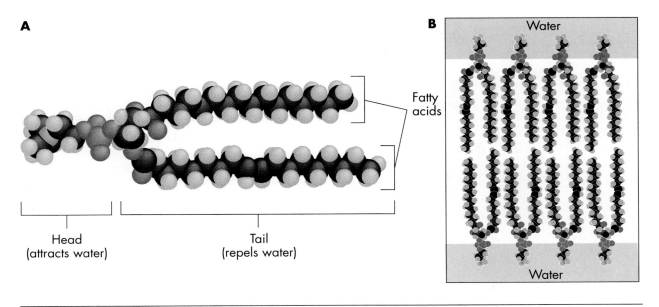

Head
(attracts water)

Tail
(repels water)

Fatty
acids

B

Water

Water

FIGURE 9   *Protein.* Protein molecules are large, complex molecules formed by a twisted and folded strand of amino acids. Each amino acid is connected to the next amino acid by covalent peptide bonds.

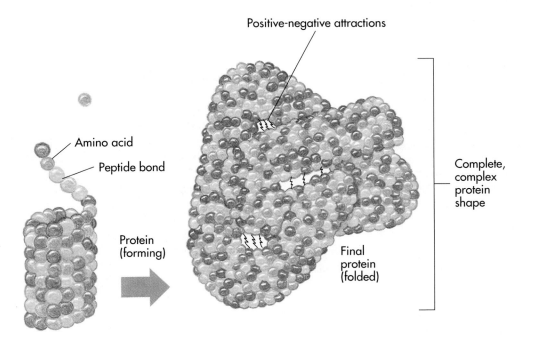

Positive-negative attractions

Amino acid

Peptide bond

Protein
(forming)

Final
protein
(folded)

Complete,
complex
protein
shape

FIGURE 10 *Enzyme Action.* Enzymes are functional proteins whose molecular shape allows them to catalyze chemical reactions. Molecules *A* and *B* are brought together by the enzyme to form a larger molecule, *AB*.

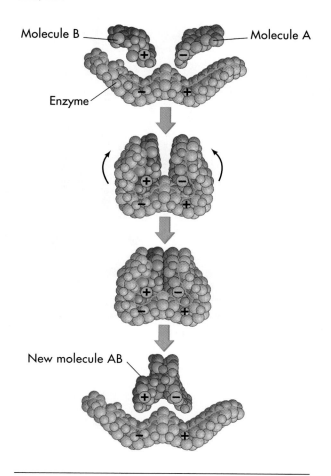

Molecule B —
Molecule A —
Enzyme —
New molecule AB —

Proteins can bond with other organic compounds and form "mixed" molecules. For example, *glycoproteins* are proteins with sugars attached. *Lipoproteins* are lipid-protein combinations.

NUCLEIC ACIDS

The two forms of nucleic acid are **deoxyribonucleic acid (DNA)** and **ribonucleic acid (RNA)**. As outlined in Chapter 2, the basic building blocks of nucleic acids are called **nucleotides.** Each nucleotide consists of a *phosphate unit,* a sugar *(ribose* or *deoxyribose),* and a *nitrogen base.* DNA nucleotide bases include **adenine, thymine, guanine,** and **cytosine.** RNA uses the same set of bases, except for the substitution of **uracil** for thymine. See Table 4.

Nucleotides bind to one another to form strands or other structures. In the DNA molecule, nucleotides are arranged in a twisted, double strand called a **double helix** (Figure 11).

The sequence of different nucleotides along the DNA double helix is the "master code" for assembling proteins and other nucleic acids. *Messenger RNA (mRNA)* molecules have a sequence that forms a temporary "working copy" of the DNA code. The code in nucleic acids ultimately directs the entire symphony of living chemistry.

> **Carbohydrates are composed of monosaccharide units and can be broken apart to yield energy. Lipids are fat molecules composed mainly of glycerol and fatty acids and have many functions. Proteins are complex strings of amino acids. Proteins comprise body structures or regulate body functions. Nucleic acids DNA and RNA are composed of nucleotides. The sequence of nucleotides serves as a code for assembling proteins.**

TABLE 4   Components of Nucleotides

NUCLEOTIDE	DNA	RNA
Sugar	Deoxyribose	Ribose
Phosphate	Phosphate	Phosphate
Nitrogen base	Cytosine	Cytosine
	Guanine	Guanine
	Adenine	Adenine
	Thymine	Uracil

FIGURE 11   *DNA.* Deoxyribonucleic acid (DNA), like all nucleic acids, is composed of units called *nucleotides*. Each nucleotide has a phosphate, a sugar, and a nitrogen base. In DNA, the nucleotides are arranged in a double helix formation.

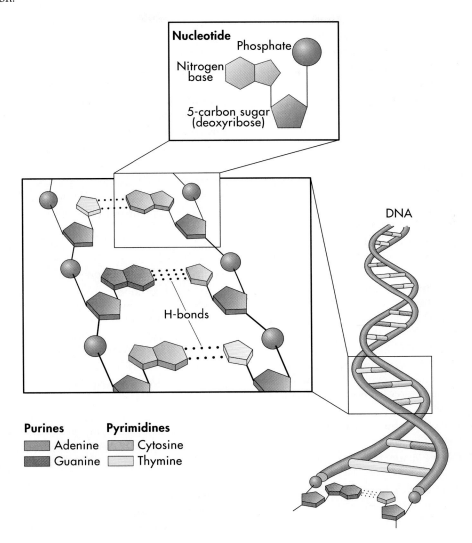

M I N I - G L O S S A R Y

acid any substance that, when dissolved in water, contributes to an excess of H⁺ ions

aqueous (AY-kwee-us) solution liquid mixture in which water is the solvent; for example, saltwater is an aqueous solution because water is the solvent

atom smallest particle of a pure substance (element) that still has the chemical properties of that substance; composed of protons, electrons, and neutrons (subatomic particles)

atomic mass combined total number of protons and neutrons in an atom

atomic number total number of protons in an atom's nucleus; atoms of each element have a characteristic atomic number

base alkaline; any substance that, when dissolved in water, contributes to an excess of OH⁻ ions

carbohydrate (KAR-bo-HYE-drate) organic molecule composed of one or more monosaccharides (containing C, H, and O in a 1:2:1 ratio)

compound (KOM-pound) substance whose molecules have more than one kind of element in them

covalent (ko-VAY-lent) bond chemical bond formed when atoms share electrons by overlapping their energy levels (electron shells)

dehydration (dee-hye-DRAY-shun) synthesis (SIN-the-sis) chemical reaction in which large molecules are formed by removing water from smaller molecules and joining them together

dissociation (dis-so-see-AY-shun) separation of ions as they dissolve in water

double helix (HE-lix) shape of DNA molecules; a double spiral

electrolyte (e-LEK-tro-lite) compound whose molecules dissolve in water to form ions

electron (e-LEK-tron) negatively charged particle orbiting the nucleus of an atom

element (EL-e-ment) pure substance, composed of only one type of atom

energy level limited region surrounding the nucleus of an atom at a certain distance containing electrons; also called a shell

enzyme (EN-zime) functional protein that catalyzes chemical reactions (helps them occur more rapidly)

glycogen (GLYE-ko-jen) polysaccharide consisting of a chain of glucose (monosaccharide) molecules

hydrolysis (hye-DROL-i-sis) chemical reaction in which water is added to a large molecule causing it to break apart into smaller molecules

inorganic compounds compounds whose molecules generally do not contain carbon

ionic (eye-ON-ic) bond chemical bond formed by the positive-negative attraction between two ions

lipid organic molecule usually composed of glycerol and fatty acid units; types include triglycerides, phospholipids, and cholesterol

matter any substance that occupies space and has mass

molecule (MOL-e-kyool) particle of matter composed of one or more smaller units called *atoms*

neutron (NOO-tron) electrically neutral particle within the nucleus of an atom

nucleic (noo-KLEE-ic) acid complex organic molecule composed of units called nucleotides that each include a phosphate, a five carbon sugar, and a nitrogen base

nucleus (of an atom) central core of the atom; contains protons and neutrons (hydrogen atoms have a single proton and no neutrons)

organic (or-GAN-ic) compounds compounds whose large molecules contain carbon, which forms C-C bonds and/or C-H bonds

peptide bond covalent bond linking amino acids within a protein molecule

pH unit expressing relative H⁺ concentration (acidity); pH value higher than 7 is basic, pH value less than 7 is acidic, pH value equal to 7 is neutral

product any substance formed as a result of a chemical reaction

protein (PRO-teen) nitrogen-containing organic compound composed of a folded strand of amino acids

proton (PRO-ton) positively charged particle within the nucleus of an atom

reactant (ree-AK-tant) any substance entering (and being changed by) a chemical reaction

solute (SOL-yoot) substance that dissolves into another substance; for example, in saltwater the salt is the solute dissolved in water

solvent (SOL-vent) substance in which other substances are dissolved; for example, in saltwater the water is the solvent for salt

Common Medical Abbreviations, Prefixes, and Suffixes

Appendix

ABBREVIATIONS

aa of each
a̅.c̅. before meals
ad lib. as much as desired
alb. albumin
AM before noon
amt. amount
ante before
aq. water
AV. average
Ba barium
b.i.d. twice a day
b.m. bowel movement
BMR basal metabolic rate
BP blood pressure
BRP bathroom privileges
BUN blood urea nitrogen
c̅ with
CBC complete blood count
CCU coronary care unit
CHF congestive heart failure
CNS central nervous system
Co cobalt
CVA cerebrovascular accident, stroke
D & C dilation and curettage
d/c discontinue
DOA dead on arrival
Dx diagnosis
ECG electrocardiogram
EDC expected date of confinement

EEG electroencephalogram
EENT ear, eye, nose, throat
EKG electrocardiogram
ER emergency room
FUO fever of undetermined origin
GI gastrointestinal
GP general practitioner
GU genitourinary
h. hour
HCT hematocrit
Hb hemoglobin
H_2O water
h.s. at bedtime
ICU intensive care unit
KUB kidney, ureter, and bladder
MI myocardial infarction
non rep. do not repeat
NPO nothing by mouth
OR operating room
p.c. after meals
per by
PH past history
PI previous illness
PM after noon
p.r.n. as needed
q. every
q.d. every day
q.h. every hour
q.i.d. four times a day
q.n.s. quantity not sufficient

q.o.d.   every other day
q.s.   quantity required or sufficient
RBC   red blood cell
℞   prescription
s̄   without
sp. gr.   specific gravity
s̄s.   one half
stat.   at once, immediately
T & A   tonsillectomy and adenoidectomy
T.B.   tuberculosis
t.i.d.   three times a day
TPR   temperature, pulse, respiration
TUR   transurethral resection
WBC   white blood cell

para-   beside
peri-   around, beyond
post-   after, behind
pre-   before, in front of
pro-   before, in front of
re-   again
retro-   backward, back
semi-   half
sub-   under, beneath
super-   above, over
supra-   above, on the upper side
syn-   with, together
trans-   across, beyond
ultra-   excessive

PREFIXES

a-   without
ab-   away from
ad-   to, toward
adeno-   glandular
amphi-   on both sides
an-   without
ante-   before, forward
anti-   against
bi-   two, double, twice
circum-   around, about
contra-   opposite, against
de-   away from, from
di-   double
dia-   across, through
dis-   separate from, apart
dys-   difficult
e-   out, away
ecto-   outside
en-   in
endo-   in, inside
epi-   on
eu-   well
ex-   from, out of, away from
exo-   outside
extra-   outside, beyond; in addition
hemi-   half
hyper-   over, excessive, above
hypo-   under, deficient
infra-   underneath, below
inter-   between, among
intra-   within, on the side
intro-   into, within
iso-   equal, like

SUFFIXES

-algia   pain, painful
-asis   condition
-blast   young cell
-cele   swelling
-centesis   puncture for aspiration
-cide   killer
-cyte   cell
-ectomy   cut out
-emia   blood
-genesis   production, development
-itis   inflammation
-kinin   motion, action
-logy   study of
-megaly   enlargement
-odynia   pain
-oid   resembling
-oma   tumor
-osis   condition
-opathy   disease
-penia   abnormal reduction
-pexy   fixation
-phagia   eating, swallowing
-phasia   speaking condition
-phobia   fear
-plasty   plastic surgery
-plegia   paralysis
-poiesis   formation
-ptosis   downward displacement
-rhaphy   suture
-scope   instrument for examination
-scopy   examination
-stomy   creation of an opening
-tomy   incision
-uria   urine

Chapter Test Answers

CHAPTER 1

1. anatomy; physiology
2. tissue
3. lateral
4. anterior; posterior
5. axial, appendicular
6. homeostasis
7. h
8. e
9. a
10. j
11. b
12. d
13. c
14. f
15. i
16. g
17. a
18. d
19. c
20. c

CHAPTER 2

1. a
2. b
3. c
4. d
5. a
6. b
7. d
8. c
9. a
10. f
11. e
12. d
13. b
14. h
15. g
16. ribosomes
17. active
18. prophase
19. epithelial
20. epithelial

CHAPTER 3

1. skin
2. bones
3. communication; integration; control; recognition
4. hormones
5. lymphatic
6. urine
7. accessory
8. alveoli
9. d
10. g
11. b
12. h
13. a
14. c
15. e
16. f

CHAPTER 4

1. epithelial; connective tissue
2. parietal; visceral
3. synovial
4. skin
5. cutaneous
6. epidermis; dermis
7. keratin
8. melanin
9. root
10. pressure
11. eccrine; apocrine
12. sebaceous
13. burns
14. d
15. g
16. e
17. a
18. c
19. b
20. f

CHAPTER 5

1. trabeculae

2. Haversian systems
3. osteoblasts; osteoclasts
4. diaphysis
5. axial; appendicular
6. sinuses
7. cervical
8. scapula; clavicle
9. radius; ulna
10. olecranon
11. carpal; phalanges
12. ileum; ischium; pubis
13. patella; tibia
14. feet
15. diarthrotic
16. F
17. T
18. F
19. F
20. T
21. F
22. T
23. T
24. F
25. T

CHAPTER 6

1. b
2. a
3. a
4. b
5. b
6. d
7. a
8. smooth or involuntary
9. tendons
10. myosin
11. sliding-filament
12. synergists
13. posture
14. oxygen debt
15. isotonic
16. plantar
17. atrophy
18. hamstrings

19. c
20. a
21. d
22. b
23. g
24. e
25. f

CHAPTER 7

1. central; peripheral
2. meninges
3. hydrocephalus
4. neurons; neuroglia
5. oligodendroglia
6. away; axon
7. synapse
8. neurotransmitters
9. dermatome
10. endorphins; enkephalins
11. hypothalamus; thalamus
12. 12; 31
13. cardiac muscle; smooth muscle; glandular epithelium
14. sympathetic; parasympathetic
15. sympathetic
16. c
17. h
18. a
19. b
20. e
21. i
22. f
23. g
24. d

CHAPTER 8

1. general
2. rods; cones
3. sclera; choroid; retina
4. iris
5. aqueous humor
6. cataracts
7. choroid
8. mechanoreceptors
9. external; middle; inner (internal)
10. malleus
11. perilymph
12. Corti
13. chemoreceptors
14. gustatory

15. g
16. a
17. i
18. c
19. b
20. h
21. d
22. f
23. e

CHAPTER 9

1. hormones
2. exocrine
3. second messenger
4. negative
5. prostaglandins
6. hypersecretion
7. thyroid stimulating
8. anterior pituitary
9. antidiuretic
10. thyroxine
11. goiter
12. glomerulosa
13. medulla
14. glucagon
15. diabetes mellitus
16. progesterone
17. testosterone
18. thymosin
19. pineal
20. estrogen
21. F
22. T
23. T
24. F
25. T
26. F
27. T
28. F
29. T
30. F
31. T
32. T
33. T
34. T
35. F

CHAPTER 10

1. b
2. b

3. d
4. a
5. c
6. d
7. a
8. b
9. serum
10. red bone marrow
11. polycythemia
12. embolus
13. antigen
14. AB
15. Rh

CHAPTER 11

1. atria; ventricles
2. myocardium
3. pericardium
4. atrioventricular
5. tricuspid
6. vena cava
7. pulmonary
8. angina pectoris
9. pacemaker
10. electrocardiogram (ECG)
11. ventricles
12. veins
13. tunica media
14. systemic
15. capillary
16. foramen ovale
17. cardiac output
18. ductus venosus
19. ductus arteriosus
20. systole
21. cardiac cycle
22. central venous pressure
23. h
24. f
25. b
26. d
27. a
28. c
29. e
30. g

CHAPTER 12

1. thoracic duct
2. lymph nodes
3. artificial
4. antibodies; complement

5. antigens
6. humoral
7. antibodies
8. lymphocytes
9. neutrophils; monocytes
10. lymphocytes
11. humoral
12. cell-mediated
13. T
14. F
15. T
16. F
17. F
18. T
19. T
20. T
21. T
22. F
23. F
24. T

CHAPTER 13

1. diffusion
2. surfactant
3. lower
4. oropharynx
5. nasopharynx
6. larynx
7. epiglottis
8. pneumothorax
9. external
10. oxyhemoglobin
11. vital capacity
12. medulla
13. alveoli
14. e
15. b
16. g
17. d
18. f
19. c
20. a

CHAPTER 14

1. mechanical; chemical
2. feces
3. accessory
4. lumen
5. mucosa
6. palates

7. crown
8. mastication
9. amylase
10. esophagus
11. chyme
12. peristalsis
13. jejunum
14. d
15. a
16. f
17. b
18. h
19. c
20. e
21. g
22. T
23. F
24. F
25. T
26. T
27. F
28. F
29. F
30. T

CHAPTER 15

1. catabolism
2. anabolism
3. metabolism
4. hepatic portal vein
5. glycolysis
6. glycogenesis
7. fats
8. vitamins
9. ATP
10. basal metabolic rate (BMR)
11. hypothalamus
12. F
13. T
14. T
15. F
16. F
17. T
18. T

CHAPTER 16

1. uremia
2. 10%
3. cortex; medulla
4. nephron
5. Bowman's capsule; glomerulus

6. proximal
7. filtration
8. osmosis
9. proximal
10. polyuria
11. pelvis
12. renal colic
13. urethra; meatus
14. incontinence
15. cystitis
16. 125
17. e
18. i
19. d
20. b
21. a
22. g
23. c
24. f
25. h

CHAPTER 17

1. intracellular
2. interstitial; plasma
3. nonelectrolytes
4. ions
5. cations; anions
6. urine
7. sodium; chloride
8. diuretic
9. edema
10. T
11. T
12. F
13. F
14. T
15. F
16. T
17. T
18. F

CHAPTER 18

1. neutral
2. alkaline; acid
3. carbonic anhydrase
4. buffers
5. carbonic
6. acidosis
7. ammonia
8. alkalosis

9. acidosis
10. homeostasis
11. f
12. b
13. j
14. d
15. a
16. h
17. c
18. e
19. g
20. i

CHAPTER 19

1. testes
2. spermatozoa
3. gonads
4. seminiferous
5. interstitial
6. prostate

7. cervix
8. progesterone
9. endometrium
10. areola
11. menarche
12. 28
13. f
14. j
15. a
16. d
17. h
18. c
19. e
20. g
21. i
22. b

CHAPTER 20

1. conception; birth
2. birth; death

3. embryology
4. zygote
5. amniotic
6. placenta
7. organogenesis
8. postnatal
9. childhood
10. arteriosclerosis
11. cataract
12. F
13. F
14. T
15. T
16. T
17. T
18. T
19. F
20. T

Glossary

abdomen (AB-doe-men) body area between the diaphragm and pelvis

abdominal cavity (ab-DOM-i-nal KAV-i-tee) the cavity containing the abdominal organs

abdominal muscles (ab-DOM-i-nal MUS-els) muscles supporting the anterior aspect of the abdomen

abdominal quadrants (ab-DOM-i-nal KWOD-rants) health professionals divide the abdomen (through the navel) into four areas to help locate specific organs

abdominal regions (ab-DOM-i-nal REE-juns) anatomists have divided the abdomen into nine regions to identify the location of organs

abdominopelvic cavity (ab-DOM-i-no-PEL-vik KAV-i-tee) term used to describe the single cavity containing the abdominal and pelvic organs

abduction (ab-DUK-shun) moving away from the midline of the body, opposite motion of adduction

absorption (ab-SORP-shun) passage of a substance through a membrane, such as skin or mucosa, into blood

accessory organ (ak-SES-o-ree OR-gan) an organ that assists other organs in accomplishing their functions

acetabulum (as-e-TAB-yoo-lum) socket in the hip bone (ox coxa or innominate bone) into which the head of the femur fits

acetylcholine (as-e-til-KO-lean) chemical neurotransmitter

acid-base balance (AS-id base BAL-ans) maintaining the concentration of hydrogen ions in body fluids

acidosis (as-i-DOE-sis) condition in which there is an excessive proportion of acid in the blood

acquired immunodeficiency syndrome (AIDS) (ah-KWIRED i-myoo-no-de-FISH-en-see SIN-drome) disease in which the HIV virus attacks the T cells, thereby compromising the body's immune system

acquired immunity (ah-KWIRED i-MYOO-ni-tee) immunity that is obtained after birth through the use of injections or exposure to a harmful agent

acromegaly (ak-ro-MEG-ah-lee) condition caused by hypersecretion of growth hormone after puberty, resulting in enlargement of facial features (for example, jaw, nose), fingers, and toes

acrosome (AK-ro-sohm) specialized structure on the sperm containing enzymes that break down the covering of the ovum to allow entry

actin (AK-tin) contractile protein found in the thin myofilaments of skeletal muscle

action potential (AK-shun po-TEN-shal) nerve impulse

active transport (AK-tiv TRANS-port) movement of a substance into and out of a living cell requiring the use of cellular energy

Addison's disease (AD-i-sons di-ZEEZ) disease of the adrenal gland resulting in low blood sugar, weight loss, and weakness

adduction (ah-DUK-shun) moving toward the midline of the body, opposite motion of abduction

adenohypophysis (ad-e-no-hye-POF-i-sis) anterior pituitary gland, which has the structure of an endocrine gland

adenoid (AD-e-noyd) literally, glandlike; adenoids, or pharyngeal tonsils, are paired lymphoid structures in the nasopharynx

adenosine diphosphate (ADP) (ah-DEN-o-sen dye-FOS-fate) molecule similar to adenosine triphosphate but containing only two phosphate groups

adenosine triphosphate (ATP) (ah-DEN-o-sen try-FOS-fate) chemical compound that provides energy for use by body cells

adipose (AD-i-pose) fat tissue

adolescence (ad-o-LES-ens) period between puberty and adulthood

adrenal cortex (ah-DREE-nal KOR-teks) outer portion of adrenal gland that secretes hormones called *corticoids*

adrenal gland (ah-DREE-nal gland) glands that rest on the top of the kidneys, made up of the cortex and medulla

adrenal medulla (ah-DREE-nal me-DUL-ah) inner portion of adrenal gland that secretes epinephrine and norepinephrine

adrenergic fibers (a-dre-NER-jik FYE-bers) axons whose terminals release norepinephrine and epinephrine

adrenocorticotropic hormone (ACTH) (ah-dree-no-kor-te-ko-TRO-pic HOR-mone) hormone that stimulates the adrenal cortex to secrete larger amounts of hormones

adulthood (ah-DULT-hood) period after adolescence

aerobic respiration (air-O-bik res-pi-RAY-shun) the stage of cellular respiration requiring oxygen

aerobic training (air-O-bik TRAIN-ing) continuous vigorous exercise requiring the body to increase its consumption of oxygen and develop the muscles' ability to sustain activity over a long period of time

afferent (AF-fer-ent) carrying or conveying toward the center (for example, an afferent neuron carries nerve impulses toward the central nervous system)

agglutinate (ah-GLOO-tin-ate) antibodies causing antigens to clump or stick together

aging process (AJ-ing PROS-es) the gradual degenerative changes that occur after young adulthood as a person ages

AIDS-related complex (ARC) (AIDS ree-LAY-ted KOM-pleks) a more mild form of AIDS that produces fever, weight loss, and swollen lymph nodes

albumin (AL-byoo-min) one of several types of proteins normally found in blood plasma, it helps thicken the blood

aldosterone (AL-doe-ste-rone) hormone that stimulates the kidney to retain sodium ions and water

alimentary canal (al-e-MEN-tar-ee kah-NAL) the digestive tract as a whole

alkalosis (al-kah-LO-sis) condition in which there is an excessive proportion of alkali in the blood, opposite of acidosis

allergy (AL-er-jee) hypersensitivity of the immune system to relatively harmless environmental antigens

all or none when stimulated, a muscle fiber will contract fully or not at all; whether a contraction occurs depends on whether the stimulus reaches the required threshold

alpha cell (AL-fah sell) pancreatic cell that secretes glucagon

alveolar duct (al-VEE-o-lar dukt) airway that branches from the smallest bronchioles; alveolar sacs arise from alveolar ducts

alveolar sac (al-VEE-o-lar sak) each alveolar duct ends in several sacs that resemble a cluster of grapes

alveolus (al-VEE-o-lus) literally, a small cavity; alveoli of lungs are microscopic saclike dilations of terminal bronchioles

amniotic cavity (am-nee-OT-ik KAV-i-tee) cavity within the blastocyst that will become a fluid-filled sac in which the embryo will float during development

amino acid (ah-MEE-no AS-id) structural units from which proteins are built

amphiarthrosis (am-fee-ar-THRO-sis) slightly movable joint such as the joint joining the two pubic bones

amylase (AM-i-lase) enzyme that digests carbohydrates

anabolic steroid (a-nah-BOL-ik STE-royd) hormones that stimulate the building of large molecules, specifically proteins in muscle and bones

anabolism (ah-NAB-o-lizm) cells making complex molecules (for example, hormones) from simpler compounds (for example, amino acids), opposite of catabolism

anal canal (AY-nal kah-NAL) terminal portion of the rectum

anaphase (AN-ah-faze) stage of mitosis; duplicate chromosomes move to poles of dividing cell

anaphylactic shock (an-ah-fi-LAK-tik shock) shock resulting from a severe allergic reaction, may be fatal

anatomical position (an-ah-TOM-i-kal po-ZISH-un) the reference position for the body, which gives meaning to directional terms

anatomy (ah-NAT-o-mee) the study of the structure of an organism and the relationships of its parts

androgen (AN-dro-jen) male sex hormone

anemia (ah-NEE-mee-ah) deficient number of red blood cells or deficient hemoglobin

anesthesia (an-es-THEE-zee-ah) loss of sensation

angina pectoris (an-JYE-nah PECK-tor-is) severe chest pain resulting when the myocardium is deprived of sufficient oxygen

angstrom (ANG-strom) unit 0.1 m (1/10,000,000,000 of a meter or about 1/250,000,000 of an inch)

anion (AN-eye-on) negatively charged particle

anorexia nervosa (an-o-REK-see-ah ner-VO-sah) a behavior involving an irrational fear of being overweight, resulting in severe weight loss due to self-starvation

antagonist muscle (an-TAG-o-nist MUS-el) those having opposing actions; for example, muscles that flex the upper arm are antagonists to muscles that extend it

antebrachial (an-tee-BRAY-kee-al) refers to the forearm

antecubital (an-tee-KYOO-bi-tal) refers to the elbow

antenatal medicine (an-tee-NAY-tal MED-i-sin) prenatal medicine

anterior (an-TEER-ee-or) front or ventral; opposite of posterior or dorsal

antibody (AN-ti-bod-ee) substance produced by the body that destroys or inactivates a specific substance (antigen) that has entered the body

antibody-mediated immunity (AN-ti-bod-ee MEE-dee-ate-ed i-MYOO-ni-tee) immunity that is produced when antibodies make antigens unable to harm the body

antidiuretic hormone (ADH) (an-ti-dye-yoo-RET-ik HOR-mone) hormone produced in the posterior pituitary gland to regulate the balance of water in the body by accelerating the reabsorption of water

antigen (AN-ti-jen) substance that, when introduced into the body, causes formation of antibodies against it

antrum (AN-trum) cavity

anuria (ah-NOO-ree-ah) absence of urine

anus (AY-nus) distal end or outlet of the rectum

aorta (ay-OR-tah) main and largest artery in the body

aortic body (ay-OR-tik BOD-ee) small cluster of chemosensitive cells that respond to carbon dioxide and oxygen levels

aortic semilunar valve (ay-OR-tic sem-i-LOO-nar valve) valve between the aorta and left ventricle that prevents blood from flowing back into the ventricle

apex (A-peks) pointed end of a conical structure

apnea (AP-nee-ah) temporary cessation of breathing

apocrine (AP-o-krin) sweat glands located in the axilla and genital regions; these glands enlarge and begin to function at puberty

appendicitis (ah-pen-di-SYE-tis) inflammation of the vermiform appendix

appendage (ah-PEN-dij) something that is attached

appendicular (a-pen-DIK-yoo-lar) refers to the upper and lower extremities of the body

appendicular skeleton (a-pen-DIK-yoo-lar SKEL-e-ton) the bones of the upper and lower extremities of the body

aqueous humor (AY-kwee-us HYOO-mor) watery fluid that fills the anterior chamber of the eye, in front of the lens

arachnoid (ah-RAK-noyd) delicate, weblike middle membrane covering the brain, the meninges

areola (ah-REE-o-lah) small space; the pigmented ring around the nipple

areolar (ah-REE-o-lar) a type of connective tissue consisting of fibers and a variety of cells embedded in a loose matrix of soft, sticky gel

arrector pili (ah-REK-tor PYE-lie) smooth muscles of the skin, which are attached to hair follicles; when contraction occurs, the hair stands up, resulting in "goose flesh"

arteriole (ar-TEER-ee-ole) small branch of an artery

artery (AR-ter-ee) vessel carrying blood away from the heart

articular cartilage (ar-TIK-yoo-lar KAR-ti-lij) cartilage covering the joint ends of bones

articulation (ar-tik-yoo-LAY-shun) joint

artificial kidney (ar-ti-FISH-al KID-nee) mechanical device that removes wastes from the blood that would normally be removed by the kidney

artificial pacemaker (ar-ti-FISH-al PAYS-may-ker) an electrical device that is implanted into the heart to treat a heart block

asexual (a-SEKS-yoo-al) one-celled plants and bacteria that do not produce specialized sex cells

assimilation (ah-sim-i-LAY-shun) when food molecules enter the cell and undergo chemical changes

association area (ah-so-shee-AY-shun AIR-ee-ah) region of the cerebral cortex of the brain that functions to put together or associate information from many parts of the brain to help make sense of or analyze the information

astrocyte (AS-tro-site)   a glial cell

atherosclerosis (ath-er-o-skle-RO-sis)   hardening of the arteries; lipid deposits lining the inside of the arteries

atrial natriuretic hormone (ANH) (AY-tree-al na-tree-yoo-RET-ik HOR-mone)   hormone secreted by the heart cells that regulates fluid and electrolyte homeostasis

atrioventricular valves (ay-tree-o-ven-TRIK-yoo-lar valves)   two valves that separate the atrial chambers from the ventricles

atrium (AY-tree-um)   chamber or cavity; for example, atrium of each side of the heart

atrophy (AT-ro-fee)   wasting away of tissue; decrease in size of a part; sometimes referred to as disuse atrophy

auditory tube (AW-di-toe-ree tube)   tube that connects the throat with the middle ear

auricle (AW-ri-kul)   part of the ear attached to the side of the head; earlike appendage of each atrium of the heart

autonomic effector (aw-toe-NOM-ik ef-FEK-tor)   tissues to which autonomic neurons conduct impulses

autonomic nervous system (ANS) (aw-toe-NOM-ik NER-vus SIS-tem)   division of the human nervous system that regulates involuntary actions

autonomic neuron (aw-toe-NOM-ik NOO-ron)   motor neurons that make up the autonomic nervous system

AV bundle (A V BUN-dul)   fibers in the heart that relay a nerve impulse from the AV node to the ventricles; also known as the bundle of His

avitaminosis (ay-vye-tah-mi-NO-sis)   vitamin deficiency

axial (AK-see-al)   refers to the head, neck, and torso, or trunk of the body

axial skeleton (AK-see-al SKEL-e-ton)   the bones of the head, neck, and torso

axilla (AK-sil-ah)   refers to the armpit

axon (AK-son)   nerve cell process that transmits impulses away from the cell body

B cells (B sells)   a lymphocyte; activated B cells develop into plasma cells, which secrete antibodies into the blood

Bartholin's glands (BAR-toe-lins glands)   gland located on either side of the vaginal outlet that secretes mucuslike lubricating fluid; also known as greater vestibular glands

basal ganglia (BAY-sal GANG-glee-ah)   see *cerebral nuclei*

basal metabolic rate (BMR) (BAY-sal met-ah-BOL-ik rate)   number of calories of heat that must be produced per hour by catabolism to keep the body alive, awake, and comfortably warm

basement membrane (BASE-ment MEM-brane)   the connective tissue layer of the serous membrane that holds and supports the epithelial cells

basophil (BAY-so-fil)   white blood cell that stains readily with basic dyes

benign prostatic hypertrophy (be-NINE pros-TAT-ik hye-PER-tro-fee)   a noncancerous enlargement of the prostate in older men

benign tumor (be-NINE TOO-mer)   a relatively harmless neoplasm

beta cell (BAY-tah sell)   pancreatic islet cell that secretes insulin

bicarbonate loading (bye-KAR-bo-nate LOHD-ing)   ingesting large amounts of sodium bicarbonate to counteract the effects of lactic acid buildup, thereby reducing fatigue; however, there are potential dangerous side effects

biceps brachii (BYE-seps BRAY-kee-eye)   the primary flexor of the forearm

biceps femoris (BYE-seps FEM-o-ris)   powerful flexor of the lower leg

bicuspid (bye-KUS-pid)   premolars

bicuspid valve (bye-KUS-pid valve)   one of the two AV valves, it is located between the left atrium and ventricle and is sometimes called the *mitral valve*

bile (bile)   substance that reduces large fat globules into smaller droplets of fat that are more easily broken down

bile duct (bile dukt)   duct that drains bile into the small intestine and is formed by the union of the comman hepatic and cystic ducts

biological filtration (bye-o-LOJ-e-kal fil-TRAY-shun)   process in which cells alter the contents of the filtered fluid

blackhead (BLACK-hed)   when sebum accumulates, darkens, and enlarges some of the ducts of the sebaceous glands; also known as a *comedo*

bladder (BLAD-der)   a sac, usually referring to the urinary bladder

blastocyst (BLAS-toe-sist)   postmorula stage of developing embryo; hollow ball of cells

blister (BLIS-ter)   a baglike point on the skin caused by some irritant, usually full of fluid

blood-brain barrier (blud brayn BARE-ee-er)   two-ply wall formed by the wall of a capillary and the surrounding extensions of a glial cell called an *astrocyte*; it functions to prevent harmful chemicals from entering vital brain tissue

blood doping (blud DOE-ping)   a practice used to improve athletic performance by removing red blood cells weeks before an event and then reinfusing them just before competition to increase the oxygen-carrying capacity of the blood

blood pressure (blud PRESH-ur)   pressure of blood in the blood vessels, expressed as systolic pressure/diastolic pressure (for example, 120/80 mmHg)

blood pressure gradient (blud PRESH-ur GRAY-dee-ent)   the difference between two blood pressures in the body

blood types (blud tipes)   the different types of blood that are identified by certain antigens in red blood cells (A, B, AB, O, and Rh-negative or Rh-positive)

body (BOD-ee)   unified and complex assembly of structurally and functionally interactive components

body composition (BOD-ee com-po-ZISH-un)   assessment to identify the percentage of the body that is lean tissue and the percentage that is fat

bolus (BO-lus)   a small, rounded mass of masticated food to be swallowed

bone (bone)   highly specialized connective tissue whose matrix is hard and calcified

bone marrow (bone MAR-o)   soft material that fills cavities of the bones; red bone marrow is vital to blood cell formation; yellow bone marrow is inactive fatty tissue

bony labyrinth (BONE-ee LAB-i-rinth)   the fluid-filled complex maze of three spaces (the vestibule, semicircular canals, and cochlea) in the temporal bone

Bowman's capsule (BO-mens KAP-sul)   the cup-shaped top of a nephron that surrounds the glomerulus

brachial (BRAY-kee-al)   pertaining to the arm

breast (brest)   anterior aspect of the chest; in females, also an accessory sex organ

bronchi (BRONG-ki)   the branches of the trachea

bronchiole (BRONG-kee-ole)   small branch of a bronchus

buccal (BUK-al)   pertaining to the cheek

buffer (BUF-er)   compound that combines with an acid or with a base to form a weaker acid or base, thereby lessening the change in hydrogen-ion concentration that would occur without the buffer

buffer pairs (BUF-er pairs)   two kinds of chemical substances that prevent a sharp change in the pH of a fluid; for example, sodium bicarbonate ($NaHCO_3$) and carbonic acid (H_2CO_3)

bulbourethral gland (BUL-bo-yoo-REE-thral gland)   small glands located just below the prostate gland whose mucus-like secretions lubricate the terminal portion of the urethra and contribute less than 5% of the seminal fluid volume; also known as *Cowper's glands*

bundle of His (BUN-dul of his)   see AV bundle

burn (bern)   an injury to tissues resulting from contact with heat, chemicals, electricity, friction, or radiant and electro-magnetic energy; classified into three categories, depending on the number of tissue layers involved

bursae (BER-see)   small, cushionlike sacs found between moving body parts, making movement easier

bursitis (ber-SYE-tis)   inflammation of a bursa

calcaneus (kal-KAY-nee-us)   heel bone; largest tarsal in the foot

calcitonin (kal-si-TOE-nin)   a hormone secreted by the thyroid that decreases calcium in the blood

calorie (c) (KAL-or-ree)   heat unit; the amount of heat needed to raise the temperature of 1 g of water 1° C

Calorie (C) (KAL-or-ree)   heat unit; kilocalorie; the amount of heat needed to raise the temperature of 1 kilogram of water 1° C

calyx (KAY-liks)   cup-shaped division of the renal pelvis

canaliculi (kan-ah-LIK-yoo-lie)   an extremely narrow tubular passage or channel in compact bone

canine tooth (KAY-nine tooth)   the tooth with the longest crown and the longest root, which is located lateral to the second incisor

capillary (KAP-i-lair-ee)   tiny vessels that connect arterioles and venules

capillary blood pressure (KAP-i-lair-ee blud PRESH-ur)   the blood pressure found in the capillary vessels

capsule (KAP-sul)   found in diarthrotic joints, holds the bones of joints together while allowing movement and is made of fibrous connective tissue lined with a smooth, slippery synovial membrane

carbaminohemoglobin (kar-bam-ee-no-hee-mo-GLO-bin)   the compound formed by the union of carbon dioxide with hemoglobin

carbohydrate (kar-bo-HYE-drate)   organic compounds containing carbon, hydrogen, and oxygen in certain specific proportions; for example, sugars, starches, and cellulose

carbohydrate loading (kar-bo-HYE-drate LOHD-ing)   the method used by athletes to increase the stores of muscle glycogen, allowing more sustained aerobic exercise

carbonic anhydrase (kar-BON-ik an-HYE-drays)   the enzyme that converts carbon dioxide into carbonic acid

cardiac (KAR-dee-ak)   refers to the heart

cardiac cycle (KAR-dee-ak SYE-kul)   each complete heart beat, including contraction and relaxation of the atria and ventricles

cardiac muscle (KAR-dee-ak MUS-el)   the specialized muscle that makes up the heart

cardiac output (KAR-dee-ak OUT-put)   volume of blood pumped by one ventricle per minute

cardiac sphincter (KAR-dee-ak SFINGK-ter)   a ring of muscle between the stomach and esophagus that prevents food from re-entering the esophagus when the stomach contracts

cardiac vein (KAR-dee-ak vane)   any vein that carries blood from the myocardial capillary beds to the coronary sinus

cardiopulmonary resuscitation (CPR) (kar-dee-o-PUL-mo-nair-ree ree-sus-i-TAY-shun)   combined external cardiac (heart) massage and artificial respiration

cardiovascular (kar-dee-o-VAS-kyoo-lar)   pertaining to the heart and blood vessels

caries (KARE-eez)   decay of teeth or of bone

carotid body (kah-ROT-id BOD-ee)   chemoreceptor located in the carotid artery that detects changes in oxygen, carbon dioxide, and blood acid levels

carpal (KAR-pal)   pertaining to the wrist

cartilage (KAR-ti-lij)   a specialized, fibrous connective tissue that has the consistency of a firm plastic or gristlelike gel

catabolism (kah-TAB-o-lizm)   breakdown of food compounds or cytoplasm into simpler compounds; opposite of anabolism, the other phase of metabolism

catalyst (KAT-ah-list)   chemical that speeds up reactions without being changed itself

cataract (KAT-ah-rakt)   opacity of the lens of the eye

catecholamines (kat-e-kol-AM-eens)   norepinephrine and epinephrine

catheterization (kath-e-ter-i-ZAY-shun)   passage of a flexible tube (catheter) into the bladder through the urethra for the withdrawal of urine (urinary catheterization)

cation (KAT-eye-on)   positively charged particle

cavity (KAV-i-tee)   hollow place or space in a tooth; dental caries

cecum (SEE-kum)   blind pouch; the pouch at the proximal end of the large intestine

cell (sell)   the basic biological and structural unit of the body consisting of a nucleus surrounded by cytoplasm and enclosed by a membrane

cell body (sell BOD-ee)   the main part of a neuron from which the dendrites and axons extend

cell-mediated immunity (sell MEE-dee-ate-ed i-MYOO-ni-tee)   resistance to disease organisms resulting from the actions of cells; chiefly sensitized T cells

cellular respiration (SELL-yoo-lar res-pi-RAY-shun)   enzymes in the mitochondrial wall and matrix using oxygen to break down glucose and other nutrients to release energy needed for cellular work

centimeter (SEN-ti-mee-ter)   1/100 of a meter; approximately 2.5 cm equal 1 inch

central nervous system (CNS) (SEN-tral NER-vus SIS-tem)   the brain and spinal cord

central venous pressure (SEN-tral VEE-nus PRESH-ur)   venous blood pressure within the right atrium that influences the pressure in the large peripheral veins

centriole (SEN-tree-ol)   one of a pair of tiny cylinders in the centrosome of a cell; believed to be involved with the spindle fibers formed during mitosis

centromere (SEN-tro-meer)   a beadlike structure that attaches one chromatid to another during the early stages of mitosis

cephalic (se-FAL-ik)   refers to the head

cerebellum (sair-e-BELL-um)   the second largest part of the human brain that plays an essential role in the production of normal movements

cerebral cortex (se-REE-bral KOR-teks)   a thin layer of gray matter made up of neuron dentrites and cell bodies that compose the surface of the cerebrum

cerebral nuclei (se-REE-bral NOOK-lee-i)   islands of gray matter located in the cerebral cortex that are responsible for automatic movements and postures

cerebrospinal fluid (CSF) (se-ree-bro-SPY-nal FLOO-id)   fluid that fills the subarachnoid space in the brain and spinal cord and in the cerebral ventricles

cerebrovascular accident (CVA) (se-ree-bro-VAS-kyoo-lar AK-si-dent)   a hemorrhage or cessation of blood flow through cerebral blood vessels resulting in destruction of neurons; commonly called a *stroke*

cerebrum (se-REE-brum)   the largest and uppermost part of the human brain that controls consciousness, memory, sensations, emotions, and voluntary movements

cerumen (se-ROO-men)   ear wax

ceruminous gland (se-ROO-mi-nus gland)   gland that produces a waxy substance called *cerumen* (ear wax)

cervical (SER-vi-kal)   refers to the neck

cervix (SER-viks)   neck; any necklike structure

chemoreceptors (kee-mo-ree-SEP-tors)   receptors that respond to chemicals and are responsible for taste and smell

chest (chest)   thorax

childhood (CHILD-hood)   from infancy to puberty

cholecystokinin (CCK) (ko-le-sis-toe-KYE-nin)   hormone secreted from the intestinal mucosa of the duodenum that stimulates the contraction of the gall bladder, resulting in bile flowing into the duodenum

chondrocyte (KON-dro-site)   cartilage cell

chordae tendineae (KOR-dee ten-DIN-ee)   stringlike structures that attach the AV valves to the wall of the heart

chorion (KO-ree-on)   develops into an important fetal membrane in the placenta

chorionic gonadotropins (ko-ree-ON-ik go-na-doe-TRO-pins)   hormones that are secreted as the uterus develops during pregnancy

chorionic villi (ko-ree-ON-ik VIL-eye)   connect the blood vessels of the chorion to the placenta

choroid (KO-royd)   middle layer of the eyeball that contains a dark pigment to prevent the scattering of incoming light rays

choroid plexus (KO-royd PLEK-sus)   a network of brain capillaries that are involved with the production of cerebrospinal fluid

chromatids (KRO-mah-tids)   a chromosome strand

cholinergic fiber (ko-lin-ER-jik FYE-ber)   axons whose terminals release acetylcholine

chromatin granules (KRO-mah-tin GRAN-yools)   deep-staining substance in the nucleus of cells; divides into chromosomes during mitosis

chromosome (KRO-mo-sohm)   DNA molecule that has coiled to form a compact mass during mitosis or meiosis; each chromosome is composed of regions called *genes,* each of which transmits hereditary information

chyme (kime)   partially digested food mixture leaving the stomach

cilia (SIL-ee-ah)   hairlike projections of cells

circulatory system (SER-kyoo-lah-tor-ee SIS-tem)   the system that supplies transportation for cells of the body

circumcision (ser-kum-SIZH-un)   surgical removal of the foreskin or prepuce

cisterna chyli (sis-TER-nah KYE-lye)   an enlarged pouch on the thoracic duct that serves as a storage area for lymph moving towards its point of entry in to the venous system

citric acid cycle (SIT-rik AS-id SYE-kul)   the second series of chemical reactions in the process of glucose metabolism; it is an aerobic process

clavicle (KLAV-i-kul)   collar bone, connects the upper extremity to the axial skeleton

cleavage furrow (KLEEV-ij FUR-o)   appears at the end of anaphase and begins to divide the cell into two daughter cells

clitoris (KLIT-o-ris)   erectile tissue located within the vestibule of the vagina

clone (klone)   any of a family of many identical cells descended from a single "parent" cell

cochlea (KOKE-lee-ah)   snail shell or structure of similar shape

cochlear duct (KOKE-lee-ar dukt)   membranous tube within the bony cochlea

collagen (KOL-ah-jen)   principle organic constituent of connective tissue

collecting tubule (ko-LEK-ting TOO-byool)   a straight part of a renal tubule formed by distal tubules of several nephrons joining together

colloid (KOL-oyd)   dissolved particles with diameters of 1 to 100 millimicrons (1 millimicron equals about 1/25,000,000 of an inch)

colon (KO-lon)   intestine

colostomy (ko-LAH-sto-me)   surgical procedure in which an artificial anus is created on the abdominal wall by cutting the colon and bringing the cut end or ends out to the surface to form an opening called a *stoma*

columnar (ko-LUM-nar)   shape in which cells are higher than they are wide

combining sites (kom-BINE-ing sites)   antigen-binding sites, antigen receptor regions on antibody molecule; shape of each combining site is complementary to shape of a specific antigen

compact bone (kom-PAKT bone)   dense bone

compensated metabolic acidosis (KOM-pen-say-ted met-ah-BOL-ik as-i-DOE-sis)   when metabolic acidosis occurs and the body is able to adjust to return the blood pH to near normal levels

complement (KOM-ple-ment)   any of several inactive enzymes normally present in blood, which when activated kill foreign cells by dissolving them

complementary base pairing (kom-ple-MEN-ta-ree base PAIR-ing)   bonding purines and pyridimes in DNA; adenine always binds with thymine, and cytosine always binds with guanine

complement fixation (KOM-ple-ment fik-SAY-shun) highly specialized antigen-antibody complexes are formed to destroy a foreign cell

concave (KON-kave) a rounded, somewhat depressed surface

concentric lamella (kon-SEN-trik lah-MEL-ah) ring of calcified matrix surrounding the Haversian canal

conchae (KONG-kee) shell-shaped structure; for example, bony projections into the nasal cavity

conduction (kon-DUK-shun) transfer of heat energy to the skin and then the external environment

cone (cone) receptor cell located in the retina that is stimulated by bright light

conjunctiva (kon-junk-TIE-vah) mucous membrane that lines the eyelids and covers the sclera (white portion)

connective tissue (ko-NEK-tiv TISH-yoo) most abundant and widely distributed tissue in the body; has numerous functions

connective tissue membrane (ko-NEK-tiv TISH-yoo MEM-brane) one of the two major types of body membranes; composed exclusively of various types of connective tissue

constipation (kon-sti-PAY-shun) retention of feces

contact dermatitis (KON-takt der-mah-TIE-tis) a local skin inflammation lasting a few hours or days after being exposed to an antigen

continuous ambulatory peritoneal dialysis (CAPD) (kon-TIN-yoo-us AM-byoo-lah-tor-ee pair-i-toe-NEE-al dye-AL-i-sis) an alternative form of treatment for renal failure rather than the more complex and expensive hemodialysis

contractile unit (kon-TRAK-til YOO-nit) the sarcomere, the basic functional unit of skeletal muscle

contractility (kon-TRAK-til-i-tee) the ability to contract a muscle

contraction (kon-TRAK-shun) ability of muscle cells to shorten or contract

convection (kon-VEK-shun) transfer of heat energy to air that is flowing away from the skin

convex (KON-veks) a rounded, somewhat elevated surface

coronal (ko-RO-nal) literally "like a crown"; a coronal plane divides the body or an organ into anterior and posterior regions

coronary artery (KOR-o-nair-ee AR-ter-ee) the first artery to branch off the aorta, supplies blood to the myocardium (heart muscle)

coronary bypass surgery (KOR-o-nair-ee BYE-pass SER-jer-ee) surgery to relieve severely restricted coronary blood flow; veins are taken from other parts of the body to bypass the partial blockage

coronary circulation (KOR-o-nair-ee ser-kyoo-LAY-shun) delivery of oxygen and removal of waste product from the myocardium (heart muscle)

coronary embolism (KOR-o-nair-ee EM-bo-lizm) blocking of a coronary blood vessel by a clot

coronary heart disease (KOR-o-nair-ee hart di-ZEEZ) disease (blockage or other deformity) of the vessels that supply the myocardium (heart muscle); one of the leading causes of death among adults in the United States

coronary sinus (KOR-o-nair-ree SYE-nus) area which receives deoxygenated blood from the coronary veins and empties into the right atrium

coronary thrombosis (KOR-o-nair-ree throm-BO-sis) formation of a blood clot in a coronary blood vessel

corpora cavernosa (KOR-por-ah kav-er-NO-sah) two columns of erectile tissue found in the shaft of the penis

corpus callosum (KOR-pus kal-LO-sum) where the right and left cerebral hemispheres are joined

corpus luteum (KOR-pus LOO-tee-um) a hormone-secreting glandular structure transformed after ovulation from a ruptured follicle; it secretes chiefly progesterone with some estrogen secreted as well

corpus spongiosum (KOR-pus spun-jee-O-sum) a column of erectile tissue surrounding the urethra in the penis

cortex (KOR-teks) outer part of an internal organ; for example, the outer part of the cerebrum and of the kidneys

corticoids (KOR-ti-koyds) hormones secreted by the three cell layers of the adrenal cortex

cotransport (ko-TRANS-port) active transport process in which two substances are moved together across a cell membrane; for example, sodium and glucose may be transported together across a membrane

coxal bone (KOKS-al bone) the pelvic bone or hipbone (also known as the os coxae or the innominate bone); formed by fusion of three distinct bones (ilium, ischium, and pubis) during skeletal development

cranial (KRAY-nee-al) toward the head

cranial cavity (KRAY-nee-al KAV-i-tee) space inside the skull that contains the brain

cranial nerve (KRAY-nee-al nerv) any of twelve pairs of nerves that attach to the undersurface of the brain and conduct impulses between the brain and structures in the head, neck, and thorax

craniosacral (kray-nee-o-SAY-kral) pertaining to parasympathetic nerves

cranium (KRAY-nee-um) bony vault made up of eight bones that encases the brain

crenation (kre-NAY-shun) abnormal notching in an erythrocyte due to shrinkage after suspension in a hypertonic solution

cretinism (KREE-tin-izm) dwarfism caused by hyposecretion of the thyroid gland

crista ampullaris (KRIS-tah am-pyoo-LAIR-is) a specialized receptor located within the semicircular canals that detects head movements

crown (krown) topmost part of an organ or other structure

crural (KROOR-al) refers to the leg

cryptorchidism (krip-TOR-ki-dizm) undescended testicles

cubital (KYOO-bi-tal) refers to the elbow

cuboid (KYOO-boyd) resembling a cube

cuboidal (KYOO-boyd-al) cell shape resembling a cube

Cushing's syndrome (KOOSH-ings SIN-drome) condition caused by the hypersecretion of glucocorticoids from the adrenal cortex

cuspid (KUS-pid) canine tooth, serves to pierce or tear food being eaten

cutaneous (kyoo-TANE-ee-us) pertaining to the skin

cutaneous membrane (ku-TANE-ee-us MEM-brane) primary organ of the integumentary system; the skin

cuticle (KYOO-ti-kul) skin fold covering the root of the nail

cyanosis (sye-ah-NO-sis) bluish appearance of the skin caused by deficient oxygenation of the blood

cyclic AMP (SIK-lik A M P)   one of several second messengers that delivers information inside the cell and thus regulates the cell's activity

cystic duct (SIS-tik dukt)   joins with the common hepatic duct to form the common bile duct

cystitis (sis-TI-tis)   inflammation of the urinary bladder

cytoplasm (SYE-toe-plazm)   the gellike substance of a cell exclusive of the nucleus and other organelles

deciduous (de-SID-yoo-us)   temporary; shedding at a certain stage of growth; for example, deciduous teeth that are commonly referred to as *baby teeth*

deep (deep)   farther away from the body's surface

deglutition (deg-loo-TISH-un)   swallowing

dehydration (dee-hye-DRAY-shun)   excessive loss of body water; the most common fluid imbalance; an abnormally low volume of one or more body fluids

deltoid (DEL-toyd)   triangular; for example, the deltoid muscle

dendrite (DEN-drite)   branching or treelike; a nerve cell process that transmits impulses toward the body

dense bone (dense bone)   bone where the outer layer is hard and dense

deoxyribonucleic acid (DNA) (dee-ok-see-rye-bo-NOO-klee-ik AS-id)   genetic material of the cell that carries the chemical "blueprint" of the body

depilatories (de-PIL-ah-toe-rees)   hair removers

depolarization (dee-po-lar-i-ZAY-shun)   the electrical activity that triggers a contraction of the heart muscle

dermal-epidermal junction (DER-mal-EP-i-der-mal JUNK-shun)   junction between the thin epidermal layer of the skin and the dermal layer providing support for the epidermis

dermal papillae (DER-mal pah-PIL-ee)   upper region of the dermis that forms part of the dermal-epidermal junction and forms the ridges and grooves of fingerprints

dermatomes (DER-mah-tohms)   skin surface areas supplied by a single spinal nerve

dermis (DER-mis)   the deeper of the two major layers of the skin, composed of dense fibrous connective tissue interspersed with glands, nerve endings, and blood vessels; sometimes called the *true skin*

developmental process (de-vel-op-MEN-tal PROSS-es)   changes and functions occurring during a human's early years as the body becomes more efficient and more effective

diabetes insipidus (dye-ah-BEE-teez in-SIP-i-dus)   condition resulting from hyposecretion of ADH in which large volumes of urine are formed and, if left untreated, may cause serious health problems

diabetes mellitus (dye-ah-BEE-teez mell-EYE-tus)   a condition resulting when the pancreatic islets secrete too little insulin, resulting in increased levels of blood glucose

dialysis (dye-AL-i-sis)   separation of smaller (diffusible) particles from larger (nondiffusable) particles through a semipermeable membrane

diaphragm (DYE-ah-fram)   membrane or partition that separates one thing from another; the flat muscular sheet that separates the thorax and abdomen and is a major muscle of respiration

diaphysis (dye-AF-i-sis)   shaft of a long bone

diarthroses (dye-ar-THRO-sis)   freely movable joint

diarrhea (dye-ah-REE-ah)   defecation of liquid feces

diastole (dye-AS-toe-lee)   relaxation of the heart, interposed between its contractions; opposite of systole

diastolic pressure (dye-ah-STOL-ik PRESH-ur)   blood pressure in arteries during diastole (relaxation) of the heart

diencephalon (dye-en-SEF-ah-lon)   "between" brain; parts of the brain between the cerebral hemispheres and the mesencephalon or midbrain

diffusion (di-FYOO-shun)   spreading; for example, scattering of dissolved particles

digestion (di-JEST-chun)   the breakdown of food materials either mechanically (i.e., chewing) or chemically (i.e., digestive enzymes)

digestive system (di-JEST-tiv SIS-tem)   organs that work together to ensure proper digestion and absorption of nutrients

digital (DIJ-i-tal)   refers to fingers and toes

discharging chambers (dis-CHARJ-ing CHAM-bers)   the two lower chambers of the heart called *ventricles*

dissection (di-SEK-shun)   cutting technique used to separate body parts for study

dissociate (di-SO-see-ate)   when a compound breaks apart in solution

distal (DIS-tal)   toward the end of a structure; opposite of proximal

distal convoluted tubule (DIS-tal KON-vo-loo-ted TOO-byool)   the part of the tubule distal to the ascending limb of the loop of Henle in the kidney

disuse atrophy (DIS-yoos AT-ro-fee)   when prolonged inactivity results in the muscles getting smaller in size

diuretic (dye-yoo-RET-ik)   a substance that promotes or stimulates the production of urine; diuretic drugs are among the most commonly used drugs in medicine

DNA replication (DNA rep-li-KAY-shun)   the unique ability of DNA molecules to make copies of themselves

dopamine (DOE-pah-meen)   chemical neurotransmitter

dorsal (DOR-sal)   referring to the back; opposite of ventral; in humans, the posterior is dorsal

dorsal body cavity (DOR-sal BOD-ee KAV-i-tee)   includes the cranial and spinal cavities

dorsiflexion (dor-si-FLEK-shun)   when the top of the foot is elevated (brought toward the front of the lower leg) with the toes pointing upward

ductless gland (DUKT-less gland)   specialized gland that secretes hormones directly into the blood

ductus arteriosus (DUK-tus ar-teer-ee-O-sus)   connects the aorta and the pulmonary artery, allowing most blood to bypass the fetus' developing lungs

ductus deferens (DUK-tus DEF-er-ens)   a thick, smooth, muscular tube that allows sperm to exit from the epididymis and pass from the scrotal sac into the abdominal cavity; also known as the *vas deferens*

ductus venosus (DUK-tus ve-NO-sus)   a continuation of the umbilical vein that shunts blood returning from the placenta past the fetus' developing liver directly into the inferior vena cava

duodenal papillae (doo-o-DEE-nal pah-PIL-ee)   ducts located in the middle third of the duodenum that empty pancreatic digestive juices and bile from the liver into the small intestine; there are two ducts, the major duodenal papillae and the minor papillae

duodenum (doo-o-DEE-num)   the first subdivision of the small intestine where most chemical digestion occurs

dura mater (DOO-rah MAH-ter)   literally "strong or hard mother"; outermost layer of the meninges

dust cells (dust sells)   macrophages that ingest particulate matter in the small air sacs of the lungs

dwarfism (DWARF-izm)   condition of abnormally small stature, sometimes resulting from hyposecretion of growth hormone

dyspnea (DISP-nee-ah)   difficult or labored breathing

eardrum (EAR-drum)   the tympanic membrane that separates the external ear and middle ear

eccrine (EK-rin)   small sweat glands distributed over the total body surface

ectoderm (EK-toe-derm)   the innermost of the primary germ layers that develops early in the first trimester of pregnancy

ectopic pregnancy (ek-TOP-ik PREG-nan-see)   a pregnancy in which the fertilized ovum implants some place other than in the uterus

edema (e-DEE-mah)   excessive fluid in the tissues

effector (ef-FEK-tor)   responding organ; for example, voluntary and involuntary muscle, the heart, and glands

efferent (EF-fer-ent)   carrying from, as neurons that transmit impulses from the central nervous system to the periphery; opposite of afferent

ejaculation (ee-jak-yoo-LAY-shun)   sudden discharging of semen from the body

ejaculatory duct (ee-JAK-yoo-lah-toe-ree dukt)   duct formed by the joining of the ductus deferens and the duct from the seminal vesicle that allows sperm to enter the urethra

electrocardiogram (ECG) (e-lek-tro-KAR-dee-o-gram)   graphic record of the heart's action potentials

electrolyte (e-LEK-tro-lite)   substance that ionizes in solution, rendering the solution capable of conducting an electric current

electrolyte balance (e-LEK-tro-lite BAL-ans)   homeostasis of electrolytes

electron transport system (e-LEK-tron TRANS-port SIS-tem)   cellular process within mitochrondria that transfers energy from high-energy electrons from glycolysis and the citric acid cycle to ATP molecules so that the energy is available to do work in the cell

embolism (EM-bo-lizm)   obstruction of a blood vessel by foreign matter carried in the bloodstream

embolus (EM-bo-lus)   a blood clot or other substance (bubble of air) that is moving in the blood and may block a blood vessel

embryo (EM-bree-o)   animal in early stages of intrauterine development; in humans, the first 3 months after conception

embryology (em-bree-OL-o-gee)   study of the development of an individual from conception to birth

embryonic phase (em-bree-ON-ik faze)   the period extending from fertilization until the end of the eighth week of gestation; during this phase the term *embryo* is used

emesis (EM-e-sis)   vomiting

emptying reflex (EMP-tee-ing REE-fleks)   the reflex that causes the contraction of the bladder wall and relaxation of the internal sphincter to allow urine to enter the urethra, which is followed by urination if the external sphincter is voluntarily relaxed

emulsify (e-MUL-se-fye)   in digestion, when bile breaks up fats

endocarditis (en-doe-kar-DYE-tis)   inflammation of the lining of the heart

endocardium (en-doe-KAR-dee-um)   thin layer of very smooth tissue lining each chamber of the heart

endochondral ossification (en-doe-KON-dral os-i-fi-KAY-shun)   the process in which most bones are formed from cartilage models

endocrine (EN-doe-krin)   secreting into the blood or tissue fluid rather than into a duct; opposite of exocrine

endocrine glands (EN-doe-krin glands)   ductless glands that are part of the endocrine system and secrete hormones into intercellular spaces

endocrine system (EN-doe-krin SIS-tem)   the series of ductless glands that are found in the body

endoderm (EN-doe-derm)   the outermost layer of the primary germ layers that develops early in the first trimester of pregnancy

endometrium (en-doe-MEE-tree-um)   mucous membrane lining the uterus

endoneurium (en-doe-NOO-ree-um)   the thin wrapping of fibrous connective tissue that surrounds each axon in a nerve

endoplasmic reticulum (ER) (en-doe-PLAS-mik re-TIK-yoo-lum)   network tubules and vesicles in cytoplasm

endorphins (en-DOR-fins)   chemical in central nervous system that influences pain perception; a natural painkiller

endosteum (en-DOS-tee-um)   a fibrous membrane that lines the medullary cavity

endothelium (en-doe-THEE-lee-um)   squamous epithelial cells that line the inner surface of the entire circulatory system and the vessels of the lymphatic system

endurance training (en-DOOR-ance TRAIN-ing)   continuous vigorous exercise requiring the body to increase its consumption of oxygen and developing the muscles' ability to sustain activity over a prolonged period of time

enkephalins (en-KEF-ah-lins)   peptide chemical in the central nervous system that acts as a natural painkiller

enzyme (EN-zime)   biochemical catalyst allowing chemical reactions to take place

eosinophil (ee-o-SIN-o-fils)   white blood cell that is readily stained by eosin

epicardium (ep-i-KAR-dee-um)   the inner layer of the pericardium that covers the surface of the heart; it is also called the *visceral pericardium*

epidermis (ep-i-DER-mis)   "false" skin; outermost layer of the skin

epididymis (ep-i-DID-i-mis)   tightly coiled tube that lies along the top and behind the testes where sperm mature and develop the ability to swim

epiglottis (ep-i-GLOT-is)   lidlike cartilage overhanging the entrance to the larynx

epinephrine (ep-i-NEF-rin)   adrenaline; secretion of the adrenal medulla

epineurium (ep-i-NOO-ree-um)   a tough fibrous sheath that covers the whole nerve

epiphyseal fracture (ep-i-FEEZ-ee-al FRAK-cher)   when the epiphyseal plate is separated from the epiphysis or diaphysis; this type of fracture can disrupt the normal growth of the bone

epiphyseal plate (ep-i-FEEZ-ee-al plate)   the cartilage plate that is between the epiphysis and the diaphysis and allows growth to occur; sometimes referred to as a *growth plate*

epiphyses (e-PIF-i-sees)   ends of a long bone

episiotomy (e-piz-ee-OT-o-mee)   a surgical procedure used during birth to prevent a laceration of the mother's perineum or the vagina

epithelial membrane (ep-i-THEE-lee-al MEM-brane)   membrane composed of epithelial tissue with an underlying layer of specialized connective tissue

epithelial tissue (ep-i-THEE-lee-al TISH-yoo)   covers the body and its parts; lines various parts of the body; forms continuous sheets that contain no blood vessels; classified according to shape and arrangement

erythroblastosis fetalis (e-rith-ro-blas-TOE-sis fee-TAL-is)   a disease that may develop when an Rh-negative mother has anti-Rh antibodies and gives birth to an Rh-positive baby and the antibodies react with the Rh positive cells of the baby

erythrocytes (e-RITH-ro-sites)   red blood cells

esophagus (e-SOF-ah-gus)   the muscular, mucous-lined tube that connects the pharynx with the stomach; also known as the *foodpipe*

essential organs (ee-SEN-shal OR-gans)   reproductive organs that must be present for reproduction to occur and are known as gonads

estrogen (ES-tro-jen)   sex hormone secreted by the ovary that causes the development and maintenance of the female secondary sex characteristics and stimulates growth of the epithelial cells lining the uterus

eupnea (YOOP-nee-ah)   normal respiration

eustachian tube (yoo-STAY-shun toob)   tube extending from inside the ear to the throat to equalize air pressure

evaporation (ee-vap-o-RAY-shun)   heat being lost from the skin by sweat being vaporized

exhalation (eks-hah-LAY-shun)   moving air out of the lungs; also known as expiration

exocrine (EK-so-krin)   secreting into a duct; opposite of endocrine

exocrine gland (EK-so-krin glands)   glands that secrete their products into ducts that empty onto a surface or into a cavity; for example, sweat glands

expiration (eks-pi-RAY-shun)   moving air out of the lungs; also known as exhalation

expiratory center (eks-PYE-rah-tor-ee SEN-ter)   one of the two most important respiratory control centers, located in the medulla

expiratory muscles (eks-PYE-rah-tor-ee MUS-els)   muscles that allow more forceful expiration to increase the rate and depth of ventilation; the internal intercostals and the abdominal muscles

expiratory reserve volume (ERV) (eks-PYE-rah-tor-ee re-ZERV VOL-yoom)   the amount of air that can be forcibly exhaled after expiring the tidal volume (TV)

extension (ek-STEN-shun)   increasing the angle between two bones at a joint

external auditory canal (eks-TER-nal AW-di-toe-ree kah-NAL)   a curved tube (approximately 2.5 cm) extending from the auricle into the temporal bone, ending at the tympanic membrane

external ear (eks-TER-nal ear)   the outer part of the ear that is made up of the auricle and the external auditory canal

external genitalia (eks-TER-nal jen-i-TAIL-yah)   external reproductive organs

external intercostals (eks-TER-nal in-ter-KOS-tals)   inspiratory muscles that enlarge the thorax, causing the lungs to expand and air to rush in

external nares (eks-TER-nal NAY-reez)   nostrils

external oblique (eks-TER-nal o-BLEEK)   the outermost layer of the anterolateral abdominal wall

external otitis (eks-TER-nal o-TIE-tis)   a common infection of the external ear; also known as *swimmer's ear*

external respiration (eks-TER-nal res-pi-RAY-shun)   the exchange of gases between air in the lungs and in the blood

extracellular fluid (ECF) (eks-trah-SELL-yoo-lar FLOO-id)   the water found outside of cells located in two compartments between cells (interstitial fluid) and in the blood (plasma)

facial (FAY-shal)   referring to the face

fallen arch (fallen arch)   when the tendons and ligaments of the foot weaken, allowing the normally curved arch to flatten out

fallopian tubes (fal-LO-pee-an toobs)   the pair of tubes that conduct the ovum from the ovary to the uterus

false ribs (fawls ribs)   the eighth, ninth, and tenth pairs of ribs that are attached to the cartilage of the seventh ribs rather than the sternum

fasciculus (fah-SIK-yoo-lus)   little bundle

fat (fat)   one of the three basic food types; primarily a source of energy

fatigue (fah-TEEG)   loss of muscle power; weakness

fat tissue (fat TISH-yoo)   adipose tissue; specialized to store lipids

feces (FEE-seez)   waste material discharged from the intestines

feedback control loop (FEED-bak kon-TROL loop)   a highly complex and integrated communication control network, classified as negative or positive; negative feedback loops are the most important and numerous homeostatic control mechanisms

femoral (FEM-or-al)   referring to the thigh

femur (FEE-mur)   the thigh bone, which is the longest bone in the body

fertilization (FER-ti-li-ZAY-shun)   the moment the female's ovum and the male's sperm cell unite

fetal alcohol syndrome (FAS) (FEE-tal AL-ko-hol SIN-drome)   a condition that may cause congenital abnormalities in a baby that results from a woman consuming alcohol during pregnancy

fetal phase (FEE-tal faze)   period extending from the eighth to the thirty-ninth week of gestation; during this phase the term *fetus* is used

fetus (FEE-tus)   unborn young, especially in the later stages; in human beings, from the third month of the intrauterine period until birth

fibers (FYE-bers)   threadlike structures; for example, nerve fibers

fibrin (FYE-brin)   insoluble protein in clotted blood

fibrinogen (fye-BRIN-o-jen)   soluble blood protein that is converted to insoluble fibrin during clotting

fibrous connective tissue (FYE-brus ko-NEK-tiv TISH-yoo) strong, nonstretchable, white collagen fibers that compose tendons

fibula (FIB-yoo-lah) the slender non-weight-bearing bone located on the lateral aspect of the leg

fight or flight syndrome (fite or flite SIN-drome) the changes produced by increased sympathetic impulses allowing the body to deal with any type of stress

filtration (fil-TRAY-shun) movement of water and solutes through a membrane by a higher hydrostatic pressure on one side

fimbriae (FIM-bree-ee) fringe

flagellum (flah-JEL-um) single projection extending from the cell surface; only example in human is the "tail" of the male sperm

flat bone (flat bone) one of the four types of bone; the frontal bone is an example of a flat bone

flat feet (flat feet) when the tendons and ligaments of the foot weaken, allowing the normally curved arch to flatten out

flexion (FLEK-shun) act of bending; decreasing the angle between two bones at the joint

floating ribs (FLOW-ting ribs) the eleventh and twelfth pairs of ribs, which are only attached to the thoracic vertebrae

fluid balance (FLOO-id BAL-ans) homeostasis of fluids; the volumes of interstitial fluid, intracellular fluid, and plasma and total volume of water remain relatively constant

fluid compartments (FLOO-id kom-PART-ments) the areas in the body where the fluid is located; for example, interstitial fluid

follicles (FOL-li-kuls) specialized structures required for hair growth

follicle-stimulating hormone (FSH) (FOL-li-kul STIM-yoo-lay-ting HOR-mone) hormone present in males and females; in males, FSH stimulates the production of sperm; in females, FSH stimulates the ovarian follicles to mature and follicle cells to secrete estrogen

fontanels (FON-tah-nels) "soft spots" on the infant's head; unossified areas in the infant skull

foramen (fo-RAY-men) small opening; for example, the vertebral foramen, which allows the spinal cord to pass through the vertebral canal

foramen ovale (fo-RAY-men o-VAL-ee) shunts blood from the right atrium directly into the left atrium, allowing most blood to bypass the baby's developing lungs

foreskin (FORE-skin) a loose-fitting retractable casing located over the glans of the penis; also know as the *prepuce*

fractal geometry (FRAK-tul jee-OM-e-tree) the study of surfaces with a seemingly infinite area, such as the lining of the small intestine

free nerve endings (free nerv END-ings) specialized receptors in the skin that respond to pain

frenulum (FREN-yoo-lum) the thin membrane that attaches the tongue to the floor of the mouth

frontal (FRON-tal) lengthwise plane running from side to side, dividing the body into anterior and posterior portions

frontal muscle (FRON-tal MUS-el) one of the muscles of facial expression; it moves the eyebrows and furrows the skin of the forehead

frontal sinusitis (FRON-tal sye-nyoo-SYE-tis) inflammation in the frontal sinus

gene (jean) one of many segments of a chromosome (DNA molecule); each gene contains the genetic code for synthesizing a protein molecule such as an enzyme or hormone

genitalia (jen-i-TAIL-yah) reproductive organs

gestation (jes-TAY-shun) the length of pregnancy, approximately 9 months in humans

gigantism (jye-GAN-tizm) a condition produced by hypersecretion of growth hormone during the early years of life; results in a child who grows to gigantic size

gland (gland) secreting structure

glandular epithelium (GLAN-dyoo-lar ep-i-THEE-lee-um) cells that are specialized for secreting activity

glans (glans) the distal end of the shaft of the penis

glaucoma (glaw-KO-mah) disorder characterized by elevated pressure in the eye

glia (GLEE-ah) supporting cells of nervous tissue; also called *neuroglia*

glioma (glee-O-mah) one of the most common types of brain tumors

globulin (GLOB-yoo-lin) a type of plasma protein that includes antibodies

glomerulus (glo-MARE-yoo-lus) compact cluster; for example, capillaries in the kidneys

glottis (GLOT-is) the space between the vocal cords

glucagon (GLOO-kah-gon) hormone secreted by alpha cells of the pancreatic islets

glucocorticoids (GCs) (gloo-ko-KOR-ti-koyds) hormones that influence food metabolism; secreted by the adrenal cortex

gluconeogenesis (gloo-ko-nee-o-JEN-e-sis) formulation of glucose or glycogen from protein or fat compounds

glucose (GLOO-kose) monosaccharide or simple sugar; the principal blood sugar

gluteal (GLOO-tee-al) of or near the buttocks

gluteus maximus (GLOO-tee-us MAX-i-mus) major extensor of the thigh and also supports the torso in an erect position

glycerol (GLIS-er-ol) product of fat digestion

glycogen (GLYE-ko-jen) polysaccharide; animal starch

glycogen loading (GLYE-ko-jen LOHD-ing) see *carbohydrate loading*

glycogenesis (glye-ko-JEN-e-sis) formation of glycogen from glucose or from other monosaccharides, fructose, or galactose

glycogenolysis (glye-ko-je-NOL-i-sis) hydrolysis of glycogen to glucose-6-phosphate or to glucose

glycolysis (glye-KOL-i-sis) the first series of chemical reactions in glucose metabolism; changes glucose to pyruvic acid in a series of anaerobic reactions

glycosuria (glye-ko-SOO-ree-ah) glucose in the urine; a sign of diabetes mellitus

goblet cells (GOB-let sells) specialized cells found in simple columnar epithelium that produce mucus

goiter (GOY-ter) enlargement of the thyroid gland

golgi apparatus (GOL-jee ap-ah-RA-tus) small sacs stacked on one another near the nucleus that makes carbohydrate compounds, combines them with protein molecules, and packages the product in a globule

golgi tendon receptors (GOL-jee TEN-don ree-SEP-tors) sensors that are responsible for proprioception

gonads (GO-nads) sex glands in which reproductive cells are formed

graafian follicle (GRAF-ee-an FOL-li-kul)   a mature ovum in its sac

gradient (GRAY-dee-ent)   a slope or difference between two levels; for example, blood pressure gradient: a difference between the blood pressure in two different vessels

gram (gram)   the unit of measure in the metric system on which mass is based (approximately 454 grams equals one pound)

granulosa cell (gran-yoo-LO-sah sell)   cell layer surrounding the oocyte

gray matter (gray MAT-er)   tissue comprising cell bodies and unmyelinated axons and dendrites

greater omentum (GRATE-er o-MEN-tum)   a pouchlike extension of the visceral peritoneum

growth hormone (growth HOR-mone)   hormone secreted by the anterior pituitary gland that controls the rate of skeletal and visceral growth

gustatory cell (GUS-tah-tor-ee sell)   cells of taste

gyrus (JYE-rus)   convoluted ridge

hair follicle (hair FOL-li-kul)   a small tube where hair growth occurs

hair papilla (hair pah-PIL-ah)   a small, cap-shaped cluster of cells located at the base of the follicle where hair growth begins

hamstring muscles (HAM-string MUS-els)   powerful flexors of the hip made up of the semimembranosus, semitendinosis, and biceps femoris muscles

Haversian canal (ha-VER-shun kah-NAL)   the canal in the Haversian system that contains a blood vessel

Haversian system (hah-VER-shun SIS-tem)   the circular arrangements of calcified matrix and cells that give bone its characteristic appearance

heart block (hart blok)   a blockage of impulse conduction from atria to ventricles so that the heart beats at a slower rate than normal

heartburn (HART-bern)   burning sensation characterized by pain and a feeling of fullness beneath the sternum caused by the esophogeal mucosa being irritated by stomach acid

Heimlich maneuver (HIME-lik mah-NOO-ver)   livesaving technique used to free the trachea of objects blocking the airway

hematocrit (he-MAT-o-krit)   volume percent of blood cells in whole blood

hemodialysis (hee-mo-dye-AL-i-sis)   use of dialysis to separate waste products from the blood

hemoglobin (hee-mo-GLO-bin)   iron-containing protein in red blood cells

hemopoiesis (hee-mo-poy-EE-sis)   blood cell formation

hemopoietic tissue (hee-mo-poy-ET-ik TISH-yoo)   specialized connective tissue that is responsible for the formation of blood cells and lymphatic system cells; found in red bone marrow, spleen, tonsils, and lymph nodes

heparin (HEP-ah-rin)   substance obtained from the liver; inhibits blood clotting

hepatic colic flexure (he-PAT-ik KOL-ik FLEK-sher)   the bend between the ascending colon and the transverse colon

hepatic ducts (he-PAT-ik dukts)   drain bile out of the liver

hepatic portal circulation (he-PAT-ik POR-tal ser-kyoo-LAY-shun)   the route of blood flow through the liver

hepatic portal vein (he-PAT-ik POR-tal vane)   delivers blood directly from the gastrointestinal tract to the liver

herpes zoster (HER-peez ZOS-ter)   "shingles," viral infection that affects the skin of a single dermatone

hiccup (HIK-up)   involuntary spasmodic contraction of the diaphragm

hip (hip)   the joint connecting the legs to the trunk

histogenesis (his-toe-JEN-e-sis)   formation of tissues from primary germ layers of embryo

homeostasis (ho-mee-o-STAY-sis)   relative uniformity of the normal body's internal environment

homeostatic mechanism (ho-mee-o-STAT-ik MEK-ah-nizm)   a system that maintains a constant environment enabling body cells to function efffectively

hormone (HOR-mone)   substance secreted by an endocrine gland

human immunodeficiency virus (HYOO-man i-myoo-no-de-FISH-en-see VYE-rus)   the retrovirus that causes acquired immunodeficiency syndrome (AIDS)

humerus (HYOO-mer-us)   the second longest bone in the body; the long bone of the arm

humoral immunity (HYOO-mor-al i-MYOO-ni-tee)   antibody-mediated immunity

hybridoma (hye-brid-O-ma)   fused or hybrid cells that continue to produce the same antibody as the original lymphocyte

hydrocephalus (hye-dro-SEF-ah-lus)   abnormal accumulation of cerebrospinal fluid; "water on the brain"

hydrocortisone (hye-dro-KOR-ti-zone)   a hormone secreted by the adrenal cortex; cortisol; compound F

hydrogen ion (HYE-dro-jen EYE-on)   found in water and water solutions; produces an acidic solution; H^+

hydrostatic pressure (hye-dro-STAT-ik PRESH-ur)   the force of a fluid pushing against some surface

hydroxide ion (hye-DROK-side EYE-on)   found in water and water solutions; produces an alkaline solution; H^+

hymen (HYE-men)   Greek for "membrane"; mucous membrane that may partially or entirely occlude the vaginal outlet

hyperacidity (hye-per-a-SID-i-tee)   excessive secretion of acid; an important factor in the formation of ulcers

hypercalcemia (hye-per-kal-SEE-mee-ah)   a condition in which there is harmful excesses of calcium in the blood

hyperglycemia (hye-per-glye-SEE-mee-ah)   higher than normal blood glucose concentration

hyperopia (hye-per-O-pee-ah)   farsightedness

hypersecretion (hye-per-se-KREE-shun)   too much of a substance is being secreted

hypertension (hye-per-TEN-shun)   abnormally high blood pressure

hyperthyroidism (hye-per-THYE-royd-izm)   oversecretion of thyroid hormones that increases metabolic rate resulting in loss of weight, increased appetite, and nervous irritability

hypertonic (hye-per-TON-ik)   a solution containing a higher level of salt (NaCl)   than is found in a living red blood cell (above 0.9% NaCl)

hypertrophy (hye-PER-tro-fee)   increased size of a part caused by an increase in the size of its cells

hyperventilation (hye-per-ven-ti-LAY-shun)   very rapid deep respirations

hypervitaminosis (hye-per-vye-tah-mi-NO-sis)   condition caused by excess amounts of vitamins; usually associated with the use of vitamin supplements

hypoglycemia (hye-po-glye-SEE-mee-ah)   lower-than-normal blood glucose concentration

hyposecretion (hye-po-se-KREE-shun)   too little of a substance is being secreted

hypothalamus (hye-po-THAL-ah-mus)   vital neuroendocrine and autonomic control center beneath the thalamus

hypothermia (hye-po-THER-mee-ah)   subnormal core body temperature below 37° C

hypothesis (hye-POTH-e-sis)   a proposed explanation of an observed phenomena

hypothyroidism (hye-po-THYE-royd-izm)   undersecretion of thyroid hormones; early in life results in cretinism; later in life results in myxedema

hypotonic (hye-po-TON-ik)   a solution containing a lower level of salt (NaCl) than is found in a living red blood cell (below 0.9% NaCl)

hypoventilation (hye-po-ven-ti-LAY-shun)   slow and shallow respirations

hysterectomy (his-te-REK-toe-mee)   surgical removal of the uterus

ileocecal valve (il-ee-o-SEE-kal valv)   the sphincterlike structure between the end of the small intestine and the beginning of the large intestine

ileum (IL-ee-um)   the distal portion of the small intestine

iliac crest (IL-ee-ak krest)   the superior edge of the illium

iliopsoas (il-ee-op-SO-as)   a flexor of the thigh and an important stabilizing muscle for posture

ilium (IL-ee-um)   one of the three separate bones that forms the ox coxa

immune system (i-MYOON SIS-tem)   the body's defense system against disease

immunization (i-myoo-ni-ZAY-shun)   deliberate artificial exposure to disease to produce acquired immunity

implantation (im-plan-TAY-shun)   when a fertilized ovum implants in the uterus

inborn immunity (IN-born i-MYOO-ni-tee)   immunity to disease that is inherited

incontinence (in-KON-ti-nens)   when an individual voids urine involuntarily

incus (IN-kus)   the anvil, the middle ear bone that is shaped like an anvil

infancy (IN-fan-see)   from birth to about 18 months of age

infant respiratory distress syndrome (IN-fant RES-pi-rah-toe-ree di-STRESS SIN-drome)   leading cause of death in premature babies, due to a lack of surfactant in the alveolar air sacs

inferior (in-FEER-ee-or)   lower; opposite of superior

inferior vena cava (in-FEER-ee-or VEE-nah KAY-vah)   one of two large veins carrying blood into the right atrium

inflammatory response (in-FLAM-ah-toe-ree re-SPONS)   nonspecific immune process produced in response to injury and resulting in redness, pain, heat, and swelling and promoting movement of white blood cells to the affected area

inguinal (ING-gwi-nal)   of the groin

inhalation (in-hah-LAY-shun)   inspiration or breathing in; opposite of exhalation or expiration

inherited immunity (in-HAIR-i-ted i-MYOO-ni-tee)   inborn immunity

inhibiting hormone (in-HIB-i-ting HOR-mone)   hormone produced by the hypothalamus that slows the release of anterior pituitary hormones

insertion (in-SER-shun)   attachment of a muscle to the bone that it moves when contraction occurs (as distinguished from its origin)

inspiration (in-spi-RAY-shun)   inhalation, moving air into the lungs; same as inhalation, opposite of exhalation or expiration

inspiratory muscle (in-SPY-rah-tor-ee MUS-el)   the muscles that increase the size of the thorax, including the diaphragm and external intercostals, and allow air to rush into the lungs

inspiratory center (in-SPY-rah-tor-ee SEN-ter)   one of the two most important control centers located in the medulla; the other is the expiratory center

inspiratory reserve volume (IRV) (in-SPY-rah-tor-ee re-SERV VOL-yoom)   the amount of air that can be forcibly inspired over and above a normal respiration

insulin (IN-suh-lin)   hormone secreted by the pancreatic islets

integument (in-TEG-yoo-ment)   refers to the skin

integumentary system (in-teg-yoo-MEN-tar-ee SIS-tem)   the skin; the largest and most important organ in the body

intercalated disks (in-TER-kah-lay-ted disks)   cross striations and unique dark bands that are found in cardiac muscle fibers

intercostal muscle (in-ter-KOS-tal MUS-el)   the respiratory muscles located between the ribs

interferon (in-ter-FEER-on)   small proteins produced by the immune system that inhibit virus multiplication

internal oblique (in-TER-nal o-BLEEK)   the middle layer of the anterolateral abdominal walls

internal respiration (in-TER-nal res-pi-RAY-shun)   the exchange of gases that occurs between the blood and cells of the body

interneuron (in-ter-NOO-ron)   nerves that conduct impulses from sensory neurons to motor neurons

interphase (IN-ter-faze)   the phase immediately before the visible stages of cell division when the DNA of each chromosome replicates itself

interstitial cell (in-ter-STISH-al sell)   small specialized cells in the testes that secrete the male sex hormone, testosterone

interstitial cell-stimulating hormone (ICSH) (in-ter-STISH-al sell STIM-yoo-lay-ting HOR-mone)   the previous name for luteinizing hormone in males; causes testes to develop and secrete testosterone

interstitial fluid (in-ter-STISH-al FLOO-id)   fluid located in the microscopic spaces between the cells

intestine (in-TES-tin)   the part of the digestive tract that is after the stomach; separated into two segments, the small and the large

intestinal gland (in-TES-ti-nal gland)   thousands of glands found in the mucous membrane of the mucosa of the small intestines; secrete intestinal digestive juices

intracellular fluid (ICF) (in-tra-SELL-yoo-lar FLOO-id)   a fluid located within the cells; largest fluid compartment

in vitro (in VEE-tro)   refers to the glass laboratory container where a mature ovum is fertilized by a sperm

involuntary muscle (in-VOL-un-tare-ee MUS-el)   smooth muscles that are not under conscious control and are found in organs such as the stomach and small intestine

involution (in-vo-LOO-shun)   return of an organ to its normal size after an enlargement; also retrograde or degenerative change

ion (EYE-on)   electrically charged atom or group of atoms

ion pump (EYE-on pump)   a specialized cellular component that moves ions from an area of low concentration to an area of high concentration

iron deficiency anemia (EYE-ern de-FISH-en-see ah-NEE-mee-ah)   when there are inadequate levels of iron in the diet so that less hemoglobin is produced; results in extreme fatigue

ischium (IS-kee-um)   one of three separate bones that forms the ox coxa

isometric (eye-so-MET-rik)   type of muscle contraction in which muscle does not shorten

isotonic (eye-so-TON-ik)   of the same tension or pressure

jaundice (JAWN-dis)   abnormal yellowing of skin, mucous membranes, and white of eyes

jejunum (je-JOO-num)   the middle third of the small intestine

joints (joynts)   articulation

Kaposi's sarcoma (KAH-po-sees sar-KO-mah)   a type of skin cancer that can take advantage of an impaired immune system

keratin (KARE-ah-tin)   protein substance found in hair, nails, outer skin cells, and horny tissues

kidney (KID-nee)   organ that cleanses the blood of waste products continually produced by metabolism

kilocalorie (Kcal) (KIL-o-kal-o-ree)   1000 calories

kinesthesia (kin-es-THEE-zee-ah)   "muscle sense"; that is, sense of position and movement of body parts

Krause's end bulb (KROWZ end bulb)   skin receptor that detects sensations of cold

Kupffer cell (KOOP-fer sell)   macrophage found in spaces between liver cells

labia majora (LAY-bee-ah ma-JO-rah)   "large lips" of the vulva

labia minora (LAY-bee-ah mi-NO-rah)   "small lips" of the vulva

labor (LAY-bor)   the process that results in the birth of the baby

lacrimal gland (LAK-ri-mal gland)   the glands that produce tears, located in the upper lateral portion of the orbit

lacteal (LAK-tee-al)   a lymphatic vessel located in each villus of the intestine; serves to absorb fat materials from the chyme passing through the small intestine

lactiferous duct (lak-TIF-er-us dukt)   the duct that drains the grapelike cluster of milk-secreting glands in the breast

lacuna (lah-KOO-nah)   space or cavity; for example, lacunae in bone contain bone cells

lambdoidal suture (LAM-doyd-al SOO-chur)   the immovable joint formed by the parietal and occipital bones

lamella (lah-MEL-ah)   thin layer, as of bone

lanugo (lah-NOO-go)   the extremely fine and soft hair found on a newborn infant

laparoscope (LAP-ah-ro-skope)   specialized optical viewing tube

laryngopharynx (lah-ring-go-FAIR-inks)   the lowest part of the pharynx

larynx (LAIR-inks)   the voice box located just below the pharynx; the largest piece of cartilage making up the larynx is the thyroid cartilage, commonly known as the *Adam's apple*

lateral (LAT-er-al)   of or toward the side; opposite of medial

latissimus dorsi (la-TIS-i-mus DOR-si)   an extensor of the upper arm

lens (lenz)   the refracting mechanism of the eye that is located directly behind the pupil

leukemia (loo-KEE-mee-ah)   blood cancer characterized by an increase in white blood cells

leukocyte (LOO-ko-site)   white blood cells

leukocytosis (loo-ko-SYE-toe-sis)   abnormally high white blood cell numbers in the blood

leukopenia (loo-ko-PEE-nee-ah)   abnormally low white blood cell numbers in the blood

ligament (LIG-ah-ment)   bond or band connecting two objects; in anatomy a band of white fibrous tissue connecting bones

lipase (LYE-pase)   fat-digesting enzymes

lithotriptor (LITH-o-trip-tor)   a specialized ultrasound generator that is used to pulverize kidney stones

liver glycogenolysis (LIV-er glye-ko-je-NOL-i-sis)   chemical process by which liver glycogen is converted to glucose

longitudinal arch (lon-ji-TOO-di-nal arch)   two arches, the medial and lateral, that extend lengthwise in the foot

loop of Henle (loop of HEN-lee)   extension of the proximal tubule of the kidney

lumbar (LUM-bar)   lower back, between the ribs and pelvis

lumbar puncture (LUM-bar PUNK-chur)   when some cerebrospinal fluid is withdrawn from the subarachnoid space in the lumbar region of the spinal cord

lumen (LOO-men)   the hollow space within a tube

lung (lung)   organ of respiration; the right lung has three lobes and the left lung has two lobes

lunula (LOO-nyoo-lah)   crescent-shaped white area under the proximal nail bed

luteinization (loo-te-ni-ZAY-shun)   the formation of a golden body (corpus luteum) in the ruptured follicle

luteinizing hormone (LH) (LOO-te-nye-zing HOR-mone)   acts in conjunction with follicle-stimulating hormone (FSH) to stimulate follicle and ovum maturation and release of estrogen and ovulation; known as the ovulating hormone; in males, causes testes to develop and secrete testosterone

lymph (limf)   specialized fluid formed in the tissue spaces that returns excess fluid and protein molecules to the blood

lymph node (limf node)   performs biological filtration of lymph on its way to the circulatory system

lymphatic capillaries (lim-FAT-ik CAP-i-lair-ees)   tiny, blind-ended tubes distributed in the tissue spaces

lymphatic duct (lim-FAT-ik dukt)   terminal vessel into which lymphatic vessels empty lymph; the duct then empties the lymph into the circulatory system

lymphatic system (lim-FAT-ik SIS-tem)   a system that plays a critical role in the functioning of the immune system, moves fluids and large molecules from the tissue spaces and fat-related nutrients from the digestive system to the blood

lymphatic tissue (lim-FAT-ik TISH-yoo)   tissue that is responsible for manufacturing lymphocytes and moncytes; found mostly in the lymph nodes, thymus, and spleen

lymphatic vessels (lim-FAT-ik VES-els)   vessels that carry lymph to its eventual return to the circulatory system

lymphocytes (LIM-fo-sites)   one type of white blood cell

lyse (lize)   disintegration of a cell

lysosome (LYE-so-sohm)   membranous organelles containing various enzymes that can dissolve most cellular compounds; hence called *digestive bags* or *suicide bags* of cells

macrophage (MAK-ro-faje)   phagocytic cells in the immune system

malignant (mah-LIG-nant)   cancerous growth

malleus (MAL-ee-us)   hammer; the tiny middle ear bone that is shaped like a hammer

mammary gland (MAM-er-ee gland)   breasts; classified as external accessory sex organs in females

mastication (mas-ti-KAY-shun)   chewing

matrix (MAY-triks)   the intracellular substance of a tissue; for example, the matrix of bone is calcified, whereas that of blood is liquid

mature follicle (mah-CHUR FOL-li-kul)   graafian follicle

maximum oxygen consumption (Vo_{2max}) (MAX-i-mum OKS-i-jen kon-SUMP-shun)   the maximum amount of oxygen taken up by the lungs, transported to the tissues and used to do work

mechanoreceptor (mek-an-o-ree-SEP-tor)   receptors that are mechanical in nature; for example, equilibrium and balance sensors in the ears

medial (MEE-dee-al)   of or toward the middle; opposite of lateral

mediastinum (mee-dee-as-TI-num)   a subdivision in the mid-portion of the thoracic cavity

medulla (me-DUL-ah)   Latin for "marrow"; hence the inner portion of an organ in contrast to the outer portion or cortex

medulla oblongata (me-DUL-ah ob-long-GAH-tah)   the lowest part of the brain stem; an enlarged extension of the spinal cord; the vitals centers are located within this area

medullary cavity (MED-yoo-lair-ee KAV-i-tee)   hollow area inside the diaphysis of the bone that contains yellow bone marrow

meiosis (my-O-sis)   nuclear division in which the number of chromosomes are reduced to half their original number; produce gametes

Meissner's corpuscle (MIZS-ners KOR-pus-ul)   a sensory receptor located in the skin close to the surface that detects light touch

melanin (MEL-ah-nin)   brown skin pigment

melanocyte (me-LAN-o-site)   specialized cells in the pigment layer that produce melanin

melanocyte-stimulating hormone (MSH) (me-LAN-o-site STIM-yoo-lay-ting HOR-mone)   responsible for a rapid increase in the synthesis and dispersion of melanin granules in specialized skin cells

melatonin (mel-ah-TOE-nin)   important hormone produced by the pineal gland that is believed to regulate the onset of puberty and the menstrual cycle; also referred to as the *third eye* because it responds to levels of light and is thought to be involved with the body's internal clock

membrane (MEM-brane)   thin layer or sheet

membranous labyrinth (MEM-brah-nus LAB-i-rinth)   a membranous sac that follows the shape of the bony labyrinth and is filled with endolymph

memory cell (MEM-o-ree sell)   cells that remain in reserve in the lymph nodes until their ability to secrete antibodies is needed

menarche (me-NAR-kee)   beginnings of the menstrual function

meninges (me-NIN-jeez)   fluid-containing membranes surrounding the brain and spinal cord

menopause (MEN-o-pawz)   termination of menstrual cycles

menses (MEN-seez)   menstrual flow

menstrual cycle (MEN-stroo-al SYE-kul)   the cyclical changes in the uterine lining

mesentery (MEZ-en-tair-ee)   a large double fold of peritoneal tissue that anchors the loops of the digestive tract to the posterior wall of the abdominal cavity

mesoderm (MEZ-o-derm)   the middle layer of the primary germ layers

messenger RNA (mRNA) (MES-en-jer RNA)   a duplicate copy of a gene sequence on the DNA that passes from the nucleus to the cytoplasm

metabolic acidosis (met-ah-BOL-ik as-i-DOE-sis)   a disturbance affecting the bicarbonate element of the bicarbonate-carbonic acid buffer pair; bicarbonate deficit

metabolic alkalosis (met-ah-BOL-ik al-kah-LO-sis)   disturbance affecting the bicarbonate element of the bicarbonate-carbonic acid buffer pair; bicarbonate excess

metabolism (me-TAB-o-lizm)   complex process by which food is used by a living organism

metacarpal (met-ah-KAR-pal)   the part of the hand between the wrist and fingers

metaphase (MET-ah-faze)   second stage of mitosis, during which the nuclear envelope and nucleolus disappear

metatarsal arch (met-ah-TAR-sal arch)   the arch that extends across the ball of the foot; also called the *transverse arch*

meter (MEE-ter)   a measure of length in the metric system; equal to about 39.5 inches

microcephaly (my-kro-SEF-ah-lee)   a congenital abnormality in which an infant is born with a small head

microglia (my-KROG-lee-ah)   one type of connective tissue found in the brain and spinal cord

micron (MY-kron)   1/1000 millimeter; 1/25,000 inch

microvilli (my-kro-VIL-eye)   the brushlike border made up of epithelial cells found on each villus in the small intestine; increases the surface area for absorption of nutrients

micturition (mik-too-RISH-un)   urination, voiding

midbrain (MID-brain)   one of the three parts of the brain stem

middle ear (MID-ul eer)   a tiny and very thin epithelium-lined cavity in the temporal bone that houses the ossicles; in the middle ear, sound waves are amplified

midsagittal (mid-SAJ-i-tal)   a cut or plane that divides the body or any of its parts into two equal halves

minerals (MIN-er-als)   inorganic elements or salts found naturally in the earth that are vital to the proper functioning of the body

mineralocorticoid (MC) (min-er-al-o-KOR-ti-koyd)   hormone that influences mineral salt metabolism; secreted by adrenal cortex; aldosterone is the chief mineralocorticoid

mitochondria (my-toe-KON-dree-ah)   threadlike structures

mitosis (my-TOE-sis)   indirect cell division involving complex changes in the nucleus

mitral valve (MY-tral valv)   also known as the *bicuspid valve;* located between the left atrium and ventricle

monoclonal antibody (mon-o-KLONE-al AN-ti-bod-ee)   specific antibody produced from a population of identical cells

monocyte (MON-o-site)   a phagocyte

mons pubis (monz PYOO-bis)   skin-covered pad of fat over the symphysis pubis in the female

morula (MOR-yoo-lah)   a solid mass of cells formed by the divisions of a fertilized egg

motor neuron (MO-tor NOO-ron) transmits nerve impulses from the brain and spinal cord to muscles and glandular epithelial tissues

motor unit (MO-tor YOO-nit) a single motor neuron with the muscle cells it innervates

mucocutaneous junction (myoo-ko-kyoo-TAY-nee-us JUNK-shun) the transitional area where the skin and mucous membrane meet

mucosa (myoo-KO-sah) mucous membrane

mucous membrane (MYOO-kus MEM-brane) epithelial membranes that line body surfaces opening directly to the exterior and secrete a thick, slippery material called *mucus*

mucus (MYOO-kus) thick, slippery material that is secreted by the mucous membrane and that keeps the membrane moist

multiple sclerosis (MS) (MUL-ti-pul skle-RO-sis) the most common primary disease of the central nervous system; a myelin disorder

muscle fiber (MUS-el FYE-ber) the specialized contractile cells of muscle tissue that are grouped together and arranged in a highly organized way

muscular system (MUS-kyoo-lar SIS-tem) the muscles of the body

muscularis (mus-kyoo-LAIR-is) two layers of muscle surrounding the digestive tube that produce wavelike, rhythmic contractions, called *peristalsis,* which move food material

myelin (MY-e-lin) lipoid substace found in the myelin sheath around some nerve fibers

myelinated fiber (MY-e-li-nay-ted FYE-ber) axons outside the central nervous system that are surrounded by a segmented wrapping of myelin

myeloid (MY-e-loyd) pertaining to bone marrow

myocardial infarction (my-o-KAR-dee-al in-FARK-shun) death of cardiac muscle cells resulting from inadequate blood supply as in coronary thrombosis

myocardium (my-o-KAR-dee-um) muscle of the heart

myofilaments (my-o-FIL-ah-ments) ultramicroscopic, thread-like structures found in myofibrils

myometrium (my-o-MEE-tree-um) muscle layer in the uterus

myopia (my-O-pee-ah) nearsightedness

myosin (MY-o-sin) contractile protein found in the thick filaments of skeletal muscle

myxedema (mik-se-DEE-mah) condition caused by deficiency of thyroid hormone in adults

nail body (nail BOD-ee) the visible part of the nail

nail root (nail root) the part of the nail that is hidden by the cuticle

nanometer (NAN-o-mee-ter) a measure of length in the metric system; one billionth of a meter

nares (NAY-reez) nostrils

nasal cavity (NAY-zal KAV-i-tee) the moist, warm cavities lined by mucosa located just beyond the nostrils; olfactory receptors are located in the mucosa

nasal septum (NAY-zal SEP-tum) a partition that separates the right and left nasal cavities

nasopharynx (nay-zo-FAIR-inks) the uppermost portion of the tube just behind the nasal cavities

neonatology (nee-o-nay-TOL-o-jee) diagnosis and treatment of disorders of the newborn infant

neoplasm (NEE-o-plazm) an abnormal mass of proliferating cells that may be either benign or malignant

nephritis (ne-FRY-tis) kidney disease; inflammation of the nephrons

nephron (NEF-ron) anatomical and functional unit of the kidney, consisting of the renal corpuscle and the renal tubule

nerve (nerv) collection of nerve fibers

nerve impulse (nerv IM-puls) signals that carry information along the nerves

nervous tissue (NER-vus TISH-yoo) consists of neurons and glia that provide rapid communication and control of body function

nucleic acids (noo-KLEE-ik AS-ids) the two nucleic acids are ribonucleic acid, found in the cytoplasm, and deoxyribonucleic acid, found in the nucleus

neurilemma (noo-ri-LEM-mah) nerve sheath

neurohypophysis (noo-ro-hye-POF-i-sis) posterior pituitary gland

neuromuscular junction (noo-ro-MUS-kyoo-lar JUNK-shun) the point of contact between the nerve endings and muscle fibers

neuron (NOO-ron) nerve cell, including its processes (axons and dendrites)

neurotransmitter (noo-ro-trans-MIT-ter) chemicals by which neurons communicate

neutrophil (NOO-tro-fils) white blood cell that stains readily with neutral dyes

nodes of Ranvier (nodes of rahn-vee-AY) indentations that are found between adjacent Schwann cells

norepinephrine (nor-ep-i-NEF-rin) hormone secreted by adrenal medulla; released by sympathetic nervous system

nose (noze) respiratory organ

nuclear envelope (NOO-klee-ar EN-vel-ope) membrane that surrounds the cell nucleus

nucleolus (noo-KLEE-o-lus) critical to protein formation because it "programs" the formation of ribosomes in the nucleus

nucleoplasm (NOO-klee-o-plazm) a special type of cytoplasm found in the nucleus

nucleus (NOO-klee-us) spherical structure within a cell; a group of neuron cell bodies in the brain or spinal cord

nutrition (noo-TRI-shun) food, vitamins, and minerals that are ingested and assimilated into the body

old age (old age) see *senescence*

olecranon fossa (o-LEK-rah-non FOS-ah) a large depression on the posterior surface of the humerus

olecranon process (o-LEK-rah-non PROSS-es) the large bony process of the ulna; commonly referred to as the tip of the elbow

olfaction (ol-FAK-shun) sense of smell

olfactory receptor (ol-FAK-tor-ee ree-SEP-tor) chemical receptors responsible for the sense of smell; located in the epithelial tissue in the upper part of the nasal cavity

oligodendroglia (ol-i-go-den-DROG-lee-ah) holds nerve fibers together and, more important, produces the myelin sheath around axons in the CNS

oliguria (ol-i-GOO-ree-ah) scanty amounts of urine

oocyte (O-o-site) immature stage of the female sex cell

oogenesis (o-o-JEN-e-sis) production of female gametes

oophorectomy (o-off-o-REK-toe-mee) surgical procedure to remove the ovaries

opposition (op-o-ZISH-un) moving the thumb to touch the tips of the fingers; the movement used to hold a pencil to write

optic disc (OP-tic disk) the area in the retina where the optic nerve fibers exit and there are no rods or cones; also known as a *blind spot*

oral cavity (OR-al KAV-i-tee) mouth

orbicularis oculi (or-bik-yoo-LAIR-is OK-yoo-lie) facial muscle that causes a squint

orbicularis oris (or-bik-yoo-LAIR-is O-ris) facial muscle that puckers the lips

organ (OR-gan) group of several tissue types that performs a special function

organelle (or-gah-NELL) cell organ; for example, the ribosome

organism (OR-gah-nizm) an individual living thing

organ of Corti (OR-gan of KOR-tee) the organ of hearing located in the cochlea and filled with endolymph

organogenesis (or-ga-no-JEN-e-sis) formation of organs from the primary germ layers of the embryo

origin (OR-i-jin) the attachment of a muscle to the bone that does not move when contraction occurs, as distinguished from insertion

oropharynx (o-ro-FAIR-inks) the portion of the pharynx that is located behind the mouth

osmosis (os-MO-sis) movement of a fluid through a semipermeable membrane

ossicles (OS-si-kls) little bones; found in the ears

osteoblast (OS-tee-o-blast) bone-forming cell

osteoclast (OS-tee-o-klast) bone-absorbing cell

osteocyte (OS-tee-o-site) bone cell

osteoporosis (os-tee-o-po-RO-sis) a bone disease where there is an excessive loss of calcified matrix and collagenous fibers from bone

otitis media (o-TIE-tis MEE-dee-ah) a middle ear infection

ova (O-vah) female sex cells (singular ovum)

oval window (O-val WIN-doe) a small, membrane-covered opening that separates the middle and inner ear

ovarian follicles (o-VARE-ee-an FOL-i-kuls) contain oocytes

ovaries (O-var-ees) female gonads that produce ova (sex cells)

overhydration (o-ver-hye-DRAY-shun) too large a fluid input that can put a burden on the heart

oviducts (O-vi-dukts) uterine or fallopian tubes

oxygen debt (OK-si-jen det) continued increased metabolism that occurs in a cell to remove excess lactic acid that resulted from exercise

oxyhemoglobin (ok-see-hee-mo-GLO-bin) hemoglobin combined with oxygen

oxytocin (ok-se-TOE-sin) hormone secreted by the posterior pituitary gland before and after delivering a baby; thought to initiate and maintain labor and it causes the release of breast milk into ducts for the baby to suck

pacemaker (PASE-may-ker) see *sinoatrial node*

pacinian corpuscle (pah-SIN-ee-an KOR-pus-ul) a receptor found deep in the dermis that detects pressure on the skin surface

palate (PAL-let) the roof of the mouth; made up of the hard (anterior portion of the mouth) and soft (posterior portion of the mouth) palates

palmar (PAHL-mar) palm of the hand

pancreas (PAN-kree-as) endocrine gland located in the abdominal cavity; contains pancreatic islets that secrete glucagon and insulin

pancreatic islets (pan-kree-AT-ik eye-LETS) endocrine portion of the pancreas; made up of alpha and beta cells among others

papillae (pah-PIL-ee) small, nipple-shaped elevations

paralysis (pah-RAL-i-sis) loss of the power of motion, especially voluntary motion

paranasal sinus (pair-ah-NAY-sal SYE-nus) four pairs of sinuses that have openings into the nose

parasympathetic nervous system (PNS) (par-ah-sim-pah-THE-tic NER-vus SIS-tem) part of the autonomic nervous system; ganglia are connected to the brain stem and the sacral segments of the spinal cord; controls many visceral effectors under normal conditions

parathyroid glands (pair-ah-THYE-royd glands) endocrine glands located in the neck on the posterior aspect of the thyroid gland; secrete parathyroid hormone

parathyroid hormone (PTH) (pair-ah-THYE-royd HOR-mone) hormone released by the parathryroid gland that increases the concentration of calcium in the blood

parietal (pah-RYE-i-tal) of the walls of an organ or cavity

parietal pericardium (pah-RYE-i-tal pair-i-KAR-dee-um) pericardium surrounding the heart like a loose-fitting sack to allow the heart enough room to beat

parietal portion (pah-RYE-i-tal POR-shun) serous membrane that lines the walls of a body cavity

parturition (par-too-RISH-un) act of giving birth

patella (pah-TEL-ah) small, shallow pan; the kneecap

pectoral girdle (PEK-toe-ral GIR-dul) shoulder girdle; the scapula and clavicle

pectoralis major (pek-tor-RAL-is MAY-jor) major flexor of the upper arm

pedal (PEED-al) foot

pelvic cavity (PEL-vik KAV-i-tee) the lower portion of the ventral cavity; the distal portion of the abdominopelvic cavity

pelvic girdle (PEL-vik GIR-dul) connects the legs to the trunk

pelvis (PEL-vis) basin or funnel-shaped structure

penis (PEE-nis) forms part of the male genitalia; when sexually aroused, becomes stiff to enable it to enter and deposit sperm in the vagina

pepsinogen (pep-SIN-o-jen) component of gastric juice that is converted into pepsin by hydrochloric acid

pericarditis (pair-i-kar-DYE-tis) when the pericardium becomes inflamed

pericardium (pair-i-KAR-dee-um) membrane that surrounds the heart

perilymph (PAIR-i-limf) a watery fluid that fills the bony labyrinth of the ear

perineal (pair-i-NEE-al) area between the anus and genitals; the perineum

perineum (pair-i-NEE-um) see *perineal*

periosteum (pair-i-OS-tee-um) tough, connective tissue covering the bone

peripheral (pe-RIF-er-al) pertaining to an outside surface

peripheral nervous system (PNS) (pe-RIF-er-al NER-vus SIS-tem)   the nerves connecting the brain and spinal cord to other parts of the body

peristalsis (pair-i-STAL-sis)   wavelike, rhythmic contractions of the stomach and intestines that move food material along the digestive tract

peritoneal space (pair-i-toe-NEE-al space)   small, fluid-filled space between the visceral and parietal layers that allows the layers to slide over each other freely in the abdominopelvic cavity

peritoneum (pair-i-toe-NEE-um)   large, moist, slippery sheet of serous membrane that lines the abdominopelvic cavity (parietal layer) and its organs (visceral layer)

peritonitis (pair-i-toe-NYE-tis)   inflammation of the serous membranes in the abdominopelvic cavity; sometimes a serious complication of an infected appendix

permeable membrane (PER-mee-ah-bul MEM-brane)   a membrane that allows passage of substances

pernicious anemia (per-NISH-us ah-NEE-mee-ah)   deficiency of red blood cells due to a lack of vitamin B_{12}

peroneal muscles (per-o-NEE-al MUS-els)   plantar flexors and evertors of the foot; the peroneus longus forms a support arch for the foot

perspiration (per-spi-RAY-shun)   transparent, watery liquid released by glands in the skin that eliminates ammonia and uric acid and helps maintain body temperature; also known as *sweat*

phagocytes (FAG-o-sites)   white blood cells that engulf microbes and digest them

phagocytosis (fag-o-sye-TOE-sis)   ingestion and digestion of articles by a cell

phalanges (fah-LAN-jeez)   the bones that make up the fingers and toes

pharynx (FAIR-inks)   organ of the digestive and respiratory system; commonly called the *throat*

phospholipid (fos-fo-LIP-id)   phosphate-containing fat molecule

photopigments (fo-toe-PIG-ments)   chemicals in retinal cells that are sensitive to light

phrenic nerve (FREN-ik nerv)   the nerve that stimulates the diaphragm to contract

physiology (fiz-ee-OL-o-jee)   the study of body function

pia mater (PEE-ah MAH-ter)   the vascular innermost covering (meninx) of the brain and spinal cord

pigment layer (PIG-ment LAY-er)   the layer of the epidermis that contains the melanocytes that produce melanin to give skin its color

pineal gland (PI-nee-al gland)   endocrine gland located in the third ventricle of the brain; produces melatonin

pinocytosis (pin-o-sye-TOE-sis)   the active transport mechanism used to transfer fluids or dissolved substances into cells

pituitary gland (pi-TOO-i-tair-ee gland)   endocrine gland located in the skull, made up of the adenohypophysis and the neurohypophysis

placenta (plah-SEN-tah)   anchors the developing fetus to the uterus and provides a "bridge" for the exchange of nutrients and waste products between the mother and developing baby

plantar (PLAN-tar)   pertaining to the sole of the foot

plantar flexion (PLAN-tar FLEK-shun)   the bottom of the foot is directed downward; this motion allows a person to stand on his or her tip-toes

plasma (PLAZ-mah)   the liquid part of the blood

plasma cells (PLAZ-mah sells)   cells that secrete copious amounts of antibody into the blood

plasma membrane (PLAZ-mah MEM-brane)   membrane that separates the contents of a cell from the tissue fluid; encloses the cytoplasm and forms the outer boundary of the cell

plasma protein (PLAZ-mah PRO-teen)   any of several proteins normally found in the plasma; includes albumins, globulins, and fibrinogen

platelet plug (PLAYT-let plug)   a temporary accumulation of platelets (thrombocytes) at the site of an injury; it precedes the formation of a blood clot

pleura (PLOOR-ah)   the serous membrane in the thoracic cavity

pleural cavity (PLOOR-al KAV-i-tee)   a subdivision of the thorax

pleural space (PLOOR-al space)   the space between the visceral and parietal pleuras filled with just enough fluid to allow them to glide effortlessly with each breath

pleurisy (PLOOR-i-see)   inflammation of the pleura

plica (PLYE-kah)   multiple circular folds

pneumocystitis (noo-mo-sis-TYE-tis)   a protozoan infection, most likely to invade the body when the immune system has been compromised

pneumothorax (noo-mo-THO-raks)   accumulation of air in the pleural space, causing collapse of the lung

polycythemia (pol-ee-sye-THEE-mee-ah)   an excessive number of red blood cells

polyuria (pol-ee-YOO-ree-ah)   unusually large amounts of urine

pons (ponz)   the part of the brain stem between the medulla oblongata and the midbrain

popliteal (pop-li-TEE-al)   behind the knee

pore (pore)   pinpoint-size openings on the skin that are outlets of small ducts from the eccrine sweat glands

posterior (pos-TEER-ee-or)   located behind; opposite of anterior

posterior pituitary gland (pos-TEER-ee-or pi-TOO-i-tair-ee gland)   neurohypophysis; hormones produced are ADH and oxytocin

posterior root ganglion (pos-TEER-ee-or GANG-lee-on)   ganglion located near the spinal cord; where the neuron cell body of the dendrites of the sensory neuron is located

postganglionic neurons (post-gang-glee-ON-ik NOO-rons)   autonomic neurons that conduct nerve impulses from a ganglion to cardiac or smooth muscle or glandular epithelial tissue

postnatal period (POST-nay-tal PEER-ee-od)   the period after birth and ending at death

postsynaptic neuron (post-si-NAP-tik NOO-ron)   a neuron situated distal to a synapse

posture (POS-chur)   position of the body

precapillary sphincter (pree-CAP-pi-lair-ee SFINGK-ter)   smooth muscle cells that guard the entrance to the capillary

preganglionic neurons (pree-gang-glee-ON-ik NOO-rons)   autonomic neurons that conduct nerve impulses between the spinal cord and a ganglion

prenatal period (PREE-nay-tal PEER-i-od) the period after conception until birth

presbyopia (pres-bee-O-pee-ah) farsightedness of old age

presynaptic neuron (pree-si-NAP-tik NOO-ron) a neuron situated proximal to a synapse

primary follicles (PRYE-mare-ee FOL-i-kuls) the follicles present at puberty; covered with granulosa cells

primary germ layers (PRYE-mar-ee jerm LAY-ers) three layers of specialized cells that give rise to definite structures as the embryo develops

primary spermatocyte (PRY-mar-ee SPER-mah-toe-site) specialized cell that undergoes meiosis to ultimately form sperm

prime mover (prime MOO-ver) the muscle responsible for producing a particular movement

progesterone (pro-JES-ter-ohn) hormone produced by the corpus luteum; stimulates secretion of the uterine lining; with estrogen, helps to initiate the menstrual cycle in girls entering puberty

prolactin (pro-LAK-tin) hormone secreted by the anterior pituitary gland during pregnancy to stimulate the breast development needed for lactation

pronate (PRO-nate) to turn the palm downward

prone (prone) used to describe the body lying in a horizontal position facing downward

prophase (PRO-faze) first stage of mitosis during which chromosomes become visible

proprioceptors (pro-pree-o-SEP-tors) receptors located in the muscles, tendons, and joints; allows the body to recognize its position

prostaglandins (PGs) (pross-tah-GLAN-dins) a group of naturally occurring fatty acids that affect many body functions

prostatectomy (pross-tah-TEK-toe-mee) surgical removal of part or all of the prostate gland

prostate gland (PROSS-tate gland) lies just below the bladder; secretes a fluid that constitutes about 30% of the seminal fluid volume; helps activate sperm and helps them maintain motility

protease (PRO-tee-ase) protein-digesting enzyme

protein (PRO-teen) one of the basic nutrients needed by the body; usually involved with anabolism

protein hormone (PRO-teen HOR-mone) first messenger, provides communication between endocrine glands and target organs; triggers second messengers to affect the cell's activity

proteinuria (pro-teen-YOO-ree-ah) presence of abnormally high amounts of plasma protein in the urine; usually an indicator of kidney disease

prothrombin (pro-THROM-bin) a protein present in normal blood that is required for blood clotting

prothrombin activator (pro-THROM-bin AK-tiv-ayt-or) a protein formed by clotting factors from damaged tissue cells and platelets; it converts prothrombin into thrombin, a step essential to forming a blood clot

proximal (PROK-si-mal) next or nearest; located nearest the center of the body or the point of attachment of a structure

proximal convoluted tubule (PROK-si-mal kon-vo-LOO-ted TOOB-yool) the first segment of a renal tubule

pseudo (SOO-doe) false

pubis (PYOO-bis) joint in the midline between the two pubic bones

pulmonary artery (PUL-mo-nair-ee AR-ter-ee) artery that carries deoxygenated blood from the right ventricle to the lungs

pulmonary circulation (PUL-mo-nair-ee ser-kyoo-LAY-shun) venous blood flow from the right atrium to the lung and returning to the left atrium

pulmonary semilunar valve (PUL-mo-nair-ee sem-i-LOO-nar valv) valve located at the beginning of the pulmonary artery

pulmonary vein (PUL-mo-nair-ee vane) any vein that carries oxygenated blood from the lungs to the left atrium

pulmonary ventilation (PUL-mo-nair-ee ven-ti-LAY-shun) breathing; process that moves air in and out of the lungs

pupil (PYOO-pil) the opening in the center of the iris that regulates the amount of light entering the eye

purkinje fibers (pur-KIN-jee FYE-bers) specialized cells located in the walls of the ventricles; relay nerve impulses from the AV node to the ventricles causing them to contract

P wave (P wave) deflection on an ECG that occurs with depolarization of the atria

pyloric sphincter (pye-LOR-ik SFINGK-ter) sphincter that prevents food from leaving the stomach and entering the duodenum

pylorus (pye-LOR-us) the small narrow section of the stomach that joins the first part of the small intestine

pyramids (PEER-ah-mids) triangular-shaped divisions of the medulla of the kidney

QRS complex (QRS KOM-pleks) deflection on an ECG that occurs as a result of depolarization of the ventricles

quadriceps femoris (KWOD-re-seps fe-MOR-is) extensor of the lower leg

quickening (KWIK-en-ing) when a pregnant woman first feels recognizable movements of the fetus

radiation (ray-dee-AY-shun) flow of heat waves away from the blood

radius (RAY-dee-us) one of the two bones in the forearm; located on the thumb side of the forearm

reabsorption (ree-ab-SORP-shun) process of absorbing again that occurs in the kidneys

receiving chambers (ree-SEE-ving CHAM-bers) atria of the heart; receive blood from the superior and inferior vena cava

receptor (ree-SEP-tor) peripheral beginning of a sensory neuron's dendrite

rectum (REK-tum) distal portion of the large intestine

rectus abdominis (REK-tus ab-DOM-i-nis) muscle that runs down the middle of the abdomen; protects the abdominal viscera and flexes the spinal column

reflex (REE-fleks) involuntary action

reflex arc (REE-fleks ark) allows an impulse to travel in only one direction

reflux (REE-fluhks) back flow, as in flow of stomach contents back into esophagus

refraction (ree-FRAK-shun) bending of a ray of light as it passes from a medium of one density to one of a different density

releasing hormones (ree-LEE-sing HOR-mones) hormone produced by the hypothalamus gland that causes the anterior pituitary gland to release its hormones

renal calculi (REE-nal KAL-kyoo-lie) kidney stones

renal colic (REE-nal KOL-ik)   pain caused by the passage of a kidney stone

renal corpuscle (REE-nal KOR-pus-ul) the part of the nephron located in the cortex of the kidney

renal pelvis (REE-nal PEL-vis)   basinlike upper end of the ureter that is located inside the kidney

renal tubule (REE-nal TOOB-yool)   one of the two principal parts of the nephron

repolarization (ree-po-lah-ri-ZAY-shun)   begins just before the relaxation phase of cardiac muscle activity

reproductive system (ree-pro-DUK-tiv SIS-tem)   produces hormones that permit the development of sexual characteristics and the propagation of the species

residual volume (RV) (re-ZID-yoo-al VOL-yoom)   the air that remains in the lungs after the most forceful expiration

respiratory acidosis (RES-pi-rah-tor-ee as-i-DOE-sis)   a respiratory disturbance that results in a carbonic acid excess

respiratory alkalosis (RES-pi-rah-tor-ee al-kah-LO-sis)   a respiratory disturbance that results in a carbonic acid deficit

respiratory arrest (RES-pi-rah-tor-ee ah-REST)   cessation of breathing without resumption

respiratory control centers (RES-pi-rah-tor-ee kon-TROL SEN-ters)   centers located in the medulla and pons that stimulate the muscles of respiration

respiratory membrane (RES-pi-rah-tor-ee MEM-brane)   the single layer of cells that makes up the wall of the alveoli

respiratory mucosa (RES-pi-rah-tor-ee myoo-KO-sah)   mucus-covered membrane that lines the tubes of the respiratory tree

respiratory muscles (RES-pi-rah-tor-ee MUS-els)   muscles that are responsible for the changing shape of the thoracic cavity that allows air to move in and out of the lungs

respiratory system (re-SPY-rah-tor-ee SIS-tem)   the organs that allow the exchange of oxygen from the air with the carbon dioxide from the blood

respiratory tract (RES-pi-rah-tor-ee trakt)   the two divisions of the respiratory system are the upper and lower respiratory tracts

reticular formation (re-TIK-yoo-lar for-MAY-shun)   located in the medulla where bits of gray and white matter mix intricately

retina (RET-i-nah)   innermost layer of the eyeball; contains rods and cones and continues posteriorly with the optic nerve

retroperitoneal (re-tro-pair-i-toe-NEE-al)   area outside of the peritoneum

Rh-negative (R H NEG-ah-tiv)   red blood cells that do not contain the antigen called *Rh factor*

RhoGAM (RO-gam)   an injection of a special protein given to an Rh-negative woman who is pregnant to prevent her body from forming anti-Rh antibodies, which may harm an Rh-positive baby

Rh-positive (R H POZ-i-tiv)   red blood cells that contain an antigen called *Rh factor*

ribonucleic acid (RNA) (rye-bo-noo-KLEE-ik AS-id)   a nucleic acid found in the cytoplasm that is crucial to protein synthesis

ribosome (RYE-bo-sohm)   organelle in the cytoplasm of cells that synthesizes proteins; also known as a protein factory

rods (rods)   receptors located in the retina that are responsible for night vision

rotation (ro-TAY-shun)   movement around a longitudinal axis; for example, shaking your head "no"

rugae (ROO-gee)   wrinkles or folds; singular: ruga (ROO-gah)

"rule of nines"   a frequently used method to determine the extent of a burn injury; the body is divided into 11 areas of 9% each to help estimate the amount of skin surface burned in an adult

sagittal (SAJ-i-tal)   longitudinal; like an arrow

salivary amylase (SAL-i-vair-ee AM-i-lase)   digestive enzyme found in the saliva that begins the chemical digestion of carbohydrates

saltatory conduction (SAL-tah-tor-ee kon-DUK-shun)   when a nerve impulse encounters myelin and "jumps" from one node of Ranvier to the next

sarcomere (SAR-ko-meer)   contractile unit of muscle; length of a myofibril between two Z bands

scapula (SKAP-yoo-lah)   shoulder blade

Schwann cells (shwon sells)   large nucleated cells that form myelin

sclera (SKLE-rah)   white outer coat of the eyeball

scrotum (SKRO-tum)   pouchlike sac that contains the testes

sebaceous gland (se-BAY-shus gland)   oil-producing glands found in the skin

sebum (SEE-bum)   secretion of sebaceous glands

secondary sexual characteristics (SEK-on-dair-ee SEK-shoo-al kair-ak-ter-IS-tiks)   sexual characteristics that appear at the onset of puberty

second messenger (SEK-und MES-en-jer)   provide communication within a hormone's target cell; for example, cyclic AMP

sella turcica (SEL-lah TER-si-kah)   small depression of the sphenoid bone that contains the pituitary gland

semen (SEE-men)   male reproductive fluid

semicircular canals (sem-i-SIR-kyoo-lar kah-NALS)   located in the inner ear; contains a specialized receptor called *crista ampullaris* that generates a nerve impulse on movement of the head

semilunar valves (sem-i-LOO-nar valvs)   valves located between the two ventricular chambers and the large arteries that carry blood away from the heart; valves found in the veins

seminal fluid (SEM-i-nal FLOO-id)   semen

seminal vesicle (SEM-i-nal VES-i-kul)   paired, pouchlike glands that contribute about 60% of the seminal fluid volume; rich in fructose, which is a source of energy for sperm

seminiferous tubule (se-mi-NIF-er-us TOOB-yool)   long, coiled structure that forms the bulk of the testicular mass

senescence (se-NES-enz)   older adulthood; aging

sensory neurons (SEN-sor-ee NOO-rons)   neurons that transmit impulses to the spinal cord and brain from all parts of the body

serosa (se-RO-sah)   outermost covering of the digestive tract; composed of the parietal pleura in the abdominal cavity

serotonin (sair-o-TOE-nin)   a neurotransmitter that belongs to a group of compounds called *catecholamines*

serous membrane (SE-rus MEM-brane)   a two-layered epithelial membrane that lines body cavities and covers the surfaces of organs

serum (SEER-um)   blood plasma minus its clotting factors, still contains antibodies

shingles (SHING-guls)   see *herpes zoster*

sickle cell anemia (SIK-ul sell ah-NEE-mee-ah)   severe, possibly fatal, hereditary disease caused by an abnormal type of hemoglobin

sickle cell trait (SIK-ul sell trate)   when only one defective gene is inherited and only a small amount of hemoglobin that is less soluble than usual is produced

sigmoid colon (SIG-moyd KO-lon)   S-shaped segment of the large intestine that terminates in the rectum

sinoatrial (SA) node (sye-no-AY-tree-al node)   the heart's pacemaker; where the impulse conduction of the heart normally starts; located in the wall of the right atrium near the opening of the superior vena cava

sinus (SYE-nus)   a space or cavity inside some of the cranial bones

sinusitis (sye-nyoo-SYE-tis)   sinus infections

skeletal muscle (SKEL-e-tal MUS-el)   also known as *voluntary muscle;* muscles under willed or voluntary control

skeletal system (SKEL-e-tal SIS-tem)   the bones, cartilage, and ligaments that provide the body with a rigid framework for support and protection

smooth muscle (smooth MUS-el)   muscles that are not under conscious control; also known as *involuntary* or *visceral;* forms the walls of blood vessels and hollow organs

sodium-potassium pump (SO-dee-um po-TAS-ee-um pump)   a system of coupled ion pumps that actively transports sodium ions out of a cell and potassium ions into the cell at the same time—found in all living cells

solute (SOL-yoot)   dissolved particles in water

somatic nervous system (so-MA-tik NER-vus SIS-tem)   the motor neurons that control the voluntary actions of skeletal muscles

specific immunity (spe-SI-fik i-MYOON-i-tee)   the protective mechanisms that provide specific protection against certain types of bacteria or toxins

sperm (sperm)   the male spermatazoon; sex cell

spermatids (SPER-mah-tids)   the resulting daughter cells from the primary spermatocyte undergoing meiosis; these cells have only half the genetic material and half the chromosomes of other body cells

spermatogenesis (sper-mah-toe-JEN-e-sis)   the production of sperm cells

spermatogonia (sper-mah-toe-GO-nee-ah)   sperm precursor cells

spermatozoa (sper-mah-tah-ZO-ah)   sperm cells (singular: spermatozoon)

sphincter (SFINGK-ter)   ring-shaped muscle

spinal cavity (SPY-nal KAV-i-tee)   the space inside the spinal column through which the spinal cord passes

spinal nerves (SPY-nal nervs)   nerves that connect the spinal cord to peripheral structures such as the skin and skeletal muscles

spinal tracts (SPY-nal tracts)   the white columns of the spinal cord that provide two-way conduction paths to and from the brain; ascending tract carries information to the brain, whereas descending tracts conduct impulses from the brain

spindle fiber (SPIN-dul FYE-ber)   a network of tubules formed in the cytoplasm between the centrioles as they are moving away from each other

spirometer (spi-ROM-e-ter)   an instrument used to measure the amount of air exchanged in breathing

spleen (spleen)   largest lymphoid organ; filters blood, destroys worn out red blood cells, salvages iron from hemoglobin, and serves as a blood resevoir

splenectomy (splen-NEK-toe-mee)   surgical removal of the spleen

splenic flexure (SPLEEN-ik FLEK-shur)   where the descending colon turns downward on the left side of the abdomen

spongy bone (SPUN-jee bone)   porous bone in the end of the long bone, may be filled with marrow

squamous (SKWAY-mus)   scalelike

squamous suture (SKWAY-mus SOO-chur)   the immovable joint between the temporal bone and the sphenoid bone

stapes (STAY-peez)   tiny, stirrup-shaped bone in the middle ear

Stensen's ducts (STEN-sens dukts)   the ducts of the parotid gland as they enter the mouth

sternoclavicular joint (ster-no-klah-VIK-yoo-lar joynt)   the direct point of attachment between the bones of the upper extremity and the axial skeleton

sternocleidomastoid (stern-o-klye-doe-MAS-toyd)   "strap" muscle located on the anterior aspect of the neck

steroid hormones (STE-royd HOR-mones)   lipid-soluble hormones that pass intact through the cell membrane of the target cell and influence cell activity by acting on specific genes

stimulus (STIM-yoo-lus)   agent that causes a change in the activity of a structure

stoma (STO-mah)   an opening, such as the opening created in a colostomy procedure

stomach (STUM-ak)   an expansion of the digestive tract between the esophagus and small intestine

stratum corneum (STRA-tum KOR-nee-um)   the tough outer layer of the epidermis; cells are filled with keratin

stratum germinativum (STRA-tum JER-mi-nah-tiv-um)   the innermost of the tightly packed epithelial cells of the epidermis; cells in this layer are able to reproduce themselves

strength training (strength TRAIN-ing)   contracting muscles against resistance to enhance muscle hypertrophy

striated muscle (STRYE-ay-ted MUS-el)   see *skeletal muscle*

stroke volume (stroke VOL-yoom)   the amount of blood that is ejected from the ventricles of the heart with each beat

subcutaneous tissue (sub-kyoo-TAY-nee-us TISH-yoo)   tissue below the layers of skin; made up of loose connective tissue and fat

submucosa (sub-myoo-KO-sah)   connective tissue layer containing blood vessels and nerves in the wall of the digestive tract

sudoriferous gland (soo-doe-RIF-er-us gland)   glands that secrete sweat; also referred as *sweat glands*

sulcus (SUL-kus)   furrow or groove

superficial (soo-per-FISH-al)   near the body surface

superior (soo-PEER-ee-or)   higher, opposite of inferior

superior vena cava (soo-PEER-ee-or VEE-nah KAY-vah)   one of two large veins returning deoxygenated blood to the right atrium

supinate (SOO-pi-nate)   to turn the palm of the hand upward; opposite of pronate

supine (SOO-pine)   used to describe the body lying in a horizontal position facing upward

supraclavicular (soo-prah-cla-VIK-yoo-lar) area above the clavicle

surfactant (sur-FAK-tant) a substance covering the surface of the respiratory membrane inside the alveolus, which reduces surface tension and prevents the alveoli from collapsing

suture (SOO-chur) immovable joint

sweat (swet) transparent, watery liquid released by glands in the skin that eliminates ammonia and uric acid and helps maintain body temperature; also known as *perspiration*

sympathetic nervous system (sim-pah-THE-tik NER-vus SIS-tem) part of the autonomic nervous system; ganglia are connected to the thoracic and lumbar regions of the spinal cord; functions as an emergency system

sympathetic postganglionic neurons (sim-pah-THE-tik post-gang-glee-ON-ik NOO-rons) dendrites and cell bodies are in sympathetic ganglia and axons travel to a variety of visceral effectors

sympathetic preganglionic neurons (sim-pah-THE-tik pree-gang-glee-ON-ik NOO-rons) dendrites and cell bodies are located in the gray matter of the thoracic and lumbar segments of the spinal cord; leaves the cord through an anterior root of a spinal nerve and terminates in a collateral ganglion

synapse (SIN-aps) junction between adjacent neurons

synaptic cleft (si-NAP-tik kleft) the space between a synaptic knob and the plasma membrane of a postsynaptic neuron

synaptic knob (si-NAP-tik nob) a tiny bulge at the end of a terminal branch of a presynaptic neuron's axon that contains vesicles with neurotransmitters

synarthrosis (sin-ar-THRO-sis) a joint in which fibrous connective tissue joins bones and holds them together tightly; commonly called *sutures*

synergist (SIN-er-jists) muscle that assists a prime mover

synovial fluid (si-NO-vee-al FLOO-id) the thick, colorless lubricating fluid secreted by the synovial membrane

synovial membrane (si-NO-vee-al MEM-brane) connective tissue membrane lining the spaces between bones and joints that secretes synovial fluid

system (SIS-tem) group of organs arranged so that the group can perform a more complex function than any one organ can perform alone

systemic circulation (sis-TEM-ik ser-kyoo-LAY-shun) blood flow from the left ventricle to all parts of the body and back to the right atrium

systole (SIS-toe-lee) contraction of the heart muscle

target organ cell (TAR-get OR-gan sell) organ or cell acted on by a particular hormone and responding to it

tarsals (TAR-sals) seven bones of the heel and back part of the foot; the calcaneus is the largest

taste buds (taste buds) chemical receptors that generate nerve impulses, resulting in the sense of taste

telophase (TEL-o-faze) last stage of mitosis in which the cell divides

temporal (TEM-po-ral) muscle that assists the masseter in closing the jaw

tendons (TEN-dons) bands or cords of fibrous connective tissue that attach a muscle to a bone or other structure

tendon sheath (TEN-don sheeth) tube-shaped structure lined with synovial membrane that encloses certain tendons

tenosynovitis (ten-o-sin-o-VYE-tis) inflammation of a tendon sheath

testes (TES-teez) male gonads that produce the male sex cells or sperm

testosterone (tes-TOS-te-rone) male sex hormone produced by the interstitial cells in the testes; the "masculinizing hormone"

tetanic contraction (te-TAN-ik kon-TRAK-shun) sustained contraction

tetanus (TET-ah-nus) sustained muscular contraction

thalamus (THAL-ah-mus) located just above the hypothalamus; its functions are to help produce sensations, associates sensations with emotions and plays a part in the arousal mechanism

thermoregulation (ther-mo-reg-yoo-LAY-shun) maintaining homeostasis of body temperature

thoracic duct (thor-AS-ik dukt) largest lymphatic vessel in the body

thorax (THOR-aks) chest

threshold stimulus (THRESH-hold STIM-yoo-lus) minimal level of stimulation required to cause a muscle fiber to contract

thrombin (THROM-bin) protein important in blood clotting

thrombocytes (THROM-bo-sites) also called *platelets*; play a role in blood clotting

thrombosis (throm-BO-sis) formation of a clot in a blood vessel

thrombus (THROM-bus) stationary blood clot

thymosin (THY-mo-sin) hormone produced by the thymus that is vital to the development and functioning of the body's immune system

thymus gland (THY-mus gland) endocrine gland located in the mediastinum; vital part of the body's immune system

thyroid gland (THY-royd gland) endocrine gland located in the neck that stores its hormones until needed; thyroid hormones regulate cellular metabolism

thyroid-stimulating hormone (TSH) (THY-royd STIM-yoo-lay-ting HOR-mone) a tropic hormone secreted by the anterior pituitary gland that stimulates the thyroid gland to increase its secretion of thyroid hormone

thyroxine (T_4) (thy-ROK-sin) thyroid hormone that stimulates cellular metabolism

tibia (TIB-ee-ah) shinbone

tibialis anterior (tib-ee-AL-is an-TEER-ee-or) dorsiflexor of the foot

tidal volume (TV) (TIE-dal VOL-yoom) amount of air breathed in and out with each breath

tissue (TISH-yoo) group of similar cells that perform a common function

tissue fluid (TISH-yoo FLOO-id) a dilute salt water solution that bathes every cell in the body

tissue hormone (TISH-yoo HOR-mone) prostaglandins; produced in a tissue and only diffuses a short distance to act on cells within the tissue

tissue typing (TISH-yoo TIE-ping) a procedure used to identify tissue compatability before an organ transplant

T-lymphocytes (T LIM-fo-sites) cells that are critical to the function of the immune system; produce cell-mediated immunity

tonic contraction (TON-ik kon-TRAK-shun) special type of skeletal muscle contraction used to maintain posture

tonsillectomy (ton-si-LEK-toe-mee) surgical procedure used to remove the tonsils

tonsillitis (ton-si-LIE-tis)   an inflammation of the tonsils

tonsils (TON-sils)   masses of lymphoid tissue; protect against bacteria; three types: palatine tonsils, located on each side of the throat; pharyngeal tonsils (adenoids), near the posterior opening of the nasal cavity; and lingual tonsils, near the base of the tongue

total metabolic rate (TMR) (TOE-tal met-ah-BOL-ik rate)   total amount of energy used by the body per day

trabeculae (trah-BEK-yoo-lee)   needlelike threads of spongy bone that surround a network of spaces

trachea (TRAY-kee-ah)   the windpipe; the tube extending from the larynx to the bronchi

transcription (trans-KRIP-shun)   when the double stranded DNA molecules unwind and form mRNA

translation (trans-LAY-shun)   the synthesis of a protein by ribosomes

transverse arch (TRANS-vers arch)   see *metatarsal arch*

transversus abdominis (trans-VER-sus ab-DOM-i-nis)   the innermost layer of the anterolateral abdominal wall

trapezium (trah-PEE-zee-um)   the carpal bone of the wrist that forms the saddle joint that allows the opposition of the thumb

trapezius (trah-PEE-zee-us)   triangular muscle in the back that elevates the shoulder and extends the head backwards

triceps brachii (TRY-seps BRAY-kee-eye)   extensor of the elbow

tricuspid valve (try-KUS-pid valv)   the valve located between the right atrium and ventricle

trigone (TRY-gon)   triangular area on the wall of the urinary bladder

triiodothyonine (T₃) (try-eye-o-doe-THY-ro-nine)   thyroid hormone that stimulates cellular metabolism

tropic hormone (TRO-pik HOR-mone)   hormone that stimulates another endocrine gland to grow and secrete its hormones

true ribs (troo ribs)   the first seven pairs of ribs that are attached to the sternum

tumor (TOO-mer)   growth of tissues in which cell proliferation is uncontrolled and progressive

tunica adventitia (TOO-ni-kah ad-ven-TISH-ah)   the outermost layer found in blood vessels

tunica albuginea (TOO-ni-kah al-byoo-JIN-ee-ah)   a tough, whitish membrane that surrounds each testis and enters the gland to divide it into lobules

tunica intima (TOO-ni-kah IN-tih-mah)   endothelium that lines the blood vessels

tunica media (TOO-ni-kah MEE-dee-ah)   the muscular middle layer found in blood vessels; the tunica media of arteries is more muscular than that of veins

T wave (T wave)   deflection on an electrocardiogram that occurs with repolarization of the ventricles

twitch (twitch)   a quick, jerky response to a single stimulus

tympanic (tim-PAN-ik)   drumlike

ulcer (UL-ser)   a necrotic open sore or lesion

ulna (UL-nah)   one of the two forearm bones; located on the little finger side

ultrasonogram (ul-tra-SOHN-o-gram)   a technique using sound to produce images

umbilical artery (um-BIL-i-kul AR-ter-ee)   two small arteries that carry oxygen-poor blood from the developing fetus to the placenta

umbilical cord (um-BIL-i-kul cord)   flexible structure connecting the fetus with the placenta, which allows the umbilical arteries and vein to pass

umbilical vein (um-BIL-i-kul vane)   a large vein carrying oxygen-rich blood from the placenta to the developing fetus

urea (yoo-REE-ah)   nitrogen-containing waste product

uremia (yoo-REE-mee-ah)   high levels of nitrogen-containing waste products in the blood; also referred to as *uremic poisoning*

uremic poisoning (yoo-REE-mik POY-zon-ing)   see *uremia*

urethra (yoo-REE-thrah)   passageway for elimination of urine; in males, also acts as a genital duct that carries sperm to the exterior

urinary meatus (YOOR-i-nair-ee mee-AY-tus)   external opening of the urethra

urinary system (YOOR-i-nair-ee SIS-tem)   system responsible for excreting liquid waste from the body

urination (yoor-i-NAY-shun)   passage of urine from the body; emptying of the bladder

urine (YOOR-in)   fluid waste excreted by the kidneys

uterus (YOO-ter-us)   hollow, muscular organ where a fertilized egg implants and grows

uvula (YOO-vyoo-lah)   cone-shaped process hanging down from the soft palate that helps prevent food and liquid from entering the nasal cavities

vagina (vah-JYE-nah)   internal tube from uterus to vulva

vas deferens (vas DEF-er-enz)   see *ductus deferens*

vastus (VAS-tus)   wide; of great size

vein (vane)   vessel carrying blood toward the heart

ventral (VEN-tral)   of or near the belly; in humans, front or anterior; opposite of dorsal or posterior

ventricles (VEN-tri-kuls)   small cavities

venule (VEN-yool)   small blood vessels that collect blood from the capillaries and join to form veins

vermiform appendix (VERM-i-form ah-PEN-diks)   a tubular structure attached to the cecum composed of lymphatic tissue

vertebrae (VER-te-bray)   bones that make up the spinal column

vertebral column (ver-TEE-bral KOL-um)   the spinal column, made up of a series of separate vertebrae that form a flexible, curved rod

vestibular nerve (ves-TIB-yoo-lar nerv)   a division of the vestibulocochlear nerve (the eighth cranial nerve)

vestibule (VES-ti-byool)   located in the inner ear; the portion adjacent to the oval window between the semicircular canals and the cochlea

villi (VIL-eye)   fingerlike folds covering the plicae of the small intestines

visceral pericardium (VIS-er-al pair-i-KAR-dee-um)   the pericardium that covers the heart

visceral portion (VIS-er-al POR-shun)   serous membrane that covers the surface of organs found in the body cavity

vital capacity (VC) (VYE-tal kah-PAS-i-tee)   largest amount of air that can be moved in and out of the lungs in one inspiration and expiration

vitamins (VYE-tah-mins)   organic molecules needed in small quantities to help enzymes operate effectively

vitreous humor (VIT-ree-us HYOO-mor)   the jellylike fluid found in the eye, posterior to the lens

voiding (VOYD-ing)   emptying of the bladder

volar (VO-lar)   palm or sole

voluntary muscle (VOL-un-tair-ee MUS-el)   see *skeletal muscle*

vulva (VUL-vah)   external genitals of the female

white matter (wite MAT-er)   nerves covered with white myelin

withdrawl reflex (with-DRAW-al REE-fleks)   a reflex that moves a body part away from an irritating stimulus

yolk sac (yoke sak)   in humans, involved with the production of blood cells in the developing embryo

zona fasciculata (ZO-nah fas-sic-yoo-LAY-tah)   middle zone of the adrenal cortes that secretes glucocorticoids

zona glomerulosa (ZO-nah glo-mare-yoo-LO-sah)   outer zone of the adrenal cortex that secretes mineralocorticoids

zona reticularis (ZO-nah re-tik-yoo-LAIR-is)   inner zone of the adrenal cortex that secretes small amounts of sex hormones

zygomaticus (zye-go-MAT-ik-us)   muscle that elevates the corners of the mouth and lips; also known as the *smiling muscle*

zygote (ZYE-gote)   a fertilized ovum

Illustration/Photo Credits

Chapter 1: 1-1, 1-4, 1-8, 1-9, Rolin Graphics; 1-2, Terry Cockerham/Synapse Media Production; 1-3, 1-5, 1-6, Joan Beck; 1-7, Cynthia Alexander Turner/Terry Cockerham/Synapse Media Production/Christine Oleksyk.

Chapter 2: 2-1, 2-2, William Ober; 2-3, Lennart Nilsson; 2-4, 2-6, Tonicity box, Table 2-2, Table 2-3, Rolin Graphics; 2-5, 2-8, Joan Beck; 2-9, 2-10, 2-11, 2-12, 2-14, 2-19, 2-20, Edward Reschke; 2-13, courtesy of S. Erlandsen (from *Color Atlas of Histology,* 1992, Mosby); 2-15, 2-16, 2-18, 2-21, Phototake; 2-17, 2-22, Robert Callentine.

Chapter 3: 3-2, 3-3, 3-4, 3-5, 3-6, 3-7, 3-8, 3-9, 3-10, 3-11, 3-12, 3-13, Joan Beck.

Chapter 4: 4-1A,B, 4-2, 4-6A,B, Rolin Graphics; 4-3, Edward Reschke; 4-4, 4-7A,B, Christine Oleksyk; 4-5, David Scharf/Peter Arnold, Inc.; 4-8, Joan Beck; Home Health Care box, courtesy of Potter/Perry (from *Fundamentals of Nursing,* 1993, Mosby).

Chapter 5: 5-1, Joan Beck; 5-2, Laurie O'Keefe/John Daugherty; 5-3, Phototake; 5-5, Stephen Oh; 5-6, Network Graphics; 5-7A,B, David J. Mascaro & Associates; 5-8, 5-9, 5-10, 5-13A,B, 5-14 (Drawing), 5-15A, 5-15C, 5-16, 5-19, Ernest W. Beck; 5-11, Ron Edwards; 5-12, 5-13C, 5-14 (Photo), 5-15B, 5-15D, courtesy of B. Vidic and F.R. Suarez (from *Photographic Atlas of the Human Body,* 1984, Mosby); 5-17, Yvonne Wylie Walston; 5-18, Christine Oleksyk; 5-20, David J. Mascaro & Associates; 5-21, The Knee Joint box (Drawing), Rolin Graphics; Epiphyseal Fracture box, J.M. Booher and G.A. Thibodeau: *Athletic Injury Assessment,* St. Louis, 1985, Mosby; Palpable Bony Landmarks box, Terry Cockerham/Synapse Media Production; The Knee Joint box (Photo), Stewart Halpernin.

Chapter 6: 6-1A-C, Christine Oleksyk; 6-2, Network Graphics; 6-3A, Barbara Cousins/Rolin Graphics; 6-3B, courtesy of Dr. H.E. Huxley; 6-4, courtesy of S. Erlandsen and J. Magney (from *Color Atlas of Histology,* 1992, Mosby); 6-5A,B, 6-9A, 6-9C, 6-10A, 6-10C, Rolin Graphics; 6-6A,B, 6-7, 6-8, Intramuscular Injections box, John V. Hagen; 6-9B, 6-10B, 6-11A, 6-11C, 6-11D, Terry Cockerham/Synapse Media Production; 6-11B, Patrick Watson.

Chapter 7: 7-2A,B, 7-13, 7-16, 7-20, Rolin Graphics; 7-2C, Edward Reschke; 7-4 (Drawing), 7-7, Laurie O'Keefe/John Daugherty; 7-4 (Photo), courtesy of H.L. McCance and S. Huether (from *Pathophysiology,* 1991, Mosby); 7-5, 7-15, Barbara Cousins; 7-6A, 7-8, 7-19, Joan Beck; 7-6B, 7-15 (inset), 7-17, Herpes Zoster box (art), Network Graphics; 7-3, Barbara Cousins; 7-6 (inset), 7-14, Scott Bodell; 7-9A, 7-10A,B, William Ober; 7-9B, 7-10 (inset), 7-11 (Photo), courtesy of B. Vidic and F.R. Suarez (from *Photographic Atlas of the Human Body,* 1984, Mosby); 7-11 (Drawing), George Wassilchenko; 7-12, Christine Oleksyk; 7-18, Raychel Ciemma; Herpes Zoster box (Photo), courtesy of Thomas P. Habif (from *Clinical Dermatology,* ed. 2, 1990, Mosby).

Chapter 8: 8-1, 8-3, Christine Oleksyk; 8-2, George Wassilchenko; 8-4, Ernest W. Beck; 8-5, Table 8-1, Rolin Graphics; 8-6, 8-7A-C, Focusing on Problems box A-D, Table 8-1 (Muscle Spindles), Table 8-1 (Golgi Tendon Receptors), Network Graphics; Color Blindness box, courtesy of Ishihara (from *Tests for Colour Blindness,* Tokyo, Japan, 1973, Kanehara Shuppan Co., Ltd. provided by Washington University Department of Ophthalmology), 8-8, Joan Beck.

Chapter 9: 9-1, Joan Beck; 9-2, 9-3, Network Graphics; 9-4, 9-8, Rolin Graphics; 9-5, Barbara Cousins; 9-6, 9-9, 9-11, Ernest W. Beck; 9-10, Graphic Works; Thyroid Hormone Abnormalities box (A), Seidel, et. al. (from *Mosby's Guide to Physical Examination,* ed. 3, 1995, Mosby); Thyroid Hormone Abnormalities box (B), L.V. Bergman & Associates, Inc., Cold Springs, New York; Adrenal Hormone Abnormalities box, courtesy of Gower Medical Publishers.

Chapter 10: 10-1, Barbara Cousins; 10-2, Sickle Cell Anemia box, courtesy of G. Bevelander and J.A. Ramaley (from *Essentials of Histology,* ed. 8, Mosby); 10-3, 10-7, Rolin Graphics; 10-4A-E, courtesy of S. Erlandsen and J. Magney (from *Color Atlas of Histology,* 1992, Mosby); 10-5, courtesy of A. Arlan Hinchee; 10-6, Laurie O'Keefe/John Daugherty; 10-8, Molly Babich/John Daugherty; Table 10-1, Ernest W. Beck.

Chapter 11: 11-1, 11-7A, Ernest W. Beck; 11-2, Rusty Jones; 11-2 (inset), 11-3A,B (bottom), George J. Wassilchenko; 11-3A,B (top), Christine Oleksyk; 11-4, 11-8, 11-9, Barbara Cousins; 11-5, Joan M. Beck; 11-6, Lisa Shoemaker/Joan M. Beck; 10-10, Karen Waldo; 11-11, Rolin Graphics; 11-12, 11-15, Network Graphics; 11-13, Molly Babich; Blood Pressure Readings box, Joan M. Beck/Donna Odle.

Chapter 12: 12-1, 12-4, Joan Beck; 12-2, G. David Brown; 12-5, George Wassilchenko; 12-6, 12-11, 12-12, Rolin Graphics; 12-8, 12-9, Network Graphics; 12-10, courtesy of Emma Shelton; 12-13, courtesy of Dr. James T. Barrett.

Chapter 13: 13-1, 13-2, Heimlich Maneuver box, Joan M. Beck; 13-4, Margaret Gerrity; 13-5, George J. Wassilchenko; 13-6A, Ernest W. Beck; 13-6B, Christine Oleksyk; 13-6C, Custom Medical Stock Photos; 13-7 (Drawing), Lisa Shoemaker/Joan M. Beck; 13-7 (Photo), courtesy of S. Erlandsen and J. Magney (from *Color Atlas of Histology*, 1992, Mosby); 13-8, 13-14 (heart and lungs), Network Graphics; 13-9, 13-12, Rolin Graphics; 13-11, John Daugherty; 13-14 (inset and brain stem), William Ober.

Chapter 14: 14-1, Lisa Shoemaker; 14-2, Bill Ober; 14-3, 14-4, Table 14-2, 14-7, Rolin Graphics; 14-5, 14-6, 14-9, 14-10, 14-12, Ernest W. Beck; 14-8, G. David Brown; 14-11, courtesy of B. Vidic and F.R. Suarez (from *Photographic Atlas of the Human Body*, 1984, Mosby); 14-13, CNRI Science Photo Library/Photo Researchers; 14-14, Michael P. Schenck; Colostomy box, Network Graphics.

Chapter 15: 15-1, 15-5, Network Graphics; 15-2, 15-3, 15-4, Rolin Graphics.

Chapter 16: 16-1A, 16-3A, Christine Oleksyk; 16-1B, courtesy of Patricia Kane, University of Indiana Medical School; 16-2, David J. Mascaro; 16-3B, Dr. Andrew P. Evan, University of Indiana; 16-4, 16-8, Ernest W. Beck; 16-5, Rolin Graphics; 16-6, courtesy of Potter/Perry (*Fundamentals of Nursing*, 1993, Mosby); 16-7, courtesy of G. Bevelander and J.A. Ramaley (from *Essentials of Histology*, ed. 8, Mosby); Artificial Kidney box (Photo), Pat Watson; Artificial Kidney box (Art), Barbara Stackhouse.

Chapter 17: 17-1 (right), 17-2, Rolin Graphics; 17-3, Joan M. Beck; 17-4, Network Graphics; 17-7, Patrick Watson (from Barkauskas, et. al., *Health and Physical Assessment*, 1994, Mosby).

Chapter 18: 18-7, Laurie O'Keefe/John Daugherty.

Chapter 19: 19-1, Ronald J. Ervin; 19-2 (Photo), Lennart Nilsson; 19-2 (Art), Ernest W. Beck; 19-4A, courtesy of S. Erlandsen and J. Magney (from *Color Atlas of Histology*, 1992, Mosby); 19-4B, Barbara Cousins; 19-5A, Carolyn B. Coulam/John A. McIntyre; 19-5B, William Ober; 19-6A, Kevin A. Sommerville; 19-6B, courtesy of B. Vidic and F.R. Suarez (from *Photographic Atlas of the Human Body*, 1984, Mosby); 19-7, 19-10, George J. Wassilchenko; 19-8, 19-9, Kevin A. Sommerville/Kathy Mitchell Gray; 19-11, David J. Mascaro & Associates; 19-13, Yvonne Wylie Walston.

Chapter 20: 20-1(Art), Rolin Graphics; 20-1(Photo), 20-6, Lennart Nilsson; 20-2, Scott Bodell; 20-3, courtesy of Lucinda L. Veek, Jones Institute for Reproductive Medicine, Norfolk, Virginia; 20-4, 20-8, Ernest W. Beck; 20-5, 20-7, Kevin A. Somerville; 20-9, courtesy of Marjorie M. Pyle, for Lifecircle, Costa Mesa, California; 20-10, Ron Edwards; 20-11, Patrick Watson; Antenatal Diagnosis and Treatment box, Network Graphics.

Chemistry Appendix: A-1, A-7, A-10, A-11, Network Graphics; A-2, A-3, Precision Graphics; A-4, A-5, A-6, A-9, Rolin Graphics; S-8, J.B. Wooskey and Associates.

Index

t indicates term found in table only; italics indicates term found in illustration only.